Apostolos Giannopoulos, Alexander Koldobsky

Tomography of Measures on Convex Bodies

Advances in Analysis and Geometry

Volume 14

Apostolos Giannopoulos, Alexander Koldobsky

Tomography of Measures on Convex Bodies

DE GRUYTER

Mathematics Subject Classification 2020
Primary: 52A20, 46B07; Secondary: 42B10, 44A12

Authors

Prof. Apostolos Giannopoulos
National Technical University of Athens
School of Applied Mathematical and Physical Sciences
Heroon Polytechniou Str. 9
157 80 Athens
Greece
apgiannop@math.ntua.gr

Prof. Alexander Koldobsky
University of Missouri-Columbia
Department of Mathematics
810 East Rollins St
Columbia, MO 65211
USA
koldobskiya@missouri.edu

ISSN 2511-0438
ISBN 978-3-11-135945-8
ISBN 978-3-11-136050-8 (PDF)
ISBN 978-3-11-136117-8 (E-PUB)
DOI https://doi.org/10.1515/9783111360508

Library of Congress Control Number: 2026943015

Bibliographic information published by the Deutsche Nationalbibliothek
The Deutsche Nationalbibliothek lists this publication in the Deutsche Nationalbibliografie; detailed bibliographic data are available on the Internet at https://dnb.dnb.de.

De Gruyter and Walter de Gruyter GmbH are part of De Gruyter Brill.
www.degruyterbrill.com

Questions about General Product Safety Regulation: productsafety@degruyterbrill.com

Printing: CPI books GmbH, Leck

Contents

Preface

In this book, we review the main results that lie at the intersection of two very active branches of convex geometry. The first is the *functional approach*, which studies what happens to classical results from convex geometry when volume is replaced by an arbitrary measure or by measures from certain classes. The second branch is *geometric tomography*, which investigates geometric properties of convex bodies based on data about their sections and projections. We call the intersection of these two branches "the tomography of measures."

A typical result in this category is a generalization of the Busemann–Petty problem. The Busemann–Petty problem, posed in 1956, asks the following:

"Let K, L be origin-symmetric convex bodies in $\mathbb{R}^n$ that satisfy the Busemann–Petty hypothesis, i. e. the $(n-1)$-dimensional volume of every central hyperplane section of K is smaller than that of the corresponding hyperplane section of L:

$$|K \cap \xi^\perp| \leqslant |L \cap \xi^\perp| \quad \text{for all } \xi \in S^{n-1}.$$

Does it necessarily follow that the n-dimensional volume satisfies $|K| \leqslant |L|$?"

Here, $\xi^\perp = \{x \in \mathbb{R}^n : \langle x, \xi \rangle = 0\}$ is the central hyperplane perpendicular to $\xi \in S^{n-1}$, and $|K|$ denotes volume in the appropriate dimension. The answer is affirmative if $n \leqslant 4$, and negative if $n \geqslant 5$; see Section 3.1 for the solution and historical details.

In 2005, Zvavitch found a generalization to measures with even continuous density. Suppose that f is an even, continuous, strictly positive function on $\mathbb{R}^n$, and K and L are origin-symmetric convex bodies in $\mathbb{R}^n$ such that

$$\int_{K \cap \xi^\perp} f\,dx \leqslant \int_{L \cap \xi^\perp} f\,dx \quad \text{for all } \xi \in S^{n-1}.$$

Does it necessarily follow that

$$\int_K f\,dx \leqslant \int_L f\,dx\,?$$

It turns out that for every function f satisfying these conditions, the answer is the same as for volume: affirmative if $n \leqslant 4$ and negative if $n \geqslant 5$. We prove this result in Section 3.2.

An *isomorphic* version of this result was proved in [324]. If K and L satisfy the above Busemann–Petty hypothesis for f, then

$$\int_K f\,dx \leqslant \sqrt{n} \int_L f\,dx.$$

https://doi.org/10.1515/9783111360508-203

Very recently, it was announced in [325] that the order of $\sqrt{n}$ is optimal up to an absolute constant. We present the proof of this result in Section 3.3. Sections 3.3 and 3.4 also include more general versions of the isomorphic result.

Another important result in the tomography of measures is the slicing inequality for measures. The slicing problem of Bourgain asks whether there exists an absolute constant $c > 0$ such that every origin-symmetric convex body of volume 1 in any dimension has a hyperplane section whose area is greater than c. The problem has recently been solved affirmatively. After a series of developments, the final step was achieved by Klartag and Lehec [295], and a few weeks later a different proof was announced by Bizeul [76]. We describe the solution in Chapter 2.

In [313, 315, 317], the problem was extended to arbitrary measures in place of volume, as follows. Suppose that K is a convex body of volume one in $\mathbb{R}^n$, and f is an even, continuous, non-negative function on K with $\int_K f = 1$. Does there exist a constant $s_n > 0$, independent of K and f, such that there exists a hyperplane H in $\mathbb{R}^n$ with

$$\int\limits_{K\cap H} f\,dx \geqslant s_n?$$

It was proved in [315] that such a constant exists, and moreover, $s_n \geqslant \frac{1}{2\sqrt{n}}$. Subsequently, it was shown in [291, 297] that $s_n \leqslant O(\frac{1}{\sqrt{n}})$, so the estimate from [315] is optimal up to an absolute constant (the original result in [291] included a double logarithmic term, which was later removed in [297]).

Let us briefly describe the contents of the book.

Chapter 1 introduces fundamental concepts in convex geometry and harmonic analysis that will be used throughout the book.

Chapter 2 is devoted to Bourgain's hyperplane problem, or the slicing conjecture, asking whether there exists an absolute constant $c > 0$ such that every convex body of unit volume in $\mathbb{R}^n$ has a hyperplane section of volume at least c. This is equivalent to uniformly bounding the isotropic constant L_K for isotropic convex bodies. The problem naturally leads to the study of isotropic log-concave measures, whose densities are centered log-concave functions with covariance matrix equal to the identity. The hyperplane problem is reduced to strong small-ball probability estimates for isotropic log-concave measures. Connections to the Kannan–Lovász–Simonovits conjecture and Cheeger constants further relate isotropic constants to Poincaré inequalities.

Recent advances rely on Eldan's stochastic localization, which decomposes an isotropic log-concave measure into a mixture of uniformly log-concave measures with controlled covariance, allowing precise small-ball and variance estimates. Using these techniques, one can establish optimal small-ball probability estimates, culminating in an affirmative answer to the hyperplane conjecture.

Chapter 3 surveys the classical Busemann–Petty problem, its generalizations to arbitrary measures, and related comparison problems. The final answer to the classical problem is clarified through Lutwak's theory of intersection bodies and techniques involving the spherical Radon transform and Fourier analysis. Extensions to arbitrary measures, isomorphic versions, and lower-dimensional sections introduce constants controlling the deviation from exact inequalities, often bounded by $\sqrt{n}$ or other functionals depending on the Banach–Mazur or outer volume ratio distance to intersection bodies. Stability and separation results quantify how small differences in sections affect volumes, while fractional Laplacians allow further refinements.

The Shephard problem, concerned with projections instead of sections, has a positive answer only for projection bodies, but quantitative volume bounds in higher dimensions can be obtained using zonoids, volume ratio distances, and affine quermassintegrals.

Functional-analytic perspectives unify these problems by linking operator inequalities to L_p-norm comparisons, with the Radon and cosine transforms recovering Busemann–Petty-type results. Finally, convex measures allow hyperplane section inequalities and other classical convex geometric results to extend beyond the Lebesgue measure.

Chapter 4 investigates the classical slicing problem and its extensions to lower-dimensional sections, general measures, p-th moments, and fractional derivatives of the Radon transform. It begins by examining whether the volume of a centered convex body can be controlled by the maximal volume of its hyperplane or $(n-k)$-dimensional sections, establishing uniform bounds via isotropic constants and generalized k-intersection bodies.

The discussion extends to measures with even continuous densities, using outer volume ratio distances to relate global integrals to sectional measures, and demonstrates sharpness through probabilistic constructions.

The chapter further develops p-th moment inequalities for convex bodies relative to L_p^n subspaces, yielding nearly optimal directional and global estimates, and connects these to Busemann–Petty-type comparisons.

Finally, fractional derivatives of the Radon transform are introduced to generalize slicing inequalities, with constants controlled by distances to L_{-1-q}^n-bodies, providing a unified framework that links geometric, analytic, and measure-theoretic methods to control integrals over convex bodies through sections, moments, and directional derivatives.

Chapter 5 develops a range of integral-geometric tools and inequalities that are central to high-dimensional convex geometry. The chapter begins with the introduction of the Blaschke–Petkantschin formulas, which express integrals over k-tuples of points in $\mathbb{R}^n$ in terms of integrals over lower-dimensional subspaces and the Haar measures on Grassmannians, weighted by the k-dimensional volume of the spanned parallelepipeds. Building on these tools, the chapter presents the Busemann–Straus inequality and its generalization by Grinberg, giving sharp bounds on the average volume of intersections of a set with lower-dimensional subspaces in terms of the volume radius, with equality characterized by origin-symmetric ellipsoids.

A variant of the classical Busemann–Petty and Shephard problems is then discussed, establishing that inequalities between all lower-dimensional projections and sections imply inequalities between full volumes, even for non-symmetric or non-convex sets, with stability and rigidity phenomena highlighted.

Finally, alternative proofs of the slicing and comparison inequalities are presented, employing multiple integration, affine isoperimetric techniques, and the outer volume ratio to relate measures of sets to the measures of their lower-dimensional sections.

Overall, the chapter illustrates the interplay between integral geometry, symmetrization, and affine invariants, providing a unified framework for volume comparison and slicing bounds.

Chapter 6 investigates volume difference inequalities, which quantify how differences in the volumes of high-dimensional bodies can be controlled by differences in their lower-dimensional sections and projections. Extending classical slicing problems, the focus is on stability: understanding when small discrepancies in sectional or projection data lead to small discrepancies in total volume, and identifying the geometric structures that make such control possible.

The chapter develops sharp inequalities for structured classes of bodies, such as intersection bodies and projection bodies, and shows that without these structures the inequalities fail dramatically.

Multiple complementary methods are employed, including Fourier analysis, Radon transforms, and integral geometry, yielding results for both volume and more general measures. Together, these findings clarify the precise role of geometric structure in section- and projection-based estimates and provide a unified perspective on stability phenomena.

Chapter 7 investigates the interplay between the volume and surface area of convex bodies in $\mathbb{R}^n$ and the geometry of their projections and sections.

Starting from classical inequalities such as the Loomis–Whitney inequality and its dual form due to Meyer, the chapter develops a unified framework that connects volume estimates to lower-dimensional projections and sections through modern extensions, including the Bollobás–Thomason uniform cover inequality, its weighted and restricted variants, and dual inequalities obtained via L_p-centroid bodies.

A central theme is the study of the surface area of projections, where sharp upper and lower bounds are established for general convex bodies, projection bodies, and zonoids, particularly when bodies are placed in classical positions such as isotropic, John, Löwner, or minimal surface area position.

The chapter then explores slicing inequalities for surface area, showing that, in contrast to volume, no universal slicing inequality can hold; extremal examples are provided by highly anisotropic ellipsoids, highlighting fundamental limitations and the role of polar duality.

The final part of the chapter addresses the average volume of sections, proving that for wide classes of convex and star bodies the average size of lower-dimensional sections can be controlled by maximal sections of fixed codimension, with constants that remain bounded when the codimension is proportional to the ambient dimension.

Chapter 8 investigates the rich interplay between convex geometry and discrete structures, focusing on how classical properties of convex bodies translate when intersected with the integer lattice.

Central themes include discrete analogues of the slicing problem, discrete Brunn–Minkowski inequalities, and correlation inequalities on abelian groups, highlighting how combinatorial, probabilistic, and geometric methods combine to reveal the subtle and sometimes surprising behavior of convexity in the lattice setting.

Chapter 9 summarizes the recent progress on the extremal sections and projections of ℓ_p^n-balls. These bodies display striking geometric phenomena when sliced or projected onto lower-dimensional subspaces.

Hyperplane sections and projections reveal that extremal volumes often occur along coordinate axes or diagonal directions, depending on the shape parameter p, while probabilistic representations and analytic techniques provide insight into these patterns.

The perimeter and volume of sections, as well as lower-dimensional projections, are governed by symmetry and majorization principles, illustrating how the interplay between dimension, shape, and orientation shapes the structure of high-dimensional spaces.

These results highlight the rich, subtle geometry underlying even seemingly simple objects like cubes and ℓ_p^n-balls.

Finally, Chapter 10 explores the interplay between functional inequalities and classical convex geometric results. It begins with the dimensional Brunn–Minkowski inequality and discusses its gaussian analogue, which has been established for origin-symmetric convex sets. This inequality neither trivially follows from nor implies Ehrhard's inequality; however, it does imply the log-concavity of gaussian measure for symmetric convex sets. Several related results also rely on convexity and symmetry, including the B-inequality of Cordero-Erausquin, Fradelizi, and Maurey, the S-inequality of Latala and Oleszkiewicz, and Royen's gaussian correlation inequality. These developments motivate further conjectures, such as the log-Brunn–Minkowski inequality for symmetric log-concave measures.

The discussion then turns to volume product inequalities, highlighting the Blaschke–Santaló inequality and its functional analogue, which extends classical bounds on the volume product of a convex body and its polar, and examines Mahler's conjecture and its functional form. Connections to transport and entropy inequalities are presented, illustrating how functional inequalities encode concentration phenomena.

Further topics that are discussed are functional Rogers–Shephard inequalities, which generalize classical bounds on difference bodies to log-concave functions, and mixed integrals that extend the theory of mixed volumes and give rise to functional Alexandrov–Fenchel- and Brunn–Minkowski-type inequalities. Additional directions and results in "functional convexity" are briefly outlined in the further reading section.

We would like to thank J. Haddad, D. Langharst, D. Ryabogin, V. Yaskin, and A. Zvavitch for reading specific chapters and helping improve the presentation. We also thank M. Pafis and N. Tziotziou for their contribution to Chapter 2. The first named author acknowledges partial support from the National Technical University of Athens. The second named author was partially supported by the US National Science Foundation under Grant No. DMS-2450745 and by the Great Books Program at the University of Missouri.

August 2026

Apostolos Giannopoulos
Alexander Koldobsky

1 Basic concepts from convex geometry and harmonic analysis

In this chapter, we introduce fundamental concepts in convex geometry and harmonic analysis that underpin the study of high-dimensional bodies. Beginning with star bodies and convex bodies, we develop key concepts such as mixed volumes, intersection bodies, and k-intersection bodies, while systematically introducing tools from harmonic analysis that connect volume, sections, and projections to analytic representations. We refer the reader to [27, 28, 113, 194, 195, 505] for more details from convex geometry and for comprehensive coverage of the literature. We also refer to the monographs [310] and [323] for details on the Fourier approach to sections and projections of convex bodies.

1.1 Star bodies

A closed bounded set $K \neq \{0\}$ in $\mathbb{R}^n$ is called a star body if for every $x \in K \setminus \{0\}$ we have that the interval $[0, x)$ is contained in the interior of K (thus, every straight line passing through the origin crosses the boundary of K at exactly two points different from the origin), and the Minkowski functional of K defined by

$$\|x\|_K = \min\{a \geqslant 0 : x \in aK\}$$

is a continuous function on $\mathbb{R}^n$. This implies that the origin is an interior point of K and $\|\cdot\|_K$ is strictly positive outside of the origin. The Minkowski functional of K is a homogeneous function of degree 1 on $\mathbb{R}^n$ and

$$K = \{x \in \mathbb{R}^n : \|x\|_K \leqslant 1\}.$$

The radial function of a star body K is defined by

$$\rho_K(x) = \|x\|_K^{-1}, \quad x \in \mathbb{R}^n,\ x \neq 0.$$

If $x \in S^{n-1}$, then $\rho_K(x)$ is the radius of K in the direction of x, i. e., the distance from the origin to the boundary of K in the direction of x.

A star body K in $\mathbb{R}^n$ is called origin-symmetric if $x \in K$ implies that $-x \in K$. The radial metric on the set of all origin-symmetric star bodies in $\mathbb{R}^n$ is defined by

$$\rho(K, L) = \max_{x \in S^{n-1}} \left|\rho_K(x) - \rho_L(x)\right|.$$

If K and L are compact sets in $\mathbb{R}^n$, then the Minkowski sum of K and L is the set

$$K + L = \{x + y : x \in K,\ y \in L\}.$$

https://doi.org/10.1515/9783111360508-001

For every $p \in \mathbb{R} \setminus \{0\}$ we define the p-sum of K and L to be the star body $K +_p L$ with Minkowski functional

$$\|x\|_{K+_pL} = \left(\|x\|_K^p + \|x\|_L^p\right)^{1/p}.$$

The p-radial sum of K and L is the star body $K\tilde{+}_pL$ defined by

$$\rho_{K\tilde{+}_pL}(x) = \left(\rho_K^p(x) + \rho_L^p(x)\right)^{1/p}.$$

The 0-sum of K and L is the so-called multiplicative sum of K and L, defined by

$$\|x\|_{K+_0L} = \sqrt{\|x\|_K \|x\|_L}.$$

Let $k \geqslant 0$. A star body K in $\mathbb{R}^n$ is called k-smooth if the restriction of $\|\cdot\|_K$ to the unit sphere S^{n-1} belongs to the space $C^k(S^{n-1})$ of k-times continuously differentiable functions on S^{n-1}. The norm of a function in the space $C^k(S^{n-1})$ is the maximum of the L_∞-norms of the function itself and all its derivatives of the orders up to k. We also say that a star body is infinitely smooth if it is k-smooth for every $k \in \mathbb{N}$.

The next lemma asserts that if K is a star body in $\mathbb{R}^n$, then $\|\cdot\|_K^{-p}$ is locally integrable on $\mathbb{R}^n$ for all $0 < p < n$. Since we are going to consider $\|\cdot\|_K^{-p}$ as a distribution, local integrability implies that the action of this distribution on test functions is by integration.

Lemma 1.1.1. *Let K be an origin-symmetric star body in $\mathbb{R}^n$. For every $0 < p < n$ the function $\|\cdot\|_K^{-p}$ is locally integrable on $\mathbb{R}^n$. Moreover, if f is a bounded integrable function on $\mathbb{R}^n$, then the function $\|\cdot\|_K^{-p} f$ is integrable on $\mathbb{R}^n$.*

Proof. Note that, by the definition of a star body, there exist $r_1, r_2 > 0$ such that $r_1 B_2^n \subseteq K \subseteq r_2 B_2^n$, where B_2^n is the Euclidean unit ball. Equivalently, there exist constants $c_1, c_2 > 0$ such that $c_1|x| \leqslant \|x\|_K \leqslant c_2|x|$ for all $x \in \mathbb{R}^n$, where $|\cdot|$ is the Euclidean norm in $\mathbb{R}^n$.

An elementary computation in polar coordinates shows that if $0 < p < n$, then the function $|\cdot|^{-p}$ is integrable on any Euclidean ball with center at the origin. This shows that $\|\cdot\|_K^{-p}$ is locally integrable. For the second assertion of the lemma, note that $\|\cdot\|_K^{-p} f$ is integrable on B_2^n because $\|\cdot\|_K^{-p}$ is locally integrable and f is bounded, and that $\|\cdot\|_K^{-p} f$ is integrable on $\mathbb{R}^n \setminus B_2^n$, this time because f is integrable and $\|\cdot\|_K^{-p}$ is bounded. □

The parallel section function of a star body K in $\mathbb{R}^n$ in the direction of $\xi \in S^{n-1}$ is the function $A_{K,\xi} : \mathbb{R} \to \mathbb{R}$ defined by

$$A_{K,\xi}(t) = \left|K \cap (\xi^\perp + t\xi)\right|,$$

where $\xi^\perp + t\xi$ is the hyperplane perpendicular to ξ at distance t from the origin, and $|\cdot|$ denotes volume in the appropriate dimension.

Let χ be the indicator function of the interval $[-1,1]$. Denote by $\langle\cdot,\cdot\rangle$ the scalar product in $\mathbb{R}^n$. Then, $\chi(\|\cdot\|_K)$ is the indicator function of the star body K, and we have

$$A_{K,\xi}(t) = \int_{\{x:\langle x,\xi\rangle=t\}} \chi(\|x\|_K)\,dx. \tag{1.1}$$

In the case $t=0$, expressing this integral in polar coordinates in the hyperplane $\langle x,\xi\rangle = 0$, we get

$$\begin{aligned} A_{K,\xi}(0) = |K\cap\xi^\perp| &= \int_{\{x:\langle x,\xi\rangle=0\}} \chi(\|x\|_K)\,dx \\ &= \int_{S^{n-1}\cap\xi^\perp} \left(\int_0^\infty r^{n-2}\chi(r\|\theta\|_K)\,dr\right) d\theta \\ &= \int_{S^{n-1}\cap\xi^\perp} \left(\int_0^{1/\|\theta\|_K} r^{n-2}\,dr\right) d\theta \\ &= \frac{1}{n-1}\int_{S^{n-1}\cap\xi^\perp} \|\theta\|_K^{-n+1}\,d\theta \\ &= \frac{1}{n-1}\int_{S^{n-1}\cap\xi^\perp} \rho_K^{n-1}(\theta)\,d\theta, \end{aligned} \tag{1.2}$$

where $d\theta$ stands for the uniform measure on the sphere with density 1. This is the polar formula for the volume of hyperplane sections of K. A similar argument shows that if H is an m-dimensional subspace of $\mathbb{R}^n$, then we may express the volume of the section of K by H as

$$|K\cap H| = \frac{1}{m}\int_{S^{n-1}\cap H} \|\theta\|_K^{-m}\,d\theta. \tag{1.3}$$

For $m=n$, we get the polar formula for the volume of a star body

$$|K| = \frac{1}{n}\int_{S^{n-1}} \|\theta\|_K^{-n}\,d\theta. \tag{1.4}$$

1.2 Convex bodies

A convex body in $\mathbb{R}^n$ is a compact, convex subset K of $\mathbb{R}^n$ with non-empty interior. We say that K is origin-symmetric if $K=-K$, and we say that K is centered if it has its barycenter $\mathrm{bar}(K)$ at 0, i. e., if

$$\int_K \langle x, \xi \rangle \, dx = 0$$

for every $\xi \in S^{n-1}$. Every convex body with the origin as an interior point is a star body. The Minkowski functional $\|\cdot\|_K$ of an origin-symmetric convex body K is a norm on $\mathbb{R}^n$.

The support function of K is defined for every $y \in \mathbb{R}^n$ by

$$h_K(y) = \max\{\langle x, y \rangle : x \in K\}.$$

The support function of an origin-symmetric convex body is the norm on $\mathbb{R}^n$ dual to $\|\cdot\|_K$.

Note that one has $\rho_K(\xi) \leqslant h_K(\xi)$ for every $\xi \in S^{n-1}$. The mean width of K is the quantity

$$w(K) = \int_{S^{n-1}} h_K(\xi) \, d\sigma(\xi),$$

where σ is the rotationally invariant probability measure on S^{n-1}. The radius of K is

$$R(K) = \max\{|x| : x \in K\}.$$

If 0 is an interior point of K, we denote by $r(K)$ the largest $r > 0$ for which $rB_2^n \subseteq K$. The volume radius of K is the quantity

$$\mathrm{vrad}(K) = \left(\frac{|K|}{\omega_n} \right)^{1/n},$$

where

$$\omega_n = |B_2^n| = \frac{\pi^{n/2}}{\Gamma(1 + \frac{n}{2})}$$

is the volume of the Euclidean unit ball. The polar body K° of a convex body K with $0 \in \mathrm{int}(K)$ is defined as follows:

$$K^\circ = \{x \in \mathbb{R}^n : \langle x, y \rangle \leqslant 1 \text{ for all } y \in K\}. \tag{1.5}$$

We write $\overline{K}$ for the multiple of $K \subseteq \mathbb{R}^n$ that has volume 1; in other words, $\overline{K} := |K|^{-1/n} K$.

1.2.1 Brunn–Minkowski inequality

The Brunn–Minkowski inequality relates Minkowski addition and volume in $\mathbb{R}^n$.

Theorem 1.2.1 (Brunn–Minkowski). *Let K and L be two non-empty compact subsets of* $\mathbb{R}^n$. *Then,*

$$|K + L|^{1/n} \geqslant |K|^{1/n} + |L|^{1/n}. \tag{1.6}$$

Theorem 1.2.1 expresses the fact that volume is a concave function with respect to Minkowski addition. For this reason, we also write it in the following form: If K and L are non-empty compact subsets of $\mathbb{R}^n$, then for every $\lambda \in (0,1)$ we have

$$\left|\lambda K + (1-\lambda)L\right|^{1/n} \geqslant \lambda|K|^{1/n} + (1-\lambda)|L|^{1/n}. \tag{1.7}$$

From (1.7) and the arithmetic-geometric means inequality we get

$$\left|\lambda K + (1-\lambda)L\right| \geqslant |K|^{\lambda}|L|^{1-\lambda}. \tag{1.8}$$

This form of the Brunn–Minkowski inequality has the advantage of being dimension free. In fact, one can show that it is equivalent to (1.6) in the sense that knowing (1.8) for all K, L, and λ we can then show that the stronger inequality (1.6) holds true.

There are many interesting proofs of the Brunn–Minkowski inequality. The first one, in chronological order, was restricted to the class of convex bodies and it was based on Brunn's concavity principle.

Theorem 1.2.2 (Brunn). *Let K be a convex body in $\mathbb{R}^n$ and let F be a k-dimensional subspace. Then, the function $f : F^{\perp} \to \mathbb{R}$ defined by*

$$f(x) = \left|K \cap (F + x)\right|^{1/k}$$

is concave on its support.

For the proof of Brunn's principle we introduce Steiner symmetrization. For any convex body K in $\mathbb{R}^n$ and any $\xi \in S^{n-1}$ we consider the set $S_{\xi}(K)$ consisting of all points of the form $x + \lambda\xi$, where x is in the projection $P_{\xi^{\perp}}(K)$ of K onto $\xi^{\perp}$ and $|\lambda| \leqslant \frac{1}{2} \times \text{length}[(x + \mathbb{R}\xi) \cap K]$. In other words, we obtain $S_{\xi}(K)$ by sliding its chords so that their midpoint will be on $\xi^{\perp}$ and take the union of all resulting chords. The set $S_{\xi}(K)$ is the Steiner symmetrization of K in the direction of ξ.

From the definition one can check a number of basic properties of Steiner symmetrization, which are summarized below:

(i) Steiner symmetrization preserves convexity: if K is a convex body, then $S_{\xi}(K)$ is also a convex body.

(ii) $S_{\xi}(K)$ can be described as follows:

$$S_{\xi}(K) = \left\{x + \frac{t_1 - t_2}{2}\xi : x \in P_{\xi^{\perp}}K,\ x + t_1\xi \in K,\ x + t_2\xi \in K\right\}.$$

(iii) Steiner symmetrization preserves volume: $|S_{\xi}(K)| = |K|$.

(iv) If K_1 and K_2 are two convex bodies, then, for every $\lambda \in (0,1)$

$$S_\xi(\lambda K_1 + (1-\lambda)K_2) \supseteq \lambda S_\xi(K_1) + (1-\lambda)S_\xi(K_2).$$

A very useful fact is that, given a convex body K in $\mathbb{R}^n$, we can find a sequence $\{\xi_j\}$ of directions so that applying successive Steiner symmetrizations with respect to ξ_j we obtain a sequence of convex bodies that converges to a Euclidean ball in the Hausdorff metric. More generally, it is a well-known fact that goes back to Steiner and Schwarz that if F is a k-dimensional subspace of $\mathbb{R}^n$, $1 \leqslant k \leqslant n$, then for every convex body K one can find a sequence of successive Steiner symmetrizations in directions $\xi \in F$ so that the limiting convex body $\tilde{K}$ has the following property: For every $x \in F^\perp$, $\tilde{K} \cap (F+x)$ is a ball with center at x and radius $r(x)$ such that $|\tilde{K} \cap (F+x)| = |K \cap (F+x)|$; see, for example, [71, Theorem 9.13.6].

Then, the proof of Theorem 1.2.2 is easily completed: using the convexity of $\tilde{K}$ we see that the function r is concave on its support, and hence f is also concave.

Proof of Theorem 1.2.1. Brunn's concavity principle implies the Brunn–Minkowski inequality for convex bodies as follows. If K and L are convex bodies in $\mathbb{R}^n$, we define

$$K_1 = K \times \{0\} \quad \text{and} \quad L_1 = L \times \{1\}$$

in $\mathbb{R}^{n+1}$ and consider their convex hull M. If

$$M(t) = \{x \in \mathbb{R}^n : (x,t) \in M\} \quad (t \in [0,1]),$$

we easily check that $M(0) = K$, $M(1) = L$ and

$$M(1/2) = \frac{K+L}{2}.$$

Then, Brunn's concavity principle for $F = \mathbb{R}^n$ shows that

$$\left|\frac{K+L}{2}\right|^{1/n} \geqslant \frac{1}{2}|K|^{1/n} + \frac{1}{2}|L|^{1/n}$$

and (1.6) is proved. □

One more proof of the Brunn–Minkowski inequality may be given using an inequality of Prékopa and Leindler; see [368, 469] or [113, Theorem 1.2.3].

Theorem 1.2.3 (Prékopa–Leindler). *Let $f,g,h : \mathbb{R}^n \to \mathbb{R}^+$ be measurable functions, and let $\lambda \in (0,1)$. We assume that f and g are integrable, and for every $x,y \in \mathbb{R}^n$*

$$h(\lambda x + (1-\lambda)y) \geqslant f(x)^\lambda g(y)^{1-\lambda}. \tag{1.9}$$

Then,

$$\int_{\mathbb{R}^n} h \geqslant \Big(\int_{\mathbb{R}^n} f\Big)^{\lambda}\Big(\int_{\mathbb{R}^n} g\Big)^{1-\lambda}.$$

The dimension-free version (1.8) of the Brunn–Minkowski inequality is a simple consequence of the Prékopa–Leindler inequality. We consider two non-empty compact subsets K and L of $\mathbb{R}^n$ and, given $\lambda \in (0,1)$, we define $f = \mathbb{1}_K$, $g = \mathbb{1}_L$, and $h = \mathbb{1}_{\lambda K+(1-\lambda)L}$, where $\mathbb{1}_A$ denotes the indicator function of a set A. We check that the assumptions of Theorem 1.2.3 are satisfied, therefore

$$|\lambda K + (1-\lambda)L| = \int_{\mathbb{R}^n} h \geqslant \Big(\int_{\mathbb{R}^n} f\Big)^{\lambda}\Big(\int_{\mathbb{R}^n} g\Big)^{1-\lambda} = |K|^{\lambda}|L|^{1-\lambda}.$$

1.2.2 Geometric inequalities

We will often use the following basic inequalities for convex bodies.

(i) *Urysohn inequality* ([529]; see also [113, Theorem 1.3.5]). If K is a convex body in $\mathbb{R}^n$, then

$$w(K) \geqslant \left(\frac{|K|}{|B_2^n|}\right)^{1/n}. \tag{1.10}$$

(ii) *Blaschke–Santaló inequality* ([78, 497]; see also [113, Theorem 1.3.4]). If K is a convex body in $\mathbb{R}^n$, then the function $|K|\,|(K-z)^{\circ}|$, defined on $\mathrm{int}(K)$, is strictly convex and has a unique point of minimum, the Santaló point $s(K)$ of K. The Blaschke–Santaló inequality states that if $s(K)=0$, then

$$|K|\,|K^{\circ}| \leqslant |B_2^n|^2 = \omega_n^2. \tag{1.11}$$

One can also prove that if $\mathrm{bar}(K)=0$, then $s(K^{\circ})=0$. Since $(K^{\circ})^{\circ}=K$, we see that if $\mathrm{bar}(K)=0$, then we have again

$$|K|\,|K^{\circ}| = \big|(K^{\circ}-s(K^{\circ}))^{\circ}\big|\,|K^{\circ}| \leqslant \omega_n^2.$$

A refined version of the Blaschke–Santaló inequality was presented by Meyer and Pajor in [412]. For any $\lambda \in (0,1)$ we say that a hyperplane $H=\{x \in \mathbb{R}^n : \langle x,u_H\rangle = a_H\}$, where $u_H \in \mathbb{R}^n \setminus \{0\}$ and $a_H \in \mathbb{R}$, is λ-separating for a convex body K in $\mathbb{R}^n$ if

$$\big|\{x \in K : \langle x,u_H\rangle \geqslant a_H\}\big| = \lambda\,|K|.$$

Note that if H is λ-separating for K, then $\mathrm{int}(K) \cap H \neq \varnothing$. With these definitions we have that if K is a convex body in $\mathbb{R}^n$ and H is a λ-separating hyperplane for K, where $\lambda \in (0,1)$, then there exists $z \in \mathrm{int}(K) \cap H$ such that

$$|K|\,|(K-z)^{\circ}| \leqslant \frac{\omega_n^2}{4\lambda(1-\lambda)}.$$

In fact, z is the unique point in $\mathrm{int}(K) \cap H$ with the property that

$$\mathrm{bar}((K-z)^{\circ}) \in \{tu_H : t \in \mathbb{R}\}.$$

(iii) *Bourgain–V. Milman inequality* ([104]; see also [113, Theorem 1.13.1]). There exists an absolute constant $0 < c < 1$ with the following property: for every $n \geqslant 1$ and any convex body K in $\mathbb{R}^n$ with $0 \in \mathrm{int}(K)$

$$|K|\,|K^{\circ}| \geqslant c^n |B_2^n|^2. \tag{1.12}$$

(iv) *Rogers–Shephard inequalities* ([480]; see also [113, Theorem 1.3.1]). If K is a convex body in $\mathbb{R}^n$, then the volume of the difference body $K-K := \{x-y : x,y \in K\}$ satisfies

$$|K-K| \leqslant \binom{2n}{n}|K|. \tag{1.13}$$

The next inequality estimates the product of the volumes of a projection of a convex body K in $\mathbb{R}^n$ and of the section of K with the orthogonal subspace (see [481] for the first and [514] for the second claim). If K is a convex body in $\mathbb{R}^n$ with $0 \in \mathrm{int}(K)$, then, for any $1 \leqslant k \leqslant n-1$ and any $F \in G_{n,k}$ we have that

$$|P_F(K)|_k\,|K \cap F^{\perp}|_{n-k} \leqslant \binom{n}{k}|K|. \tag{1.14}$$

If $\mathrm{bar}(K) = 0$, then we also have that

$$|K| \leqslant |P_F(K)|_k\,|K \cap F^{\perp}|_{n-k}. \tag{1.15}$$

1.2.3 Mixed volumes

The theory of mixed volumes appeared in [427, 428]. We refer to the book [505] for the proofs and more details. Let $\mathcal{K}^n$ denote the class of non-empty, compact, convex subsets of $\mathbb{R}^n$. Minkowski's fundamental theorem states that if $K_1,\dots,K_m \in \mathcal{K}^n$, then the volume of $t_1K_1 + \cdots + t_mK_m$ is a homogeneous polynomial of degree n in $t_i > 0$. That is,

$$|t_1K_1 + \cdots + t_mK_m| = \sum_{1\leqslant i_1,\dots,i_n\leqslant m} V(K_{i_1},\dots,K_{i_n})t_{i_1}\cdots t_{i_n},$$

where the coefficients $V(K_{i_1},\dots,K_{i_n})$ are chosen to be invariant under permutations of their arguments. The coefficient $V(K_1,\dots,K_n)$ is the mixed volume of $K_1,\dots,K_n$. In particular, if K and L are two convex bodies in $\mathbb{R}^n$ then the function $|K+tL|$ is a polynomial in $t \in [0,\infty)$

$$|K+tL| = \sum_{j=0}^{n} \binom{n}{j} V_{n-j}(K,L)\, t^j,$$

where $V_{n-j}(K,L) = V((K,n-j),(L,j))$ is the j-th mixed volume of K and L (we use the notation (L,j) for the j-tuple $(L,\dots,L)$). If $L = B_2^n$, then we set $W_j(K) := V_{n-j}(K,B_2^n) = V((K,n-j),(B_2^n,j))$; this is the j-th quermassintegral of K. Note that

$$V_{n-1}(K,L) = \frac{1}{n}\lim_{t\to 0^+}\frac{|K+tL|-|K|}{t}$$

and by the Brunn–Minkowski inequality we see that

$$V_{n-1}(K,L) \geqslant |K|^{\frac{n-1}{n}}|L|^{\frac{1}{n}} \tag{1.16}$$

for all K and L (this is Minkowski's first inequality). The mixed volume $V_{n-1}(K,L)$ can be expressed as

$$V_{n-1}(K,L) = \frac{1}{n}\int_{S^{n-1}} h_L(\xi)\, dS(K,\xi), \tag{1.17}$$

where $S(K,\cdot)$ is the surface area measure of K; this is the Borel measure on S^{n-1} defined by

$$S(K,A) = \lambda(\{x \in \mathrm{bd}(K) : \text{the outer normal to } K \text{ at } x \text{ exists and belongs to } A\}),$$

where λ is the Hausdorff measure on $\mathrm{bd}(K)$. In particular, the surface area $S(K) := S(K,S^{n-1})$ of K satisfies

$$S(K) = nW_1(K).$$

The volume of K can be written as

$$|K| = \frac{1}{n}\int_{S^{n-1}} h_K(\xi)\, dS(K,\xi). \tag{1.18}$$

The well-known Cauchy projection formula expresses the volume of the projection of K to the central hyperplane perpendicular to a vector θ in terms of the surface area measure

$$|P_{\theta^\perp}(K)| = \frac{1}{2}\int_{S^{n-1}} |\langle\theta,\xi\rangle|\, dS(K,\xi), \quad \theta \in S^{n-1}. \tag{1.19}$$

Kubota's integral formula expresses the quermassintegral $W_j(K)$ as the average of the volumes of $(n-j)$-dimensional projections of K

$$W_j(K) = \frac{\omega_n}{\omega_{n-j}} \int_{G_{n,n-j}} |P_H(K)|\, d\nu_{n,n-j}(H),$$

where $\nu_{n,k}$ is the Haar probability measure on the Grassmann manifold of k-dimensional subspaces of $\mathbb{R}^n$, and $P_H(K)$ is the orthogonal projection of K to H. Applying this formula for $j = n-1$ we see that

$$W_{n-1}(K) = \omega_n w(K).$$

It is convenient to work with a normalized variant of $W_{n-k}(K)$. If we set

$$Q_k(K) = \left(\frac{W_{n-k}(K)}{\omega_n}\right)^{\frac{1}{k}} = \left(\frac{1}{\omega_k}\int_{G_{n,k}} |P_H(K)|\, d\nu_{n,k}(H)\right)^{\frac{1}{k}}, \tag{1.20}$$

then $k \mapsto Q_k(K)$ is decreasing. This is a consequence of the Aleksandrov–Fenchel inequality (see [116] and [505]). In particular, for every $1 \leqslant k \leqslant n-1$ we have

$$\mathrm{vrad}(K) = \left(\frac{|K|}{\omega_n}\right)^{\frac{1}{n}} \leqslant \left(\frac{1}{\omega_k}\int_{G_{n,k}} |P_H(K)|\, d\nu_{n,k}(H)\right)^{\frac{1}{k}} \leqslant w(K). \tag{1.21}$$

A convex body K is said to have curvature function

$$f_K : S^{n-1} \to \mathbb{R},$$

if its surface area measure $S(K,\cdot)$ is absolutely continuous with respect to Lebesgue measure σ on S^{n-1}, and

$$\frac{dS(K,\cdot)}{d\sigma} = f_K \in L_1(S^{n-1}),$$

so f_K is the density of $S(K,\cdot)$. The volume of a body can be expressed in terms of its support function and curvature function

$$|K| = \frac{1}{n} \int_{S^{n-1}} h_K(x) f_K(x)\, dx. \tag{1.22}$$

The mixed volume $V_{n-1}(K,L)$ can also be expressed in terms of the support and curvature functions

$$V_{n-1}(K,L) = \frac{1}{n} \int_{S^{n-1}} h_L(x) f_K(x)\, dx. \tag{1.23}$$

Let K be an origin-symmetric convex body in $\mathbb{R}^n$. The projection body ΠK of K is defined as an origin-symmetric convex body in $\mathbb{R}^n$ whose support function in every direction is equal to the volume of the hyperplane projection of K to this direction: for every $\xi \in S^{n-1}$

$$h_{\Pi K}(\xi) = \left|P_{\xi^\perp}(K)\right|. \tag{1.24}$$

If L is the projection body of some convex body, we simply say that L is a projection body. The Minkowski (vector) sum of projection bodies is also a projection body. Every projection body is the limit in the Hausdorff metric of Minkowski sums of symmetric intervals. An origin-symmetric convex body in $\mathbb{R}^n$ is a projection body if and only if its polar body is the unit ball of an n-dimensional subspace of L_1. In particular, by the definition (1.24) and the Cauchy formula (1.19), L is a projection body if and only if its support function can be written as

$$h_L(\xi) = \int_{S^{n-1}} \left|\langle x, \xi\rangle\right| d\mu_L(x), \quad \xi \in S^{n-1}, \tag{1.25}$$

where μ_L is some non-negative finite Borel measure on S^{n-1}. We refer the reader to [505, 194, 310] for proofs and more properties of projection bodies.

It is well known that one can approximate any convex body in $\mathbb{R}^n$ in the radial metric by a sequence of infinitely smooth convex bodies. This can be proved by a simple convolution argument; see [505, Theorem 3.3.1].

Theorem 1.2.4. *Let K be an origin-symmetric convex body with support function h_K and containing a Euclidean ball with center at the origin and radius $1/R$. Also, let $\phi_\varepsilon : [0,\infty) \to [0,\infty)$ be an infinitely differentiable function supported in $[\varepsilon/2, \varepsilon]$ and such that*

$$\int_{\mathbb{R}^n} \phi_\varepsilon(|x|)\, dx = 1.$$

Define a function f on $\mathbb{R}^n$ by

$$f_\varepsilon(x) = \int_{\mathbb{R}^n} h_K(x + |x|z)\phi_\varepsilon(|z|)\, dz.$$

Then, the function f_ε is infinitely differentiable on $\mathbb{R}^n \setminus \{0\}$ and is the support function of a convex body K_ε so that for every $x \in \mathbb{R}^n$

$$\left|h_K(x) - h_{K_\varepsilon}(x)\right| < R\varepsilon.$$

Since the Minkowski functional of a convex body is the support function of the polar body, the latter theorem gives an approximation of any convex body by infinitely smooth convex bodies in the radial metric.

Also, considering bodies K_ε with Minkowski functionals

$$\|x\|_{K_\varepsilon} = \|x\|_K + \varepsilon|x|$$

one can approximate a convex body K in the radial metric by convex bodies with strictly positive curvature. More results on approximation with special properties can be found in [236, Section 5].

An argument similar to that of Theorem 1.2.4 shows that one can approximate bodies in the norm of the space $C^k(S^{n-1})$ in the following sense: any k-smooth star body K can be approximated by a sequence of infinitely smooth star bodies K_m so that the radial functions ρ_{K_m} converge to ρ_K in the metric of the space $C^k(S^{n-1})$.

1.2.4 Some facts from the local theory of normed spaces

We will also need some basic facts from the local theory of normed spaces. Let K be an origin-symmetric convex body in $\mathbb{R}^n$. Recall that the function $\|\cdot\|_K : \mathbb{R}^n \to \mathbb{R}^+$ defined by

$$\|x\|_K = \inf\{t > 0 : x \in tK\}$$

is a norm on $\mathbb{R}^n$. The space $(\mathbb{R}^n, \|\cdot\|_K)$ will be denoted by X_K. Conversely, if $X = (\mathbb{R}^n, \|\cdot\|)$ is a normed space, then the unit ball $K = \{x \in \mathbb{R}^n : \|x\| \leqslant 1\}$ of X is an origin-symmetric convex body.

Let X, Y be two n-dimensional normed spaces. The Banach–Mazur distance from X to Y is defined as follows:

$$d(X, Y) = \inf\{\|T\| \cdot \|T^{-1}\| \mid T : X \to Y \text{ linear isomorphism}\}. \tag{1.26}$$

In a geometric language, the Banach–Mazur distance has the following description: if $X = X_K$ and $Y = X_L$ (i. e., the unit balls of X, Y are the convex bodies K, L, respectively), then the distance $d(X, Y)$ is the smallest $d \geqslant 1$ such that

$$K \subseteq T(L) \subseteq dK \tag{1.27}$$

for some $T \in GL_n$. From John's theorem [271] we know that $d(X, \ell_2^n) \leqslant \sqrt{n}$ for every n-dimensional normed space X.

We also often use the geometric distance $d_G(K, L)$ between two symmetric convex bodies K and L in $\mathbb{R}^n$, or more generally two convex bodies having the origin as an interior point, which is the smallest $d \geqslant 1$ for which there exist $a, b > 0$ with $ab \leqslant d$ such that

$$\frac{1}{a}K \subseteq L \subseteq bK. \tag{1.28}$$

We define

$$M(K) := \int_{S^{n-1}} \|\xi\|_K \, d\sigma(\xi).$$

Since $\|x\|_K = h_{K^\circ}(x)$ for every $x \in \mathbb{R}^n$, we see that $M(K) = w(K^\circ)$ and that

$$M(K)^{-1} \leqslant \mathrm{vrad}(K) \leqslant w(K) = M(K^\circ).$$

The left-hand side inequality is easily checked if we express the volume of K as an integral in polar coordinates and use the inequalities of Hölder and Jensen, while the right-hand side inequality is an immediate consequence of Urysohn's inequality.

Combining the results of Lewis [371], Figiel and Tomczak-Jaegermann [169], and Pisier [461, 464] (see [113, Section 1.11.2]) one has the following fact: If K is a centered convex body in $\mathbb{R}^n$, then there exists an origin-symmetric and positive definite $T \in GL_n$ such that the position $\tilde{K} = T(K)$ of K satisfies

$$w(\tilde{K}) \leqslant c\sqrt{n}\log n\,|\tilde{K}|^{1/n}, \tag{1.29}$$

where $c > 0$ is an absolute constant.

Another classical result in the local theory of normed spaces is Milman's M^*-inequality [420, 421]. If K is an origin-symmetric convex body in $\mathbb{R}^n$, then, for every $1 \leqslant k \leqslant n$, a random subspace $F \in G_{n,k}$ satisfies

$$R(K \cap F) \leqslant c\sqrt{\frac{n}{n-k}}\, w(K),$$

with probability greater than $1 - \exp(-c_2(n-k))$, where $c_1, c_2 > 0$ are absolute constants.

1.2.5 Covering numbers

Let K and L be two convex bodies in $\mathbb{R}^n$. The covering number $N(K,L)$ of K by L is the least integer N for which there exist N translates of L whose union covers K

$$N(K,L) = \min\left\{N \in \mathbb{N} : \exists x_1,\dots,x_N \in \mathbb{R}^n \text{ such that } K \subseteq \bigcup_{j=1}^{N}(x_j + L)\right\}.$$

A variant of this notion is defined as follows:

$$\overline{N}(K,L) = \min\left\{N \in \mathbb{N} : \exists x_1,\dots,x_N \in K \text{ such that } K \subseteq \bigcup_{j=1}^{N}(x_j + L)\right\}.$$

From the definition we see that $N(K,L) \leqslant \overline{N}(K,L)$. One can also easily check that $\overline{N}(K,L-L) \leqslant N(K,L)$. In particular, if L is convex and symmetric, then $\overline{N}(K,2L) \leqslant N(K,L)$.

If K, L are convex bodies in $\mathbb{R}^n$, and L is symmetric, for every $t > 0$ we define

$$S_t(K,L) = \max\{m \in \mathbb{N} : \exists x_1,\dots,x_m \in K \text{ such that } \|x_i - x_j\|_L > t \text{ for } i \neq j\},$$

where $\|\cdot\|_L$ is the norm induced by L on $\mathbb{R}^n$. From the definition we easily check that

$$\overline{N}(K,tL) \leqslant S_t(K,L) \leqslant \overline{N}\left(K,\frac{t}{2}L\right).$$

We shall use some basic theorems for covering numbers. The first one is Sudakov's inequality [517].

Theorem 1.2.5. *If K is a convex body in $\mathbb{R}^n$, then for every $t > 0$ one has*

$$N(K,tB_2^n) \leqslant 2\exp(cn(w(K)/t)^2), \tag{1.30}$$

where $c > 0$ is an absolute constant.

The next theorem is due to Artstein-Avidan, V. Milman, and Szarek [30].

Theorem 1.2.6 (duality of entropy). *There exist absolute positive constants α and β such that for any $n \geqslant 1$ and any symmetric convex body K in $\mathbb{R}^n$*

$$N(B_2^n, \alpha^{-1}K^\circ)^{\frac{1}{\beta}} \leqslant N(K,B_2^n) \leqslant N(B_2^n,\alpha K^\circ)^\beta. \tag{1.31}$$

V. Milman [422] proved that there exists an absolute constant $\beta > 0$ such that every centered convex body K in $\mathbb{R}^n$ has a linear image $\tilde{K}$ that satisfies $|\tilde{K}| = |B_2^n|$ and

$$\max\{N(\tilde{K},B_2^n), N(B_2^n,\tilde{K}), N(\tilde{K}^\circ,B_2^n), N(B_2^n,\tilde{K}^\circ)\} \leqslant \exp(\beta n). \tag{1.32}$$

We say that a convex body K that satisfies this estimate is in M-position with constant β.

Pisier has proposed a different approach to this result, which allows one to find a whole family of M-positions and to give more detailed information on the behavior of the corresponding covering numbers. The precise statement is as follows; see [463].

Theorem 1.2.7 (Pisier). *For every $0 < \alpha < 2$ and every symmetric convex body K in $\mathbb{R}^n$ there exists a linear image $\tilde{K}$ of K such that*

$$\max\{N(\tilde{K}, tB_2^n), N(B_2^n, t\tilde{K}), N(\tilde{K}^\circ, tB_2^n), N(B_2^n, t\tilde{K}^\circ)\} \leqslant \exp\Big(\frac{c(\alpha)n}{t^\alpha}\Big)$$

for every $t \geqslant c(\alpha)^{1/\alpha}$, where $c(\alpha)$ depends only on α, and $c(\alpha) = O((2-\alpha)^{-\alpha/2})$ as $\alpha \to 2$.

The dual Sudakov inequality of Pajor and Tomczak-Jaegermann provides an upper bound for the covering numbers $N(B_2^n, tK)$ in terms of the parameter $M(K)$; see [450].

Theorem 1.2.8. *Let K be an origin-symmetric convex body in $\mathbb{R}^n$. For every $t > 0$*

$$\log N(B_2^n, tK) \leqslant cn(M(K)/t)^2, \tag{1.33}$$

where $c > 0$ is an absolute constant.

We refer the reader to the book [505] for the theory of convex bodies and to the books [27, 425], and [464] for the local theory of normed spaces and asymptotic convex geometry.

1.3 Radon transform

The Radon transform and its connection with the Fourier transform play an important role in this book. We refer the reader to the books by Helgason [254, 255] for a systematic study of the Radon transform.

Let ϕ be an integrable function on $\mathbb{R}^n$, which is also integrable on every hyperplane. The spatial Radon transform of the function ϕ is defined as a function of $(\xi; t)$, $\xi \in S^{n-1}$, $t \in \mathbb{R}$

$$\mathcal{R}\phi(\xi; t) = \int\limits_{\{x:\langle x,\xi\rangle = t\}} \phi(x)\,dx. \tag{1.34}$$

We use a well-known connection between the Radon and Fourier transforms:

Lemma 1.3.1. *For a fixed ξ, the Fourier transform of the function $g(t) = \mathcal{R}\phi(\xi; t)$, $t \in \mathbb{R}$ is equal to the function $z \mapsto \widehat{\phi}(z\xi)$, $z \in \mathbb{R}$.*

Proof. By the Fubini theorem, making the change of variables $\langle x, \xi \rangle = t$ we get

$$\widehat{\phi}(z\xi) = \int_{\mathbb{R}^n} \phi(x) e^{-iz\langle x,\xi\rangle}\, dx = \int_{\mathbb{R}} e^{-izt} \Big(\int_{\{x:\langle x,\xi\rangle = t\}} \phi(x)\, dx \Big)\, dt = \widehat{g}(z). \qquad \square$$

The spherical Radon transform $R : C(S^{n-1}) \to C(S^{n-1})$ is the linear operator defined by

$$Rf(\xi) = \int_{S^{n-1}\cap\xi^\perp} f(x)\, dx, \quad \xi \in S^{n-1}$$

for every function $f \in C(S^{n-1})$. Using (1.2) one can express the volume of central hyperplane sections in terms of the spherical Radon transform: for every origin-symmetric star body K in $\mathbb{R}^n$ and every $\xi \in S^{n-1}$

$$A_{K,\xi}(0) = |K \cap \xi^\perp| = \frac{1}{n-1} R(\|\cdot\|_K^{-n+1})(\xi). \tag{1.35}$$

Let $G_{n,m}$ be the Grassmann manifold of m-dimensional subspaces of $\mathbb{R}^n$, and let dH be the rotationally invariant probability measure on $G_{n,m}$. The m-dimensional spherical Radon transform is the operator $R_m : C(S^{n-1}) \to C(G_{n,m})$ defined by

$$R_m f(H) = \int_{S^{n-1}\cap H} f(x)\, dx.$$

The dual transform R_m^* (see [255, p. 144] or [507, pp. 146–150]) is an operator from $C(G_{n,m})$ to $C(S^{n-1})$ such that, for every $F \in C(G_{n,m})$

$$R_m^* F(\xi) = \int_{\xi\in H} F(H)\, dH, \quad \xi \in S^{n-1}, \tag{1.36}$$

where the integral on the right-hand side is equal to

$$\int_{\xi\in H} F(H)\, dH = \frac{\kappa_m}{\kappa_n} \int_{G_\xi(n-1,m-1)} F(\mathrm{span}(H,\xi))\, dH. \tag{1.37}$$

Here, we denote by $G_\xi(n-1, m-1)$ the Grassmannian of $(m-1)$-dimensional linear subspaces of the $(n-1)$-dimensional subspace $\xi^\perp$ and we still denote by dH the rotationally invariant probability measure on this Grassmannian. We write κ_m for the surface area of the Euclidean unit sphere S^{m-1}, i. e.

$$\kappa_m := |S^{m-1}| = m\omega_m = \frac{2\pi^{m/2}}{\Gamma(m/2)}.$$

For any $f \in C(S^{n-1})$ and $F \in C(G_{n,m})$ we have the duality relation

$$\begin{aligned}\langle R_m f, F\rangle &= \int_{G_{n,m}} \Big(\int_{S^{n-1}\cap H} f(\xi)\, d\xi \Big) F(H)\, dH \\ &= \int_{S^{n-1}} f(\xi) \Big(\int_{\xi\in H} F(H)\, dH \Big)\, d\xi = \langle f, R_m^* F\rangle. \end{aligned} \tag{1.38}$$

For a proof of (1.38), let X be the set of pairs (ξ, H), where $\xi \in S^{n-1}$ and $H \in G_{n,m}$ is such that $\xi \in H$. The orthogonal group $O(n)$ acts transitively on X, so X admits a unique Haar probability measure μ. For any continuous function g on X we have that

$$\begin{aligned}\int_X g\, d\mu &= c_1 \int_{G_{n,m}} \Big(\int_{\xi\in H} g(\xi, H)\, d\xi \Big)\, dH \\ &= c_2 \int_{S^{n-1}} \Big(\int_{\xi\in H} g(\xi, H)\, dH \Big)\, d\xi \end{aligned} \tag{1.39}$$

for some constants $c_i > 0$. In fact, all three integrals define continuous $O(n)$-invariant functionals on $C(X)$, and hence they all correspond to constant multiples of the Haar measure. Consider any two functions $f \in C(S^{n-1})$ and $F \in C(G_{n,m})$. Setting $g(\xi, H) = f(\xi)F(H)$ and computing the integrals in (1.39), we obtain (1.38).

Putting $F(H) \equiv 1$ in (1.38) we see that, for any function $f \in C(S^{n-1})$

$$\int_{G_{n,m}} \Big(\int_{S^{n-1}\cap H} f(\xi)\, d\xi \Big)\, dH = \frac{\kappa_m}{\kappa_n} \int_{S^{n-1}} f(\xi)\, d\xi. \tag{1.40}$$

Note that, while the measure on the Grassmannian is a probability measure, the uniform measures on the spheres are not normalized in (1.38) and (1.40), which explains the appearance of the constant on the right-hand side of (1.40). This constant can be computed by putting $f \equiv 1$.

The following lemma was proved in [309, Lemma 1].

Lemma 1.3.2. *Let $1 \leqslant m < n$, and $f, g \in C(S^{n-1})$. Then,*

$$\begin{aligned}&\int_{S^{n-1}} f(\xi) \Big(\int_{S^{n-1}\cap\xi^\perp} g(x)\, dx \Big)\, d\xi \\ &= \frac{\kappa_n \kappa_{n-1}}{\kappa_m \kappa_{n-m}} \int_{G_{n,m}} \Big(\int_{S^{n-1}\cap H} f(\xi)\, d\xi \Big) \Big(\int_{S^{n-1}\cap H^\perp} g(x)\, dx \Big)\, dH. \end{aligned}$$

Proof. Applying (1.38) to the functions f and

$$F(H) = \int_{S^{n-1}\cap H^\perp} g(x)\,dx,$$

we get

$$\int_{G_{n,m}} \Big(\int_{S^{n-1}\cap H} f(\xi)\,d\xi\Big)\Big(\int_{S^{n-1}\cap H^\perp} g(x)\,dx\Big)\,dH$$
$$= c\int_{S^{n-1}} f(\xi)\Big(\int_{\xi\in H}\Big(\int_{S^{n-1}\cap H^\perp} g(x)\,dx\Big)\,dH\Big)\,d\xi, \tag{1.41}$$

where c is a constant to be determined later. Making the change of variables $H \to H^\perp$ and applying (1.37) we see that

$$\int_{\xi\in H}\Big(\int_{S^{n-1}\cap H^\perp} g(x)\,dx\Big)\,dH = c_1 \int_{G_\xi(n-1,n-m)}\Big(\int_{S^{n-1}\cap H} g(x)\,dx\Big)\,dH.$$

Applying (1.40) to the unit sphere $S^{n-1}\cap\xi^\perp$ of the hyperplane $\xi^\perp$, we see that the latter integral is equal to

$$c_2 \int_{S^{n-1}\cap\xi^\perp} g(x)\,dx,$$

which proves the lemma up to a constant. To compute the constant, put $f \equiv 1$ and $g \equiv 1$. □

Applying Lemma 1.3.2 with $m = n-1$ we see that the spherical Radon transform $R = R_{n-1}$ is self-dual (see [237, Lemma 1.3.3] for a simple proof not involving Grassmannians).

Lemma 1.3.3. *For any functions $f, g \in C(S^{n-1})$*

$$\int_{S^{n-1}} Rf(\xi)\,g(\xi)\,d\xi = \int_{S^{n-1}} f(\xi)\,Rg(\xi)\,d\xi. \tag{1.42}$$

Finally, we extend the spherical Radon transform to measures. Let μ be a finite Borel measure on S^{n-1}. We define the spherical Radon transform of μ as a functional $R\mu$ on the space $C(S^{n-1})$ acting by

$$\langle R\mu, f\rangle = \langle \mu, Rf\rangle = \int_{S^{n-1}} Rf(x)\,d\mu(x).$$

F. Riesz's characterization of continuous linear functionals on the space $C(S^{n-1})$, see [153, p. 262], implies that $R\mu$ is also a finite Borel measure on S^{n-1}. If μ has continuous density g, then, by Lemma 1.3.3, the Radon transform of μ has density Rg.

1.4 Fourier transform and positive definite distributions

1.4.1 Fourier transform of distributions

In this section, we collect necessary definitions and facts from the theory of distributions. We refer the reader to [205, 491, 310] for details.

We denote by $\mathcal{S} = \mathcal{S}(\mathbb{R}^n)$ the space of complex-valued functions $\phi \in C^\infty(\mathbb{R}^n)$ rapidly decreasing at infinity together with their derivatives of all orders, i. e., for every $k \in \mathbb{N}$, the seminorm

$$p_k(\phi) = \sup_{\alpha_1+\cdots+\alpha_n \leqslant k} \sup_{x \in \mathbb{R}^n} (1 + |x|)^k |D^\alpha \phi(x)|$$

is finite. Here, $\alpha = (\alpha_1, \ldots, \alpha_n)$, $\alpha_k \in \mathbb{N} \cup \{0\}$ is a multiindex, and D^α is the differential operator defined for every $f \in C^{\alpha_1+\cdots+\alpha_n}(\mathbb{R}^n)$ by

$$D^\alpha f = \frac{\partial^{\alpha_1+\cdots+\alpha_n} f}{\partial x_1^{\alpha_1} \cdots \partial x_n^{\alpha_n}}.$$

The elements of the space $\mathcal{S}$ will be called (Schwartz) test functions.

We denote by $\mathcal{S}'$ the space of linear functionals on $\mathcal{S}$ continuous in the topology generated by the sequence of seminorms p_k, $k \geqslant 0$. We call the elements of $\mathcal{S}'$ distributions over $\mathcal{S}$. Suppose that f is a locally integrable complex-valued function on $\mathbb{R}^n$ with power growth at infinity, i. e., there exists a number $\beta > 0$ so that

$$\lim_{|x|\to\infty} \frac{f(x)}{|x|^\beta} = 0.$$

Then, f represents a distribution acting by integration: for every $\phi \in \mathcal{S}$

$$\langle f, \phi \rangle = \int_{\mathbb{R}^n} f(x)\phi(x)\,dx.$$

For a test function ϕ we define the Fourier transform of ϕ by

$$\mathcal{F}\phi(x) = \widehat{\phi}(x) = \int_{\mathbb{R}^n} \phi(\xi) e^{-i\langle x,\xi\rangle}\,d\xi.$$

The Fourier transform is invertible on $\mathcal{S}$, and the inverse operator is given by

$$\check{\mathcal{F}}\phi(\xi) = (2\pi)^{-n} \int_{\mathbb{R}^n} \phi(x) e^{i\langle x,\xi\rangle}\, dx.$$

It follows that, for every $\phi \in \mathcal{S}$

$$\mathcal{F}\mathcal{F}\phi(\xi) = (2\pi)^n \phi(-\xi).$$

An easy application of the Fubini theorem is the following Parseval's formula for test functions. If ϕ, ψ are test functions, then

$$\int_{\mathbb{R}^n} \widehat{\phi}(x)\, \psi(x)\, dx = \int_{\mathbb{R}^n} \phi(\xi)\, \widehat{\psi}(\xi)\, d\xi. \tag{1.43}$$

Since test functions are dense in the spaces $L_1(\mathbb{R}^n)$ and $L_2(\mathbb{R}^n)$, Parseval's formula can be extended to the cases where both functions ϕ and ψ belong to $L_1(\mathbb{R}^n)$ or both belong to $L_2(\mathbb{R}^n)$.

Accordingly, we define the Fourier transform of a distribution $f \in \mathcal{S}'$ as a distribution $\widehat{f}$ acting by

$$\langle \widehat{f}, \phi \rangle = \langle f, \widehat{\phi} \rangle$$

for every test function ϕ. Since $\phi \mapsto \widehat{\phi}$ is an isomorphism of $\mathcal{S}(\mathbb{R}^n)$, the Fourier transform of a distribution is well defined, and $\widehat{f} = \widehat{g}$ implies that $f = g$. When f is given by a longer expression, we write $(f)^\wedge$ for the Fourier transform of f. If a test function ϕ is even, we have

$$\mathcal{F}\mathcal{F}\phi = (\widehat{\phi})^\wedge = (2\pi)^n \phi \quad \text{and} \quad \langle \widehat{f}, \widehat{\phi} \rangle = (2\pi)^n \langle f, \phi \rangle$$

for every $f \in \mathcal{S}'$ and $\phi \in \mathcal{S}$.

A distribution f is called even, homogeneous of degree $p \in \mathbb{R}$ if

$$\langle f(x), \phi(x/t) \rangle = |t|^{n+p} \langle f, \phi \rangle \tag{1.44}$$

for every test function ϕ and every $t \in \mathbb{R}$, $t \neq 0$.

Lemma 1.4.1. *The Fourier transform of an even, homogeneous distribution of degree p is an even, homogeneous distribution of degree $-n-p$.*

Proof. A simple computation shows that, for any test function ϕ and any $t \neq 0$

$$\widehat{\phi(x/t)}(\xi) = |t|^n \widehat{\phi}(t\xi).$$

By (1.44)

$$\langle \hat{f}, \phi(x/t) \rangle = \langle f, \widehat{\phi(x/t)} \rangle = \langle f, |t|^n \hat{\phi}(t\cdot) \rangle$$
$$= |t|^{-p} \langle f, \hat{\phi} \rangle = |t|^{n+(-n-p)} \langle \hat{f}, \phi \rangle,$$

which means that $\hat{f}$ is an even, homogeneous distribution of degree $-n-p$. □

For any open set $Q \subset \mathbb{C}$ we say that a family of distributions f_q, $q \in Q$ is analytic on Q if for every test function ϕ, the function

$$q \to \langle f_q, \phi \rangle$$

is analytic on Q. This concept is important for us, because we prove certain formulas for $q \in (-1, 0)$ and then use analytic extension to show that these formulas hold for other values of q. We use the following fact; see, for example, [310, Lemma 2.22].

Lemma 1.4.2. *For any star body K in $\mathbb{R}^n$ the family of distributions $\|x\|_K^q$ is analytic in the domain $q \in \mathbb{C}$, $\mathbb{R}(q) > -n$. The family of Fourier transforms $(\|\cdot\|_K^q)^\wedge$ is also analytic in the same domain.*

Proof. By Lemma 1.1.1, the function $\|x\|_K^q$ is locally integrable on $\mathbb{R}^n$. Therefore, for any test function ϕ

$$h(q) = \langle \|x\|_K^q, \phi \rangle = \int_{\mathbb{R}^n} \|x\|_K^q \phi(x)\, dx.$$

Differentiating under the integral, it is easy to see that the derivative $h'(q)$ exists everywhere in the domain $\mathbb{R}(q) > -n$ and is equal to

$$h'(q) = \int_{\mathbb{R}^n} \|x\|_K^q \log(\|x\|)\phi(x)\, dx.$$

Note that an easy way of proving the possibility of differentiation by q under the integral is to do it in the opposite direction – integrate by q the equality for the derivatives. □

For any multiindex α, the derivative of the order α of a distribution f is defined by

$$\langle D^\alpha f, \phi \rangle = (-1)^{\alpha_1+\cdots+\alpha_n} \langle f, D^\alpha \phi \rangle.$$

The Fourier transform is related to differentiation as follows:

$$(D^\alpha f)^\wedge = i^{\alpha_1+\cdots+\alpha_n} x_1^{\alpha_1} \cdots x_n^{\alpha_n} \hat{f} \tag{1.45}$$

and

$$D^\alpha \hat{f} = (-i)^{\alpha_1+\cdots+\alpha_n} (x_1^{\alpha_1} \cdots x_n^{\alpha_n} f(x))^\wedge,$$

where the product of a polynomial and a distribution is defined by

$$\langle x_1^{\alpha_1}\cdots x_n^{\alpha_n} f,\phi\rangle = \langle f, x_1^{\alpha_1}\cdots x_n^{\alpha_n}\phi\rangle.$$

We now compute the Fourier transform of powers of the Euclidean norm. For the definition and properties of the Γ-function, see, for example, [310, Section 2.4].

Lemma 1.4.3. *Let $0 < p < n$. Then, the Fourier transform of $|x|^{-p}$ in $\mathbb{R}^n$ is the distribution represented by the locally integrable function*

$$(|x|^{-p})^\wedge(\xi) = \frac{2^{n-p}\pi^{\frac{n}{2}}\Gamma(\frac{n-p}{2})}{\Gamma(\frac{p}{2})}|\xi|^{-n+p}, \quad \xi \in \mathbb{R}^n \setminus \{0\}.$$

Proof. First, suppose that $p \in (0,1)$. Making the substitution $u = z^2|x|^2/2$ in the convergent integral and using the definition of the Γ-function, we get

$$|x|^{-p} = \frac{2^{-p/2+1}}{\Gamma(p/2)} \int_0^\infty z^{p-1} e^{-z^2|x|^2/2}\, dz.$$

The Fourier transform of the standard gaussian density is well known (and easily computable)

$$(e^{-|x|^2/2})^\wedge(\xi) = (2\pi)^{n/2} e^{-|\xi|^2/2}, \quad \xi \in \mathbb{R}^n.$$

Now, for any even test function ϕ we get

$$\begin{aligned}\langle (|x|^{-p})^\wedge, \phi\rangle &= \langle |x|^{-p}, \hat{\phi}\rangle = \int_{\mathbb{R}^n} |x|^{-p}\hat{\phi}(x)\, dx \\ &= \frac{2^{-p/2+1}}{\Gamma(p/2)} \int_{\mathbb{R}^n}\left(\int_0^\infty z^{p-1}e^{-z^2|x|^2/2}\, dz\right)\hat{\phi}(x)\, dx \\ &= \frac{2^{-p/2+1}}{\Gamma(p/2)} \int_0^\infty z^{p-1}\left(\int_{\mathbb{R}^n} e^{-(|zx|)^2/2}\hat{\phi}(x)\, dx\right) dz.\end{aligned}$$

Applying Parseval's identity (1.43) to the integral over $\mathbb{R}^n$, using

$$(f(zx))^\wedge(\xi) = z^{-n}\hat{f}(\xi/z) \tag{1.46}$$

and making the substitution $u = |\xi|^2/2z^2$ in the integral over $[0,\infty)$, we get

$$\langle (|x|^{-p})^\wedge, \phi\rangle = \frac{2^{-p/2+1}}{\Gamma(p/2)} \int_0^\infty z^{p-1}\left(\int_{\mathbb{R}^n} (2\pi)^{n/2} z^{-n} e^{-|\xi|^2/2z^2}\phi(\xi)\, d\xi\right) dz$$

$$= \frac{(2\pi)^{n/2} 2^{-p/2+1}}{\Gamma(p/2)} \int_{\mathbb{R}^n} \left(\int_0^\infty z^{-n+p-1} e^{-|\xi|^2/2z^2}\, dz \right) \phi(\xi)\, d\xi$$

$$= \frac{2^{n-p}\pi^{n/2}\Gamma((n-p)/2)}{\Gamma(p/2)} \int_{\mathbb{R}} |\xi|^{-n+p} \phi(\xi)\, d\xi.$$

This shows that for every even test function ϕ

$$\langle (|x|^{-p})^\wedge, \phi \rangle = \frac{2^{n-p}\pi^{n/2}\Gamma((n-p)/2)}{\Gamma(p/2)} \langle |\xi|^{-n+p}, \phi \rangle, \tag{1.47}$$

which proves the lemma for every $p \in (0,1)$. The distributions $|x|^{-p}$ are defined in such a way that both sides of (1.47) are analytic functions of p everywhere in $p \in \mathbb{C}$, $-n < \Re p < 0$; see Lemma 1.4.2. By the uniqueness of analytic continuation, we can extend the formula from the interval $(0,1)$ to all such p. □

In the case $n = 1$ the coefficient in the formula of Lemma 1.4.3 was written in a different form in [310, Lemma 2.23].

Corollary 1.4.4. *If $n = 1$, $q \in (-1,0)$, then for every $\xi \in \mathbb{R}$, $\xi \neq 0$*

$$(|t|^q)^\wedge(\xi) = -2\Gamma(1+q)\sin(\pi q/2)\, |\xi|^{-q-1}.$$

The results of Theorem 1.4.3 and Corollary 1.4.4 can be extended to other values of p and q. We refer the reader to [205] and [310, p. 36].

1.4.2 Positive definite distributions

We say that a distribution f is positive definite if its Fourier transform is a positive distribution, i. e., $\langle \hat{f}, \phi \rangle \geqslant 0$ for every non-negative test function ϕ. L. Schwartz's generalization of Bochner's theorem states that a distribution is positive definite if and only if it is the Fourier transform of a tempered measure on $\mathbb{R}^n$; see [206]. Recall that a (non-negative, not necessarily finite) measure μ is called tempered if

$$\int_{\mathbb{R}^n} (1+|x|)^{-\beta}\, d\mu(x) < \infty$$

for some $\beta > 0$.

We mostly deal with positive definite distributions that are at the same time homogeneous. The corresponding tempered measures are also homogeneous as distributions and can be written in polar coordinates, as was proved in [310, Lemma 2.24]:

Lemma 1.4.5. *Let μ be a tempered measure on $\mathbb{R}^n$ that is, at the same time, an even, homogeneous distribution of degree $-n+p$, $p \in (0,n)$. Then, there exists a finite Borel measure μ_0 on the sphere S^{n-1} such that, for every even test function ϕ*

$$\langle \mu, \phi \rangle = \int_{S^{n-1}} \left(\int_0^\infty t^{p-1} \phi(t\xi)\, dt \right) d\mu_0(\xi). \tag{1.48}$$

We also consider measures that are homogeneous distributions of degree $-n-p$, where $p > 0$. In this case, the integral in (1.48) may diverge, so we consider only test functions ϕ supported outside of the origin.

Let D be an open set in $\mathbb{R}^n$, and let $\mathcal{S}(D)$ be the space of test functions with compact support in D. We say that two distributions f, g are equal on D if $\langle f, \phi \rangle = \langle g, \phi \rangle$ for every $\phi \in \mathcal{S}(D)$.

A distribution f is said to be positive in D if $\langle f, \phi \rangle \geqslant 0$ for every non-negative test function $\phi \in \mathcal{S}(D)$.

The following fact was proved in [310, Lemma 2.25].

Lemma 1.4.6. *Let $0 < a < b < \infty$. Suppose that F is a positive distribution in the domain $(a,b) \times S^{n-1}$. Then, for every interval $[c,d] \subset (a,b)$ there exists a finite Borel measure μ on $[c,d] \times S^{n-1}$ such that $F = \mu$ on $[c,d] \times S^{n-1}$.*

If $p > 0$ and $F \in \mathcal{S}(\mathbb{R}^n)$ is a positive outside of the origin even, homogeneous distribution of degree $-n-p$, then there exists a finite Borel measure μ_0 on S^{n-1} such that for every even test function ϕ with compact support outside of the origin,

$$\langle F, \phi \rangle = \int_{S^{n-1}} \left(\int_0^\infty t^{-1-p} \phi(t\xi)\, dt \right) d\mu_0(\xi).$$

We are going to use Lemmas 1.4.5 and 1.4.6 in the following form; see [310, Corollary 2.26]. Note that the constant $\Gamma(-p/2)$ in the second part does not play any role here and is needed only for future applications.

Proposition 1.4.7. (i) *Let K be an origin-symmetric star body in $\mathbb{R}^n$ and $p \in (0,n)$. The function $\|\cdot\|_K^{-p}$ represents a positive definite distribution on $\mathbb{R}^n$ if and only if there exists a finite Borel measure μ_0 on S^{n-1} such that for every even test function ϕ*

$$\int_{\mathbb{R}^n} \|x\|_K^{-p} \phi(x)\, dx = \int_{S^{n-1}} \left(\int_0^\infty t^{p-1} \widehat{\phi}(t\xi)\, dt \right) d\mu_0(\xi). \tag{1.49}$$

(ii) *Suppose that $p > 0$ is not an even integer. The distribution $\Gamma(-p/2)(\|\cdot\|_K^p)^\wedge$ is positive on $\mathbb{R}^n \setminus \{0\}$ if and only if there exists a finite Borel measure μ_0 on S^{n-1} such that, for* *every even test function ϕ whose Fourier transform is supported in $\mathbb{R}^n \setminus \{0\}$,*

$$\int_{\mathbb{R}^n} \|x\|_K^p \phi(x)\,dx = \frac{1}{\Gamma(-p/2)} \int_{S^{n-1}} \left(\int_0^\infty t^{-1-p} \widehat{\phi}(t\xi)\,dt \right) d\mu_0(\xi). \tag{1.50}$$

Proof. The "if" parts in both statements are straightforward. In fact, if we have (1.49), then for any non-negative even test function ϕ

$$\begin{aligned}\langle (\|x\|^{-p})^\wedge, \phi \rangle &= \langle \|x\|^{-p}, \widehat{\phi} \rangle = \int_{\mathbb{R}^n} \|x\|^{-p} \widehat{\phi}(x)\,dx \\ &= (2\pi)^n \int_{S^{n-1}} \left(\int_0^\infty t^{p-1} \phi(t\xi)\,dt \right) d\mu_0(\xi) \geqslant 0.\end{aligned}$$

The second statement is similar; we start with a non-negative test function ϕ supported on $\mathbb{R}^n \setminus \{0\}$ and apply (1.50) to the function $\widehat{\phi}$.

Now, we prove the "only if" implications. For the first part, the function $\|\cdot\|_K^{-p}$ is the Fourier transform of a tempered measure μ that, by Lemma 1.4.1, is an even, homogeneous distribution of degree $-n+p$. By Lemma 1.4.5, there exists a finite Borel measure μ_0 on S^{n-1} such that for any even test function ϕ

$$\int_{\mathbb{R}^n} \|x\|_K^{-p} \phi(x)\,dx = \langle \widehat{\mu}, \phi \rangle = \langle \mu, \widehat{\phi} \rangle = \int_{S^{n-1}} \left(\int_0^\infty t^{p-1} \widehat{\phi}(t\xi)\,dt \right) d\mu_0(\xi).$$

For the second part, the function $\Gamma(-p/2)\|\cdot\|_K^p$ is the Fourier transform of the distribution

$$F = \frac{\Gamma(-p/2)}{(2\pi)^n} (\|\cdot\|_K^p)^\wedge,$$

which is positive in $\mathbb{R}^n \setminus \{0\}$ and, again by Lemma 1.4.1, even, homogeneous of degree $-n-p$. For any even test function ϕ whose Fourier transform is supported outside of the origin, we have

$$\int_{\mathbb{R}^n} \|x\|_K^p \phi(x)\,dx = \frac{1}{\Gamma(-p/2)} \langle F, \widehat{\phi} \rangle$$

and the result follows from Lemma 1.4.6. □

1.4.3 Fourier transform of homogeneous functions

In this subsection, we establish some properties of the Fourier transform of homogeneous functions. If f is an even, continuous function on S^{n-1} and $p \in \mathbb{R}$, we denote by

$$f \cdot r^p = f(\theta)r^p = f(x/|x|)|x|^p$$

an even, homogeneous of degree p function of the variable $x = r\theta \in \mathbb{R}^n$, $r \in \mathbb{R}$, $\theta \in S^{n-1}$. When we write a function in the form $f \cdot r^p$, we actually mean a function of the variable x, as in the formula above. If $p \neq -n-k$ for even, non-negative integers k, then the action of the distribution $f \cdot r^p$ on a test function ϕ can be written as

$$\langle f \cdot r^p, \phi \rangle = \langle f(x/|x|)|x|^p, \phi(x) \rangle = \frac{1}{2} \int_{S^{n-1}} f(\theta) \langle |z|^{n+p-1}, \phi(z\theta) \rangle \, d\theta.$$

If $p > -n$ or if ϕ is supported outside of the origin, then the right-hand side of this equation takes the form

$$\frac{1}{2} \int_{S^{n-1}} f(\theta) \left(\int_0^\infty |z|^{n+p-1} \phi(z\theta) \, dz \right) d\theta.$$

We know from Lemma 1.4.1 that the Fourier transform of a homogeneous function of degree p is a homogeneous distribution of degree $-n-p$. We want to know when this Fourier transform is a continuous function on $\mathbb{R}^n \setminus \{0\}$, and to study the convergence of such Fourier transforms.

Lemma 1.4.8. (i) *Suppose that $p > -n$ and let f_k, $k \in \mathbb{N}$ and f be even, continuous functions on S^{n-1} such that $f_k \to f$ in $C(S^{n-1})$. Then, for every even test function ϕ*

$$\lim_{k\to\infty} \langle (f_k \cdot r^p)^\wedge, \phi \rangle = \langle (f \cdot r^p)^\wedge, \phi \rangle.$$

(ii) *Let $\delta > 0$ and let $\{p_k\}_{k=1}^\infty$ be a sequence of numbers such that $p_k > -n + \delta$ for all k and $\lim_{k\to\infty} p_k = p$. Then, for every $f \in C(S^{n-1})$ and every even test function ϕ*

$$\lim_{k\to\infty} \langle (f \cdot r^{p_k})^\wedge, \phi \rangle = \langle (f \cdot r^p)^\wedge, \phi \rangle.$$

Proof. The functions f_k are uniformly bounded by some constant C. Since $p > -n$, arguing as in Lemma 1.1.1 we see that the functions $f_k \cdot r^p$ are locally integrable on $\mathbb{R}^n$. The test function $\widehat{\phi} \in \mathcal{S}(\mathbb{R}^n)$ decreases infinitely faster than any power of the Euclidean norm, so the functions $f_k \cdot r^p \widehat{\phi}(x)$ are integrable on $\mathbb{R}^n$. These functions are bounded by an integrable function $C \cdot r^p |\widehat{\phi}(x)|$ and converge pointwise to $f \cdot r^p \widehat{\phi}(x)$. By the Lebesgue dominated convergence theorem

$$\begin{aligned} \lim_{k\to\infty} \langle (f_k \cdot r^p)^\wedge, \phi \rangle &= \lim_{k\to\infty} \int_{\mathbb{R}^n} f_k \cdot r^p \widehat{\phi}(x) \, dx = \int_{\mathbb{R}^n} f \cdot r^p \widehat{\phi}(x) \, dx \\ &= \langle (f \cdot r^p)^\wedge, \phi \rangle, \end{aligned}$$

which proves the first part.

Now, for the second part, the functions $f \cdot r^{p_k}$ are locally integrable, since each $p_k > -n$. Hence, the functions $f \cdot r^{p_k}\widehat{\phi}(x)$ are integrable on $\mathbb{R}^n$. There exists a constant $s > -n$ such that $p_k < s$ for each k. Now, the functions r^{p_k} are bounded by the function $h(r) = \max(r^s, r^{-n+\delta})$. Therefore, the functions $f \cdot r^{p_k}\widehat{\phi}(x)$ are bounded by the integrable function $|f| \cdot h(r)|\widehat{\phi}(x)|$, and they converge pointwise to $f \cdot r^p\widehat{\phi}(x)$. We can finish the proof in the same way as in the first part. □

We shall use the following well-known formula.

Lemma 1.4.9. *For every $q > -1$ and every $x \in S^{n-1}$, we have*

$$\int_{S^{n-1}} |\langle x, \xi\rangle|^q \, d\xi = \frac{2\Gamma((q+1)/2)\pi^{(n-1)/2}}{\Gamma((n+q)/2)}.$$

Proof. We use the formula

$$\int_{S^{n-1}} \Phi(\langle x, \xi\rangle) \, d\xi = \kappa_{n-1} \int_{-1}^{1} \Phi(t)(1-t^2)^{(n-3)/2} \, dt,$$

which holds true for any integrable function Φ on $[-1,1]$ and every $x \in S^{n-1}$. We may apply this formula for $\Phi(t) = |t|^q$, because $q > -1$ and hence Φ is integrable. Using the formulas (see, for example, [310, p. 31–32])

$$\frac{\Gamma((q+1)/2)\Gamma((n-1)/2)}{\Gamma((n+q)/2)} = \int_{-1}^{1} |t|^q (1-t^2)^{(n-3)/2} dt$$

and

$$\kappa_{n-1} = \frac{2\pi^{(n-1)/2}}{\Gamma((n-1)/2)},$$

we conclude the proof. □

Lemma 1.4.10. *Let u be a continuous function on S^{n-1}. For every $q > -1$ the function*

$$h(x) = \int_{S^{n-1}} |\langle x, \xi\rangle|^q u(\xi) \, d\xi$$

is continuous on S^{n-1}, and

$$\|h\|_{C(S^{n-1})} \leqslant \frac{2\Gamma((q+1)/2)\pi^{(n-1)/2}}{\Gamma((n+q)/2)} \|u\|_{C(S^{n-1})}.$$

Proof. Let (x_k) be a sequence of points in S^{n-1} converging to $x \in S^{n-1}$. We can find a sequence of rotations T_k such that $T_k(x) = x_k$ and

$$\|u \circ T_k - u\|_{C(S^{n-1})} \to 0,$$

as $k \to \infty$, because u is uniformly continuous on the compact set S^{n-1}. Then,

$$h(x_k) - h(x) = \int_{S^{n-1}} |\langle x, \xi \rangle|^q (u(T_k(\xi)) - u(\xi))\, d\xi$$

and the lemma follows from Lemma 1.4.9. □

Our next lemma is a combination of Corollary 1.4.4 and Lemma 1.3.1.

Lemma 1.4.11. *Let $q > -1$, where q is not an even integer. Then, for every even test function $\phi \in \mathcal{S}(\mathbb{R}^n)$ supported outside of the origin and every fixed vector $\xi \in \mathbb{R}^n$, $\xi \neq 0$, we have*

$$\int_{\mathbb{R}^n} |\langle x, \xi \rangle|^q \hat{\phi}(x)\, dx = -2(2\pi)^{n-1} \Gamma(q+1) \sin(\pi q/2) \int_{\mathbb{R}} |t|^{-1-q} \phi(t\xi)\, dt. \tag{1.51}$$

Proof. First, let $q \in (-1, 0)$. Because of homogeneity, it is enough to prove the lemma for $\xi \in S^{n-1}$. Since ϕ is an even function, we have $(\hat{\phi})^\wedge = (2\pi)^n \phi$. By Lemma 1.3.1, the function $t \to (2\pi)^n \phi(t\xi)$ is the Fourier transform of the function $z \to \int_{\{x: \langle x,\xi \rangle = z\}} \hat{\phi}(x)\, dx$. Using the Fubini theorem and Corollary 1.4.4 we get

$$\begin{aligned}
\int_{\mathbb{R}^n} |\langle x, \xi \rangle|^q\, \hat{\phi}(x)\, dx &= \int_{\mathbb{R}} |z|^q \Big(\int_{\{x: \langle x,\xi \rangle = z\}} \hat{\phi}(x)\, dx \Big)\, dz \\
&= \Big\langle |z|^q, \int_{\{x: \langle x,\xi \rangle = z\}} \hat{\phi}(x)\, dx \Big\rangle \\
&= \frac{1}{2\pi} \langle (-2\Gamma(1+q) \sin(\pi q/2)) |t|^{-1-q}, (2\pi)^n \phi(t\xi) \rangle \\
&= (2\pi)^{n-1} (-2\Gamma(1+q) \sin(\pi q/2)) \int_{\mathbb{R}} |t|^{-1-q} \phi(t\xi)\, dt.
\end{aligned}$$

Now, since both sides of (1.51) are analytic functions of q in the domain $\Re q > -1$, we can use analytic continuation by q. □

Corollary 1.4.12. *Suppose that $q > -1$ is not an even integer and let u be an even, continuous function on S^{n-1}. Define a function F on $\mathbb{R}^n$ by*

$$F(x) = \int_{S^{n-1}} |\langle x, \xi \rangle|^q\, u(\xi)\, d\xi.$$

Then, the Fourier transform $\widehat{F}$ is a homogeneous of degree $-n-q$, continuous function on $\mathbb{R}^n \setminus \{0\}$ such that for every $y \in \mathbb{R}^n \setminus \{0\}$

$$\widehat{F}(y) = -4(2\pi)^{n-1}\Gamma(q+1)\sin(\pi q/2)\, u \cdot r^{-n-q}.$$

Proof. For any even test function ϕ supported outside of the origin, using Lemma 1.4.11 we get

$$\begin{aligned}\langle \widehat{F}, \phi\rangle = \langle F, \widehat{\phi}\rangle &= \int_{S^{n-1}} u(\xi)\Big(\int_{\mathbb{R}^n} |\langle x,\xi\rangle|^q \widehat{\phi}(x)\,dx\Big)\,d\xi \\ &= -2(2\pi)^{n-1}\Gamma(q+1)\sin(\pi q/2)\int_{S^{n-1}} u(\xi)\Big(\int_{\mathbb{R}} |t|^{-1-q}\phi(t\xi)\,dt\Big)\,d\xi \\ &= -4(2\pi)^{n-1}\Gamma(q+1)\sin(\pi q/2)\,\langle u\cdot r^{-n-q}, \phi\rangle.\end{aligned}$$ □

The next lemma shows that the Fourier transform of a smooth homogeneous function is also smooth.

Lemma 1.4.13. *Let $k \geqslant 0$ and $f \in C^{2k}(S^{n-1})$ be an even function. Let $q \leqslant 2k$, where q is not an odd integer. Then, we have the following.*

(i) *The Fourier transform of the distribution $f \cdot r^{-n+q+1}$ is a homogeneous of degree $-(q+1)$, continuous function on $\mathbb{R}^n \setminus \{0\}$. If $q < 2k$, then, for every $x \in \mathbb{R}^n$*

$$|x|^{2k}(f\cdot r^{-n+q+1})^\wedge(x) = c_{k,q}\int_{S^{n-1}} |\langle x,\xi\rangle|^{2k-q-1}\Delta^k(f\cdot r^{-n+q+1})(\xi)\,d\xi, \tag{1.52}$$

where

$$c_{k,q} = \frac{(-1)^k\pi}{-2\Gamma(2k-q)\sin(\pi(2k-q-1)/2)}.$$

If $q = 2k$, then

$$|x|^{2k}(f\cdot r^{-n+q+1})^\wedge(x) = (-1)^k\pi|x|^{-1}\int_{S^{n-1}} \Delta^k(f\cdot r^{-n+q+1})(\xi)\,d\xi. \tag{1.53}$$

Here, $\Delta = \sum_{i=1}^n \partial^2/\partial x_i^2$ is the Laplace operator on $\mathbb{R}^n$.

(ii) *If $f \in C^\infty(S^{n-1})$, then there exists an even function $g \in C^\infty(S^{n-1})$ such that, for every $x = t\xi \in \mathbb{R}^n$, where $t \neq 0$ and $\xi \in S^{n-1}$*

$$(f\cdot r^{-n+q+1})^\wedge(x) = g(\xi)t^{-1-q}$$

and hence the Fourier transform of $f \cdot r^{-n+q+1}$ is an infinitely smooth function on $\mathbb{R}^n \setminus \{0\}$.

Proof. For any even test function ϕ such that $\widehat{\phi}$ is supported outside of the origin, by the connection between differentiation and the Fourier transform,

$$\langle |x|^{2k}(f\cdot r^{-n+q+1})^{\wedge}(x), \phi(x)\rangle = (-1)^k\langle \Delta^k(f\cdot r^{-n+q+1}), \widehat{\phi}\rangle.$$

On the other hand, if $q < 2k$, where q is not an odd integer, then, by Corollary 1.4.12

$$\begin{aligned}
&\Big\langle \int_{S^{n-1}} |\langle x,\xi\rangle|^{2k-q-1}\Delta^k(f\cdot r^{-n+q+1})(\xi)\,d\xi, \phi(x)\Big\rangle \\
&= \frac{1}{(2\pi)^n}\Big\langle \Big(\int_{S^{n-1}} |\langle x,\xi\rangle|^{2k-q-1}\Delta^k(f\cdot r^{-n+q+1})(\xi)\,d\xi\Big)^{\wedge}(u), \widehat{\phi}(u)\Big\rangle \\
&= \frac{-4(2\pi)^{n-1}\Gamma(2k-q)\sin(\pi(2k-q-1)/2)}{(2\pi)^n}\langle \Delta^k(f\cdot r^{-n+q+1}), \widehat{\phi}\rangle,
\end{aligned}$$

because $\Delta^k(f\cdot r^{-n+q+1})$ is a homogeneous function of degree $-n-2k+q+1$, which is continuous on $\mathbb{R}^n\setminus\{0\}$. This shows that the functions on both sides of (1.52), considered as distributions, coincide on test functions ϕ whose Fourier transform is supported outside of the origin. Therefore, they can differ only by a polynomial. But this polynomial must be zero because both functions are even, homogeneous of degree $2k-q-1$ that is not an even integer.

If $q = 2k$ the formula (1.53) follows from Lemma 1.5.1, because the function $\Delta^k(f\cdot r^{-n+q+1})$ is homogeneous of degree $-n+1$, and then the argument is the same as in the case $q < 2k$.

Finally, if f is infinitely smooth, then we can choose k in (1.52) as large as we want. The function of x in the right-hand side of (1.52) is $[2k-q-3]$ times differentiable with respect to the variable x. To see this, first note that for $p > 1$

$$\frac{\partial}{\partial x_i}|\langle x,\xi\rangle|^p = p\xi_i|\langle x,\xi\rangle|^{p-1}\mathrm{sgn}(\langle x,\xi\rangle)$$

and

$$\frac{\partial}{\partial x_i}|\langle x,\xi\rangle|^p\mathrm{sgn}(\langle x,\xi\rangle) = p\xi_i|\langle x,\xi\rangle|^{p-1}$$

and then use a standard justification of differentiation under the integral (the best way to do it is backwards, by integration of the derivative). □

An immediate consequence of Lemma 1.4.13 and Lemma 1.4.10 is the following proposition.

Proposition 1.4.14. *Let $k \geqslant 0$ and $q \leqslant 2k$. If $f \in C^{2k}(S^{n-1})$, then the Fourier transform $(f\cdot r^{-n+q+1})^{\wedge}$ is a homogeneous of degree $-q-1$, continuous function on $\mathbb{R}^n\setminus\{0\}$.*

Moreover, if $f_m, f \in C^{2k}(S^{n-1})$ and $f_m \to f$ in $C^{2k}(S^{n-1})$, then the Fourier transforms $(f_m \cdot r^{-n+q+1})^\wedge$ converge to $(f \cdot r^{-n+q+1})^\wedge$ in the space $C(S^{n-1})$.

Throughout the text, we use the notation $(f \cdot r^{-n+q+1})^\wedge(\xi)$, $\xi \in S^{n-1}$ when it is clear that $(f \cdot r^{-n+q+1})^\wedge$ is a continuous function on the sphere extended to a homogeneous function on $\mathbb{R}^n \setminus \{0\}$.

1.5 Volume of sections and the Fourier transform

The Fourier analytic approach to sections of convex bodies is based on a number of formulas expressing volume in terms of the Fourier transform. Our first result applies to arbitrary origin-symmetric star bodies. It was proved in [304]; see also [310, Lemma 3.7].

Lemma 1.5.1. *Let f be an even, continuous function on S^{n-1}. Then, the Fourier transform of $f \cdot r^{-n+1}$ is equal to $\pi Rf \cdot r^{-1}$, where Rf is the spherical Radon transform of f.*

Proof. Since f is even, we may consider only even test functions ϕ. We have

$$\langle (f \cdot r^{-n+1})^\wedge, \phi \rangle = \int_{\mathbb{R}^n} f \cdot r^{-n+1}(x)\hat{\phi}(x)\,dx = \int_{S^{n-1}} f(\theta)\left(\int_0^\infty \hat{\phi}(t\theta)\,dt\right)d\theta. \tag{1.54}$$

Using Lemma 1.3.1 and the fact that ϕ is even, we see that for every fixed $\theta \in S^{n-1}$ the Fourier transform of the function $t \mapsto \hat{\phi}(t\theta)$ is equal to the function

$$z \mapsto 2\pi \int_{\{x:\langle x,\theta\rangle = z\}} \phi(x)\,dx,$$

therefore

$$2\int_0^\infty \hat{\phi}(t\theta)\,dt = \int_{-\infty}^\infty \hat{\phi}(t\theta)\,dt = (\hat{\phi}(t\theta))^\wedge(0) = 2\pi \int_{\{x:\langle x,\theta\rangle = 0\}} \phi(x)\,dx.$$

Substituting the latter integral in (1.5.1) and writing it in polar coordinates of the hyperplane $\langle x, \theta\rangle = 0$, we get that the expression in (1.5.1) is equal to

$$\begin{aligned}&\pi \int_{S^{n-1}} f(\theta)\left(\int_{S^{n-1}\cap\theta^\perp}\left(\int_0^\infty r^{n-2}\phi(r\xi)\,dr\right)d\xi\right)d\theta\\ &= \pi \int_{S^{n-1}} f(\theta) R\left(\int_0^\infty r^{n-2}\phi(r\xi)\,dr\right)(\theta)\,d\theta.\end{aligned} \tag{1.55}$$

We now use the self-duality of the spherical Radon transform (Lemma 1.3.3) with the functions $f(\theta)$ and $g(\xi) = \int_0^\infty r^{n-2}\phi(r\xi)\,dr$. We get that the expression in (1.55) is equal to

$$\pi \int_{S^{n-1}} \left(\int_0^\infty r^{n-2}\phi(r\xi)\,dr \right) Rf(\xi)\,d\xi$$
$$= \pi \int_{S^{n-1}} \left(\int_0^\infty r^{n-2}\phi(r\xi)\,dr \right) \left(\int_{S^{n-1}\cap\xi^\perp} f(\theta)\,d\theta \right) d\xi$$
$$= \pi \int_{\mathbb{R}^n} |x|^{-1} \left(\int_{S^{n-1}\cap(x/|x|)^\perp} f(\theta)\,d\theta \right) \phi(x)\,dx,$$

where the last equality can be verified by moving to the polar coordinates $x = r\xi$.

We see now that the distribution $\hat{f}$ is equal to the locally integrable continuous on $\mathbb{R}^n \setminus \{0\}$ function

$$\pi|x|^{-1} \int_{S^{n-1}\cap(x/|x|)^\perp} f(\theta)\,d\theta = \pi Rf \cdot r^{-1}.$$

In particular, for every unit vector $\xi \in S^{n-1}$, we get $\hat{f}(\xi) = \pi Rf(\xi)$. □

For a lower-dimensional version of this formula, see Lemma 1.6.7.

As an application of Lemma 1.5.1, we obtain a formula expressing the volume of central hyperplane sections of origin-symmetric star bodies in terms of the Fourier transform. This formula was proved in [304].

Theorem 1.5.2. *Let K be an origin-symmetric star body in $\mathbb{R}^n$. The Fourier transform of the function $\|\cdot\|_K^{-n+1}$ is a homogeneous of degree -1 function on $\mathbb{R}^n$, continuous on $\mathbb{R}^n \setminus \{0\}$ and such that, for every $\xi \in S^{n-1}$*

$$A_{K,\xi}(0) = |K \cap \xi^\perp| = \frac{1}{\pi(n-1)} (\|\cdot\|_K^{-n+1})^\wedge(\xi).$$

Proof. Apply Lemma 1.5.1 to the function $f(\theta) = \|\theta\|_K^{-n+1}$, $\theta \in S^{n-1}$ and use (1.35). We get that $(\|\cdot\|_K^{-n+1})^\wedge$ is a continuous function on $\mathbb{R}^n \setminus \{0\}$ and, for every $\xi \in S^{n-1}$

$$(\|\cdot\|_K^{-n+1})^\wedge(\xi) = \pi R(\|\cdot\|_K^{-n+1})(\xi) = \pi(n-1)|K \cap \xi^\perp|.$$ □

Theorem 1.5.2 provides a proof of the Minkowski–Funk uniqueness theorem for sections.

Corollary 1.5.3. *Let K, L be origin-symmetric star bodies in $\mathbb{R}^n$ such that, for every direction $\xi \in S^{n-1}$*

$$|K \cap \xi^\perp| = |L \cap \xi^\perp|.$$

Then, $K = L$.

Proof. By Theorem 1.5.2, we get that the Fourier transforms of the (homogeneous of degree $-n+1$) distributions $\|\cdot\|_K^{-n+1}$ and $\|\cdot\|_L^{-n+1}$ are equal on $\mathbb{R}^n$. This implies that $\|x\|_K = \|x\|_L$ for every $x \in \mathbb{R}^n$, so $K = L$. □

We can also prove a more general uniqueness theorem for the spherical Radon transform in arbitrary dimensions.

Corollary 1.5.4. *Let $1 \leqslant m < n$ and $f, g \in C(S^{n-1})$ be two even functions such that for any m-dimensional subspace H of $\mathbb{R}^n$*

$$\int_{S^{n-1}\cap H} f(\theta)\, d\theta = \int_{S^{n-1}\cap H} g(\theta)\, d\theta. \tag{1.56}$$

Then, $f = g$.

Proof. Fix any $\xi \in S^{n-1}$ and denote by $G_\xi(n-1, m)$ the Grassmannian of m-dimensional subspaces of $\xi^\perp$. By (1.40), integrating both sides of (1.56) over $G_\xi(n-1, m)$ we get

$$\int_{S^{n-1}\cap\xi^\perp} f(\theta)\, d\theta = \int_{S^{n-1}\cap\xi^\perp} g(\theta)\, d\theta.$$

Extend f, g to homogeneous functions of degree $-n+1$ on the whole $\mathbb{R}^n$. By Corollary 1.5.1, the Fourier transforms of the distributions $f \cdot r^{-n+1}$ and $g \cdot r^{-n+1}$ are equal. By the uniqueness theorem for the Fourier transform, $f = g$. □

An important role in what follows is played by formulas expressing the derivatives of the parallel section functions in terms of the Fourier transform. We start with the definition of fractional derivatives of a smooth function at zero. We refer the reader to [310, Section 2.6] for details and to [496] for more about fractional derivatives.

Let $m \in \mathbb{N} \cup \{0\}$ and suppose that h is an even, continuous function on $\mathbb{R}$ that is m times continuously differentiable in some neighborhood of zero. For $q \in \mathbb{C}$, $-1 < \Re(q) < m$, $q \neq 0, 1, \ldots, m-1$, the fractional derivative of the order q of the function h at zero is defined as the action of the distribution $t_+^{-1-q}/\Gamma(-q)$ on the function h, as follows:

$$h^{(q)}(0) = \frac{1}{\Gamma(-q)} \int_0^1 t^{-1-q}\left(h(t) - h(0) - \cdots - h^{(m-1)}(0)\frac{t^{m-1}}{(m-1)!}\right) dt$$

$$+\frac{1}{\Gamma(-q)}\int_1^\infty t^{-1-q}h(t)\,dt+\frac{1}{\Gamma(-q)}\sum_{k=0}^{m-1}\frac{h^{(k)}(0)}{k!(k-q)}. \tag{1.57}$$

It can be seen that for a fixed q the definition does not depend on the choice of $m > \Re(q)$, as long as h is m times continuously differentiable. Note that without dividing by $\Gamma(-q)$ the expression for the fractional derivative represents an analytic function in the domain $\{q \in \mathbb{C} : \Re(q) > -1\}$ not including integers, and has simple poles at integers. The function $\Gamma(-q)$ is analytic in the same domain and also has simple poles at non-negative integers, so after the division we get an analytic function in the whole domain $\{q \in \mathbb{C} : m > \Re(q) > -1\}$, which also defines fractional derivatives of integer orders. Moreover, computing the limit as $q \to k$, where k is a non-negative integer, we see that the fractional derivatives of integer orders coincide with usual derivatives up to a sign (when we compute the limit the first two summands in the right-hand side of (1.57) converge to zero, since $\Gamma(-q) \to \infty$, and the limit in the third summand can be computed using the property $\Gamma(x+1) = x\Gamma(x)$ of the Γ-function)

$$h^{(k)}(0) = (-1)^k \frac{d^k}{dt^k}h(t)\Big|_{t=0}. \tag{1.58}$$

The sign does not matter, because h is an even function, and its derivatives of odd orders at the origin are equal to zero. Also, in the case where h is even, for $m-2 < \Re q < m$ the expression (1.57) becomes

$$h^{(q)}(0) = \frac{1}{\Gamma(-q)}\int_0^\infty t^{-1-q}\left(h(t) - \sum_{j=0}^{(m-2)/2}\frac{t^{2j}}{(2j)!}h^{(2j)}(0)\right)dt. \tag{1.59}$$

We also note that if $-1 < q < 0$, then

$$h^{(q)}(0) = \frac{1}{\Gamma(-q)}\int_0^\infty t^{-1-q}h(t)\,dt. \tag{1.60}$$

To establish the existence of fractional derivatives of parallel section functions and more general Radon transforms of functions supported on a convex body K, we use a slight generalization of Lemma 2.4 from [310]. This lemma was proved in [310] under the assumption that the body K is convex. The convexity of K was needed only to make sure that $K \cap H$ is a star body for hyperplanes H close enough to the origin so that we can use the polar formula for volume. Yaskin [548] observed that convexity is not needed here, due to the following argument.

Lemma 1.5.5. *Let K be a star body in $\mathbb{R}^n$ with a C^1 radial function. Then, there is a ball centered at the origin such that K is a star body with respect to any point in this ball.*

Proof. Since the boundary of K is a C^1 hypersurface in $\mathbb{R}^n$, we have a unique tangent hyperplane at each point of $\mathrm{bd}(K)$. We will first prove that the origin O does not belong to any of the tangent planes to $\mathrm{bd}(K)$. Suppose this is not true, i. e., there is a point $p \in \mathrm{bd}(K)$ such that the tangent plane Π at this point contains the origin. Consider the 2-dimensional plane spanned by the normal vector to Π and the vector p (here and below we use the same letter to denote a point and the vector from the origin to this point). Without loss of generality, we can assume that Π is the x_1x_2-coordinate plane. Let us denote by $\rho(\phi)$, $\phi \in [0, 2\pi)$, the radial function (parameterized by the polar angle) of the planar curve $\Pi \cap \mathrm{bd}(K)$, so that the curve is given by the equations

$$\begin{cases} x_1 = \rho(\phi)\cos\phi, \\ x_2 = \rho(\phi)\sin\phi. \end{cases}$$

By our assumption the tangent line at p is parallel to p. On the other hand, the tangent vector equals $(\rho'(\phi)\cos\phi - \rho(\phi)\sin\phi, \rho'(\phi)\sin\phi + \rho(\phi)\cos\phi)$. If this vector is parallel to $(\cos\phi, \sin\phi)$, for some ϕ, then it is orthogonal to $(-\sin\phi, \cos\phi)$ and therefore

$$-\sin\phi(\rho'(\phi)\cos\phi - \rho(\phi)\sin\phi) + \cos\phi(\rho'(\phi)\sin\phi + \rho(\phi)\cos\phi) = \rho(\phi) = 0,$$

which is impossible.

Let n_p be the outward normal vector to $\mathrm{bd}(K)$ at $p \in \mathrm{bd}(K)$. Observe that this means that for any $\xi \in S^{n-1}$ such that $\langle \xi, n_p \rangle < 0$ (respectively, $\langle \xi, n_p \rangle > 0$) there is $\varepsilon_0 > 0$ such that for all $\varepsilon \in (0, \varepsilon_0)$ the points of the form $p + \varepsilon\xi$ are interior points of K (respectively, exterior points of K). As was shown above, $\langle p, n_p \rangle \neq 0$ for all $p \in \mathrm{bd}(K)$. Moreover, $\langle p, n_p \rangle > 0$, since the segment $[0, p)$ is in the interior of K. By continuity the following minimum is also positive:

$$\eta = \min_{p \in \mathrm{bd}(K)} \langle p, n_p \rangle > 0.$$

Consider the convex set

$$M = \bigcap_{p \in \mathrm{bd}(K)} \{x \in \mathbb{R}^n : \langle x, n_p \rangle < \langle p, n_p \rangle\}.$$

The set has a non-empty interior, as it contains the open ball B of radius η centered at the origin. Moreover, $M \subset K$. Indeed, if we had a point $q \in K \setminus M$, then the segment $[0, q]$ would contain a boundary point of K, say y, and the half-space

$$\{x \in \mathbb{R}^n : \langle x, n_y \rangle < \langle y, n_y \rangle\}$$

would not contain q (as it contains 0).

We claim that the body K is star-shaped with respect to any point $z \in B$. Indeed, suppose that this is not true and there is a point $z \in B$ and a direction $\xi \in S^{n-1}$ such that the ray $\gamma = \{x \in \mathbb{R}^n : x = z + t\xi, t \geqslant 0\}$ intersects the boundary of K in at least two points. Let u be the point on $\gamma \cap \mathrm{bd}(K)$ closest to z. Since the segment $[z, u)$ consists entirely of interior points of K, there is $\varepsilon_0 > 0$ such that for all $0 < \varepsilon < \varepsilon_0$ the points $u + \varepsilon\xi$ are not in K. Denote by v the point of $\gamma \cap \mathrm{bd}(K)$ closest to u (and different from u). Since the interval (u, v) lies outside of K, the half-space

$$\{x \in \mathbb{R}^n : \langle x, n_v\rangle < \langle v, n_v\rangle\}$$

does not contain the interval (u, v), and therefore does not contain 0. This is a contradiction to the construction of M.

Finally, we observe that the radial function of K with respect to any point $z \in B$ is a continuous function on the sphere (it is actually of class C^1, by the implicit function theorem). □

Let f be a function supported in a star body K. Our next lemma proves the differentiability of the function

$$t \to \mathcal{R}f(\xi, t) = \int_{K\cap\{x:\langle x,\xi\rangle=t\}} f(x)\, dx$$

in a neighborhood of zero, which assures that the fractional derivatives of this function at zero exist and are analytic with respect to the order of differentiation.

Lemma 1.5.6. *Let K be a C^m-smooth star body in $\mathbb{R}^n$, for some $m \geqslant 1$. Suppose that f is a C^m function on $\mathbb{R}^n$. Then, the function $t \to \mathcal{R}f(\xi, t)$ is m times continuously differentiable function of t (uniformly with respect to ξ) is some neighborhood U of zero. Moreover, for every $t \in U$, the derivative $(\mathcal{R}f(\xi, t))^{(m)}(t)$ is a continuous function of $\xi \in S^{n-1}$.*

Proof. We proceed as in [310, Lemma 2.4]. Fix a direction $\xi \in S^{n-1}$. For every t consider the hyperplane $H_t = \{x : \langle x, \xi\rangle = t\}$. As was shown in Lemma 1.5.5, K contains a Euclidean ball with radius r and center at the origin, such that K is a star body with respect to any point in this ball. We assume that $|t| < r$. Let S^{n-2} be the unit Euclidean sphere in H_t with center at the orthogonal projection of the origin to H_t (which is inside of $K \cap H_t$ since $|t| < r$), and let $\rho_{K\cap H_t}$ be the radial function of the star body $K \cap H_t$ in H_t with respect to this center. Moving to polar coordinates $x = t\xi + r\theta$ in H_t, we obtain

$$\mathcal{R}f(\xi, t) = \int_{K\cap H_t} f(x)\, dx = \int_{S^{n-2}} \int_0^{\rho_{K\cap H_t}} r^{n-2} f(t\xi + r\theta)\, dr\, d\theta. \tag{1.61}$$

Since f is uniformly differentiable (up to order m) on K, the question is reduced to uniform differentiability (with respect to θ and ξ) of $\rho_{K\cap H_t}(\theta)$ by the variable t. Fix θ and consider the two-dimensional plane passing through the origin and spanned by ξ and θ. Let D be the section of K by this plane, and let ρ_D be the radial function of D defined on $[0, 2\pi]$. Since K is C^m, the function ρ_D is C^m. Let us now get an implicit formula expressing $\rho_{K\cap H_t}(\theta)$ in terms of ρ_D. The line connecting the origin with the point on the boundary corresponding to $\rho_{K\cap H_t}(\theta)$ is at the angle $\phi = \arctan(\frac{t}{\rho_{K\cap H_t}(\theta)})$ with respect to the hyperplane $\xi^\perp$. Therefore, the radius at this point is equal to $\rho_D(\arctan(\frac{t}{\rho_{K\cap H_t}(\theta)}))$. From the right triangle

$$\rho_{K\cap H_t}(\theta) = \sqrt{\rho_D^2\left(\arctan\left(\frac{t}{\rho_{K\cap H_t}(\theta)}\right)\right) - t^2}. \tag{1.62}$$

This is an implicit formula that we are now going to use to compute the derivatives of $y(t) = \rho_{K\cap H_t}(\theta)$. Formula (1.62) now has the form

$$y = \sqrt{\rho_D^2\left(\arctan\left(\frac{t}{y}\right)\right) - t^2}.$$

Differentiating both sides by t and separating the derivative of y, we get

$$y'(t) = \frac{\rho_D(\arctan(\frac{t}{y}))\rho_D'(\arctan(\frac{t}{y}))\frac{1}{y^2+t^2}(y - ty'(t)) - t}{y},$$

therefore

$$y'(t) = \frac{\rho_D(\arctan(\frac{t}{y}))\rho_D'(\arctan(\frac{t}{y}))(1/(y^2+t^2))y - t}{y + t\rho_D(\arctan(\frac{t}{y}))\rho_D'(\arctan(\frac{t}{y}))\frac{1}{y^2+t^2}}.$$

Again, K contains a Euclidean ball with center at the origin, so ρ_D and y are separated from zero for small t with a bound independent from θ and ξ. Also, ρ_K is $C^m(S^{n-1})$, so ρ_D and all its derivatives up to the order m are bounded from above uniformly with respect to θ and ξ. This shows that the denominator in the expression for y' is positive for small t, so the derivative exists in a neighborhood of zero uniformly with respect to θ and ξ. Computing derivatives of higher orders is similar (note that we never get zero in the denominator if t is small). This shows that y is m times uniformly differentiable in some neighborhood of zero.

It follows from (1.61) that the uniform differentiability of $y = \rho_{K\cap H_t}(\theta)$ implies that $\mathcal{R}f(\xi, t)$ is uniformly (with respect to ξ) differentiable in a neighborhood of zero. This also implies the last statement of the lemma by induction by m. Given the result for $m-1$, we use the fact that, for every t in some neighborhood of zero, the function $\xi \to (\mathcal{R}f(\xi, t))^{(m)}(t)$ is the limit in $C(S^{n-1})$ (i. e., uniformly by ξ) of the continuous functions

$$\xi \to \frac{(\mathcal{R}f(\xi, t))^{(m-1)}(t+h) - (\mathcal{R}f(\xi, t))^{(m-1)}(t)}{h}$$

as $h \to 0$. □

Lemma 1.5.6 shows that if K is infinitely smooth and f is an infinitely smooth function on K, then fractional derivatives of the Radon transform of all orders exist at the origin.

The following formulas express the fractional derivatives of the parallel section function in terms of the Fourier transform. These formulas were proved in [202] (see also [310, Theorem 3.18]) and served as one of the main ingredients of the Fourier analytic solution of the Busemann–Petty problem; see Chapter 3.

Theorem 1.5.7. *Let K be an infinitely smooth, origin-symmetric star body in $\mathbb{R}^n$, $\xi \in S^{n-1}$. Then, for every $q \in (-1, \infty)$, $q \neq n-1$, the fractional derivative of order q at zero of the parallel section function $A_{K,\xi}$ is equal to*

$$A_{K,\xi}^{(q)}(0) = \frac{\cos(\pi q/2)}{\pi(n-q-1)}(\|\cdot\|_K^{-n+q+1})^\wedge(\xi). \tag{1.63}$$

In particular, if $k \geqslant 0$, $k \neq n-1$ is an even integer, then we get the usual derivative of the order k

$$(\|\cdot\|_K^{-n+k+1})^\wedge(\xi) = (-1)^{k/2}\pi(n-k-1)A_{K,\xi}^{(k)}(0). \tag{1.64}$$

If $k \geqslant 1$, $k \neq n-1$ is an odd integer, then

$$\begin{aligned}(\|\cdot\|_K^{-n+k+1})^\wedge(\xi) &= (-1)^{(k+1)/2}2(n-k-1)k! \\ &\times \int_0^\infty \frac{A_{K,\xi}(z) - A_{K,\xi}(0) - \cdots - A_{K,\xi}^{(k-1)}(0)\frac{z^{k-1}}{(k-1)!}}{z^{k+1}}\, dz.\end{aligned} \tag{1.65}$$

A generalization of Theorem 1.5.7 to functions was proved in [233].

Theorem 1.5.8. *Let K be an infinitely smooth, origin-symmetric star body in $\mathbb{R}^n$, let f be an even infinitely smooth function on K, and let $q \in (-1, n-1)$. Then, for every fixed $\xi \in S^{n-1}$*

$$(\mathcal{R}f(\xi,t))_t^{(q)}(0) = \frac{\cos(\pi q/2)}{\pi}\left(|x|^{-n+q+1}\left(\int_0^{\frac{|x|}{\|x\|_K}} r^{n-q-2} f\left(r\frac{x}{|x|}\right) dr\right)\right)_x^\wedge(\xi). \tag{1.66}$$

Proof. Let $-1 < q < 0$. By (1.60) and the Fubini theorem

$$\begin{aligned}(\mathcal{R}f(\xi,t))_t^{(q)}(0) &= \frac{1}{2\Gamma(-q)}\int_{-\infty}^{\infty} |t|^{-1-q}\left(\int_{K\cap\{x:\langle x,\xi\rangle=t\}} f(x)\,dx\right) dt \\ &= \frac{1}{2\Gamma(-q)}\int_{\mathbb{R}^n} |\langle x,\xi\rangle|^{-1-q} f(x)\mathbb{1}_K(x)\,dx\end{aligned}$$

$$= \frac{1}{2\Gamma(-q)} \int_{S^{n-1}} |\langle \theta, \xi \rangle|^{-1-q} \left(\int_0^{\|\theta\|_K^{-1}} r^{n-q-2} f(r\theta)\, dr \right) d\theta.$$

Consider the latter as a homogeneous of degree $-1-q$ function of $\xi \in \mathbb{R}^n \setminus \{0\}$, apply it to an even test function ϕ and use Lemma 1.4.11:

$$\begin{aligned}
&\langle (Rf(\xi,t))_t^{(q)}(0), \phi \rangle \\
&= \frac{1}{2\Gamma(-q)} \int_{\mathbb{R}^n} \phi(\xi) \left(\int_{S^{n-1}} |\langle \theta, \xi \rangle|^{-1-q} \left(\int_0^{\|\theta\|_K^{-1}} r^{n-q-2} f(r\theta)\, dr \right) d\theta \right) d\xi \\
&= \frac{\cos(q\pi/2)}{\pi} \int_{S^{n-1}} \left(\int_0^{\infty} t^q \hat{\phi}(t\theta)\, dt \right) \left(\int_0^{\|\theta\|_K^{-1}} r^{n-q-2} f(r\theta)\, dr \right) d\theta.
\end{aligned}$$

On the other hand, if we apply the function in the right-hand side of (1.66) to the test function ϕ we get

$$\begin{aligned}
&\left\langle \frac{\cos(\pi q/2)}{\pi} \left(|x|^{-n+q+1} \left(\int_0^{\frac{|x|}{\|x\|_K}} r^{n-q-2} f\left(\frac{rx}{|x|} \right) dr \right) \right)_x^{\wedge} (\xi), \phi(\xi) \right\rangle \\
&= \frac{\cos(\pi q/2)}{\pi} \int_{\mathbb{R}^n} |x|^{-n+q+1} \left(\int_0^{\frac{|x|}{\|x\|_K}} r^{n-q-2} f\left(\frac{rx}{|x|} \right) dr \right) \hat{\phi}(x)\, dx \\
&= \frac{\cos(\pi q/2)}{\pi} \int_{S^{n-1}} \left(\int_0^{\infty} t^q \hat{\phi}(t\theta)\, dt \right) \left(\int_0^{\|\theta\|_K^{-1}} r^{n-q-2} f(r\theta)\, dr \right) d\theta.
\end{aligned}$$

Since the right-hand sides in the last two computations coincide, we see that for any even test function ϕ,

$$\begin{aligned}
&\langle (Rf(\xi,t))_t^{(q)}(0), \phi(\xi) \rangle \\
&= \left\langle \frac{\cos(\pi q/2)}{\pi} \left(|x|^{-n+q+1} \left(\int_0^{\frac{|x|}{\|x\|_K}} r^{n-q-2} f\left(\frac{rx}{|x|} \right) dr \right) \right)_x^{\wedge} (\xi), \phi(\xi) \right\rangle. \qquad (1.67)
\end{aligned}$$

Since both distributions are even, this proves the theorem for $-1 < q < 0$. By the analyticity of fractional derivatives with respect to q and an argument similar to that in Lemma 1.4.2, one can see that both sides of (1.67) are analytic functions of q in the domain $-1 < \Re q < n-1$. By analytic extension, (1.67) holds for all $-1 < q < n-1$, which completes the proof. □

We move on to a formula for the volume of projections via the Fourier transform. Let μ be a finite Borel measure on the unit sphere S^{n-1}. We extend μ to a homogeneous distribution of degree $-n-1$. A distribution μ_e is called the extended measure of μ if, for every even test function $\phi \in \mathcal{S}(\mathbb{R}^n)$

$$\langle \mu_e, \phi\rangle = \frac{1}{2}\int_{S^{n-1}} \langle r^{-2}, \phi(r\xi)\rangle \, d\mu(\xi). \tag{1.68}$$

In most cases, we are only interested in test functions supported outside of the origin, for which $\langle r^{-2}, \phi(r\xi)\rangle = \int_{\mathbb{R}} r^{-2}\phi(r\xi)\,dr$.

If μ is absolutely continuous with density $g \in L_1(S^{n-1})$, we define the extension $g(x), x \in \mathbb{R}^n \setminus \{0\}$ as a homogeneous function of degree $-n-1$: $g(x) = |x|^{-n-1} g(x/|x|)$ and identify μ_e with g.

Lemma 1.5.9. *The Fourier transform $\widehat{\mu}_e$ is a continuous, homogeneous of degree 1, even function on $\mathbb{R}^n$. For every $\xi \in S^{n-1}$*

$$\widehat{\mu_e}(\xi) = -\frac{\pi}{2}\int_{S^{n-1}} |\langle \xi, y\rangle| \, d\mu(y).$$

Proof. Let $\phi \in \mathcal{S}(\mathbb{R}^n)$ be an even test function such that $0 \notin \operatorname{supp}(\widehat{\phi})$. Then, by the definition of μ_e

$$\langle \widehat{\mu_e}, \phi\rangle = \langle \mu_e, \widehat{\phi}\rangle = \frac{1}{2}\int_{S^{n-1}} \left(\int_{-\infty}^{\infty} r^{-2}\widehat{\phi}(r\theta)\,dr\right) d\mu_e(\theta).$$

By Lemma 1.4.9

$$\int_{\mathbb{R}^n} |\langle \theta, x\rangle| \widehat{\psi}(x)\,dx = (2\pi)^{n-1}\frac{4\sqrt{\pi}}{\Gamma(-1/2)}\int_{\mathbb{R}} r^{-2}\psi(r\theta)\,dr$$

for any even $\psi \in \mathcal{S}(\mathbb{R}^n)$ with $0 \notin \operatorname{supp}(\psi)$.

This gives (with $\widehat{\psi} = \phi$)

$$\begin{aligned}\langle \widehat{\mu_e}, \phi\rangle &= -\frac{\pi}{2}\int_{S^{n-1}} \left(\int_{R^n} |\langle \theta, \xi\rangle| \phi(\xi)\,d\xi\right) d\mu(\theta) \\ &= -\frac{\pi}{2}\int_{R^n} \phi(\xi)\left(\int_{S^{n-1}} |\langle \theta, \xi\rangle|\,d\mu(\theta)\right) d\xi \\ &= \left\langle -\frac{\pi}{2}\int_{S^{n-1}} |\langle \theta, \xi\rangle|\,d\mu(\theta), \phi\right\rangle.\end{aligned}$$

Since ϕ is an arbitrary even test function with $0 \notin \operatorname{supp}(\hat{\phi})$, the distributions $\widehat{\mu_e}$ and $-\pi/2 \int_{S^{n-1}} |\langle \theta, \xi \rangle| \, d\mu(\theta)$ can differ by a polynomial only. But both distributions are even and homogeneous of degree one, so this polynomial must be equal to zero, and hence $\widehat{\mu_e}$ coincides with the continuous function $-\pi/2 \int_{S^{n-1}} |\langle \theta, \xi \rangle| \, d\mu(\theta)$. □

The next theorem immediately follows from the Cauchy formula and Lemma 1.5.9. We denote by $S_e(L)$ the extended measure of the surface area measure $S(L, \cdot)$. We also extend the curvature function f_L to $\mathbb{R}^n \setminus \{0\}$ as a homogeneous function of degree $-n-1$. The following formula was proved in [333].

Theorem 1.5.10. *Let L be an origin-symmetric convex body in $\mathbb{R}^n$. Then, for all $\xi \in S^{n-1}$*

$$\widehat{S_e(L)}(\xi) = -\pi |P_{\xi^\perp}(L)|.$$

In particular, if the body L has a curvature function, then, for all $\xi \in S^{n-1}$

$$(f_L \cdot r^{-n-1})^\wedge(\xi) = -\pi |P_{\xi^\perp}(L)|. \tag{1.69}$$

1.6 Parseval's formula on the sphere

In this section, we describe a spherical version of Parseval's formula. This formula was proved in [307] in the infinitely smooth case and later extended in [310] to a slightly more general setting; see Proposition 1.6.5. Below, we reproduce the proof from [310, Section 3.4]. Shorter proofs involving spherical harmonics can be found in [310, Section 5.3] and [414]. For a generalization to the non-symmetric case, see [223, 224].

For an origin-symmetric star body K in $\mathbb{R}^n$, define a function

$$\mu_K(\xi) = (\exp(-\|x\|_K^4))^\wedge(\xi).$$

Lemma 1.6.1. *Let K be an infinitely smooth origin-symmetric star body K in $\mathbb{R}^n$. Then, the function μ_K is continuous and integrable on $\mathbb{R}^n$.*

Proof. Let n be an even integer. Consider the function

$$F(x) = \Delta^{(n+2)/2} \exp(-\|x\|_K^4) = h(x) \exp(-\|x\|_K^4),$$

where Δ is the Laplace operator. The function h that appears in front of the exponential after differentiation is the sum of homogeneous functions of degrees $-n+2$ and higher, which are continuous on the sphere S^{n-1}, because K is infinitely smooth (after each differentiation the degree of homogeneity decreases by 1; this explains the choice of the power 4). Clearly, h is locally integrable, and, therefore, F is integrable on $\mathbb{R}^n$. Using the connection between the Fourier transform and differentiation, given by (1.45), we get

$$\hat{F}(x) = (-1)^{(n+2)/2} |x|^{n+2} \mu_K(x).$$

The Fourier transform of an integrable function F is a bounded function on $\mathbb{R}^n$. Thus, there exists a constant C such that $|\mu_K(x)| \leqslant C|x|^{-n-2}$ for every $x \in \mathbb{R}^n$. The function $|x|^{-n-2}$ is integrable outside of the unit ball in $\mathbb{R}^n$. Also, since $\exp(-\|x\|_K^4)$ is integrable on $\mathbb{R}^n$, μ_K is a continuous function and is locally integrable, so the result follows. A similar argument (with $\Delta^{(n+1)/2}$) works in the case where n is an odd integer. □

Lemma 1.6.2. *Let $0 < p < n$ and let K be an infinitely smooth, origin-symmetric star body in $\mathbb{R}^n$. Then, for every $\theta \in S^{n-1}$*

$$\frac{\Gamma((n-p)/4)}{4}(\|x\|_K^{-n+p})^\wedge(\theta) = \int_0^\infty t^{p-1}\mu_K(t\theta)\,dt.$$

Proof. For any even test function $\phi \in \mathcal{S}(\mathbb{R}^n)$, consider the integral

$$\int_{\mathbb{R}^n} \mu_K(x)\left(\int_0^\infty t^{n-p-1}\phi(tx)\,dt\right)dx. \tag{1.70}$$

This integral converges absolutely, because by Lemma 1.6.1, the function μ_K is continuous and integrable, and the inner integral represents a continuous on $\mathbb{R}^n \setminus \{0\}$ and homogeneous of degree $-n+p$ function of x, so we can apply the argument of Lemma 1.1.1.

First, let us write the integral (1.70) in spherical coordinates (r,θ) (we also make a substitution $z = rt$ in the inner integral). We get that

$$(1.70) = \int_{S^{n-1}} \left(\int_0^\infty r^{p-1}\mu_K(r\theta)\,dr\right)\left(\int_0^\infty z^{n-p-1}\phi(z\theta)\,dz\right)d\theta.$$

Now, let us write the same integral in a different way. Note that, by Proposition 1.4.14, $(\|x\|_K^{-n+p})^\wedge$ is a continuous function on S^{n-1}. This function is homogeneous of degree $-p$ on $\mathbb{R}^n$. Also, $(\exp(-t^4\|x\|_K^4))^\wedge(\xi) = t^{-n}\mu_K(\xi/t)$. We have

$$\begin{aligned}
(1.70) &= \int_0^\infty t^{n-p-1}\left(\int_{\mathbb{R}^n} \phi(tx)\mu_K(x)\,dx\right)dt \\
&= \int_0^\infty t^{n-p-1}\left(\int_{\mathbb{R}^n} \phi(\xi)\mu_K(\xi/t)t^{-n}\,d\xi\right)dt \\
&= \int_0^\infty t^{n-p-1}\left(\int_{\mathbb{R}^n} \hat{\phi}(y)\exp(-t^4\|y\|_K^4)\,dy\right)dt \\
&= \int_{\mathbb{R}^n} \hat{\phi}(y)\left(\int_0^\infty t^{n-p-1}\exp(-t^4\|y\|_K^4)\,dt\right)dy
\end{aligned}$$

$$= \frac{\Gamma((n-p)/4)}{4} \int_{\mathbb{R}^n} \|y\|_K^{-n+p} \widehat{\phi}(y)\, dy$$
$$= \frac{\Gamma((n-p)/4)}{4} \langle \|\cdot\|_K^{-n+p}, \widehat{\phi}\rangle$$
$$= \frac{\Gamma((n-p)/4)}{4} \langle (\|\cdot\|_K^{-n+p})^\wedge, \phi\rangle$$
$$= \frac{\Gamma((n-p)/4)}{4} \int_{\mathbb{R}^n} (\|\cdot\|_K^{-n+p})^\wedge(\xi)\phi(\xi)\, d\xi$$
$$= \frac{\Gamma((n-p)/4)}{4} \int_{S^{n-1}} (\|\cdot\|_K^{-n+p})^\wedge(\theta)\left(\int_0^\infty r^{n-p-1}\phi(r\theta)\, dr\right) d\theta.$$

Now, if we put $\phi(x) = u(r)v(\theta)$ in both expressions for the integral (1.70), where v is any infinitely differentiable function on the sphere and u is a non-negative test function on $\mathbb{R}$ with compact support, we get

$$\int_{S^{n-1}} \left(\int_0^\infty r^{p-1}\mu_K(r\theta)\, dr\right) v(\theta)\, d\theta = \frac{\Gamma((n-p)/4)}{4} \int_{S^{n-1}} (\|x\|_K^{-n+p})^\wedge(\theta)v(\theta)\, d\theta$$

for every $v \in C^\infty(S^{n-1})$. The result follows. □

Next, we prove Parseval's formula on $\mathbb{R}^n$ in the case where one of the functions is homogeneous of degree $-p$, $p \in (-n, 0)$.

Lemma 1.6.3. *Let $0 < p < n$ and let K and D be infinitely smooth origin-symmetric star bodies in $\mathbb{R}^n$. Then,*

$$(2\pi)^n \int_{\mathbb{R}^n} \|x\|_K^{-p} \exp(-\|x\|_D^4)\, dx = \int_{\mathbb{R}^n} (\|\cdot\|_K^{-p})^\wedge(\xi)\, \mu_D(\xi)\, d\xi. \tag{1.71}$$

Proof. By Lemma 1.1.1, the integrals in both sides of (1.71) converge absolutely. By Lemma 1.6.2 (with p in place of $n-p$; we extend both sides of the equation to the whole $\mathbb{R}^n$ as homogeneous functions of degree $-n+p$), the right-hand side of (1.71) is equal to

$$\frac{4}{\Gamma(p/4)} \int_{\mathbb{R}^n} \left(\int_0^\infty t^{n-p-1}\mu_K(t\xi)\, dt\right) \mu_D(\xi)\, d\xi.$$

By the Fubini theorem, the latter is equal to

$$\frac{4}{\Gamma(p/4)} \int_0^\infty t^{n-p-1} \left(\int_{\mathbb{R}^n} \mu_K(t\xi) \mu_D(\xi)\, d\xi \right) dt.$$

We can apply Parseval's formula for L_1-functions (see (1.43)) to the inner integral, because μ_K and μ_D are even, continuous integrable functions on $\mathbb{R}^n$, which are the Fourier transforms of the functions $\exp(-\|\cdot\|_K^4)$ and $\exp(-\|\cdot\|_D^4)$. We continue the calculation

$$\begin{aligned}
&= \frac{4(2\pi)^n}{\Gamma(p/4)} \int_0^\infty t^{-p-1} \left(\int_{\mathbb{R}^n} \exp(-\|x\|_K^4/t^4) \exp(-\|x\|_D^4)\, dx \right) dt \\
&= \frac{4(2\pi)^n}{\Gamma(p/4)} \int_{\mathbb{R}^n} \left(\int_0^\infty t^{-p-1} \exp(-\|x\|_K^4/t^4)\, dt \right) \exp(-\|x\|_D^4)\, dx \\
&= (2\pi)^n \int_{\mathbb{R}^n} \|x\|_K^{-p} \exp(-\|x\|_D^4)\, dx. \qquad \square
\end{aligned}$$

Now, we can prove Parseval's formula on the sphere in the infinitely smooth case.

Proposition 1.6.4. *Let K and D be infinitely smooth, origin-symmetric star bodies in $\mathbb{R}^n$, and let $0 < p < n$. Then,*

$$\int_{S^{n-1}} (\|\cdot\|_K^{-p})^\wedge(\theta) (\|\cdot\|_D^{-n+p})^\wedge(\theta)\, d\theta = (2\pi)^n \int_{S^{n-1}} \|\theta\|_K^{-p} \|\theta\|_D^{-n+p}\, d\theta.$$

Proof. Moving to spherical coordinates we get

$$\begin{aligned}
\int_{\mathbb{R}^n} \|x\|_K^{-p} \exp(-\|x\|_D^4)\, dx &= \int_{S^{n-1}} \|\theta\|_K^{-p} \left(\int_0^\infty r^{n-p-1} \exp(-r^4 \|\theta\|_D^4)\, dr \right) d\theta \\
&= \frac{\Gamma((n-p)/4)}{4} \int_{S^{n-1}} \|\theta\|_K^{-p} \|\theta\|_D^{-n+p}\, d\theta. \qquad (1.72)
\end{aligned}$$

Now, use Lemma 1.6.3 to get a different expression for the integral in (1.72)

$$\begin{aligned}
\int_{\mathbb{R}^n} \|x\|_K^{-p} \exp(-\|x\|_D^4)\, dx &= \frac{1}{(2\pi)^n} \int_{\mathbb{R}^n} (\|x\|_K^{-p})^\wedge(\xi) \mu_D(\xi)\, d\xi \\
&= \frac{1}{(2\pi)^n} \int_{S^{n-1}} (\|x\|_K^{-p})^\wedge(\theta) \left(\int_0^\infty r^{p-1} \mu_D(r\theta)\, dr \right) d\theta.
\end{aligned}$$

The result follows from Lemma 1.6.2. □

We can use an approximation argument to weaken the smoothness assumptions on the bodies K and D.

Proposition 1.6.5. *Let $1 \leqslant k \leqslant n-1$, and let K and D be origin-symmetric star bodies in $\mathbb{R}^n$. Suppose that D is $(k-1)$-smooth if k is odd and that D is k-smooth if k is even. Suppose also that $\|\cdot\|_K^{-k}$ is a positive definite distribution, and let μ_0 be the finite Borel measure on S^{n-1} that corresponds to $\|\cdot\|_K^{-k}$ according to Proposition* 1.4.7. *Then,*

$$\int_{S^{n-1}} (\|\cdot\|_D^{-n+k})^\wedge(\theta)\, d\mu_0(\theta) = \int_{S^{n-1}} \|\theta\|_K^{-k} \|\theta\|_D^{-n+k}\, d\theta. \tag{1.73}$$

Proof. Since D is $(k-1)$-smooth, we can approximate D by a sequence D_m of infinitely smooth symmetric star bodies in such a way that the functions $\|\cdot\|_{D_m}$ converge to $\|\cdot\|_D$ in the metric of the space $C^{k-1}(S^{n-1})$. Then, by Proposition 1.4.14, the Fourier transforms $(\|\cdot\|_{D_m}^{-n+k})^\wedge$ are continuous functions on the sphere that converge in $C(S^{n-1})$ to another continuous function $(\|\cdot\|_D^{-n+k})^\wedge$, as $m \to \infty$.

Also, consider any sequence of infinitely smooth symmetric star bodies $K_j \subset K$, $j \in \mathbb{N}$ approximating K in the radial metric. For every even test function ϕ the sequence of integrable (by Lemma 1.1.1) functions $\|x\|_{K_j}^{-k} |\widehat{\phi}(x)|$ is bounded by the integrable function $\|x\|_K^{-k} |\widehat{\phi}(x)|$. By the dominated convergence theorem and the first part of Proposition 1.4.7

$$\begin{aligned}\int_{\mathbb{R}^n} (\|x\|_{K_j}^{-k})^\wedge(y)\phi(y)\, dy &= \int_{\mathbb{R}^n} \|x\|_{K_j}^{-k} \widehat{\phi}(x)\, dx \\ &\to \int_{\mathbb{R}^n} \|x\|_K^{-k} \widehat{\phi}(x)\, dx \\ &= (2\pi)^n \int_{S^{n-1}} \left(\int_0^\infty r^{k-1}\phi(r\theta)\, dr \right) d\mu_0(\theta),\end{aligned} \tag{1.74}$$

as $j \to \infty$. Since the bodies D_m are infinitely smooth

$$\phi_m(x) = u(r)(\|x\|_{D_m}^{-n+k})^\wedge(\theta)$$

is a test function, where u is any non-negative test function on $\mathbb{R}$ with compact support. Substituting ϕ_m in (1.74) and moving to spherical coordinates, we get

$$\lim_{j\to\infty} \int_{S^{n-1}} (\|x\|_{K_j}^{-k})^\wedge(\theta)(\|x\|_{D_m}^{-n+k})^\wedge(\theta)\, d\theta = (2\pi)^n \int_{S^{n-1}} (\|x\|_{D_m}^{-n+k})^\wedge(\theta)\, d\mu_0(\theta). \tag{1.75}$$

By Proposition 1.4.14, the functions $(\|x\|_{D_m}^{-n+k})^\wedge(\theta)$ converge in $C(S^{n-1})$ to the function $(\|x\|_D^{-n+k})^\wedge(\theta)$, as $m \to \infty$, so the integral on the right-hand side of (1.75) converges to the integral on the left-hand side of (1.73). Now, we get the result if we apply Lemma 1.6.4 to the left-hand side of (1.75) and let $m, j \to \infty$. □

Sometimes, we use Parseval's formula on the sphere in a slightly different form. Since K, D are arbitrary star bodies, their Minkowski functionals can be written as

$$\|\cdot\|_K = f \cdot r, \ \|\cdot\|_D = g \cdot r,$$

where f, g are arbitrary, strictly positive continuous, even functions on the sphere. If f, g are infinitely smooth, strictly positive functions on the sphere, the formula of Proposition 1.6.4 can be written for $p \in (0, n)$ in the form

$$\int_{S^{n-1}} (f^{-p} \cdot r^{-p})^\wedge(\theta)(g^{-n+p} \cdot r^{-n+p})^\wedge(\theta)\, d\theta = (2\pi)^n \int_{S^{n-1}} f^{-p}(\theta)\, g^{-n+p}(\theta)\, d\theta.$$

If we denote by $h = f^{-p}$, $s = g^{-n+p}$, we get

$$\int_{S^{n-1}} (h \cdot r^{-p})^\wedge(\theta)(s \cdot r^{-n+p})^\wedge(\theta)\, d\theta = (2\pi)^n \int_{S^{n-1}} h(\theta)s(\theta)\, d\theta. \tag{1.76}$$

This formula can be extended to sign-changing infinitely smooth functions. Also, using Lemma 1.4.13 and the argument of Lemma 1.4.2 one can prove that both sides of (1.76) are analytic functions of $p \in \mathbb{C}$ as long as $p \neq 0, 2, 4...$ and $p \neq -n, -n-2, -n-4...$, so (1.76) can be extended to these values of p.

In particular, if K, L are origin-symmetric convex bodies in $\mathbb{R}^n$ such that the support function h_K of K and the curvature function f_L of L are infinitely smooth on the sphere S^{n-1}, then one has

$$\int_{S^{n-1}} (h_K \cdot r)^\wedge(\theta)(f_L \cdot r^{-n-1})^\wedge(\theta)\, d\theta = (2\pi)^n \int_{S^{n-1}} h_K(\theta) f_L(\theta)\, d\theta. \tag{1.77}$$

Independent proofs of the latter equality can be found in [333] and [310, Lemma 8.8].

The technique of this section allows us to prove a lower-dimensional version of the Fourier formula for the volume of sections.

Lemma 1.6.6. *Let $1 \leqslant k < n$, let ϕ be an even, continuous, integrable function on $\mathbb{R}^n$, and let H be an $(n-k)$-dimensional subspace of $\mathbb{R}^n$. Suppose that ϕ is integrable on all translations of H and that the Fourier transform $\widehat{\phi}$ is integrable on $H^\perp$. Then,*

$$(2\pi)^k \int_H \phi(x)\, dx = \int_{H^\perp} \widehat{\phi}(x)\, dx.$$

Proof. Let $\xi_1, \dots, \xi_k$ be an orthonormal basis in $H^\perp$. For every $t \in \mathbb{R}^k$, making the change of variables $u_1 = \langle x, \xi_1\rangle, \dots, u_k = \langle x, \xi_k\rangle$ we get

$$\widehat{\phi}(t_1\xi_1 + \cdots + t_k\xi_k) = \int_{\mathbb{R}^n} \phi(x) \exp(-i\langle x, t_1\xi_1 + \cdots + t_k\xi_k\rangle)\, dx$$

$$= \int_{\mathbb{R}^n} \phi(x) \exp(-it_1\langle x, \xi_1\rangle + \cdots + t_k\langle x, \xi_k\rangle)\, dx$$

$$= \int_{\mathbb{R}^k} \exp(-i(t_1u_1 + \cdots + t_ku_k))\Big(\int_{H+u_1\xi_1+\cdots+u_k\xi_k} \phi(x)\, dx\Big)\, du.$$

This means that the function $f(t) = \widehat{\phi}(t_1\xi_1 + \cdots + t_k\xi_k)$ is the Fourier transform of the function $g(u) = \int_{H+u_1\xi_1+\cdots+u_k\xi_k} \phi(x)\, dx$. The function g is integrable on $\mathbb{R}^k$ because ϕ is integrable on $\mathbb{R}^n$. Also, since $\widehat{\phi}$ is even integrable on $H^\perp$, the function f is integrable and even on $\mathbb{R}^k$. By the inversion theorem for the Fourier transform

$$\int_{\mathbb{R}^k} f(t)\, dt = (\widehat{g})^\wedge(0) = (2\pi)^k g(0),$$

which immediately implies the result, because the left-hand side is exactly $\int_{H^\perp} \widehat{\phi}$. □

The following lemma provides a Fourier transform formula for the volume of lower-dimensional sections. This formula extends the result of Theorem 1.5.2 and first appeared in [307, 309].

Lemma 1.6.7. *Suppose $1 \leqslant k < n$ is an integer, and L is an origin-symmetric star body in $\mathbb{R}^n$, so that L is $(k-1)$-smooth if k is odd, and it is k-smooth if k is even. Then, for every $(n-k)$-dimensional subspace H of $\mathbb{R}^n$ we have*

$$|L \cap H| = \frac{1}{(2\pi)^k(n-k)} \int_{S^{n-1}\cap H^\perp} (\|x\|_L^{-n+k})^\wedge(\theta)\, d\theta.$$

Proof. We first assume that L is infinitely smooth. By Lemma 1.6.2

$$\int_{S^{n-1}\cap H^\perp} (\|x\|_L^{-n+k})^\wedge(\theta)\, d\theta$$

$$= (4/\Gamma((n-k)/4)) \int_{S^{n-1}\cap H^\perp} \Big(\int_0^\infty t^{k-1}\mu_L(t\theta)\, dt\Big)\, d\theta$$

$$= (4/\Gamma((n-k)/4)) \int_{H^\perp} \mu_L(x)\, dx.$$

Using Lemma 1.6.6 with $\phi(x) = \exp(-\|x\|_L^4)$, and writing the integral in polar coordinates, we get

$$(4(2\pi)^k/\Gamma((n-k)/4)) \int_H \exp(-\|x\|_L^4)\, dx = (2\pi)^k \int_{S^{n-1}\cap H} \|\theta\|_L^{-n+k}\, d\theta$$

$$= (2\pi)^k(n-k)|L \cap H|,$$

by (1.3). The assumption that L is infinitely smooth can be removed by approximation in the same way, as was done in Proposition 1.6.5. □

1.7 Intersection bodies and generalizations

The concept of an intersection body was introduced by Lutwak [392] and played the crucial role in the solution of the Busemann–Petty problem.

Let K, L be origin-symmetric star bodies in $\mathbb{R}^n$. Following Lutwak [392], we say that K is the intersection body of L if the radius of K in every direction is equal to the $(n-1)$-dimensional volume of the section of L by the central hyperplane orthogonal to this direction, i. e., for every $\xi \in S^{n-1}$

$$\rho_K(\xi) = \|\xi\|_K^{-1} = |L \cap \xi^\perp|. \tag{1.78}$$

It follows from Corollary 1.5.3 that if two origin-symmetric star bodies L_1, L_2 generate the same intersection body, then $L_1 = L_2$.

By (1.2), the right-hand side of (1.78) can be written as the spherical Radon transform of a positive continuous function, as follows:

$$\|\xi\|_K^{-1} = \frac{1}{n-1} \int_{S^{n-1}\cap\xi^\perp} \|\theta\|_L^{-n+1}\, d\theta = \frac{1}{n-1} R(\|\cdot\|_L^{-n+1})(\xi).$$

This allows us to introduce a more general class of bodies. Following Goodey, Lutwak, and Weil [222], we say that a star body K in $\mathbb{R}^n$ is an intersection body if there exists a finite Borel measure μ on S^{n-1} such that $\|\cdot\|_K^{-1} = R\mu$ as functionals on $C(S^{n-1})$, i. e., for every continuous function f on S^{n-1}

$$\int_{S^{n-1}} \|x\|_K^{-1} f(x)\, dx = \int_{S^{n-1}} Rf(x)\, d\mu(x). \tag{1.79}$$

We denote the class of intersection bodies in $\mathbb{R}^n$ by $\mathcal{I}_n$.

Suppose that K is the intersection body of some symmetric star body L. Then, using Theorem 1.5.2, for every $x \in S^{n-1}$, we get

$$\|x\|_K^{-1} = |L \cap x^\perp| = \frac{1}{\pi(n-1)} (\|\cdot\|_L^{-n+1})^\wedge(x).$$

We extend this equality to the whole $\mathbb{R}^n$, as an equality of homogeneous functions of degree -1. Since K is symmetric, the Fourier transform is self-invertible (up to a constant), so

$$(\|\cdot\|_K^{-1})^\wedge(\xi) = \frac{(2\pi)^n}{\pi(n-1)} \|\xi\|_L^{-n+1} > 0,$$

which means that the locally integrable function $\|\cdot\|_K^{-1}$ represents a positive definite distribution.

On the other hand, suppose that $\|\cdot\|_K^{-1}$ is positive definite and its Fourier transform is a continuous positive function on S^{n-1}. Then, define an origin-symmetric star body L by

$$\|\xi\|_L = \left(\frac{\pi(n-1)}{(2\pi)^n}(\|\cdot\|_K^{-1})^\wedge(\xi)\right)^{-1/(n-1)}, \quad \xi \in S^{n-1}.$$

It is easy to reverse the argument above and see that K is the intersection body of L. Thus, we have proved that an origin-symmetric star body K in $\mathbb{R}^n$ is an intersection body of a star body if and only if the Fourier transform of $\|\cdot\|_K^{-1}$ is a continuous strictly positive function on the sphere.

This Fourier analytic characterization holds for the whole class of intersection bodies, as proved in [305]; see also [310, Theorem 4.1].

Theorem 1.7.1. *An origin-symmetric star body K in $\mathbb{R}^n$ is an intersection body if and only if $\|x\|_K^{-1}$ represents a positive definite distribution on $\mathbb{R}^n$.*

Proof. By Lemma 1.3.1, for every even test function ϕ and every fixed $\theta \in S^{n-1}$ the function $f(t) = (2\pi)^n\phi(t\theta)$ is the Fourier transform of the function $g(z) = \mathcal{R}\widehat{\phi}(\theta; z)$, where $t, z \in \mathbb{R}$ and $\mathcal{R}$ is the Radon transform. Inverting the Fourier transform we get $\int_{\mathbb{R}} f = 2\pi g(0)$, which can be written as

$$(2\pi)^n \int_{\mathbb{R}} \phi(t\theta)\,dt = 2\pi\mathcal{R}\widehat{\phi}(\theta; 0) = 2\pi \int_{\theta^\perp} \widehat{\phi}(x)\,dx.$$

Writing the integral in the right-hand side in polar coordinates of the hyperplane $\theta^\perp$ we get

$$2\pi \int_{S^{n-1}\cap\theta^\perp} \left(\int_0^\infty t^{n-2}\widehat{\phi}(t\xi)\,dt\right) d\xi = 2\pi R\left(\int_0^\infty t^{n-2}\widehat{\phi}(t\xi)\,dt\right)(\theta),$$

where on the right-hand side we have the spherical Radon transform of a function of the variable $\xi \in S^{n-1}$ at the point θ. Combining the last two formulas, we get

$$\int_{\mathbb{R}} \phi(t\theta)\,dt = \frac{1}{(2\pi)^{n-1}} R\left(\int_0^\infty t^{n-2}\widehat{\phi}(t\xi)\,dt\right)(\theta). \tag{1.80}$$

Suppose that K is an intersection body, and let μ be the measure on S^{n-1} for which $R\mu = \|\cdot\|_K^{-1}$ in the sense of equality (1.79). By Lemma 1.1.1, $\|\cdot\|_K^{-1}$ is a locally integrable function on $\mathbb{R}^n$, so, for every non-negative even test function ϕ, we have

$$\langle (\|\cdot\|_K^{-1})^\wedge, \phi\rangle = \langle \|\cdot\|_K^{-1}, \hat{\phi}\rangle = \int_{\mathbb{R}^n} \|x\|_K^{-1}\hat{\phi}(x)\,dx.$$

We need to prove that the latter integral is non-negative. Writing this integral in polar coordinates we get that it is equal to

$$\int_{S^{n-1}} \|\xi\|_K^{-1}\left(\int_0^\infty t^{n-2}\hat{\phi}(t\xi)\,dt\right)d\xi,$$

which, by (1.79), equals

$$\int_{S^{n-1}} R\left(\int_0^\infty t^{n-2}\hat{\phi}(t\xi)\,dt\right)(\theta)\,d\mu(\theta).$$

Applying (1.80), we write the latter integral as

$$(2\pi)^{n-1}\int_{S^{n-1}}\left(\int_{\mathbb{R}} \phi(t\theta)\,dt\right)d\mu(\theta) \geqslant 0,$$

because ϕ is a non-negative test function. Therefore, $\|x\|_K^{-1}$ represents a positive definite distribution.

Now, suppose that $\|x\|_K^{-1}$ represents a positive definite distribution. By Proposition 1.4.7, there exists a finite Borel measure μ_0 on S^{n-1} such that for every even test function ϕ

$$\langle \|x\|_K^{-1}, \phi\rangle = \int_{S^{n-1}}\left(\int_{\mathbb{R}} \hat{\phi}(t\theta)\,dt\right)d\mu_0(\theta).$$

By (1.80) with $\hat{\phi}$ in place of ϕ, the latter expression is equal to

$$2\pi \int_{S^{n-1}} R\left(\int_0^\infty t^{n-2}\phi(t\xi)\,dt\right)(\theta)\,d\mu_0(\theta). \tag{1.81}$$

On the other hand

$$\langle \|x\|_K^{-1}, \phi\rangle = \int_{S^{n-1}} \|\theta\|_K^{-1}\left(\int_0^\infty t^{n-2}\phi(t\theta)\,dt\right)d\theta. \tag{1.82}$$

For any infinitely differentiable function v on S^{n-1}, consider a test function $\phi \in \mathcal{S}(\mathbb{R}^n)$ defined by $\phi(x) = u(t)v(\theta)$ for every $x = t\theta \in \mathbb{R}^n$, $t > 0$, $\theta \in S^{n-1}$, where $u \in \mathcal{S}(\mathbb{R})$ is a non-negative even function supported outside of zero and $\int_{\mathbb{R}} t^{n-2}u(t)\,dt = 1$. We have

$$v(\theta) = \int_0^\infty t^{n-2}\phi(t\theta)\,dt.$$

Substituting this ϕ into (1.81) and (1.82), we get that for any $v \in C^\infty(S^{n-1})$

$$\int_{S^{n-1}} \|\theta\|_K^{-1} v(\theta)\,d\theta = 2\pi \int_{S^{n-1}} Rv(\theta)\,d\mu_0(\theta).$$

Since infinitely differentiable functions are dense in $C(S^{n-1})$, the latter equality holds for any continuous function v on the sphere, so the body K satisfies the definition (1.79) with the measure $\mu = 2\pi\mu_0$. □

The result of Theorem 1.7.1 turned out to be rather effective in identifying bodies that are and are not intersection bodies. We will review these examples below in the more general setting of k-intersection bodies and embedding in L_{-p}. At this time, we will just mention a result that was used in the solution of the Busemann–Petty problem.

Theorem 1.7.2. *Every origin-symmetric convex body D in $\mathbb{R}^n$, $n = 2, 3, 4$ is an intersection body. On the other hand, for every $n \geqslant 5$ there exists an origin-symmetric convex body D in $\mathbb{R}^n$ that is not an intersection body.*

The first example of a non-intersection body in dimension 5 was given by Gardner [191]. The case $n = 3$ was also established by Gardner [193]. For $n = 4$, different proofs were given by Zhang in [554] and in [202]. The proof from [202] is unified in the sense that it works for all dimensions and clearly shows the reason for the transition between dimensions 4 and 5. This proof also appears in [310].

Proof. Let $n = 4$. By Lemma 1.7.13 we can assume that the body D is infinitely smooth. Then, by Proposition 1.4.13, the Fourier transform of $\|\cdot\|_D^{-1}$ is an infinitely smooth function on the sphere extended to a homogeneous of degree $-n+1$ function on $\mathbb{R}^n \setminus \{0\}$. Putting $m = 2$ in (1.64) we get

$$A''_{D,\xi}(0) = -\pi(\|x\|_D^{-1})^\wedge(\xi), \quad \xi \in S^{n-1}. \tag{1.83}$$

We have $A''_{D,\xi}(0) \leqslant 0$, because by Theorem 1.2.2 the central section of an origin-symmetric convex body is maximal among all sections perpendicular to a given direction. So, $(\|x\|_D^{-1})^\wedge(\xi) \geqslant 0$ for every ξ, and by Theorem 1.7.1 we conclude that D is an intersection body. For $n = 3$, the argument is the same except that we use (1.65) instead of (1.64). The case $n = 2$ is obvious.

If the dimension $n = 5$, we need $m = 3$ to get $(\|x\|_D^{-1})^\wedge$ in (1.65). However, the third derivative is not controlled by convexity, and one can construct a non-intersection body in $\mathbb{R}^5$ using (1.65). This consideration is implicitly used in the following example.

Let $N > 0$, and define $f_N(z) = (1 - z^2 - Nz^4)^{1/4}$, and let a_N be such that $f_N(a_N) = 0$ and $1 - x^2 - Nx^4 > 0$ on $(0, a_N)$. Put

$$D = \left\{ (x_1, \dots, x_5) \in \mathbb{R}^5 : x_5 \in [-a_N, a_N], \left(\sum_{i=1}^{4} x_i^2 \right)^{1/2} \leqslant f_N(|x_5|) \right\}.$$

A simple computation, which can be found in [310, Corollary 4.4], shows that for large N the Fourier transform of $\|x\|_D^{-1}$ is negative at the point $(0,0,0,0,1)$. By Theorem 1.7.1, D is not an intersection body. □

1.7.1 k-intersection bodies and embeddings in L_{-p}

Let $1 \leqslant k < n$, and let D and L be origin-symmetric star bodies in $\mathbb{R}^n$. Following [308, 309], we say that D is a k-intersection body of L if for every $(n-k)$-dimensional subspace H of $\mathbb{R}^n$

$$|D \cap H^\perp| = |L \cap H|. \tag{1.84}$$

In the case where $k = 1$ we get intersection bodies. However, if $k > 1$ it is not immediately clear when a k-intersection body exists for a given L.

By the polar formula (1.3) for the volume, equality (1.84) can be written in the form

$$\int_{S^{n-1}\cap H^\perp} \|\xi\|_D^{-k}\, d\xi = \frac{k}{n-k} \int_{S^{n-1}\cap H} \|x\|_L^{-n+k}\, dx. \tag{1.85}$$

Let us first consider the case where L is infinitely smooth. Then, by Corollary 1.4.14, $(\|\cdot\|_L^{-n+k})^\wedge$ is a continuous function on the sphere. Moreover, we can apply Lemma 1.6.7 to the right-hand side of (1.85): for every $(n-k)$-dimensional subspace H of $\mathbb{R}^n$

$$\int_{S^{n-1}\cap H^\perp} \|\xi\|_D^{-k}\, d\xi = \frac{k}{(n-k)(2\pi)^k} \int_{S^{n-1}\cap H^\perp} (\|\cdot\|_L^{-n+k})^\wedge(\xi)\, d\xi.$$

Now, by Corollary 1.5.4, the two functions of the variable ξ are equal everywhere on S^{n-1}, and we get the following.

Lemma 1.7.3. *Let L be an infinitely smooth origin-symmetric star body in $\mathbb{R}^n$, and let $1 \leqslant k < n$. The k-intersection body D of L exists if and only if the function $(\|\cdot\|_L^{-n+k})^\wedge$ is strictly positive on S^{n-1}. The body D is then given by*

$$\|\xi\|_D^{-k} = \frac{k}{(n-k)(2\pi)^k} (\|\cdot\|_L^{-n+k})^\wedge(\xi)$$

for every $\xi \in S^{n-1}$.

On the other hand, reversing the Fourier transform in the latter formula, one can see that if a body D is the k-intersection body of some infinitely smooth symmetric star body, then $(\|\cdot\|_D^{-k})^\wedge$ is a strictly positive continuous function on the sphere. We are going to prove a more general fact by removing the smoothness assumption.

Theorem 1.7.4. *Let $1 \leqslant k < n$, and suppose that D is an origin-symmetric star body in $\mathbb{R}^n$ that is the k-intersection body of an origin-symmetric star body L. Then, $(\|x\|_D^{-k})^\wedge$ is a strictly positive continuous function on $\mathbb{R}^n \setminus \{0\}$. The corresponding body L satisfies*

$$\|\theta\|_L^{-n+k} = \frac{n-k}{(2\pi)^{n-k}k}(\|\cdot\|_D^{-k})^\wedge(\theta), \quad \theta \in S^{n-1}.$$

Proof. Let $D_m \subset D$, $m \in \mathbb{N}$ be infinitely smooth symmetric star bodies in $\mathbb{R}^n$ approximating D in the uniform metric on the sphere. Then, as $m \to \infty$

$$\int_{S^{n-1}\cap H^\perp} \|x\|_{D_m}^{-k}\,dx \to \int_{S^{n-1}\cap H^\perp} \|x\|_D^{-k}\,dx$$

uniformly with respect to $H \in G_{n,n-k}$. Now, we apply the definition of a k-intersection body of a star body in the form of (1.85) to the right-hand side of the latter formula, and we also apply Lemma 1.6.7 to the left-hand side of this formula. We get that

$$\int_{S^{n-1}\cap H} (\|\cdot\|_{D_m}^{-k})^\wedge(\theta)\,d\theta \to \frac{(2\pi)^{n-k}k}{n-k} \int_{S^{n-1}\cap H} \|x\|_L^{-n+k}\,dx \tag{1.86}$$

uniformly with respect to H. Denote the constant on the right-hand side by c. Let $\psi \in C^\infty(S^{n-1})$ be an even infinitely differentiable function on the sphere. Then, by Lemma 1.4.13, $(\psi(\theta)r^{-1})^\wedge$ is an infinitely differentiable function on $\mathbb{R}^n \setminus \{0\}$. The uniform convergence in (1.7.13) implies that

$$\lim_{m\to\infty} \int_{G_{n,n-k}} \Big(\int_{S^{n-1}\cap H} (\|\cdot\|_{D_m}^{-k})^\wedge\Big)\Big(\int_{S^{n-1}\cap H^\perp} (\psi(\theta)r^{-1})^\wedge\Big)\,dH$$
$$= c \int_{G_{n,n-k}} \Big(\int_{S^{n-1}\cap H} \|x\|_L^{-n+k}\Big)\Big(\int_{S^{n-1}\cap H^\perp} (\psi(\theta)r^{-1})^\wedge\Big)\,dH.$$

Applying Lemma 1.3.2 to both sides of the latter equality, we get

$$\lim_{m\to\infty} \int_{S^{n-1}} (\|\cdot\|_{D_m}^{-k})^\wedge(\xi)\Big(\int_{S^{n-1}\cap\xi^\perp} (\psi(\theta)r^{-1})^\wedge\Big)\,d\xi$$
$$= c \int_{S^{n-1}} \|\xi\|_L^{-n+k}\Big(\int_{S^{n-1}\cap\xi^\perp} (\psi(\theta)r^{-1})^\wedge\Big)\,d\xi.$$

Since the function $(\psi(\theta)r^{-1})^\wedge$ is homogeneous of degree $-n+1$, by Proposition 1.5.1

$$\int_{S^{n-1}\cap\xi^{\perp}} (\psi(\theta)r^{-1})^{\wedge} = \frac{1}{\pi}((\psi(\theta)r^{-1})^{\wedge})^{\wedge}(\xi) = \frac{(2\pi)^n}{\pi}\psi(\xi)$$

for every $\xi \in S^{n-1}$. Therefore,

$$\int_{S^{n-1}} (\|\cdot\|_{D_m}^{-k})^{\wedge}(\xi)\psi(\xi)\,d\xi \to c\int_{S^{n-1}} \|\xi\|_L^{-n+k}\psi(\xi)\,d\xi, \tag{1.87}$$

as $m \to \infty$.

Let $\phi \in \mathcal{S}(\mathbb{R}^n)$ be any even test function. For every m

$$\|x\|_{D_m}^{-k}|\widehat{\phi}(x)| \leqslant \|x\|_D^{-k}|\widehat{\phi}(x)|$$

and the latter function belongs to $L_1(\mathbb{R}^n)$ by Lemma 1.4.8, as the product of a locally integrable, bounded outside of the origin function $\|x\|_D^{-k}$ and an L_1-function $|\widehat{\phi}|$. By the dominated convergence theorem and formula (1.87) applied to the function $\psi(\xi) = \int_0^{\infty} t^{k-1}\phi(t\xi)\,dt$,

$$\begin{aligned}
\langle(\|\cdot\|_D^{-k})^{\wedge},\phi\rangle &= \int_{\mathbb{R}^n} \|x\|_D^{-k}\widehat{\phi}(x)\,dx \\
&= \lim_{m\to\infty}\int_{\mathbb{R}^n} \|x\|_{D_m}^{-k}\widehat{\phi}(x)\,dx \\
&= \lim_{m\to\infty}\int_{\mathbb{R}^n} (\|\cdot\|_{D_m}^{-k})^{\wedge}(y)\phi(y)\,dy \\
&= \lim_{m\to\infty}\int_{S^{n-1}} (\|\cdot\|_{D_m}^{-k})^{\wedge}(\xi)\left(\int_0^{\infty} t^{k-1}\phi(t\xi)\,dt\right)d\xi \\
&= c\int_{S^{n-1}} \|\xi\|_L^{-n+k}\left(\int_0^{\infty} t^{k-1}\phi(t\xi)\,dt\right)d\xi \\
&= c\int_{\mathbb{R}^n} \|y\|_L^{-n+k}\phi(y)\,dy.
\end{aligned}$$

Since ϕ is an arbitrary even test function, we get

$$(\|\cdot\|_D^{-k})^{\wedge} = c\|\cdot\|_L^{-n+k}. \qquad \square$$

Suppose that D is a k-intersection body of a star body. By Theorem 1.7.4, $(\|x\|_D^{-k})^{\wedge}$ is a positive continuous, homogeneous of degree $-n+k$ function on $\mathbb{R}^n \setminus \{0\}$. For every even test function $\phi \in \mathcal{S}(\mathbb{R}^n)$, we have

$$\int_{\mathbb{R}^n} \|x\|_D^{-k}\phi(x)\,dx = \frac{1}{(2\pi)^n}\int_{\mathbb{R}^n} (\|\cdot\|_D^{-k})^{\wedge}(y)\widehat{\phi}(y)\,dy$$

$$= \frac{1}{(2\pi)^n} \int_{S^{n-1}} (\|\cdot\|_D^{-k})^\wedge(\xi) \left(\int_0^\infty t^{k-1}\widehat{\phi}(t\xi)\, dt \right) d\xi.$$

Similar to how it was done in the case $k = 1$ for intersection bodies, we define a more general class of bodies by replacing the positive continuous function $(\|\cdot\|_D^{-k})^\wedge$ in the latter equality by any measure on the sphere S^{n-1}.

Definition 1.7.5. Let $1 \leqslant k < n$. We say that an origin-symmetric star body D in $\mathbb{R}^n$ is a k-intersection body if there exists a finite Borel measure μ on S^{n-1} such that for every even test function ϕ in $\mathbb{R}^n$

$$\int_{\mathbb{R}^n} \|x\|_D^{-k} \phi(x)\, dx = \int_{S^{n-1}} \left(\int_0^\infty t^{k-1}\widehat{\phi}(t\xi)\, dt \right) d\mu(\xi). \tag{1.88}$$

By the calculation before the definition, every k-intersection body of a star body is also a k-intersection body in the general sense.

We now present a Fourier analytic characterization of k-intersection bodies similar to Theorem 1.7.1.

Theorem 1.7.6. *Let $1 \leqslant k < n$. An origin-symmetric star body D in $\mathbb{R}^n$ is a k-intersection body if and only if the function $\|\cdot\|_D^{-k}$ represents a positive definite distribution on $\mathbb{R}^n$.*

Proof. Suppose that D is a k-intersection body. Consider a non-negative even test function ϕ. Then, by (1.88) and the fact that $(\widehat{\phi})^\wedge = (2\pi)^n\phi$, we get

$$\begin{aligned}\langle (\|\cdot\|_D^{-k})^\wedge, \phi\rangle = \langle \|\cdot\|_D^{-k}, \widehat{\phi}\rangle &= \int_{\mathbb{R}^n} \|x\|_D^{-k}\widehat{\phi}(x)\, dx \\ &= (2\pi)^n \int_{S^{n-1}} \left(\int_0^\infty t^{k-1}\phi(t\xi)\, dt \right) d\mu(\xi) \geqslant 0,\end{aligned}$$

so the function $\|\cdot\|_D^{-k}$ represents a positive definite distribution.

On the other hand, if $\|\cdot\|_D^{-k}$ is a positive definite distribution, by Proposition 1.4.7, there exists a measure μ_0 on the sphere S^{n-1} such that for any test function ϕ

$$\int_{\mathbb{R}^n} \|x\|_D^{-k} \phi(x)\, dx - \int_{S^{n-1}} \left(\int_0^\infty t^{k-1}\widehat{\phi}(t\xi)\, dt \right) d\mu_0(\xi),$$

which means exactly that D is a k-intersection body. □

The concept of embedding of finite-dimensional normed spaces in L_p with negative p was introduced in [308, 309].

Definition 1.7.7. Let $0 < p < n$. Let D be an origin-symmetric star body D in $\mathbb{R}^n$. We say that the body D belongs to the class L^n_{-p} and, in other words, the space $(\mathbb{R}^n, \|\cdot\|_D)$ embeds in L_{-p} if there exists a finite Borel measure μ on S^{n-1} such that for every even test function ϕ in $\mathbb{R}^n$

$$\int_{\mathbb{R}^n} \|x\|_D^{-p}\phi(x)\,dx = \int_{S^{n-1}} \left(\int_0^\infty t^{p-1}\hat{\phi}(t\xi)\,dt \right) d\mu(\xi). \tag{1.89}$$

A consequence of Proposition 1.4.7 is the next fact.

Proposition 1.7.8. *Let $0 < p < n$. An origin-symmetric star body D belongs to the class L^n_{-p} if and only if the function $\|\cdot\|_D^{-p}$ represents a positive definite distribution on $\mathbb{R}^n$.*

This shows that the concept of embedding in L_{-p} can be considered as an analytic extension of embedding in L_p with $p > 0$. By L_p, $p > 0$ we mean the L_p-space of functions on $[0,1]$ with Lebesgue measure. It was shown in [303] that an n-dimensional normed space embeds isometrically in L_p with $p > 0$ and not an even integer, if and only if the Fourier transform (in the sense of Schwartz distributions) of the function $\Gamma(-p/2)\|\cdot\|^p$ is a non-negative distribution outside of the origin in $\mathbb{R}^n$.

Comparing Definition 1.7.7 with Theorem 1.7.6 we get the following.

Proposition 1.7.9. *An origin-symmetric star body D in $\mathbb{R}^n$ is a k-intersection body if and only if $D \in L^n_{-k}$, or, equivalently, the space $(\mathbb{R}^n, \|\cdot\|_D)$ embeds in L_{-k}.*

The advantage of the concept of embedding in L_{-p} is that we can extend different results about L_p-spaces with positive p to negative values of p, and this immediately gives new information about k-intersection bodies.

First, note that, by Lemma 1.4.3, the Euclidean ball B_2^n belongs to the class L^n_{-p} for every $p \in (0, n)$. Also, origin-symmetric ellipsoids belong to L^n_{-p} for every $p \in (0, n)$. To see this, one can use the connection between linear transformations and the Fourier transform: for every $T \in GL_n$

$$(|Tx|^{-p})^\wedge(\xi) = c_{n,p}\frac{1}{|\det T|}|T^{-1}\xi|^{-n+p} > 0.$$

In particular, origin-symmetric ellipsoids are k-intersection bodies in $\mathbb{R}^n$ for every $k = 1, \dots, n-1$.

Bretagnolle, Dacunha-Castelle, and Krivine [114] proved that for $0 < p < q \leqslant 2$ the space L_q embeds isometrically in L_p. This fact was extended to negative p in [305] and also appears in [310, Theorem 4.11].

Proposition 1.7.10. *The unit ball of any n-dimensional subspace of L_q with $0 < q \leqslant 2$ belongs to the class L^n_{-p} for every $0 < p < n$. In particular, the unit ball of any n-dimensional subspace of L_q, $0 < q \le 2$ is a k-intersection body for every $0 < k < n$.*

Proof. By a well-known result of P. Lévy [369], for every n-dimensional subspace $B = (\mathbb{R}^n, \|\cdot\|)$ of L_q with $0 < q \leqslant 2$, the function $\exp(-\|x\|^q)$ is positive definite on $\mathbb{R}^n$. By Bochner's theorem [86], this function is the Fourier transform of a finite Borel measure μ on $\mathbb{R}^n$.

Using the change of variables $z = t\|x\|$ and the fact that, for all $p, q > 0$

$$\int_0^\infty x^{p-1} e^{-x^q}\, dx \sum = \frac{\Gamma(p/q)}{q}$$

for every $x \in \mathbb{R}^n$ we get

$$\|x\|^{-p} = \frac{q}{\Gamma(p/q)} \int_0^\infty t^{p-1} \exp(-t^q \|x\|^q)\, dt.$$

By Lemma 1.1.1, for any even test function ϕ, the function $\|x\|^{-p}\widehat{\phi}(x)$ is integrable on $\mathbb{R}^n$. Therefore, we can use the Fubini theorem in the following calculation:

$$\begin{aligned}
\langle (\|\cdot\|^{-p})^\wedge, \phi\rangle &= \int_{\mathbb{R}^n} \|x\|^{-p}\widehat{\phi}(x)\, dx \\
&= \frac{q}{\Gamma(p/q)} \int_{\mathbb{R}^n} \widehat{\phi}(x) \left(\int_0^\infty t^{p-1} \exp(-t^q\|x\|^q)\, dt \right) dx \\
&= \frac{q}{\Gamma(p/q)} \int_0^\infty t^{p-1} \left(\int_{\mathbb{R}^n} \widehat{\phi}(x) \exp(-t^q \|x\|^q)\, dx \right) dt \\
&= \frac{q}{\Gamma(p/q)} \int_0^\infty t^{p-1} \left(\int_{\mathbb{R}^n} t^{-n}\widehat{\phi}(y/t) \exp(-\|y\|^q)\, dy \right) dt. \qquad (1.90)
\end{aligned}$$

The function $y \to t^{-n}\widehat{\phi}(y/t)$ is the Fourier transform of the function $\xi \to \phi(t\xi)$. Therefore, since $\widehat{\mu} = \exp(-\|\cdot\|^q)$ and $\langle \widehat{f}, \widehat{\phi}\rangle = (2\pi)^n \langle f, \phi\rangle$, we have

$$\begin{aligned}
\int_{\mathbb{R}^n} t^{-n}\widehat{\phi}(y/t) \exp(-\|y\|^q)\, dy &= \langle \exp(-\|y\|^q), t^{-n}\widehat{\phi}(y/t)\rangle \\
&= (2\pi)^n \langle \mu, \phi(t\xi)\rangle = (2\pi)^n \int_{\mathbb{R}^n} \phi(tx)\, d\mu(x).
\end{aligned}$$

Now, (1.90) can be written as

$$\langle (\|\cdot\|^{-p})^\wedge, \phi\rangle = \frac{(2\pi)^n q}{\Gamma(p/q)} \int_0^\infty t^{k-1} \left(\int_{\mathbb{R}^n} \phi(tx)\, d\mu(x) \right) dt.$$

The latter equality shows that $\langle (\|\cdot\|^{-p})^\wedge, \phi\rangle \geq 0$ if ϕ is any even non-negative test function, so $(\|\cdot\|^{-p})^\wedge$ is a positive distribution. If K is the unit ball of the space B, then $\|\cdot\|$ is the Minkowski functional of K, so K belongs to L^n_{-p} by Theorem 1.7.6. □

Another example originates from the fact that every two-dimensional normed space embeds isometrically in L_1; see, for example, [310, Corollary 6.8]. The extension is that every n-dimensional normed space embeds in L_{-n+3}. This follows from (1.64) if we put $k = 2$ and use Brunn's theorem; see [310, p. 127]. For $n = 4$, this means that every origin-symmetric convex body in $\mathbb{R}^4$ is an intersection body.

Yet another example is the factorization theorem of Maurey [402] and Nikishin [442]. If $0 < p < q < 1$, then any Banach subspace of L_p is isomorphic to a subspace of L_q. It was proved in [273] (see also [310, Theorem 6.18]) that this result can be extended to negative p.

Proposition 1.7.11. *For every $k \in \mathbb{N}$ and every $0 < q < 1$, there exists a constant $c(k,q)$ such that for every $n \in \mathbb{N}$, $n > k$ and every origin-symmetric convex k-intersection body D in $\mathbb{R}^n$ there exists an n-dimensional subspace of $L_q([0,1])$ whose unit ball L satisfies $L \subset D \subset c(k,q)L$.*

It follows from Theorem 1.5.7, that every origin-symmetric convex body in $\mathbb{R}^n$ is a k-intersection body for $k = n-3, n-2$ and every origin-symmetric star body is an $(n-1)$-intersection body; see [310, Corollary 4.9]. On the other hand, the unit balls of the spaces ℓ^n_q with $2 < q \leq \infty$, $n \geq 3$ are not k-intersection bodies for all k, $0 < k < n-3$; see [310, Theorems 4.13, 4.15]. In particular, the unit balls of ℓ^n_q, $2 < q \leq \infty$ are not intersection bodies if $n \geq 5$ (this was first proved in [306]). Moreover, we have the following.

Proposition 1.7.12. *The spaces ℓ^n_q, $2 < q \leq \infty$ embed in L_{-p} if and only if $n-3 \leq p < n$.*

We use the following approximation lemma.

Lemma 1.7.13. *Let $1 \leq k < n$ and let D be an origin-symmetric convex body in $\mathbb{R}^n$ that is not a k-intersection body. Then, there exists a sequence of origin-symmetric convex bodies D_m such that $D_m \to D$ in the radial metric, each D_m is infinitely smooth and has positive curvature, and each D_m is not a k-intersection body.*

Proof. Consider a sequence of origin-symmetric infinitely smooth convex bodies D_m with positive curvature that approximate D in the radial metric. Suppose that all the bodies D_m, starting from some m_0, are k-intersection bodies. Then, $\|\cdot\|_{D_m}^{-k}$ is a positive definite distribution for each $m > m_0$. By Lemma 1.4.8, for every even test function ϕ

$$\langle (\|\cdot\|_D^{-k})^\wedge, \phi\rangle = \lim_{m\to\infty} \langle (\|\cdot\|_{D_m}^{-k})^\wedge, \phi\rangle \geq 0,$$

which implies that D is a k-intersection body by Theorem 1.7.6. This is a contradiction that proves our statement. □

We would like to mention at the end of this section that the concept of a k-projection body has been introduced by Dorrek and Schuster [152] in a way similar to the definition of k-intersection bodies. A Fourier characterization in terms of the area measures is remarkably similar to the Fourier characterization of k-intersection bodies. It would be interesting to find functional analytic connections similar to the connection of k-intersection bodies with L_{-k}.

1.7.2 Generalized intersection bodies

The classes $\mathcal{BP}_k^n$ of generalized k-intersection bodies were introduced by Zhang [553]. Recall that for $1 \leqslant k \leqslant n-1$, the $(n-k)$-dimensional spherical Radon transform $R_{n-k} : C(S^{n-1}) \to C(G_{n,n-k})$ is the linear operator defined by

$$R_{n-k}g(H) = \int_{S^{n-1}\cap H} g(\xi)\,d\xi, \quad H \in G_{n,n-k} \tag{1.91}$$

for every function $g \in C(S^{n-1})$. Denote the image of the operator R_{n-k} by X

$$R_{n-k}\big(C(S^{n-1})\big) = X \subset C(G_{n,n-k}).$$

Let $M^+(X)$ be the space of positive linear functionals on X, i. e., for every $\mu \in M^+(X)$ and any non-negative function $f \in X$ we have $\mu(f) \geqslant 0$. We say that an origin-symmetric star body D in $\mathbb{R}^n$ is a generalized k-intersection body, and write $D \in \mathcal{BP}_k^n$, if there exists a functional $\mu_D \in M^+(X)$ such that for every $g \in C(S^{n-1})$

$$\int_{S^{n-1}} \|\xi\|_D^{-k} g(\xi)\,d\xi = \mu_D(R_{n-k}g). \tag{1.92}$$

The class $\mathcal{BP}_1^n$ is the original class of intersection bodies $\mathcal{I}_n$ introduced by Lutwak.

A geometric characterization of generalized intersection bodies was given by Goodey and Weil for $k = 1$ [225] and by Grinberg and Zhang [236, Theorem 6.4] for $k > 1$. Suppose that we have the unit Euclidean ball in $\mathbb{R}^n$ and are allowed to do three operations – linear transformations, k-radial sums, and closure in the radial metric. As a result, we get the class of generalized k-intersection bodies.

Proposition 1.7.14. *The class $\mathcal{BP}_k^n$ is the closure in the radial metric of k-radial sums of origin-symmetric ellipsoids.*

Clearly, the classes $\mathcal{BP}_k^n$ contain origin-symmetric ellipsoids and are invariant with respect to linear transformations and k-radial sums. It was proved by Grinberg and Zhang [236] that intersection bodies are generalized k-intersection bodies for every k, i. e., $\mathcal{I}_n = \mathcal{BP}_1^n \subset \mathcal{BP}_k^n$. E. Milman [414] proved a more general result.

Proposition 1.7.15. *If $k, m \in \mathbb{N}$, $1 \leqslant m < k < n$ and k is divisible by m, then $\mathcal{BP}_m^n \subset \mathcal{BP}_k^n$ and $L_{-m}^n \subset L_{-k}^n$.*

Since ellipsoids are contained in L_{-k}^n for every k and the classes L_{-k}^n are closed with respect to k-radial sums, Proposition 1.7.14 implies that every generalized k-intersection body is a k-intersection body. This was first proved in [309]. We include here the Fourier analytic proof from [310, Theorem 4.23] that does not depend on Proposition 1.7.14.

Proposition 1.7.16. *Every generalized k-intersection body in $\mathbb{R}^n$ is a k-intersection body, i. e., $\mathcal{BP}_k^n \subset L_{-k}^n$ for every $1 \leqslant k < n$.*

Proof. Suppose that $D \in \mathcal{BP}_n^k$, and μ_D is the positive functional corresponding to D by (1.92). For any non-negative test function ϕ

$$\langle (\|\cdot\|_D^{-k})^\wedge, \phi\rangle = \int_{\mathbb{R}^n} \|x\|_D^{-k}\widehat{\phi}(x)\,dx$$
$$= \int_{S^{n-1}} \|\xi\|_D^{-k}\left(\int_0^\infty r^{n-k-1}\widehat{\phi}(r\xi)\,dr\right)d\xi. \tag{1.93}$$

By (1.92) applied to the function

$$g(\xi) = \int_0^\infty r^{n-k-1}\widehat{\phi}(r\xi)\,dr,$$

the expression on the right-hand side of (1.93) is equal to

$$\mu\left(R_{n-k}\left(\int_0^\infty r^{n-k-1}\widehat{\phi}(r\xi)\,dr\right)\right)$$
$$= \mu\left(\int_{S^{n-1}\cap H}\left(\int_0^\infty r^{n-k-1}\widehat{\phi}(r\xi)\,dr\right)d\xi\right)$$
$$= \mu\left(\int_H \widehat{\phi}\right) = \mu\left((2\pi)^{n-k}\int_{H^\perp}\phi\right) \geqslant 0,$$

because the function ϕ is non-negative. Here, we used the fact that $\int_H \widehat{\phi} = (2\pi)^{n-k}\int_{H^\perp}\phi$ by Lemma 1.6.6. So, $\|\cdot\|_D^{-k}$ is a positive definite distribution, and we have $D \in L_{-k}^n$ by Theorem 1.7.6. □

The reverse inclusion does not hold in general, as shown by E. Milman in [415]. The body providing a counterexample in [415] is not convex, so for convex bodies the equivalence question is still open.

We use Proposition 1.7.14 to extend the concept of a generalized intersection body to the case where k is not necessarily an integer.

Definition 1.7.17. For every real number q, $0 < q < n$, the class $\mathcal{BP}_q^n$ is the closure in the radial metric of q-radial sums of origin-symmetric ellipsoids in $\mathbb{R}^n$.

Since the classes L_{-q}^n contain ellipsoids and are closed with respect to q-radial sums, we immediately get the following.

Proposition 1.7.18. *For every $0 < q < n$, the class $\mathcal{BP}_q^n$ is contained in L_{-q}^n.*

1.7.3 Outer volume ratio estimates

Let K be an origin-symmetric convex body in $\mathbb{R}^n$ and let $\mathcal{A}$ be a class of star bodies. We define the outer volume ratio distance of K from $\mathcal{A}$ by

$$d_{\mathrm{ovr}}(K, \mathcal{A}) = \inf\{\mathrm{ovr}(K, C),\ C \in \mathcal{A}\},$$

where

$$d_{\mathrm{ovr}}(K, C) := \inf\left\{\left(\frac{|T(C)|}{|K|}\right)^{1/n} : K \subseteq T(C),\ T \in GL_n\right\}.$$

Clearly, if $\mathcal{A}_1 \subset \mathcal{A}_2$, then $d_{\mathrm{ovr}}(K, \mathcal{A}_2) \leqslant d_{\mathrm{ovr}}(K, \mathcal{A}_1)$. Also, if K is origin-symmetric convex and $\mathcal{A}$ contains the Euclidean ball, then by John's theorem [271] we have $d_{\mathrm{ovr}}(K, \mathcal{A}) \leqslant \sqrt{n}$.

The latter estimate applied to $\mathcal{A} = \mathcal{BP}_q^n$ yields $d_{\mathrm{ovr}}(K, \mathcal{BP}_q^n) \leqslant \sqrt{n}$. A different estimate that is better for large q was established in [329]. The result was formulated in [329] for integers q, but the proof holds for all real numbers $1 \leqslant q \leqslant n-1$ without any changes.

Theorem 1.7.19. *Let L be an origin-symmetric convex body in $\mathbb{R}^n$, and let $q \in \mathbb{R}$, $1 \leqslant q \leqslant n-1$. Then,*

$$d_{\mathrm{ovr}}(L, \mathcal{BP}_q^n) \leqslant c_0 \sqrt{n/q}\, \log(en/q),$$

where c_0 is an absolute positive constant.

The main idea of the proof of Theorem 1.7.19 is to cover a given convex body by Euclidean balls and then show that, if the number of balls is not too big, one can approximate the union of the balls by a body in $\mathcal{BP}_q^n$, where q depends on the covering number of K. We shall use Theorem 1.2.7: For every $0 < \alpha < 2$ and every origin-symmetric convex body K in $\mathbb{R}^n$ there exists a linear image K_α of K such that

$$\max\{N(K_\alpha, tB_2^n), N(B_2^n, tK_\alpha), N(K_\alpha^\circ, tB_2^n), N(B_2^n, tK_\alpha^\circ)\} \leqslant \exp\left(\frac{c(\alpha)n}{t^\alpha}\right)$$

for every $t \geqslant c(\alpha)^{1/\alpha}$, where $c(\alpha) = O((2-\alpha)^{-\alpha/2})$ as $\alpha \to 2$.

Theorem 1.2.7 implies the following reverse Brunn–Minkowski inequality.

Proposition 1.7.20. *Let $\alpha \in [1,2)$, let K be an origin-symmetric convex body in $\mathbb{R}^n$, and let K_α be the position of K established in Theorem 1.2.7. Then, for every $t \geqslant 1$*

$$|K_\alpha + tB_2^n|^{1/n} \leqslant 2e^2 c(\alpha)^{1/\alpha}\, t\, |K_\alpha|^{1/n},$$

where $c(\alpha)$ is the constant in Theorem 1.2.7.

Proof. Using the estimate for $N(B_2^n, tK_\alpha)$ with $t = c(\alpha)^{1/\alpha}$ in Theorem 1.2.7 we get

$$|B_2^n|^{1/n} \leqslant c(\alpha)^{1/\alpha} |K_\alpha|^{1/n} N(B_2^n, tK_\alpha)^{1/n} \leqslant e\, c(\alpha)^{1/\alpha}\, |K_\alpha|^{1/n}.$$

Then, for every $t \geqslant 1$, using the estimate for $N(K_\alpha, tB_2^n)$ in Theorem 1.2.7 we see that

$$\begin{aligned}\frac{|K_\alpha + tB_2^n|^{1/n}}{2t|K_\alpha|^{1/n}} &\leqslant e\, c(\alpha)^{1/\alpha} \frac{|K_\alpha + tB_2^n|^{1/n}}{2t|B_2^n|^{1/n}} \leqslant e\, c(\alpha)^{1/\alpha}\, N(K_\alpha + tB_2^n, 2tB_2^n)^{1/n} \\ &\leqslant e\, c(\alpha)^{1/\alpha}\, N(K_\alpha, tB_2^n)^{1/n} \leqslant e^2 c(\alpha)^{1/\alpha}.\end{aligned}$$

□

We will also need the following elementary lemma.

Lemma 1.7.21. *Let $z \in \mathbb{R}^n$ and $t > 0$. There exists a centered ellipsoid $\mathcal{E}$ such that*

$$z + tB_2^n \subseteq \mathcal{E} \subseteq \mathrm{conv}(\{2z + 2\sqrt{2}tB_2^n, -2z + 2\sqrt{2}tB_2^n\}).$$

Proof. We may assume that $z = z_1 e_1$. Let C be the "cylinder" defined as

$$C := \{(s,y) \in \mathbb{R} \times \mathbb{R}^{n-1} : |s| \leqslant z_1 + t,\ |y| \leqslant t\}.$$

Then, one can check that

$$z + tB_2^n \subseteq C \subseteq \mathrm{conv}(\{z + \sqrt{2}tB_2^n, -z + \sqrt{2}tB_2^n\}) =: K. \tag{1.94}$$

Let $Q := \{(s,y) \in \mathbb{R}^n : |s| \leqslant 1,\ |y| \leqslant 1\}$. We have $C = T(Q)$ for some $T \in GL_n$. Define $\mathcal{E}_1 := T(B_2^n)$. Then,

$$B_2^n \subseteq Q \subseteq \sqrt{2}B_2^n, \text{ or, equivalently, } \mathcal{E}_1 \subseteq C \subseteq \sqrt{2}\mathcal{E}_1.$$

From (1.94) it follows that

$$z + tB_2^n \subseteq C \subseteq \sqrt{2}\mathcal{E}_1 \subseteq 2C \subseteq 2K$$

and we obtain the lemma with $\mathcal{E} := \sqrt{2}\mathcal{E}_1$. □

By definition, the class $\mathcal{BP}_q^n$ is closed under q-radial sums, and ellipsoids belong to this class for all $1 \leqslant q \leqslant n-1$.

Let V_1, V_2 be two star bodies in $\mathbb{R}^n$. We define the distance between V_1 and V_2 as

$$d(V_1, V_2) := \sup_{x \neq 0} \max\left\{\frac{\rho_{V_1}(x)}{\rho_{V_2}(x)}, \frac{\rho_{V_2}(x)}{\rho_{V_1}(x)}\right\}.$$

Note that the definition implies that

$$\frac{1}{d}V_1 \subseteq V_2 \subseteq dV_1,$$

where $d := d(V_1, V_2)$.

Lemma 1.7.22. *Let $1 \leqslant q \leqslant n-1$ and $N \in \mathbb{N}$ with $\log N \leqslant q$. Let V_i, $1 \leqslant i \leqslant N$, be symmetric star bodies in $\mathcal{BP}_q^n$ and let $V = \bigcup_{i=1}^N V_i$. There exists $C \in \mathcal{BP}_q^n$ such that*

$$d(V, C) \leqslant e.$$

Proof. For any $1 \leqslant i \leqslant N$ let ρ_{V_i} be the radial function of V_i. Then,

$$\rho_V := \max_{1 \leqslant i \leqslant N} \rho_{V_i}(x), \quad x \in \mathbb{R}^n \setminus \{0\}.$$

Since $\log N \leqslant q$, for any $a \in \mathbb{R}^N$ we have that

$$\|a\|_\infty \leqslant \|a\|_q \leqslant N^{\frac{1}{q}}\|a\|_\infty \leqslant e\|a\|_\infty. \tag{1.95}$$

Let C be the star body defined by

$$\rho_C^q(x) := \sum_{i=1}^N \rho_{V_i}^q(x), \quad 0 \neq x \in \mathbb{R}^n.$$

Then, $C \in \mathcal{BP}_q^n$ and by (1.95) we also have that $d(V, C) \leqslant e$. □

Proof of Theorem 1.7.19. Let $a := 2\frac{1}{\log(en/q)}$. Recall that

$$d_{\mathrm{ovr}}(T(K), \mathcal{BP}_q^n) = d_{\mathrm{ovr}}(K, \mathcal{BP}_q^n)$$

for any $T \in GL_n$. So, we may assume that K is in the position described in Theorem 1.2.7. Then, for every $t \geqslant c(a)^{1/a}$ there exists N such that $\log N \leqslant \frac{c(a)n}{t^a}$ and $z_1, \ldots, z_N \in K$ such that

$$K \subseteq \bigcup_{i=1}^{N} (z_i + tB_2^n).$$

For every z_i consider an ellipsoid $\mathcal{E}_i$ as in Lemma 1.7.21. Then, we have that

$$K \subseteq \bigcup_{i=1}^{N} \mathcal{E}_i \subseteq 2(\mathrm{conv}\{z_1, \ldots, z_n\} + \sqrt{2}tB_2^n) \subseteq 2(K + \sqrt{2}tB_2^n).$$

By the choice of α we have that $c(\alpha)^{1/\alpha} = O(\sqrt{\log(en/q)})$. Choose t so that $t^\alpha = c(\alpha)\frac{n}{q}$, which means that $t \approx \sqrt{n/q}\,\sqrt{\log(en/q)}$. Then, $\log N \leqslant q$ and hence, by Lemma 1.7.22, there exists $C \in \mathcal{BP}_q^n$ such that $d(C, V) \leqslant e$, where $V = \bigcup_{i=1}^{N} \mathcal{E}_i$. It follows that

$$e^{-1}K \subseteq e^{-1}V \subseteq C \subseteq eV \subseteq 2e(K + \sqrt{2}tB_2^n). \tag{1.96}$$

Recall that $t \approx \sqrt{n/q}\,\sqrt{\log(en/q)}$. Then, (1.96) becomes

$$e^{-1}K \subseteq C \subseteq 2e(K + c_1\sqrt{n/q}\sqrt{\log(en/q)}B_2^n).$$

Moreover, by Proposition 1.7.20, we have that

$$|K + tB_2^n|^{1/n} \leqslant 2e^2 c(\alpha)^{1/\alpha}\, t\, |K|^{1/n},$$

which implies that

$$|C|^{1/n} \leqslant 4\sqrt{2}e^3 c(\alpha)^{1/\alpha}\, t\, |K|^{1/n} \leqslant c_2\sqrt{n/q}\, \log(en/q)|K|^{1/n}.$$

Let $C_1 := eC$. We have that $C_1 \in \mathcal{BP}_q^n$, $K \subseteq C_1$ and

$$\frac{|C_1|^{1/n}}{|K|^{1/n}} \leqslant c_0\sqrt{n/q}\, \log(en/q).$$

This completes the proof. □

1.8 Further reading

1.8.1 Fixed points of the intersection body operator

Denote by $\mathcal{I}K$ the intersection body of a star body K in $\mathbb{R}^n$; see Section 1.7. Lutwak [394, Problem 12.8] and Gardner [194, Problem 8.6] asked whether a star body K in $\mathbb{R}^n$, $n \geqslant 3$ satisfying $\mathcal{I}^2K = cK$, where $c > 0$, is necessarily an ellipsoid. Moreover, Gardner [194, Problem 8.7] asked whether the equality $\mathcal{I}K = cK$ for some $c > 0$ implies that K is a Euclidean ball in dimensions 3 and higher. Fish, Nazarov, Ryabogin, and Zvavitch [174] confirmed these conjectures in the case where K is in a small (in the Banach–Mazur distance) neighborhood of the Euclidean ball. Recently, E. Milman, Shabelman, and Yehudayoff [419] answered both questions in the affirmative in full generality.

1.8.2 Klee's problems on maximal sections

For a convex body K in $\mathbb{R}^n$ containing the origin, the inner section function $m_K : S^{n-1} \to \mathbb{R}$ is defined by

$$m_K(u) = \max_{t\in\mathbb{R}} \big|K \cap (u^\perp + tu)\big|.$$

It is well known that, for origin-symmetric convex bodies K, the function m_K determines K uniquely. A question of Klee [301] is whether for non-symmetric bodies the function m_K is constant only when K is a Euclidean ball. Nazarov, Ryabogin, and Zvavitch [439] showed that the answer is negative in all dimensions.

Another question of Klee [301] is whether the function m_K determines the body K up to translation and reflection in the origin. Gardner, Ryabogin, Yaskin, and Zvavitch [203] constructed a centrally symmetric convex body L and a non-symmetric convex K with $m_K = m_L$, which gives a negative answer to Klee's question.

For more results of this kind, see the surveys [11, 495] and the papers [12, 438, 547].

1.8.3 Problems of Grünbaum and Loewner on the centroids of sections

Let K be a convex body in $\mathbb{R}^n$. The centroid of K is defined as

$$\mathrm{bar}(K) = \frac{1}{|K|} \int_K x\, dx.$$

Let $\mu(K)$ be the number of hyperplane sections of K passing through $c(K)$ whose centroid coincides with $\mathrm{bar}(K)$. Loewner [168, Problem 28] asked what is the minimum of $\mu(K)$ for all convex bodies in $\mathbb{R}^n$. Grünbaum [242] asked whether $\mu(n) \geqslant n + 1$ for every n. Both Grünbaum [242] and Loewner [168] proved that $\mu(2) \geqslant 3$. Grünbaum also showed that every interior point of a convex body is the centroid of at least one hyperplane section, so $\mu(n) \geqslant 1$. Myroshnychenko, Tatarko, and Yaskin [432] proved that for every $n \geqslant 5$ there exists a convex body K such that $\mu(K) = 1$, thus disproving the conjecture of Grünbaum in dimensions 5 and higher. The proof in [432] is Fourier analytic, and it makes use of the existence of non-intersection bodies in dimensions 5 and higher. Very recently, a complete solution was given in [537], employing the hemispherical transform and Morse theory.

1.8.4 Ulam's problem on floating bodies

Let K be a convex body of constant density that is partially submerged in water of density 1. The density of the body K is denoted by $\delta \in (0,1)$, and it is defined as the ratio of the volume of the submerged part of K and the volume of whole K. The body K is in equilibrium if the centroid of K and the centroid of the submerged part of K lie on the same vertical line.

In 1935, Ulam asked in the Scottish Book [528, Problem 19] whether a solid of uniform density that will float in water in equilibrium in every position is necessarily a Euclidean ball. In 1938, Auerbach [34] found counterexamples in dimension 2, with $\delta = \frac{1}{2}$. Counterexamples in dimension 2 with some other densities were given by Wegner [540, 541]. Bracho, Montejano, and Oliveros [107] gave an affirmative answer in dimension 2 for certain classes of bodies. In 1970, Schneider [504] answered Ulam's question in the affirmative for centrally symmetric convex bodies of density $\delta = \frac{1}{2}$ (in 1983, this was also proved by Falconer [167]). In 2012, Várkonyi [531] constructed three-dimensional nonconvex counterexamples with density $\frac{1}{2}$. Finally, in 2022, Ryabogin [494] solved Ulam's problem in the negative in all dimensions $n \geqslant 3$. His counterexamples are strictly convex bodies of revolution with density $\frac{1}{2}$.

For more details and related results, see the survey [14] and the paper [13].

1.8.5 Algebraically integrable bodies

Newton wrote in his Principia [441, Lemma 28] (translated from Latin): "There is no oval figure, whose area cut-off by right lines at pleasure can be universally found by means of equations of any number of terms and dimensions."

Arnold [22] noted that for Newton's argument to work the domain has to be infinitely smooth and convex. He formulated the problem in modern terms, as follows. Let D be a convex body in $\mathbb{R}^n$. The *cut-off function* $V_D(\xi, t)$, $\xi \in S^{n-1}$, $t \in [h_D(-\xi), h_D(\xi)]$, is defined as the n-dimensional volume of the part of D cut by the hyperplane $(x, \xi) = t$ as follows:

$$V_D(\xi, t) = \left|D \cap \{x \in \mathbb{R}^n : \langle x, \xi\rangle \leqslant t\}\right|.$$

A domain D in $\mathbb{R}^n$ is called algebraically integrable if there exists a polynomial Q in $n+2$ variables so that $Q(\xi_1, \dots, \xi_n, t, V_D(\xi, t)) = 0$ for all $\xi \in S^{n-1}$ and $t \in [h_D(-\xi), h_D(\xi)]$. In other words, the cut-off function V_D is algebraic with respect to the parameters of the cutting hyperplane.

It is easy to see that ellipsoids in odd dimensions n are algebraically integrable, but in even dimensions they are not. Newton's Lemma 28 means that there are no algebraically integrable infinitely smooth convex bodies in $\mathbb{R}^2$. Arnold [21, Problems 1990-27, 1987-14] asked the following question.

Problem 1.8.1. Is it true that the only algebraically integrable infinitely smooth convex bodies are ellipsoids in $\mathbb{R}^n$ with odd n?

Vassiliev [532] proved that there are no algebraically integrable infinitely smooth convex bodies in $\mathbb{R}^n$ with even n, extending the result of Newton. In odd dimensions, the problem is still open.

Recall that the parallel section function of D in the direction $\xi \in S^{n-1}$ is defined by

$$A_{D,\xi}(t) = \left|D \cap \{x \in \mathbb{R}^n : \langle x, \xi\rangle = t\}\right|, \quad t \in [h_D(-\xi), h_D(\xi)].$$

The parallel section function is the derivative of the cut-off function by t.

Following Agranovsky [1], we say that a domain D in $\mathbb{R}^n$ is polynomially integrable of degree N if for every $\xi \in S^{n-1}$

$$A_{D,\xi}(t) = \sum_{k=1}^{N} a_k(\xi)t^k, \quad \text{for all } t \in [h_D(-\xi), h_D(\xi)], \tag{1.97}$$

where the coefficients are arbitrary functions of ξ, and a_N is not identically zero. Agranovsky asked in the same paper whether the only infinitely smooth polynomially integrable domains are ellipsoids in odd dimensions. He proved that every such domain must be convex and that there are no infinitely smooth polynomially integrable convex domains in even dimensions. Agranovsky's conjecture was confirmed in full in [328]. The proof is based on Theorem 1.5.7.

For details and further results, see the survey [5] and papers [2–4, 6, 89, 90, 533, 534, 546, 549].

2 Bourgain's hyperplane problem

Bourgain's hyperplane problem asks if there exists an absolute constant $c > 0$ such that

$$\max\{|K \cap \xi^{\perp}| : \xi \in S^{n-1}\} \geqslant c$$

for every convex body K of volume 1 in $\mathbb{R}^n$ that has its barycenter at the origin. A natural framework for the study of this problem is the isotropic position of a convex body. More generally, a log-concave probability measure μ on $\mathbb{R}^n$ with density f_μ is called isotropic if its barycenter is at the origin and the covariance matrix $\mathrm{Cov}(\mu)$ of μ is the identity matrix I_n. As we will see, an equivalent formulation of the hyperplane problem is to ask if there exists an absolute constant $C > 0$ such that

$$\|f_\mu\|_\infty^{1/n} \leqslant C$$

for every $n \geqslant 1$ and every isotropic log-concave probability measure on $\mathbb{R}^n$.

A second main question in asymptotic convex geometry is the Kannan–Lovász–Simonovits conjecture about the isoperimetric constant χ_μ of an isotropic log-concave probability measure μ, which is the largest constant $\chi \geqslant 0$ such that

$$\mu^+(A) \geqslant \chi \min\{\mu(A), 1 - \mu(A)\}$$

for every Borel subset A of $\mathbb{R}^n$, where $\mu^+(A)$ is the Minkowski content of A. If we set $\psi_\mu = 1/\chi_\mu$, then the "KLS conjecture" is the question of whether there exists an absolute constant $C > 0$ such that

$$\psi_\mu \leqslant C$$

for every $n \geqslant 1$ and every isotropic log-concave probability measure on $\mathbb{R}^n$. An equivalent way to formulate the KLS conjecture is to ask whether the Poincaré inequality holds for every isotropic log-concave probability measure μ on $\mathbb{R}^n$ with a constant that does not depend on the measure or the dimension n.

Spectacular recent developments by Guan [243] and Klartag–Lehec [295] established an affirmative answer to Bourgain's hyperplane problem; at the same time, Klartag showed that $\psi_\mu \leqslant c\sqrt{\log n}$ for every isotropic log-concave probability measure on $\mathbb{R}^n$. Moreover, Klartag and Lehec confirmed the thin-shell conjecture, which asks if there exists an absolute constant $C > 0$ such that, for any $n \geqslant 1$ and any isotropic log-concave probability measure μ on $\mathbb{R}^n$, one has

$$\mathbb{E}_\mu(|x| - \sqrt{n})^2 \leqslant C^2.$$

This inequality implies that most of the mass of μ is concentrated on a thin spherical cell and it also implies that high-dimensional log-concave distributions have approximately gaussian marginals.

https://doi.org/10.1515/9783111360508-002

The key for the developments that we shall discuss in this chapter is Eldan's stochastic localization. We introduce the questions and present the tools and the main ideas that lead to the positive answers.

2.1 Isotropic convex bodies and Bourgain's hyperplane problem

A convex body K in $\mathbb{R}^n$ is called isotropic if it has volume $|K| = 1$, it is centered, and there is a constant $\alpha > 0$ such that

$$\int_K \langle x, y\rangle^2 \, dx = \alpha^2 |y|^2 \tag{2.1}$$

for all $y \in \mathbb{R}^n$. It is useful to note that the isotropic condition (2.1) is equivalent to the fact that

$$\int_K x_i x_j \, dx = \alpha^2 \delta_{ij} \tag{2.2}$$

for every $i, j = 1, \dots, n$, where $x_j = \langle x, e_j\rangle$ are the coordinates of x with respect to any given orthonormal basis $\{e_1, \dots, e_n\}$ of $\mathbb{R}^n$. This is, in turn, equivalent to the fact that for every linear map $T : \mathbb{R}^n \to \mathbb{R}^n$ (we write $T \in L(\mathbb{R}^n)$)

$$\int_K \langle x, Tx\rangle \, dx = \alpha^2 \mathrm{tr}(T). \tag{2.3}$$

Another consequence of the isotropic condition (2.1) is that

$$\int_K |x|^2 \, dx = \sum_{i=1}^n \int_K \langle x, e_i\rangle^2 \, dx = n\alpha^2.$$

Also, it is easily checked that if K is an isotropic convex body in $\mathbb{R}^n$, then $U(K)$ is also isotropic for every $U \in O(n)$.

The next proposition shows that every centered convex body has a linear image that satisfies the isotropic condition.

Proposition 2.1.1. *Let K be a centered convex body in $\mathbb{R}^n$. There exists $T \in GL_n$ such that $T(K)$ is isotropic.*

Proof. The linear operator $M : \mathbb{R}^n \to \mathbb{R}^n$ defined by $M(y) = \int_K \langle x, y\rangle x \, dx$ is symmetric and positive definite; therefore, it has a symmetric and positive definite square root S. Consider the linear image $\tilde{K} = S^{-1}(K)$ of K. Then, for every $y \in \mathbb{R}^n$ we have

$$\int_{\tilde{K}} \langle x, y\rangle^2 \, dx = |\mathrm{det} S|^{-1} \int_K \langle S^{-1}x, y\rangle^2 \, dx = |\mathrm{det} S|^{-1} \int_K \langle x, S^{-1}y\rangle^2 \, dx$$

$$= |\det S|^{-1}\Big\langle \int_K \langle x, S^{-1}y\rangle x\, dx, S^{-1}y\Big\rangle = |\det S|^{-1}\langle MS^{-1}y, S^{-1}y\rangle$$
$$= |\det S|^{-1}|y|^2.$$

Normalizing the volume of $\tilde{K}$ we get the result. □

We say that $\tilde{K}$ is a position of K if $\tilde{K} = T(K)$ for some $T \in GL_n$. Proposition 2.1.1 shows that every centered convex body K in $\mathbb{R}^n$ has a position $\tilde{K}$ that is isotropic. The next theorem shows that the isotropic position of a convex body is uniquely determined up to orthogonal transformations, and arises as a solution of a minimization problem.

Theorem 2.1.2. *Let K be a centered convex body of volume 1 in $\mathbb{R}^n$. Define*

$$\Delta(K) = \inf\Big\{ \int_{TK} |x|^2\, dx : T \in SL_n \Big\}. \tag{2.4}$$

Then, a position K_1 of K, of volume 1, is isotropic if and only if

$$\int_{K_1} |x|^2\, dx = \Delta(K). \tag{2.5}$$

Furthermore, if K_1 and K_2 are isotropic positions of K, then there exists $U \in O(n)$ such that $K_2 = U(K_1)$.

Proof. Fix an isotropic position K_1 of K. We know that there exists $\alpha > 0$ such that

$$\int_{K_1} \langle x, Tx\rangle\, dx = \alpha^2(\mathrm{tr}T)$$

for every $T \in L(\mathbb{R}^n)$. Then, for every $T \in SL_n$ we have

$$\int_{TK_1} |x|^2\, dx = \int_{K_1} |Tx|^2\, dx = \int_{K_1} \langle x, T^*Tx\rangle\, dx = \alpha^2\mathrm{tr}(T^*T) \geqslant n\alpha^2 = \int_{K_1} |x|^2\, dx, \tag{2.6}$$

where we have used the arithmetic-geometric means inequality in the form

$$\mathrm{tr}(T^*T) \geqslant n[\det(T^*T)]^{1/n}.$$

This shows that K_1 satisfies (2.5). In particular, the infimum in (2.4) is a minimum.

Note also that if we have equality in (2.6), then $T^*T = I_n$, and hence $T \in O(n)$. This shows that any other position $\tilde{K}$ of K that satisfies (2.5) is an orthogonal image of K_1, therefore it is isotropic.

Finally, if K_2 is some other isotropic position of K, then the first part of the proof shows that K_2 satisfies (2.5). By the previous step, we must have $K_2 = U(K_1)$ for some $U \in O(n)$. □

We can now give the following definition for the isotropic constant of a general convex body K in $\mathbb{R}^n$.

Definition 2.1.3. Let $K \subset \mathbb{R}^n$ be a convex body. Its isotropic constant L_K is defined by

$$L_K^2 = \frac{1}{n}\min\left\{\frac{1}{|T\tilde{K}|^{1+\frac{2}{n}}}\int_{T\tilde{K}} |x|^2\,dx : T \in GL_n\right\},$$

where $\tilde{K} = K - \mathrm{bar}(K)$ is the centered translate of K.

Note that L_K depends only on the affine class of K. Note also that if K is isotropic, then for all $\xi \in S^{n-1}$ we have

$$\int_K \langle x, \xi\rangle^2\,dx = L_K^2.$$

Problem 2.1.4 (isotropic constant problem). There exists an absolute constant $C > 0$ such that for any $n \geqslant 1$ and any convex body $K \subset \mathbb{R}^n$ we have

$$L_K \leqslant C.$$

Equivalently, if K is an isotropic convex body in $\mathbb{R}^n$, then

$$\int_K \langle x, \xi\rangle^2\,dx \leqslant C^2$$

for every $\xi \in S^{n-1}$.

The moments of inertia of a centered convex body are closely related with the volume of its hyperplane sections that pass through the origin. In the isotropic case this relation takes the following form.

Theorem 2.1.5. *Let K be a centered convex body of volume* 1 *in $\mathbb{R}^n$. For every $\xi \in S^{n-1}$ we have*

$$\frac{c_1}{|K \cap \xi^\perp|} \leqslant \left(\int_K \langle x, \theta\rangle^2\,dx\right)^{1/2} \leqslant \frac{c_2}{|K \cap \xi^\perp|}, \tag{2.7}$$

where $c_1, c_2 > 0$ are absolute constants. In particular, if K is isotropic, then for every $\xi \in S^{n-1}$ we have

$$\frac{c_1}{L_K} \leqslant |K \cap \xi^\perp| \leqslant \frac{c_2}{L_K}. \tag{2.8}$$

For the proof, given $\xi \in S^{n-1}$ we consider the function $f(t) = f_{K,\xi}(t) = |K \cap \{x : \langle x,\xi\rangle = t\}|$, $t \in \mathbb{R}$. It is a consequence of the Brunn–Minkowski inequality that $\log f$ is a concave function on its support.

We restrict our attention to the symmetric case. Then, f is even and $\|f\|_\infty = f(0)$. For the proof in the general case, which is more or less the same, we need an additional fact (that can be found in Fradelizi [177]), which shows that hyperplane sections through the center of mass are, up to an absolute constant, maximal: If K is a centered convex body of volume 1 in $\mathbb{R}^n$, then, for every $\xi \in S^{n-1}$

$$\|f\|_\infty \leqslant e f(0) = e\,|K \cap \xi^\perp|.$$

Proof of Theorem 2.1.5 *(symmetric case).* Let $f := f_{K,\xi}$. To prove the left-hand side of (2.7) we set $b = \int_0^{+\infty} f(t)\,dt = \frac{1}{2}$ and define

$$g(t) = \|f\|_\infty \mathbb{1}_{[0,b/\|f\|_\infty]}(t).$$

Since $g \geqslant f$ on the support of g, we have

$$\int_0^s f(t)\,dt \leqslant \int_0^s g(t)\,dt$$

for every $0 \leqslant s \leqslant b/\|f\|_\infty$. The integrals of f and g on $[0,+\infty)$ are both equal to b. It follows that

$$\int_s^\infty g(t)\,dt \leqslant \int_s^\infty f(t)\,dt$$

for every $s \geqslant 0$. Then,

$$\begin{aligned}\int_0^\infty t^2 f(t)\,dt &= \int_0^\infty\int_0^t 2sf(t)\,ds\,dt = \int_0^\infty 2s\Big(\int_s^\infty f(t)\,dt\Big)\,ds\\ &\geqslant \int_0^\infty 2s\Big(\int_s^\infty g(t)\,dt\Big)\,ds = \int_0^\infty t^2 g(t)\,dt\\ &= \int_0^{b/\|f\|_\infty} t^2\|f\|_\infty\,dt = \frac{b^3}{3\|f\|_\infty^2}.\end{aligned}$$

This shows that

$$\int_K \langle x,\xi\rangle^2\,dx = 2\int_0^\infty t^2 f(t)\,dt \geqslant \frac{2b^3}{3\|f\|_\infty^2} = \frac{1}{12f(0)^2}.$$

To prove the right-hand side inequality of (2.7) we distinguish two cases. Assume first that there exists $s > 0$ such that $f(s) = \frac{1}{2}f(0)$. Then,

$$\frac{1}{2} = \int_0^\infty f(t)\,dt \geqslant \int_0^s f(t)\,dt \geqslant sf(s) = \frac{1}{2}sf(0),$$

because, since $\log f$ is even and concave, we have that $f(t) \geqslant f(s)$ on $[0,s]$. On the other hand, if $t > s$, then

$$f(s) \geqslant (f(0))^{1-\frac{s}{t}}(f(t))^{\frac{s}{t}},$$

which implies that $f(t) \leqslant f(0)2^{-t/s}$. We now write

$$\int_0^\infty t^2 f(t)\,dt = \int_0^s t^2 f(t)\,dt + \int_s^\infty t^2 f(t)\,dt \leqslant f(0)\int_0^s t^2\,dt + \int_s^\infty t^2 f(0)2^{-t/s}\,dt$$
$$= f(0)\left(\frac{s^3}{3} + s^3\int_1^\infty u^2 2^{-u}\,du\right) \leqslant c_0 f(0)s^3 \leqslant c_0/(f(0))^2.$$

Now, assume that, for every $s > 0$ on the support of f, we have $f(s) > \frac{1}{2}f(0)$. Then, the role of s is played by $s_0 = \sup\{s > 0 : f(s) > 0\}$. We have $\frac{1}{2} \geqslant \frac{1}{2}f(0)s_0$ and

$$\int_{-\infty}^\infty t^2 f(t)\,dt = 2\int_0^\infty t^2 f(t)\,dt = 2\int_0^{s_0} t^2 f(t)\,dt \leqslant \frac{2f(0)s_0^3}{3} \leqslant \frac{2}{3(f(0))^2}.$$

Thus, we get the same estimate as before, without using the fact that $\log f$ is concave. □

Theorem 2.1.5 reveals a close connection between the isotropic constant problem and the hyperplane problem.

Problem 2.1.6 (The hyperplane problem). There exists an absolute constant $c > 0$ with the following property: if K is a convex body in $\mathbb{R}^n$ with volume 1 and barycenter at the origin, there exists $\xi \in S^{n-1}$ such that

$$|K \cap \xi^\perp| \geqslant c. \tag{2.9}$$

Problem 2.1.6 appears for the first time in the work of Bourgain [99] on high-dimensional maximal functions associated with arbitrary convex bodies. Bourgain was interested in bounds for the L_p-norm of the maximal function

$$\mathcal{M}_K f(x) = \sup\left\{\frac{1}{|tK|}\int_{tK} |f(x+y)|\,dy \;\middle|\; t > 0\right\}$$

of $f \in L^1_{\mathrm{loc}}(\mathbb{R}^n)$, where K is an origin-symmetric convex body in $\mathbb{R}^n$. Let $C_p(K)$ denote the best constant such that

$$\|\mathcal{M}_K f\|_p \leqslant C_p(K)\|f\|_p$$

and $C_{1,1}(K)$ the best constant so that the weak type inequality

$$\|\mathcal{M}_K f\|_{1,\infty} \leqslant C_{1,1}(K)\|f\|_1$$

is satisfied. Stein proved in [515] that if $K = B_2^n$ is the Euclidean unit ball, then $C_p(B_2^n)$ is bounded independently of the dimension for all $p > 1$. Bourgain showed that there exists an absolute constant $C > 0$ (independent of n and K) such that

$$\|\mathcal{M}_K\|_{L^2(\mathbb{R}^n)\to L^2(\mathbb{R}^n)} \leqslant C.$$

Around the same time, the result for $\|\mathcal{M}_K\|_{2\to 2}$ was generalized to all $p > 3/2$ by Bourgain [100] and, independently, by Carbery [124]. By the definition of $\mathcal{M}_K$ it is clear that in order to obtain a uniform bound on $\|\mathcal{M}_K\|_{2\to 2}$ one can start with any suitable position $T(K)$ of K. Bourgain used the isotropic position; the property that played an important role in his argument was that when K is isotropic, then $L_K|K \cap \xi^\perp| \approx 1$ for all $\xi \in S^{n-1}$. Bourgain [99] mentions the fact that $L_K \geqslant c$ and asks whether a reverse inequality might hold true.

We will show that the two problems are equivalent. One direction is simple, by the previous discussion; assume that the hyperplane problem has an affirmative answer. If K is isotropic, Theorem 2.1.5 shows that *all* sections $K \cap \xi^\perp$ have volume bounded from above by c_2/L_K. Since (2.9) must be true for at least one $\xi \in S^{n-1}$, we get $L_K \leqslant c_2/c$.

Conversely, we will show that if there exists an absolute bound C for the isotropic constant, then the hyperplane problem has an affirmative answer. One way to see this is through the Binet ellipsoid of inertia.

Let K be a centered convex body of volume 1 in $\mathbb{R}^n$. Consider the ellipsoid $\mathcal{E}_B(K)$ defined by

$$\|y\|^2_{\mathcal{E}_B(K)} = \langle M_K y, y\rangle = \int_K \langle x, y\rangle^2\, dx,$$

where M_K is the matrix of inertia of K. The ellipsoid $\mathcal{E}_B(K)$ is the Binet ellipsoid of K. Note that K is in an isotropic position if and only if $\mathcal{E}_B(K) = L_K^{-1}B_2^n$. It is not hard to see that the volume of $\mathcal{E}_B(K)$ is invariant under the action of SL_n. From the definition of the Binet ellipsoid we check that if $T \in SL_n$, then $\|y\|_{\mathcal{E}_B(T(K))} = \|T^* y\|_{\mathcal{E}_B(K)}$ for all $y \in \mathbb{R}^n$, therefore

$$\mathcal{E}_B(T(K)) = (T^*)^{-1}(\mathcal{E}_B(K)).$$

It follows that

$$|\mathcal{E}_B(T(K))| = |\mathcal{E}_B(K)|$$

for every $T \in SL_n$. Now, we use the fact that $T_0(K)$ is isotropic for some $T_0 \in SL_n$ and this implies that

$$|\mathcal{E}_B(T(K))| = |\mathcal{E}_B(T_0(K))| = |L_K^{-1} B_2^n| = \omega_n L_K^{-n}.$$

A consequence is the next fact.

Proposition 2.1.7. *Let K be a centered, convex body of volume 1 in $\mathbb{R}^n$. There exists $\xi \in S^{n-1}$ such that*

$$\int_K \langle x, \xi\rangle^2 \, dx \leqslant L_K^2.$$

Proof. Note that by integration in polar coordinates

$$L_K^{-n} = \frac{|\mathcal{E}_B(K)|}{\omega_n} = \int_{S^{n-1}} \|\xi\|_{\mathcal{E}_B(K)}^{-n} \, d\sigma(\xi).$$

It follows that $\min_{\xi \in S^{n-1}} \|\xi\|_{\mathcal{E}_B(K)} \leqslant L_K$. □

Assume that the isotropic constant problem has an affirmative answer and let K be a centered convex body of volume 1 in $\mathbb{R}^n$. According to Proposition 2.1.7, there exists a direction $\xi \in S^{n-1}$ such that

$$\int_K \langle x, \xi\rangle^2 \, dx \leqslant L_K^2 \leqslant C^2.$$

Then, from (2.7) we see that

$$|K \cap \xi^\perp| \geqslant c_1 \Big(\int_K \langle x, \xi\rangle^2 \, dx \Big)^{-1/2} \geqslant c_1/C.$$

This gives an affirmative answer to the hyperplane conjecture.

2.2 Isotropic log-concave probability measures

A Borel measure μ on $\mathbb{R}^n$ is called log-concave if $\mu(H) < 1$ for every hyperplane H in $\mathbb{R}^n$ and $\mu(\lambda A + (1-\lambda)B) \geqslant \mu(A)^\lambda \mu(B)^{1-\lambda}$ for any compact subsets A, B of $\mathbb{R}^n$ and any $\lambda \in (0,1)$. A theorem of Borell [92] shows that under these assumptions, μ has a log-concave density f_μ. A function $f : \mathbb{R}^n \to [0,\infty)$ is called log-concave if its support $\{f > 0\}$ is a convex set in $\mathbb{R}^n$ and the restriction of $\log f$ to it is concave. If f has a finite positive integral, then there exist constants $A, B > 0$ such that $f(x) \leqslant Ae^{-B|x|}$ for all $x \in \mathbb{R}^n$ (see [113, Lemma 2.2.1]). In particular, f has finite moments of all orders. We say that μ is even if $\mu(-B) = \mu(B)$ for every Borel subset B of $\mathbb{R}^n$ and that μ is centered if

$$\int_{\mathbb{R}^n} \langle x, \xi \rangle \, d\mu(x) = \int_{\mathbb{R}^n} \langle x, \xi \rangle f_\mu(x) \, dx = 0$$

for all $\xi \in S^{n-1}$. We shall use the fact that if μ is a centered log-concave probability measure on $\mathbb{R}^n$, then

$$\|f_\mu\|_\infty \leqslant e^n f_\mu(0). \tag{2.10}$$

This is a result of Fradelizi; for a proof see [113, Theorem 2.2.2]. Note that if K is a convex body in $\mathbb{R}^n$, then the Brunn–Minkowski inequality implies that the indicator function $\mathbb{1}_K$ of K is the density of a log-concave measure, the Lebesgue measure on K.

Let $f : \mathbb{R}^n \to \mathbb{R}$ be a measurable function. For any $\alpha \geqslant 1$ we define the ψ_α-norm of f as follows:

$$\|f\|_{\psi_\alpha} := \inf\left\{t > 0 : \int_\Omega \exp((|f|/t)^\alpha) \, d\mu \leqslant 2\right\},$$

provided that the set on the right-hand side is non-empty. Note that the ψ_α-norm is exactly the Orlicz norm corresponding to the function $t \in \mathbb{R} \to e^{|t|^\alpha} - 1$. An equivalent expression for the ψ_α-norm in terms of the L_p-norms is that

$$\|f\|_{\psi_\alpha} \approx \sup_{p \geqslant \alpha} \frac{\|f\|_{L_p(\mu)}}{p^{1/\alpha}},$$

up to some absolute constants.

A well-known lemma of Borell (see [113, Lemma 2.4.5] for a proof) asserts that if μ is a log-concave probability measure on $\mathbb{R}^n$, then, for any origin-symmetric convex set A in $\mathbb{R}^n$ with $\mu(A) = \alpha \in (0, 1)$ and any $t > 1$ we have

$$1 - \mu(tA) \leqslant \alpha \left(\frac{1-\alpha}{\alpha}\right)^{\frac{t+1}{2}}. \tag{2.11}$$

Using Borell's lemma one can show that there exists an absolute constant $c_0 > 0$ such that for every log-concave probability measure μ on $\mathbb{R}^n$, for any seminorm $f : \mathbb{R}^n \to \mathbb{R}$ and any $q > p \geqslant 1$, we have

$$\left(\int_{\mathbb{R}^n} |f|^p \, d\mu\right)^{1/p} \leqslant \left(\int_{\mathbb{R}^n} |f|^q \, d\mu\right)^{1/q} \leqslant c_0 \frac{q}{p} \left(\int_{\mathbb{R}^n} |f|^p \, d\mu\right)^{1/p} \tag{2.12}$$

(a proof is given in [113, Theorem 2.4.6]). In particular, $\|f\|_{\psi_1} \leqslant c_1 \|f\|_1$, where $c_1 > 0$ is an absolute constant.

If μ is a log-concave measure on $\mathbb{R}^n$ with density f_μ, we define the isotropic constant of μ by

$$L_\mu := \left(\frac{\sup_{x\in\mathbb{R}^n} f_\mu(x)}{\int_{\mathbb{R}^n} f_\mu(x)\,dx} \right)^{\frac{1}{n}} [\det \mathrm{Cov}(\mu)]^{\frac{1}{2n}}, \tag{2.13}$$

where $\mathrm{Cov}(\mu)$ is the covariance matrix of μ with entries

$$\mathrm{Cov}(\mu)_{ij} := \frac{\int_{\mathbb{R}^n} x_i x_j f_\mu(x)\,dx}{\int_{\mathbb{R}^n} f_\mu(x)\,dx} - \frac{\int_{\mathbb{R}^n} x_i f_\mu(x)\,dx}{\int_{\mathbb{R}^n} f_\mu(x)\,dx} \frac{\int_{\mathbb{R}^n} x_j f_\mu(x)\,dx}{\int_{\mathbb{R}^n} f_\mu(x)\,dx}.$$

We say that a log-concave probability measure μ on $\mathbb{R}^n$ is isotropic if it is centered and $\mathrm{Cov}(\mu) = I_n$, where I_n is the identity $n \times n$ matrix. In this case, $L_\mu = \|f_\mu\|_\infty^{1/n}$. For every μ there exists an affine transformation T such that $T_*\mu$ is isotropic, where $T_*\mu$ is the push-forward of μ defined by $T_*\mu(A) = \mu(T^{-1}(A))$. Note that a convex body K of volume 1 is isotropic if and only if the log-concave probability measure with density $L_K^n \mathbb{1}_{K/L_K}$ is isotropic.

Let μ and ν be two log-concave probability measures on $\mathbb{R}^n$. Let $T : \mathbb{R}^n \to \mathbb{R}^n$ be a measurable function that is defined μ-almost everywhere and satisfies

$$\nu(B) = \mu(T^{-1}(B))$$

for every Borel subset B of $\mathbb{R}^n$. We then say that T pushes forward μ to ν and write $T_*\mu = \nu$. It is not hard to check that for every log-concave probability measure μ on $\mathbb{R}^n$ there exists an invertible affine transformation T such that the log-concave probability measure $T_*\mu$ is isotropic, and $L_{T_*\mu} = L_\mu$.

One can prove that the isotropic constants of all log-concave measures are uniformly bounded from below by a constant $c > 0$ that is independent of the dimension. If μ is an isotropic log-concave probability measure, then

$$L_\mu = \|f_\mu\|_\infty^{1/n} \approx [f_\mu(0)]^{1/n} \geqslant c, \tag{2.14}$$

where $c > 0$ is an absolute constant (see [113, Proposition 2.3.12]). The isotropic constant problem for log-concave measures can now be stated as follows:

> Let μ be an isotropic log-concave probability measure on $\mathbb{R}^n$. Then, $L_\mu = \|f_\mu\|_\infty^{1/n} \leqslant C$, where $C > 0$ is an absolute constant.

We can define

$$\tilde{L}_n := \max\{L_\mu : \mu \text{ is an isotropic log-concave probability measure on } \mathbb{R}^n\}.$$

Let λ_K denote the uniform measure on a convex body K. Since $L_K = L_{\lambda_K} \leqslant \tilde{L}_n$ for every centered convex body K in $\mathbb{R}^n$, it is clear that $L_n \leqslant \tilde{L}_n$.

An important family of convex bodies associated with a given centered log-concave probability measure μ on $\mathbb{R}^n$ was introduced by K. Ball, who also established their convexity in [40]: for every $p > 0$, we define

$$K_p(\mu) = \left\{ x \in \mathbb{R}^n : \int_0^\infty r^{p-1} f_\mu(rx)\, dr \geqslant \frac{f_\mu(0)}{p} \right\}.$$

From the definition it follows that the radial function of $K_p(\mu)$ is given by

$$\varrho_{K_p(\mu)}(x) = \left(\frac{1}{f_\mu(0)} \int_0^\infty p r^{p-1} f_\mu(rx)\, dr \right)^{1/p} \tag{2.15}$$

for $x \neq 0$. It is easily checked that if K is a convex body in $\mathbb{R}^n$ with $0 \in K$, then $K_p(\mathbb{1}_K) = K$ for all $p > 0$. A very useful observation is that

$$f_\mu(0)\, |K_n(\mu)| = 1.$$

To see this, we write

$$\begin{aligned} |K_n(\mu)| &= \int_{K_n(\mu)} \mathbf{1}\, dx = n\omega_n \int_{S^{n-1}} \int_0^{\rho_{K_n(\mu)}(\xi)} r^{n-1}\, dr\, d\sigma(\xi) \\ &= \frac{n\omega_n}{f_\mu(0)} \int_{S^{n-1}} \int_0^\infty r^{n-1} f_\mu(r\xi)\, dr\, d\sigma(\xi) = \frac{1}{f_\mu(0)} \int_{\mathbb{R}^n} f_\mu(x)\, dx = \frac{1}{f_\mu(0)}, \end{aligned}$$

using (2.15) and integration in spherical coordinates.

Another important observation is that, for every $0 < p \leqslant q$ we have that

$$\frac{\Gamma(p+1)^{\frac{1}{p}}}{\Gamma(q+1)^{\frac{1}{q}}} K_q(\mu) \subseteq K_p(\mu) \subseteq e^{\frac{n}{p} - \frac{n}{q}} K_q(\mu). \tag{2.16}$$

A proof is given in [113, Proposition 2.5.7]. As a consequence, we obtain an approximate formula for the volume of $K_{n+p}(\mu)$. When $p > 0$, we have

$$e^{-1} \leqslant f_\mu(0)^{\frac{1}{n} + \frac{1}{p}} |K_{n+p}(\mu)|^{\frac{1}{n} + \frac{1}{p}} \leqslant e \frac{n+p}{n}, \tag{2.17}$$

while for $-n < p < 0$ we have

$$e^{-1} \leqslant f_\mu(0)^{\frac{1}{-p} - \frac{1}{n}} |K_{n+p}(\mu)|^{\frac{1}{-p} - \frac{1}{n}} \leqslant e. \tag{2.18}$$

If μ is an even log-concave probability measure on $\mathbb{R}^n$, then $T = K_{n+2}(\mu)$ is an origin-symmetric convex body that satisfies

$$c_1 L_\mu \leqslant L_T \leqslant c_2 L_\mu,$$

where $c_1, c_2 > 0$ are absolute constants (see [113, Proposition 2.5.9]). Furthermore, if μ is isotropic, then $\overline{T} = |T|^{-1/n} T$ is an isotropic convex body. In the case where the measure μ is centered, but not necessarily symmetric, we prefer to work with the convex body $K_{n+1}(\mu)$ instead of $K_{n+2}(\mu)$. The reason for this is that $T = K_{n+1}(\mu)$ is a centered convex body in $\mathbb{R}^n$ and we still have that

$$c_1 L_\mu \leqslant L_T \leqslant c_2 L_\mu,$$

where $c_1, c_2 > 0$ are absolute constants.

We briefly sketch the proof (for more details see [113, Proposition 2.5.12]). First, we check that T is centered and satisfies

$$\int_T |\langle x, \xi\rangle|\, dx = \frac{1}{f_\mu(0)} \int |\langle x, \xi\rangle| f_\mu(x)\, dx$$

for all $\xi \in S^{n-1}$. Borell's lemma implies that for every $y \in \mathbb{R}^n$

$$\begin{aligned} \left(\frac{1}{|T|}\int_T \langle x, y\rangle^2\, dx\right)^{1/2} &\approx \frac{1}{|T|}\int_T |\langle x, y\rangle|\, dx = \frac{1}{f_\mu(0)\,|T|}\int |\langle x, y\rangle| f_\mu(x)\, dx \\ &\approx \frac{1}{f_\mu(0)\,|T|}\left(\int \langle x, y\rangle^2 f_\mu(x)\, dx\right)^{1/2}, \end{aligned}$$

which, combined with the fact that T and f_μ are both centered, implies that there exist absolute constants $c_1, c_2 > 0$ such that as positive definite matrices

$$c_1 \mathrm{Cov}(\mathbb{1}_T) \leqslant (|T| f_\mu(0))^{-2} \mathrm{Cov}(\mu) \leqslant c_2 \mathrm{Cov}(\mathbb{1}_T).$$

Therefore,

$$(\det \mathrm{Cov}(\mathbb{1}_T))^{1/n} \approx (|T| f_\mu(0))^{-2} (\det \mathrm{Cov}(\mu))^{1/n}. \tag{2.19}$$

From the definition of the isotropic constant it follows that

$$\begin{aligned} L_T &= \frac{1}{|T|^{1/n}} (\det \mathrm{Cov}(\mathbb{1}_T))^{\frac{1}{2n}} \approx |T|^{-1/n} (f_\mu(0)\,|T|)^{-1} (\det \mathrm{Cov}(\mu))^{\frac{1}{2n}} \\ &\approx (f_\mu(0)\,|T|)^{1 \;\frac{1}{n}} L_\mu, \end{aligned}$$

where we have also used the fact that $\|f_\mu\|_\infty^{1/n} \approx f_\mu(0)^{1/n}$ by Fradelizi's inequality (2.10). Finally, applying (2.17) with $p = 1$ we get that

$$e^{-1} \leqslant (f_\mu(0)\,|T|)^{1+\frac{1}{n}} \leqslant e\frac{n+1}{n} \leqslant 2e \tag{2.20}$$

and this completes the proof. This discussion shows that

$$\tilde{L}_n \approx L_n.$$

2.3 L_p-centroid bodies

Let μ be a centered log-concave probability measure on $\mathbb{R}^n$. For any $p \geqslant 1$ we define the L_p-centroid body $Z_p(\mu)$ of μ as the convex body whose support function is

$$h_{Z_p(\mu)}(y) := \left(\int_{\mathbb{R}^n} |\langle x, y\rangle|^p f_\mu(x)\, dx \right)^{1/p}, \quad y \in \mathbb{R}^n.$$

The convex bodies $Z_p(\mu)$ are always origin-symmetric, and $Z_p(T_*\mu) = T(Z_p(\mu))$ for every $T \in GL_n$ and $p \geqslant 1$. Note that if μ is isotropic, then $Z_2(\mu) = B_2^n$. Also, from (2.12) it follows that, for every $1 \leqslant p < q$

$$Z_p(\mu) \subseteq Z_q(\mu) \subseteq \frac{cq}{p} Z_p(\mu), \tag{2.21}$$

where $c > 0$ is an absolute constant.

If $\mu = \lambda_K$ is the Lebesgue measure on a centered convex body K of volume 1 in $\mathbb{R}^n$, then we denote $Z_p(K) := Z_p(\lambda_K)$. In this case, we have some additional results. For every $p \geqslant n$ we have that

$$Z_p(K) \supseteq c_2 Z_\infty(K), \tag{2.22}$$

where $c_2 > 0$ is an absolute constant and $Z_\infty(K) = \mathrm{conv}\{K, -K\}$. This is a consequence of the inequality

$$\int_K |\langle x, \xi\rangle|^p\, dx \geqslant \frac{\Gamma(p+1)\Gamma(n)}{2e\Gamma(p+n+1)} \max\{h_K^p(\xi), h_K^p(-\xi)\},$$

which holds true for all $\xi \in S^{n-1}$ and $p \geqslant 1$. Therefore, if $p \geqslant n$ we see that

$$\|\langle \cdot, \xi\rangle\|_{L_p(K)} \approx \max\{h_K(\xi), h_K(-\xi)\}$$

and hence $Z_p(K) \supseteq cZ_\infty(K)$. In particular, the Rogers–Shephard inequality $|K - K| \leqslant \binom{2n}{n}|K|$ implies that

$$c \leqslant |Z_n(K)|^{1/n} \leqslant |K - K|^{1/n} \leqslant 4 \tag{2.23}$$

for some absolute constant $c > 0$.

A nonsymmetric variant of the L_p-centroid bodies of μ is defined as follows. For every $p \geqslant 1$ we consider the convex body $Z_p^+(\mu)$ with support function

$$h_{Z_p^+(\mu)}(y) = \left(\int_{\mathbb{R}^n} \langle x, y\rangle_+^p f_\mu(x)\,dx \right)^{1/p}, \quad y \in \mathbb{R}^n,$$

where $a_+ = \max\{a, 0\}$. When f_μ is even, we have that $Z_p^+(\mu) = 2^{-1/p} Z_p(\mu)$. In any case, it is clear that $Z_p^+(\mu) \subseteq Z_p(\mu)$. One can also check that if $1 \leqslant p < q$, then

$$\left(\frac{4}{e}\right)^{\frac{1}{p}-\frac{1}{q}} Z_p^+(\mu) \subseteq Z_q^+(\mu) \subseteq c_1 \left(\frac{4(e-1)}{e}\right)^{\frac{1}{p}-\frac{1}{q}} \frac{q}{p} Z_p^+(\mu) \tag{2.24}$$

(for a proof see [244], where the family of bodies $\tilde{Z}_p^+(\mu) = 2^{1/p} Z_p^+(\mu)$ is considered).

The L_p-centroid bodies were introduced, under a different normalization, by Lutwak and Zhang in [395], while in [452] for the first time, and in [453] later, Paouris used the geometric properties of them to acquire detailed information about the distribution of the Euclidean norm with respect to μ. A basic observation of Paouris is the next asymptotic formula that appears in [453].

Theorem 2.3.1. *Let μ be a centered log-concave probability measure on $\mathbb{R}^n$. Then,*

$$c_1 \leqslant \left(f_\mu(0)\,|Z_n(\mu)|\right)^{1/n} \leqslant c_2, \tag{2.25}$$

where $c_1, c_2 > 0$ are absolute constants.

Proof. Direct computation shows that, for every $p \geqslant 1$

$$\begin{aligned} |K_{n+p}(\mu)|^{1+\frac{p}{n}} \int_{\overline{K_{n+p}(\mu)}} |\langle x, \xi\rangle|^p\,dx &= \int_{K_{n+p}(\mu)} |\langle x, \xi\rangle|^p\,dx \\ &= \frac{1}{f_\mu(0)} \int_{\mathbb{R}^n} |\langle x, \xi\rangle|^p f_\mu(x)\,dx \end{aligned}$$

for all $\xi \in S^{n-1}$, or equivalently

$$Z_p(\overline{K_{n+p}(\mu)})\,|K_{n+p}(\mu)|^{\frac{1}{p}+\frac{1}{n}} f_\mu(0)^{\frac{1}{p}} = Z_p(\mu). \tag{2.26}$$

Now, let $1 \leqslant p \leqslant n$. Using also (2.17) we see that

$$\frac{1}{e} Z_p(\overline{K_{n+p}(\mu)}) \subseteq f_\mu(0)^{1/n} Z_p(\mu) \subseteq e\frac{n+p}{n} Z_p(\overline{K_{n+p}(\mu)}) \subseteq 2e Z_p(\overline{K_{n+p}(\mu)}). \tag{2.27}$$

On the other hand, using (2.16) we can check that

$$h_{Z_p(\overline{K_{n+p}(\mu)})}(\xi) \approx h_{Z_p(\overline{K_{n+1}(\mu)})}(\xi)$$

for every $\xi \in S^{n-1}$. This shows that $Z_p(\overline{K_{n+p}}(\mu)) \approx Z_p(\overline{K_{n+1}}(\mu))$. Therefore, for all $1 \leqslant p \leqslant n$ we get

$$c_1 f_\mu(0)^{1/n} Z_p(\mu) \subseteq Z_p(\overline{K_{n+1}}(\mu)) \subseteq c_2 f_\mu(0)^{1/n} Z_p(\mu), \tag{2.28}$$

where $c_1, c_2 > 0$ are absolute constants.

Now, recall that, since μ is centered, the body $K_{n+1}(\mu)$ is also centered. Applying (2.23) for the body $\overline{K_{n+1}}(\mu)$ we see that

$$|Z_n(\overline{K_{n+1}}(\mu))|^{1/n} \approx 1$$

and hence, by (2.28)

$$(f_\mu(0)\,|Z_n(\mu)|)^{1/n} \approx |Z_n(\overline{K_{n+1}}(\mu))|^{1/n} \approx 1.$$

This completes the proof. □

Let μ be a log-concave probability measure on $\mathbb{R}^n$. For any $1 \leqslant k \leqslant n-1$ and $F \in G_{n,k}$ we define the marginal of μ with respect to F setting

$$\pi_F(\mu)(A) := \mu(P_F^{-1}(A))$$

for every Borel subset A of F. One can check that $\pi_F(\mu)$ is a log-concave probability measure on F, whose log-concave density $f_{\pi_F(\mu)}$ agrees almost everywhere with the function $\pi_F(f_\mu)$ defined by

$$\pi_F(f_\mu)(x) = \int_{x+F^\perp} f_\mu(y)\,dy. \tag{2.29}$$

Then, for every measurable function $g : F \to \mathbb{R}$ we have

$$\int_{\mathbb{R}^n} g(P_F x) f_\mu(x)\,dx = \int_F g(x)\pi_F(f_\mu)(x)\,dx.$$

It follows that if f_μ is centered, then, for every $F \in G_{n,k}$ we have that $\pi_F(f_\mu)$ is also centered, and if f is isotropic, then $\pi_F(f_\mu)$ is also isotropic.

A basic observation of Paouris from [452] is that any projection of the L_p-centroid body of μ coincides with the L_p-centroid body of the corresponding marginal of μ.

Theorem 2.3.2. *Let μ be a centered log-concave probability measure on $\mathbb{R}^n$. For every $1 \leqslant k \leqslant n$ and any $F \in G_{n,k}$ and $p \geqslant 1$, we have that*

$$P_F(Z_p(\mu)) = Z_p(\pi_F(\mu)). \tag{2.30}$$

The proof is a direct application of Fubini's theorem.

Applying Theorem 2.3.1 to $\pi_F(\mu)$ and taking into account (2.30) we obtain the following result.

Theorem 2.3.3. *Let μ be a centered log-concave probability measure on $\mathbb{R}^n$. Then, for every $1 \leqslant k \leqslant n-1$ and any $F \in G_{n,k}$ we have*

$$c_1 \leqslant \left(\pi_F(f_\mu)(0)\,|P_F(Z_k(f))|_k\right)^{1/k} \leqslant c_2, \tag{2.31}$$

where $c_1, c_2 > 0$ are absolute constants.

The next theorem gives a very useful estimate for the volume of sections of an isotropic convex body through its barycenter.

Theorem 2.3.4. *Let K be an isotropic convex body in $\mathbb{R}^n$. Then, for every $1 \leqslant k \leqslant n-1$ and any $F \in G_{n,k}$*

$$|K \cap F^\perp|^{1/k} \approx 1. \tag{2.32}$$

Proof. We denote by μ_K the isotropic log-concave probability measure with density $L_K^n \mathbb{1}_{\frac{K}{L_K}}$ and write f_K for the density of μ_K. Fix $1 \leqslant k \leqslant n-1$ and $F \in G_{n,k}$. We know that $\pi_F(\mu_K)$ is isotropic. Using (2.28) with $p = 2$ we get

$$\begin{aligned} L_{\overline{K_{k+1}(\pi_F(\mu_K))}} &= \left(\frac{|Z_2(\overline{K_{k+1}(\pi_F(\mu_K))})|}{|B_F|}\right)^{1/k} \approx \pi_F(f_K)(0)^{1/k}\left(\frac{|Z_2(\pi_F(\mu_K))|}{|B_F|}\right)^{1/k} \\ &= \pi_F(f_K)(0)^{1/k}\left(\frac{|P_F(Z_2(\mu_K))|}{|B_F|}\right)^{1/k}, \end{aligned}$$

where we have also used the identity $Z_2(\pi_F(\mu_K)) = P_F(Z_2(\mu_K))$ from Theorem 2.3.2. Since K is isotropic, we get

$$Z_2(\mu_K) = L_K^{-1} Z_2(K) = B_2^n \quad \text{and hence } P_F(Z_2(\mu_K)) = B_F.$$

Moreover, we have

$$\pi_F(f_K)(0) = \int_{F^\perp} f_K(y)\,dy = L_K^n\left|\left(\frac{1}{L_K}K \cap F^\perp\right)\right| = L_K^k |K \cap F^\perp|.$$

Combining the above, we conclude that $L_{\overline{K_{k+1}(\pi_F(\mu_K))}} \approx L_K\,|K \cap F^\perp|^{1/k}$, which is the assertion of the theorem. □

2.4 A reduction of the hyperplane conjecture

Let μ be a centered log-concave probability measure on $\mathbb{R}^n$. For every $p \in (-n, \infty), p \neq 0$, we define

$$I_p(\mu) = \left(\int_{\mathbb{R}^n} |x|^p \, d\mu(x) \right)^{1/p}.$$

From (2.12) we see that for all $y \in \mathbb{R}^n$ and $p, q \geqslant 1$ we have

$$I_{pq}(\mu) \leqslant c_1 p I_q(\mu).$$

In particular, we have

$$I_p(\mu) \leqslant c_2 p I_2(\mu) \tag{2.33}$$

for all $p \geqslant 2$. Paouris [452] proved the following.

Theorem 2.4.1. *There exist absolute constants $c_1, c_2 > 0$ such that if μ is an isotropic log-concave probability measure on $\mathbb{R}^n$, then*

$$I_p(\mu) \leqslant c_1 I_2(\mu) \tag{2.34}$$

for all $p \leqslant c_2 \sqrt{n}$.

Assuming that we have proved Theorem 2.4.1, for any isotropic log-concave probability measure μ in $\mathbb{R}^n$ we apply Markov's inequality to get

$$\mu(\{|x| \geqslant e^3 I_p(\mu)\}) \leqslant e^{-3p}$$

for every $p \geqslant 2$. Then, Borell's lemma gives

$$\mu(\{|x| \geqslant e^3 I_p(\mu) s\}) \leqslant (1 - e^{-3p}) \left(\frac{e^{-3p}}{1 - e^{-3p}} \right)^{(s+1)/2} \leqslant e^{-ps}$$

for every $s \geqslant 1$. Choosing $p = c_2 \sqrt{n}$, and using (2.34), we see that

$$\mu(\{|x| \geqslant c_1 e^3 I_2(\mu) s\}) \leqslant \exp(-c_2 \sqrt{n} s)$$

for all $s \geqslant 1$. Since μ is isotropic, we have $I_2(\mu) = \sqrt{n}$. This proves the following deviation inequality.

Theorem 2.4.2. *Let μ be an isotropic, log-concave probability measure in $\mathbb{R}^n$. Then,*

$$\mu(\{x \in \mathbb{R}^n : |x| \geqslant ct\sqrt{n}\}) \leqslant \exp(-t\sqrt{n}) \tag{2.35}$$

for every $t \geqslant 1$, where $c > 0$ is an absolute constant.

We shall actually prove a stronger theorem, which is again due to Paouris [453].

Theorem 2.4.3. *Let μ be an isotropic, log-concave probability measure on $\mathbb{R}^n$. For every integer $1 \leqslant k \leqslant c_2\sqrt{n}$ we have*

$$I_{-k}(\mu) \approx I_k(\mu).$$

Theorem 2.4.3 shows, in particular, that for every $1 \leqslant k \leqslant c_2\sqrt{n}$ we have $I_k(\mu) \leqslant c_1 I_2(\mu)$, where $c_1 > 0$ is an absolute constant. Thus, we immediately obtain the assertion of Theorem 2.4.1.

The proof of Theorem 2.4.3 requires a number of basic results from the asymptotic theory of convex bodies. We write $k_*(C)$ for the largest integer $k \leqslant n$ that satisfies

$$\nu_{n,k}\left(\left\{F \in G_{n,k} : \frac{w(C)}{2}|x| \leqslant h_C(x) \leqslant 2w(C)|x|,\ x \in F\right\}\right) \geqslant \frac{n}{n+k}.$$

In the terminology of [27] this is the so-called Dvoretzky dimension of C°, i. e., the largest k for which the majority of k-dimensional sections of C° are 4-Euclidean. The next theorem of V. Milman and Schechtman from [426] shows that the dimension $k_*(C)$ is determined by the parameters $w(C)$ and $R(C)$ up to an absolute constant.

Theorem 2.4.4. *For every origin-symmetric convex body C in $\mathbb{R}^n$*

$$c_1 n\frac{w(C)^2}{R(C)^2} \leqslant k_*(C) \leqslant c_2 n\frac{w(C)^2}{R(C)^2},$$

where $c_1, c_2 > 0$ are absolute constants.

A parameter that was originally central in the work of Paouris is $q_*(\mu)$, which is defined for every centered log-concave probability measure μ in $\mathbb{R}^n$, as follows:

$$q_*(\mu) = \max\{p \in [2,n] : k_*(Z_p(\mu)) \geqslant p\}.$$

The next proposition provides a lower bound for $q_*(\mu)$.

Proposition 2.4.5. *There exists an absolute constant $c > 0$ with the following property: if μ is a centered log-concave probability measure on $\mathbb{R}^n$, then*

$$q_*(\mu) \geqslant c\sqrt{k_*(Z_2(\mu))}.$$

Proof. We set $q_* := q_*(\mu)$. From Theorem 2.4.7, Lemma 2.4.9, Hölder's inequality, and the simple observation that $I_2(\mu) = \sqrt{n}w_2(Z_2(\mu))$, we get

$$\begin{aligned} w(Z_{q_*}(\mu)) &\geqslant c_1 w_{q_*}(Z_{q_*}(\mu)) \geqslant c_2\sqrt{q_*/n}I_{q_*}(\mu) \\ &\geqslant c_2\sqrt{q_*/n}I_2(\mu) = c_2\sqrt{q_*/n}\,\sqrt{n}w_2(Z_2(\mu)). \end{aligned}$$

In other words

$$w(Z_{q_*}(\mu)) \geqslant c_2\sqrt{q_*}\,w(Z_2(\mu)). \tag{2.36}$$

Since $R(Z_{q_*}(\mu)) \leqslant c_3 q_*\, R(Z_2(\mu))$, using the definition of q_* and Theorem 2.4.4 we write

$$2q_* \geqslant k_*(Z_{q_*}(\mu)) \geqslant c_4 n\left(\frac{w(Z_{q_*}(\mu))}{R(Z_{q_*}(\mu))}\right)^2 \geqslant c_4 n\frac{c_2^2 q_*}{c_3^2 q_*^2}\frac{w^2(Z_2(\mu))}{R^2(Z_2(\mu))} \geqslant c_5\frac{k_*(Z_2(\mu))}{q_*}. \tag{2.37}$$

This shows that $q_*(\mu) \geqslant c\,\sqrt{k_*(Z_2(\mu))}$ for some absolute constant $c > 0$. □

Note that if μ is isotropic, then $k_*(Z_2(\mu)) = n$. Therefore, in the isotropic case we have the following.

Corollary 2.4.6. *There exists an absolute constant $c > 0$ with the following property: for every isotropic log-concave probability measure μ on $\mathbb{R}^n$*

$$q_*(\mu) \geqslant c\sqrt{n}.$$

The second tool that we need is the behavior of the p-widths of an origin-symmetric convex body C on $\mathbb{R}^n$. For every $p \neq 0$ we define

$$w_p(C) = \left(\int_{S^{n-1}} h_C(\theta)^p\,d\sigma(\theta)\right)^{1/p}.$$

Note that $w_1(C) = w(C)$. The parameters w_p, $p \geqslant 1$ were studied by Litvak, V. Milman, and Schechtman in [379].

Theorem 2.4.7. *Let C be an origin-symmetric convex body in $\mathbb{R}^n$. Then,*

$$\max\left\{w(C), c_1\frac{R(C)\sqrt{p}}{\sqrt{n}}\right\} \leqslant w_p(C) \leqslant \max\left\{2w(C), c_2\frac{R(C)\sqrt{p}}{\sqrt{n}}\right\}$$

for all $1 \leqslant p \leqslant n$, where $c_1, c_2 > 0$ are absolute constants.

Note that the behavior of $w_p(C)$ changes when $p \approx n(w(C)/R(C))^2$. This value of p is roughly equal to the "dual Dvoretzky dimension" $k_*(C)$ of C. One also has $w_n(C) \approx R(C)$, and since $w_p(C) \leqslant R(C)$ for every $p \geqslant 1$ we conclude that $w_p(C) \approx R(C)$ for all $p \geqslant n$.

Concerning $w_p(C)$ for negative values of p, Klartag and Vershynin [300] established the next important result.

Theorem 2.4.8. *Let C be an origin-symmetric convex body in $\mathbb{R}^n$. Then, $w_p(C) \approx w_{-p}(C)$ for all $1 \leqslant p \leqslant ck_*(C)$, where $c > 0$ is an absolute constant.*

We start the proof of Theorem 2.4.3 with the next lemma, which relates the p-moment of the Euclidean norm with respect to μ with the parameters w_p and the L_p-centroid bodies of μ.

Lemma 2.4.9. *Let μ be a log-concave probability measure in $\mathbb{R}^n$. For every $p \geqslant 1$ we have*

$$w_p(Z_p(\mu)) = a_{n,p}\sqrt{\frac{p}{p+n}}I_p(\mu),$$

where $a_{n,p} \approx 1$.

Proof. Direct computation shows that for every $x \in \mathbb{R}^n$ we have

$$\Big(\int_{S^{n-1}} |\langle x, \xi\rangle|^p \, d\sigma(\xi)\Big)^{1/p} = a_{n,p}\frac{\sqrt{p}}{\sqrt{p+n}}|x|,$$

where $a_{n,p} \approx 1$. Since

$$w_p(Z_p(\mu)) = \Big(\int_{S^{n-1}}\int_{\mathbb{R}^n} |\langle x, \xi\rangle|^p \, d\mu(x)\sigma(d\xi)\Big)^{1/p},$$

the lemma follows. □

We also need two identities.

Claim 2.4.10. *If μ is a centered, log-concave probability measure on $\mathbb{R}^n$ and $1 \leqslant k \leqslant n-1$ is a positive integer, then*

$$I_{-k}(\mu) = c_{n,k}\Big(\int_{G_{n,k}} \pi_F(f_\mu)(0)\, d\nu_{n,k}(F)\Big)^{-1/k}, \tag{2.38}$$

where

$$c_{n,k} = \Big(\frac{(n-k)\omega_{n-k}}{n\omega_n}\Big)^{1/k} \approx \sqrt{n}.$$

Proof. Let $1 \leqslant k \leqslant n-1$. Then, we have

$$\begin{aligned}
&\int_{G_{n,k}} \pi_F(f_\mu)(0)\, d\nu_{n,k}(F) \\
&= \int_{G_{n,n-k}} \pi_{E^\perp}(f_\mu)(0)\, d\nu_{n,n-k}(E) = \int_{G_{n,n-k}}\int_E f_\mu(y)\, dy\, d\nu_{n,n-k}(E) \\
&= \int_{G_{n,n-k}} (n-k)\omega_{n-k}\int_{S_E}\int_0^\infty r^{n-k-1} f_\mu(r\xi)\, dr\, d\sigma_E(\xi)\, d\nu_{n,n-k}(E) \\
&= \frac{(n-k)\omega_{n-k}}{n\omega_n} n\omega_n \int_{S^{n-1}}\int_0^\infty r^{n-k-1} f_\mu(r\xi)\, dr\, d\sigma(\xi)
\end{aligned}$$

$$= \frac{(n-k)\omega_{n-k}}{n\omega_n} \int_{\mathbb{R}^n} |x|^{-k} f_\mu(x)\, dx = \frac{(n-k)\omega_{n-k}}{n\omega_n} I_{-k}^{-k}(\mu).$$

It follows that

$$I_{-k}(\mu) = \left(\frac{(n-k)\omega_{n-k}}{n\omega_n}\right)^{1/k} \left(\int_{G_{n,k}} \pi_F(f_\mu)(0)\, d\nu_{n,k}(F)\right)^{-1/k}.$$

Finally, we check that $c_{n,k} = (\frac{(n-k)\omega_{n-k}}{n\omega_n})^{1/k} \approx \sqrt{n}$. □

Claim 2.4.11. *If C is an origin-symmetric convex body in $\mathbb{R}^n$ and $1 \leqslant k \leqslant n-1$ is a positive integer, then*

$$w_{-k}(C) \approx \sqrt{k}\left(\int_{G_{n,k}} |P_F(C)|^{-1}\, d\nu_{n,k}(F)\right)^{-\frac{1}{k}}. \tag{2.39}$$

Proof. Using the Blaschke–Santaló and the Bourgain–Milman inequalities, we write

$$\begin{aligned} w_{-k}^{-1}(C) &= \left(\int_{S^{n-1}} \frac{1}{h_C^k(\xi)}\, d\sigma(\xi)\right)^{1/k} = \left(\int_{G_{n,k}} \int_{S_F} \frac{1}{\|\xi\|_{(P_F C)^\circ}^k}\, d\sigma(\xi)\, d\nu_{n,k}(F)\right)^{1/k} \\ &= \left(\int_{G_{n,k}} \frac{|P_F(C)^\circ|}{|B_2^k|}\, d\nu_{n,k}(F)\right)^{1/k} \approx \left(\int_{G_{n,k}} \frac{|B_2^k|}{|P_F(C)|}\, d\nu_{n,k}(F)\right)^{1/k} \end{aligned}$$

and the result follows. □

Now, consider a centered log-concave probability measure μ on $\mathbb{R}^n$ and an integer $1 \leqslant k \leqslant n-1$. Recall that from Theorem 2.3.3 we also have

$$\frac{1}{|P_F(Z_k(\mu))|^{1/k}} \approx \pi_F(f_\mu)(0)^{1/k}$$

for every subspace $F \in G_{n,k}$. Combining Claim 2.4.10 and Claim 2.4.11 we get

$$\sqrt{k/n}\, I_{-k}(\mu) \approx w_{-k}(Z_k(\mu)). \tag{2.40}$$

Proof of Theorem 2.4.3. We shall show that, more generally, if μ is a centered log-concave probability measure on $\mathbb{R}^n$, then for every integer $1 \leqslant k \leqslant cq_*(\mu)$ we have

$$I_{-k}(\mu) \approx I_k(\mu).$$

Let $1 \leqslant k \leqslant n-1$. We know that $w_k(Z_k(\mu)) \approx \sqrt{k/n}\, I_k(\mu)$ and (2.40) shows that we also have $w_{-k}(Z_k(\mu)) \approx \sqrt{k/n}\, I_{-k}(\mu)$.

We set $k_0 = \lfloor q_* \rfloor$, where $q_* = q_*(\mu)$. Then,

$$k_*(Z_{k_0}(\mu)) \approx k_*(Z_{q_*}(\mu)) \geqslant c_1 q_* \geqslant c_1 k_0. \tag{2.41}$$

From Theorem 2.4.8 we have

$$w_{-k}(Z_{k_0}(\mu)) \approx w_k(Z_{k_0}(\mu)) \tag{2.42}$$

for every $1 \leqslant k \leqslant c_2 k_*(Z_{k_0}(\mu))$, and (2.41) shows that (2.42) holds for every $k \leqslant c_3 q_*(\mu)$. Setting $k_1 = \lfloor c_3 q_*(\mu) \rfloor \approx k_0$, and using the fact that $Z_{k_0}(\mu) \approx Z_{k_1}(\mu)$, we get

$$w_{-k_1}(Z_{k_1}(\mu)) \approx w_{k_1}(Z_{k_1}(\mu)). \tag{2.43}$$

It is now clear that $I_{-k_1}(\mu) \approx I_{k_1}(\mu)$ and since $k_1 \approx q_*(\mu)$ we see that $q \mapsto I_q(\mu)$ is "constant" in the range $1 \leqslant |q| \leqslant c q_*(\mu)$. □

A consequence of Theorem 2.4.3 is the next small ball probability estimate.

Theorem 2.4.12. *Let μ be an isotropic log-concave probability measure on $\mathbb{R}^n$. Then, for every $0 < \varepsilon < \varepsilon_0$ we have*

$$\mu(\{x \in \mathbb{R}^n : |x| \leqslant \varepsilon\sqrt{n}\}) \leqslant \varepsilon^{c\sqrt{n}}, \tag{2.44}$$

where $\varepsilon_0, c > 0$ are absolute constants.

Proof. We know that $I_2(\mu) \leqslant c_1 I_{-p}(\mu)$ for all $1 \leqslant p \leqslant c_2\sqrt{n}$. It follows that

$$\begin{aligned}\mu(\{x \in \mathbb{R}^n : |x| \leqslant \varepsilon I_2(\mu)\}) &\leqslant \mu(\{x : |x| \leqslant c_1 \varepsilon I_{-p}(\mu)\}) \\ &\leqslant (c_1\varepsilon)^p \leqslant \varepsilon^{p/2}\end{aligned}$$

for every $0 < \varepsilon < c_1^{-2}$ and $p \leqslant c_2\sqrt{n}$. This gives the result with $\varepsilon_0 = c_1^{-2}$ and $c = c_2/2$. □

2.4.1 Reduction of the hyperplane conjecture

Let μ be a centered log-concave probability measure on $\mathbb{R}^n$. Given any $\zeta \geqslant 1$, we define the parameter

$$q_{-c}(\mu, \zeta) := \max\{p \geqslant 1 : I_2(\mu) \leqslant \zeta I_{-p}(\mu)\}. \tag{2.45}$$

For an isotropic convex body K in $\mathbb{R}^n$ we set $I_p(K) = I_p(\mu_K)$.

The next theorem of Dafnis and Paouris [143] shows that if the hyperplane conjecture is correct, then there are absolute constants $\tau, \zeta_0 > 0$ such that, for every isotropic convex body K in $\mathbb{R}^n$, one has $q_{-c}(\mu_K, \zeta_0) \geqslant \tau n$.

Theorem 2.4.13. *There exists an absolute constant $c > 0$ such that, for every n and every isotropic convex body K in $\mathbb{R}^n$*

$$q_{-c}(\mu_K, cL_n) \geqslant n - 1.$$

Proof. Using Theorem 2.3.4, for every $1 \leqslant s \leqslant n-1$ we write

$$\begin{aligned} I_{-s}(K) &\approx \sqrt{n}\Big(\int_{G_{n,s}} |K \cap F^{\perp}|\, d\nu_{n,s}(F)\Big)^{-1/s} \\ &\approx \sqrt{n}\Big(\int_{G_{n,s}} \Big(\frac{L_{\overline{K_{s+1}}(\pi_F(\mu_K))}}{L_K}\Big)^s d\nu_{n,s}(F)\Big)^{-1/s} \geqslant \frac{\sqrt{n}L_K}{L_s}. \end{aligned}$$

This shows that

$$I_{-s}(K) \geqslant \frac{c_1\sqrt{n}L_K}{L_s} \geqslant \frac{c_2\sqrt{n}L_K}{L_n},$$

because it is known that $L_s \leqslant c_3 L_n$ for all integers $1 \leqslant s \leqslant n-1$ (see [106]). Since $I_2(K) = \sqrt{n}L_K$, we get

$$q_{-c}(K, c_2^{-1}L_n) := \max\{p \geqslant 1 : I_2(K) \leqslant c_2^{-1}L_n I_{-p}(K)\} \geqslant n - 1.$$

This is the claim of the theorem. □

We shall show that the hyperplane conjecture is equivalent to the following statement:

> There exist absolute constants $c, \zeta_0 > 0$ such that $q_{-c}(K, \zeta_0) \geqslant cn$ for every isotropic convex body K in $\mathbb{R}^n$.

We already know that there exists a parameter $q_* := q_*(K)$, related to the L_p-centroid bodies of K, with the following properties:

(i) $q_*(K) \geqslant c\sqrt{n}$;
(ii) $q_{-c}(K, \zeta_0) \geqslant q_*(K)$ for some absolute constant $\zeta_0 \geqslant 1$, and hence, $I_2(K) \leqslant \zeta_0 I_{-q_*}(K)$.

What is not clear is the behavior of $I_{-p}(K)$ when p lies in the interval $[q_*, n]$.

The main idea of Dafnis and Paouris in [143] is to start with an "extremal" isotropic convex body K in $\mathbb{R}^n$ with maximal isotropic constant $L_K \approx L_n$ that is at the same time in α-regular M-position. Their starting point, which has a rather technical proof, is the following precise statement.

Theorem 2.4.14. *There exist absolute constants $\kappa, \tau > 1$ and $\delta > 0$ such that, for every $\alpha \in [1, 2)$, we can find an isotropic convex body K_α in $\mathbb{R}^n$ with the following properties:*

(i) $L_{K_\alpha} \geqslant \delta L_n$;

(ii) *for every* $t \geqslant \tau(2-\alpha)^{-3/2}$

$$\log N(K_\alpha, t\sqrt{n}B_2^n) \leqslant \frac{\kappa n}{(2-\alpha)^{2\alpha}\, t^\alpha}. \tag{2.46}$$

Sketch of the proof. We shall give a very rough sketch of the proof of Theorem 2.4.14 for the case $\alpha = 1$. We should note here that the idea of considering convex bodies with "maximal" isotropic constant and their M-ellipsoids had been previously used by Bourgain, Klartag, and V. Milman in [106].

We may assume that $n = 2m$ is even, and we start with an isotropic convex body K_0 that has isotropic constant $L_{K_0} \geqslant \delta_0 L_{2m}$, where $\delta_0 \in (0,1)$. Then, for every k-codimensional subspace F of $\mathbb{R}^{2m}$ we have

$$|K \cap F^\perp|_{2m-k}^{1/k} \approx \frac{L_{\overline{K_{k+1}(\pi_F(\mu_K))}}}{L_K} \leqslant c_1(\delta_0),$$

where $c_1(\delta_0)$ is a constant depending only on δ_0. If $\mathcal{E}_1$ is a 1-regular M-ellipsoid for K_0, then we see that for every $F \in G_{2m,m}$ we have

$$|P_F(K_0)|_m^{1/m} \leqslant c_1\, |P_F(\mathcal{E}_1)|_m^{1/m}.$$

From the Rogers–Shephard/Spingarn inequality we know that

$$\left(|K_0 \cap F|_m\, |P_{F^\perp}(K_0)|_m\right)^{1/m} \approx \left(|\mathcal{E}_1 \cap F|_m\, |P_{F^\perp}(\mathcal{E}_1)_M)\right)^{1/m} \approx 1.$$

Combining the above with the fact that, for any $1 \leqslant k \leqslant n-1$, the maximal/minimal volume k-dimensional section and projection of an ellipsoid coincide, we conclude that

$$c_2(\delta_0) \leqslant |K \cap F^\perp|_m^{1/m} \leqslant c_1(\delta_0)$$

for every $F \in G_{2m,m}$. Using this fact we can check that if $\lambda_1 \geqslant \cdots \geqslant \lambda_{2m} > 0$ are the semi-axes of $\mathcal{E}_1$, then $\lambda_m \approx_\delta \sqrt{n}$. Now, we use the standard fact that there exists $F_0 \in G_{2m,m}$ such that $P_{F_0}(\mathcal{E}_1) = \lambda_m B_{F_0}$, and hence

$$c_3(\delta_0)\overline{B}_{F_0} \subseteq P_{F_0}(\mathcal{E}_1) \subseteq c_4(\delta_0)\overline{B}_{F_0}.$$

Then, we define $W = \overline{K_{m+1}(\pi_{F_0}(\mu_{K_0}))}$ and $K_1 = W \times U(W)$, where $U \in O(2m)$ satisfies $U(F_0) = F_0^\perp$. The convex body K_1 satisfies the assertion of the theorem for $\alpha = 1$. □

Then, Dafnis and Paouris try taking advantage of the fact that small ball estimates are closely related to estimates for covering numbers. The key lemma is the following.

Lemma 2.4.15. *Let K be a centered convex body of volume* 1 *in* $\mathbb{R}^n$. *Assume that, for some* $s > 0$

$$r_s := \log N(K, sB_2^n) < n. \tag{2.47}$$

Then,

$$I_{-r_s}(K) \leqslant 3es.$$

Proof. Let $z_0 \in \mathbb{R}^n$ be such that $|K \cap (-z_0 + sB_2^n)| \geqslant |K \cap (z + sB_2^n)|$ for every $z \in \mathbb{R}^n$. It follows that

$$\big|(K + z_0) \cap sB_2^n\big| N(K, sB_2^n) \geqslant |K| = 1. \tag{2.48}$$

Let $q = r_s$. Then, using Markov's inequality, the definition of $I_{-q}(K + z_0)$ and (2.47), we get

$$\big|(K + z_0) \cap 3^{-1} I_{-q}(K + z_0) B_2^n\big| \leqslant 3^{-q} < e^{-q} = e^{-r_s} \leqslant \frac{1}{N(K, sB_2^n)}$$

and hence

$$\big|(K + z_0) \cap 3^{-1} I_{-q}(K + z_0) B_2^n\big| < \big|(K + z_0) \cap sB_2^n\big|,$$

which implies that

$$3^{-1} I_{-q}(K + z_0) \leqslant s.$$

Finally, we can check that $I_{-k}(K + z) \geqslant e^{-1} I_{-k}(K)$ for any $1 \leqslant k \leqslant n - 1$ and $z \in \mathbb{R}^n$. Indeed, using Claim 2.4.10 and the fact that K is centered (more precisely, Fradelizi's inequality (2.10) for the centered log-concave function $\pi_{F^\perp}(\mathbb{1}_K)$, where $F \in G_{n,k}$) we write

$$\begin{aligned} I_{-k}(K + z) &= c_{n,k} \Big(\int_{G_{n,k}} \big|(K + z) \cap F^\perp\big| \, d\nu_{n,k}(F) \Big)^{-1/k} \\ &\geqslant \frac{c_{n,k}}{e} \Big(\int_{G_{n,k}} \big|K \cap F^\perp\big| \, d\nu_{n,k}(F) \Big)^{-1/k} = \frac{1}{e} I_{-k}(K). \end{aligned}$$

This proves the lemma. □

We can now prove the main theorem of [143].

Theorem 2.4.16. *Assume that $q_{-c}(K, \zeta) \geqslant \beta n$ for some $\zeta \geqslant 1$, some $\beta \in (0, 1)$, and every isotropic convex body K in $\mathbb{R}^n$. Then,*

$$L_n \leqslant \frac{C\zeta}{\sqrt{\beta}} \log^2(e/\beta), \tag{2.49}$$

where $C > 0$ is an absolute constant.

Proof. We set $\alpha := 2 - \log(e/\beta)^{-1}$ and, for this value of α, we apply Theorem 2.4.14 to find an isotropic convex body K_α that satisfies its conclusion: for some absolute constants κ, $\tau \geqslant 1$ and $\delta > 0$ it holds that $L_{K_\alpha} \geqslant \delta L_n$ and

$$\log N(K_\alpha, t\sqrt{n}B_2^n) \leqslant \frac{\kappa n}{(2-\alpha)^{2\alpha} t^\alpha} \quad \text{for all } t \geqslant \tau \log^{3/2}(e/\beta).$$

We may clearly assume that $\tau^2 \leqslant e\kappa$ as well. We choose

$$t_1 = (e\kappa)^{1/\alpha} \frac{1}{\sqrt{\beta}} \log^2(e/\beta).$$

Note that $\tau \leqslant \sqrt{e\kappa} \leqslant (e\kappa)^{1/\alpha}$, and hence $t_1 \geqslant \tau \frac{1}{\sqrt{\beta}} \log^2(e/\beta) \geqslant \tau \log^{3/2}(e/\beta)$. Therefore, (2.46) is valid for $t = t_1$, and this shows that

$$r_1 := \log N(K_\alpha, t_1\sqrt{n}B_2^n) \leqslant \frac{\kappa n}{(2-\alpha)^{2\alpha} t_1^\alpha} \leqslant \frac{1}{e}(\sqrt{\beta})^\alpha n \leqslant \beta n.$$

Then, from Lemma 2.4.15 we get

$$I_{-r_1}(K_\alpha) \leqslant 3et_1\sqrt{n}.$$

On the other hand, since $q_{-c}(K_\alpha, \zeta) \geqslant \beta n$ and $r_1 \leqslant \beta n$, we have that

$$\sqrt{n} L_{K_\alpha} = I_2(K_\alpha) \leqslant \zeta I_{-r_1}(K_\alpha).$$

It follows that

$$L_{K_\alpha} \leqslant 3e\zeta t_1 = 3e\zeta (e\kappa)^{1/\alpha} \frac{1}{\sqrt{\beta}} \log^2(e/\beta) \leqslant \frac{3e^2\zeta\kappa}{\sqrt{\beta}} \log^2(e/\beta).$$

Since $L_{K_\alpha} \geqslant \delta L_n$, the result follows. □

Remark 2.4.17. Since $q_{-c}(K, \zeta_0) \geqslant q_*(K) \geqslant c\sqrt{n}$ for some absolute constants $\zeta_0 \geqslant 1$ and $c > 0$, we may apply Theorem 2.4.16 with $\zeta = \zeta_0$ and $\beta = c/\sqrt{n}$ to get

$$L_n \leqslant \frac{C\zeta}{\sqrt{c}} \sqrt[4]{n} \log^2\left(\frac{e\sqrt{n}}{c}\right) \leqslant c_1 \sqrt[4]{n} (\log n)^2, \tag{2.50}$$

where $c_1 > 0$ is an absolute constant. This is a non-trivial estimate for L_n, slightly weaker but close to the estimate $L_n = O(\sqrt[4]{n} \log n)$ of Bourgain [102] and the estimate $L_n = O(\sqrt[4]{n})$ of Klartag [284].

We close this section with a much more direct reduction of the problem but with a much stronger assumption. We shall use the following corollary of Theorem 2.4.14.

Proposition 2.4.18. *For every $n \geqslant 1$ there exists an isotropic convex body K in $\mathbb{R}^n$ such that $L_K \geqslant c_1 L_n$ and*

$$|K \cap \sqrt{c_2 n} B_2^n| > c_0^n,$$

where $c_i > 0$ are absolute constants.

Proof. We apply Theorem 2.4.14 with $\alpha = 1$ to find an isotropic convex body K in $\mathbb{R}^n$ such that $L_K \geqslant c_1 L_n$ and

$$\log N(K, t\sqrt{n} B_2^n) \leqslant \frac{c_3 n}{t}$$

for all $t \geqslant c_3$. Then, it is clear that

$$1 = |K| \leqslant e^n N(K, \sqrt{c_2 n} B_2^n)\, |K \cap \sqrt{c_2 n} B_2^n| \leqslant e^{2n} |K \cap \sqrt{c_2 n} B_2^n|,$$

if $c_2 = c_3^2$, and the result follows. □

From Proposition 2.4.18 we deduce the direct reduction of the isotropic constant conjecture to small ball estimates.

Theorem 2.4.19. *Assume that there exist $\varepsilon_0, c_0 > 0$ such that for any $0 < \varepsilon \leqslant \varepsilon_0$ and every isotropic convex body K in $\mathbb{R}^n$ we have that*

$$|\{x \in K : |x| \leqslant \varepsilon \sqrt{n} L_K\}| \leqslant \varepsilon^{c_0 n}.$$

Then, $L_n \leqslant C$ for all $n \geqslant 1$, where $C = C(\varepsilon_0, c_0)$ depends only on ε_0 and c_0.

Proof. Proposition 2.4.18 shows that there exists an isotropic convex body K in $\mathbb{R}^n$ with $L_n \leqslant C_1 L_K$ and $|K \cap \sqrt{C_2 n} B_2^n| > c_2^n$ for some absolute constants $C_1, C_2, c_2 > 0$.

On the other hand, applying the hypothesis we see that for any $0 < \varepsilon \leqslant \varepsilon_0$

$$|K \cap \varepsilon \sqrt{n} L_K\, B_2^n| \leqslant \varepsilon^{c_0 n}.$$

Choosing $\varepsilon_1 = \min\{\varepsilon_0, c_2^{\frac{1}{c_0}}\}$, and comparing the above inequalities, we get

$$L_n^2 \leqslant C_1^2 L_K^2 \leqslant \frac{C_1^2 C_2}{\varepsilon_1}.$$

Thus, L_n is bounded. □

2.5 Kannan–Lovász–Simonovits conjecture

Let μ be a Borel probability measure on $\mathbb{R}^n$. For every Borel set $A \subseteq \mathbb{R}^n$ we define the Minkowski content $\mu^+(A)$ as follows:

$$\mu^+(A) = \liminf_{t\to 0^+} \frac{\mu(A_t) - \mu(A)}{t},$$

where $A_t = \{x \in \mathbb{R}^n : \operatorname{dist}(x, A) < t\}$ is the t-extension of A. The isoperimetric ratio of A is defined by $\chi_\mu(A) := \mu^+(A)/\min\{\mu(A), 1-\mu(A)\}$. Then, the (reciprocal) Cheeger constant ψ_μ of μ is the quantity

$$\psi_\mu := \sup_A \frac{1}{\chi_\mu(A)} = \sup_A \frac{\min\{\mu(A), 1-\mu(A)\}}{\mu^+(A)},$$

where the supremum is over all open sets $A \subset \mathbb{R}^n$ with smooth boundary and $0 < \mu(A) < 1$. It follows from the results of Rothaus, Cheeger, and Maz'ya (see [82]) that if a_μ is the smallest constant with the property that for every integrable, locally Lipschitz function $f : \mathbb{R}^n \to \mathbb{R}$ one has

$$\int_{\mathbb{R}^n} |f(x) - \mathbb{E}_\mu(f)|\, d\mu(x) \leqslant a_\mu \int_{\mathbb{R}^n} |\nabla f(x)|\, d\mu(x),$$

then $a_\mu/2 \leqslant \psi_\mu \leqslant 2a_\mu$.

The isoperimetric profile $I_\mu : [0,1] \to [0, +\infty]$ of μ is the function

$$I_\mu(t) = \inf\{\mu^+(A) : A \text{ Borel},\ \mu(A) = t\}.$$

Note that

$$\psi_\mu = \sup_{0<t\leqslant 1/2} \frac{t}{\min\{I_\mu(t), I_\mu(1-t)\}}.$$

The next important result of E. Milman [416], whose proof employs deep tools from geometric measure theory, is very useful.

Theorem 2.5.1. *Let μ be a log-concave probability measure on $\mathbb{R}^n$. Then, the isoperimetric profile I_μ of μ is concave on $(0,1)$, and for every $t \in (0,1)$ we have $I_\mu(t) = I_\mu(1-t)$. As a consequence*

$$\psi_\mu = \sup_{0<t\leqslant 1/2} \frac{t}{I_\mu(t)} = \frac{1}{2I_\mu(1/2)}.$$

The concavity of the isoperimetric profile on a convex domain was first obtained by Sternberg and Zumbrun in [516]. They showed that if $n \geqslant 2$ and K is a convex body in $\mathbb{R}^n$, then I_{λ_K} is concave on $[0,1]$, where λ_K is the uniform probability measure on K. Kuwert later noted in [351] that $I_{\lambda_K}^{n/(n-1)}$ is also concave on $[0,1]$. This is precisely the correct power to use. Kuwert's result was extended to convex domains in Riemannian manifolds with non-negative Ricci curvature by Bayle and Rosales [67]. E. Milman showed in [416] that the concavity of I_μ remains valid for log-concave measures on $\mathbb{R}^n$.

Theorem 2.5.1 asserts that we can calculate the Cheeger constant of a log-concave probability measure μ by looking only at Borel sets A with $\mu(A) = 1/2$. In fact, E. Milman proved in [416] that one can determine the order of the constant ψ_μ just by testing 1-Lipschitz functions of the form $x \mapsto d(x, A)$.

Theorem 2.5.2. *Let μ be a log-concave probability measure on $\mathbb{R}^n$. Then,*

$$\psi_\mu \approx \sup\left\{\int d(x,A)\,d\mu(x) : \mu(A) \geqslant \frac{1}{2}\right\}. \tag{2.51}$$

2.5.1 Kannan–Lovász–Simonovits conjecture

Kannan, Lovász, and Simonovits conjectured in [275] that the isoperimetric ratio of any Borel set A with respect to the uniform measure λ_K on a convex body K in $\mathbb{R}^n$ should be, up to an absolute constant, at least as large as the minimal isoperimetric ratio over all half-spaces. Recall that when K is an isotropic convex body in $\mathbb{R}^n$, then $1/L_K$ is approximately equal to the $(n-1)$-dimensional volume of the section of K with any hyperplane passing through the origin: we know that $|K \cap \xi^\perp| \approx 1/L_K$ for every $\xi \in S^{n-1}$. On the other hand, $|K \cap \xi^\perp|$ is the Minkowski content of the intersection of K with the half-space $H_\xi^+ := \{x : \langle x, \xi\rangle \geqslant 0\}$ or $H_\xi^- := \{x \in K : \langle x, \xi\rangle \leqslant 0\}$. Therefore, the KLS conjecture amounts to asking whether $\psi_{\lambda_K} \leqslant cL_K$. Taking into account the isotropic constant conjecture and the different normalization that we have chosen in the definition of isotropic log-concave measures, we can restate the KLS conjecture in the more general setting of isotropic log-concave probability measures as follows.

Conjecture 2.5.3 (KLS conjecture). *If μ is an isotropic log-concave probability measure on $\mathbb{R}^n$, then the reciprocal Cheeger constant of μ satisfies $\psi_\mu \leqslant C$ for some absolute constant $C > 0$. In other words*

$$\psi_n := \sup\{\psi_\mu : \mu \text{ is isotropic log-concave measure on } \mathbb{R}^n\} \leqslant C.$$

Kannan, Lovász, and Simonovits [275] gave an upper bound for ψ_{λ_K} in terms of the quantity

$$J_1(K) := \frac{1}{|K|}\int_K |x - \mathrm{bar}(K)|\,dx,$$

where $\mathrm{bar}(K)$ is the barycenter of K.

Theorem 2.5.4. *For every convex body K in $\mathbb{R}^n$ one has*

$$\psi_{\lambda_K} \leqslant cJ_1(K). \tag{2.52}$$

The proof of Theorem 2.5.4 exploited the localization lemma of Lovász and Simonovits from [387], a very useful tool that allows one to reduce inequalities for log-concave functions in $\mathbb{R}^n$ to one-dimensional inequalities. Note that, if K is an isotropic convex body in $\mathbb{R}^n$, then

$$J_1(K) \leqslant \Big(\int_K |x|^2\,dx\Big)^{1/2} = \sqrt{n}L_K$$

and hence (2.52) implies that

$$\psi_{\lambda_K} \leqslant c\sqrt{n}L_K, \tag{2.53}$$

where $c > 0$ is an absolute constant.

The KLS-conjecture is stronger than the isotropic constant conjecture. Ball and Nguyen [47] showed that $L_n \leqslant \exp(c\psi_n^2)$ for an absolute constant $c > 0$. In fact, from their work one can conclude that for each individual isotropic log-concave probability measure μ, a bound for the KLS-constant implies a bound for the isotropic constant: one has that $L_\mu \leqslant \exp(c\psi_\mu^2)$. An important result of Eldan and Klartag [158] establishes a linear dependence of L_n on ψ_n.

Theorem 2.5.5. *There exists an absolute constant $C > 0$ such that, for every $n \geqslant 1$*

$$L_n \leqslant C\psi_n.$$

Theorem 2.5.5 shows that any upper bound for ψ_n implies an equivalent, up to an absolute constant, upper bound for L_n. In fact, as we will see in this section, Klartag has established in [290] the upper bound $\psi_n \leqslant c\sqrt{\log n}$.

The Cheeger constant is closely related to another isoperimetric constant associated to μ, the Poincaré constant, defined to be the best, i. e., the smallest, constant ϑ_μ with the property that

$$\mathrm{Var}_\mu(f) = \int_{\mathbb{R}^n} f^2\,d\mu - \Big(\int_{\mathbb{R}^n} f\,d\mu\Big)^2 \leqslant \vartheta_\mu^2 \int_{\mathbb{R}^n} |\nabla f|^2\,d\mu$$

for all locally Lipschitz functions $f : \mathbb{R}^n \to \mathbb{R}$ with $\int |\nabla f|^2\,d\mu < \infty$. We shall use the fact that the Poincaré constant is subadditive: if μ, ν are probability measures, then $\vartheta_{\mu*\nu}^2 \leqslant \vartheta_\mu^2 + \vartheta_\nu^2$, where $\mu * \nu$ is the convolution measure of μ and ν; this follows by a classical variance decomposition and the convexity of $t \mapsto t^2$ (see, e.g., [97]).

A theorem of Maz'ya [405, 406] and Cheeger [129] shows that the Poincaré constant ϑ_μ of μ is bounded by the reciprocal Cheeger constant ψ_μ.

Theorem 2.5.6. *Let μ be a Borel probability measure with reciprocal Cheeger constant ψ_μ. Then, its Poincaré constant ϑ_μ satisfies*

$$\vartheta_\mu \leqslant 2\psi_\mu. \tag{2.54}$$

On the other hand, the assumption that μ is log-concave implies a reverse inequality with a constant that does not depend on the dimension. The next theorem is due to Buser [121] (see also Ledoux [360]).

Theorem 2.5.7. *Let μ be a log-concave probability measure on $\mathbb{R}^n$. Then, $\psi_\mu \leqslant c\,\vartheta_\mu$, where $c > 0$ is an absolute constant.*

2.5.2 Improved log-concave Lichnerowicz inequality

One of the crucial ingredients in Klartag's estimate for ψ_n is an improved log-concave Lichnerowicz inequality that we now describe.

In what follows, we assume that μ is a log-concave probability measure with a density $\varrho = e^{-\varphi}$, where φ is a convex function on $\mathbb{R}^n$. There are some issues that we shall ignore in our discussion. What we will do requires that μ is regular. We assume that the density ϱ of μ is smooth and positive on $\mathbb{R}^n$, and there exists $\delta > 0$ such that $\delta I_n \leqslant \nabla^2\varphi \leqslant \delta^{-1}I_n$, where $\nabla^2\varphi$ denotes the Hessian of φ. Also, we assume that φ and all its partial derivatives grow at most polynomially at infinity. One can obtain a reduction to this case, because of the following lemma: For every μ and $\varepsilon > 0$ we may find a regular ν that satisfies $\vartheta_\nu^2 \geqslant \vartheta_\mu^2 - \varepsilon$ and $\|\mathrm{Cov}(\nu) - \mathrm{Cov}(\mu)\|_{\mathrm{op}} \leqslant \varepsilon$. A proof of this fact is given in [290, Lemma 2.1].

We consider the class $\mathcal{F}_\mu$ of all functions u on $\mathbb{R}^n$ that are smooth and have subexponential decay relative to ϱ, i. e., there exist $C, a > 0$ such that

$$|u(x)| \leqslant \frac{C}{\sqrt{\varrho(x)}} e^{-a|x|},$$

where ϱ is the density of μ. We also require that the same is true for the partial derivatives of u. The Laplace–Beltrami operator is defined for $u \in \mathcal{F}_\mu$ by

$$L_\mu u = \Delta u - \langle \nabla\varphi, \nabla u\rangle.$$

The Laplace–Beltrami operator satisfies the basic identity

$$\int_{\mathbb{R}^n} (L_\mu u)\, v \, d\mu = -\int_{\mathbb{R}^n} \langle \nabla u, \nabla v\rangle \, d\mu \tag{2.55}$$

for all $u, v \in \mathcal{F}_\mu$. We will also use Bochner's formula

$$\int_{\mathbb{R}^n} (L_\mu u)^2\, d\mu = \int_{\mathbb{R}^n} \|\nabla^2 u\|_{\mathrm{HS}}^2\, d\mu + \int_{\mathbb{R}^n} \langle (\nabla^2\varphi)\nabla u, \nabla u\rangle\, d\mu, \tag{2.56}$$

which holds true for every $u \in \mathcal{F}_\mu$. Note that $-L_\mu$ is essentially self-adjoint and positive semi-definite in $\mathcal{F}_\mu \subset L^2(\mu)$, and has a discrete spectrum $\lambda_0 \leqslant \lambda_1 \leqslant \cdots \leqslant \lambda_j \leqslant \cdots$. The minimal eigenvalue λ_0 of L_μ is 0, a simple eigenvalue with constant eigenfunctions. Let $\lambda_\mu = \lambda_1$ denote the smallest positive eigenvalue of L_μ. If $-L_\mu g = \lambda_\mu g$ with $\int g\, d\mu = 0$, then

$$\lambda_\mu \int_{\mathbb{R}^n} g^2\, d\mu = \int_{\mathbb{R}^n} (-L_\mu g)\, g\, d\mu = \int_{\mathbb{R}^n} |\nabla g|^2\, d\mu.$$

This shows that $1/\lambda_\mu \leqslant \vartheta_\mu^2$. In fact, we have equality by the next lemma.

Lemma 2.5.8. *For every log-concave probability measure μ on $\mathbb{R}^n$ we have that*

$$1/\lambda_\mu = \vartheta_\mu^2.$$

Proof. Define the energy functional

$$\mathcal{E}(f,g) = \int_{\mathbb{R}^n} \langle \nabla f, \nabla g\rangle\, d\mu$$

for $f, g \in \mathcal{F}_\mu$. Note that $\mathcal{E}(\cdot,\cdot)$ is linear in each of its arguments and $\mathcal{E}(f,f) \geqslant 0$ for all $f \in \mathcal{F}_\mu$. We can write any $f \in \mathcal{F}_\mu$ as

$$f = \langle f, w_0\rangle w_0 + \sum_{j=1}^{\infty} \langle f, w_j\rangle w_j,$$

where $-L_\mu w_j = \lambda_j w_j$ and w_j form an orthonormal sequence in $L^2(\mu)$. Note that $w_0 \equiv 1$, and hence $\langle f, w_0\rangle = \int f\, d\mu$. For every $s \geqslant 1$ we have that

$$\begin{aligned} 0 &\leqslant \mathcal{E}\Big(f - \sum_{j=1}^{s} \langle f, w_j\rangle w_j, f - \sum_{k=1}^{s} \langle f, w_k\rangle w_k\Big) \\ &= \mathcal{E}(f,f) - 2\sum_{j=1}^{s} \langle f, w_j\rangle \mathcal{E}(w_j, f) + \sum_{j,k=1}^{s} \langle f, w_j\rangle \langle f, w_k\rangle \mathcal{E}(w_j, w_k). \end{aligned}$$

On observing that

$$\mathcal{E}(w_j, w_k) = \lambda_j \int_{\mathbb{R}^n} w_j w_k\, d\mu = \lambda_j \delta_{j,k}$$

and

$$\mathcal{E}(w_j,f)=\int_{\mathbb{R}^n}\langle\nabla w_j,\nabla f\rangle\,d\mu=\int_{\mathbb{R}^n}(-L_\mu w_j)f\,d\mu=\lambda_j\int_{\mathbb{R}^n}w_jf\,d\mu=\lambda_j\langle f,w_j\rangle$$

for all $1\leqslant j,k\leqslant s$, we conclude that

$$0\leqslant\mathcal{E}(f,f)-\sum_{j=1}^{s}\lambda_j\langle f,w_j\rangle^2$$

and this shows that

$$\begin{aligned}\lambda_1\mathrm{Var}_\mu(f)=\lambda_1\Big(\int_{\mathbb{R}^n}f^2\,d\mu-\Big(\int_{\mathbb{R}^n}f\,d\mu\Big)^2\Big)&\leqslant\sum_{j=1}^{\infty}\lambda_j\langle f,w_j\rangle^2\\&\leqslant\mathcal{E}(f,f)=\int_{\mathbb{R}^n}|\nabla f|^2\,d\mu.\end{aligned}$$

It follows that $\vartheta_\mu^2\leqslant 1/\lambda_1$, and this completes the proof. □

Definition 2.5.9. We say that a Borel probability measure μ on $\mathbb{R}^n$ with density $\varrho=e^{-\varphi}$ is t-uniformly log-concave if the function $\varrho(x)e^{t|x|^2/2}$ is log-concave. Then, $\varphi(x)-\frac{t|x|^2}{2}$ is convex, that is $\nabla^2\varphi(x)\geqslant tI_n$ for all $x\in\mathbb{R}^n$.

A theorem of Bakry and Ledoux [35] asserts that if μ is t-uniformly log-concave, then

$$\vartheta_\mu^2=\frac{1}{\lambda_\mu}\leqslant\frac{1}{t}.\tag{2.57}$$

This inequality is known as the log-concave Lichnerowicz inequality, because it is analogous to some investigations of Lichnerowicz [374] in Riemannian geometry (see [36] and [361]). The following stronger "improved log-concave Lichnerowicz inequality" was proved by Klartag in [290].

Theorem 2.5.10. *Let $t>0$ and let μ be a t-uniformly log-concave probability measure on $\mathbb{R}^n$. Then,*

$$\vartheta_\mu^2=\frac{1}{\lambda_\mu}\leqslant\sqrt{\frac{\|\mathrm{Cov}(\mu)\|_{\mathrm{op}}}{t}}.\tag{2.58}$$

Note that (2.58) is stronger than (2.57), since $\|\mathrm{Cov}(\mu)\|_{\mathrm{op}}\leqslant 1/\lambda_\mu$ (see Remark 2.5.12 below), and hence $\sqrt{\frac{\|\mathrm{Cov}(\mu)\|_{\mathrm{op}}}{t}}\leqslant\frac{1}{\sqrt{\lambda_\mu t}}$. So, from (2.58) we get $\frac{1}{\lambda_\mu}\leqslant\frac{1}{\sqrt{\lambda_\mu t}}$, which in turn gives (2.57). On the other hand, the Kannan–Lóvasz–Simonovits conjecture asks for the estimate

$$\frac{1}{\lambda_\mu}\leqslant c\,\|\mathrm{Cov}(\mu)\|_{\mathrm{op}}$$

for an absolute constant $c > 0$. Therefore, in the context of t-uniformly log-concave measures, (2.58) stands between the Bakry–Ledoux theorem and the Kannan–Lóvasz–Simonovits conjecture.

Equality in Theorem 2.5.10 is attained when μ is a gaussian measure, with any covariance matrix. In this case, we have that ϑ_μ^2 and $\|\mathrm{Cov}(\mu)\|_{\mathrm{op}}$ coincide, and they also coincide with the inverse lower bound on the Hessian of the potential. If $\gamma_n^{(s)}$ is the distribution of a gaussian random vector of mean zero and covariance matrix sI_n in $\mathbb{R}^n$, then $\gamma_n^{(s)}$ satisfies the assumptions of Theorem 2.5.10 for $t = 1/s$, while $\vartheta^2_{\gamma_n^{(s)}} = \|\mathrm{Cov}(\gamma_n^{(s)})\|_{\mathrm{op}} = s$.

The proof of Theorem 2.5.10 is based on the next lemma.

Lemma 2.5.11. *Let μ be a regular log-concave probability measure on $\mathbb{R}^n$ with density $\varrho = e^{-\varphi}$, where φ is a convex function. If w_1 is an eigenfunction corresponding to $\lambda_1 = \lambda_\mu$, with $\|w_1\|_{L^2(\mu)} = 1$, then*

$$\int_{\mathbb{R}^n} \langle (\nabla^2\varphi)\nabla w_1, \nabla w_1\rangle \, d\mu \leqslant \lambda_1^3 \|\mathrm{Cov}(\mu)\|_{\mathrm{op}}.$$

Proof. Applying Bochner's formula (2.56) for w_1 we get

$$\lambda_1^2 = \int_{\mathbb{R}^n} (L_\mu w_1)^2 \, d\mu = \int_{\mathbb{R}^n} \langle (\nabla^2\varphi)\nabla w_1, \nabla w_1\rangle \, d\mu + \int_{\mathbb{R}^n} \|\nabla^2 w_1\|_{\mathrm{HS}}^2 \, d\mu.$$

Note that

$$\begin{aligned}
\int_{\mathbb{R}^n} \|\nabla^2 w_1\|_{\mathrm{HS}}^2 \, d\mu &= \sum_{i=1}^n \int_{\mathbb{R}^n} |\nabla \partial^i w_1|^2 \, d\mu \\
&\geqslant \lambda_1 \Big(\sum_{i=1}^n \int_{\mathbb{R}^n} |\partial^i w_1|^2 \, d\mu - \sum_{i=1}^n \Big| \int_{\mathbb{R}^n} \partial^i w_1 \, d\mu \Big|^2 \Big) \\
&= \lambda_1 \Big(\int_{\mathbb{R}^n} |\nabla w_1|^2 \, d\mu - \Big| \int_{\mathbb{R}^n} \nabla w_1 \, d\mu \Big|^2 \Big) \\
&\geqslant \lambda_1^2 \int_{\mathbb{R}^n} w_1^2 \, d\mu - \lambda_1 \Big| \int_{\mathbb{R}^n} \nabla w_1 \, d\mu \Big|^2 \\
&= \lambda_1^2 - \lambda_1 \Big| \int_{\mathbb{R}^n} \nabla w_1 \, d\mu \Big|^2 .
\end{aligned}$$

Therefore,

$$\lambda_1^2 \geqslant \int_{\mathbb{R}^n} \langle (\nabla^2\varphi)\nabla w_1, \nabla w_1\rangle \, d\mu + \lambda_1^2 - \lambda_1 \Big| \int_{\mathbb{R}^n} \nabla w_1 \, d\mu \Big|^2 ,$$

which implies that

$$\int_{\mathbb{R}^n} \langle (\nabla^2\varphi)\nabla w_1, \nabla w_1\rangle\, d\mu \leqslant \lambda_1 \Big|\int_{\mathbb{R}^n} \nabla w_1\, d\mu\Big|^2. \tag{2.59}$$

Now, let $\xi \in S^{n-1}$ and consider the function $g_\xi(x) = \langle x, \xi\rangle$. Set $E_\xi = \int g_\xi\, d\mu$. We may write

$$\begin{aligned}\Big|\Big\langle \int_{\mathbb{R}^n} \nabla w_1\, d\mu, \xi\Big\rangle\Big|^2 &= \Big|\int_{\mathbb{R}^n} \langle \nabla w_1, \nabla g_\xi\rangle\, d\mu\Big|^2 = \Big|\int_{\mathbb{R}^n} (-L_\mu w_1) g_\xi\, d\mu\Big|^2 \\ &= \lambda_1^2 \Big|\int_{\mathbb{R}^n} w_1 g_\xi\, d\mu\Big|^2 = \lambda_1^2 \Big|\int_{\mathbb{R}^n} w_1 (g_\xi - E_\xi)\, d\mu\Big|^2 \\ &\leqslant \lambda_1^2 \int_{\mathbb{R}^n} w_1^2\, d\mu \cdot \int_{\mathbb{R}^n} (g_\xi - E_\xi)^2\, d\mu = \lambda_1^2 \langle \mathrm{Cov}(\mu)\xi, \xi\rangle,\end{aligned}$$

taking into account the fact that $\int w_1\, d\mu = 0$. It follows that

$$\begin{aligned}\Big|\int_{\mathbb{R}^n} \nabla w_1\, d\mu\Big|^2 &= \sup_{\xi\in S^{n-1}} \Big|\Big\langle \int_{\mathbb{R}^n} \nabla w_1\, d\mu, \xi\Big\rangle\Big|^2 \leqslant \lambda_1^2 \sup_{\xi\in S^{n-1}} \langle \mathrm{Cov}(\mu)\xi, \xi\rangle \\ &= \lambda_1^2 \|\mathrm{Cov}(\mu)\|_{\mathrm{op}}.\end{aligned} \tag{2.60}$$

Combining (2.59) and (2.60) we obtain the assertion of the lemma. □

Remark 2.5.12. For every $\xi \in S^{n-1}$ we have $\nabla g_\xi = \xi$, and hence

$$\langle \mathrm{Cov}(\mu)\xi, \xi\rangle = \mathrm{Var}_\mu(g_\xi) \leqslant \frac{1}{\lambda_1}\int_{\mathbb{R}^n} |\nabla g_\xi|^2\, d\mu = \frac{1}{\lambda_1}.$$

This shows that

$$\|\mathrm{Cov}(\mu)\|_{\mathrm{op}} \leqslant \frac{1}{\lambda_1}. \tag{2.61}$$

Proof of Theorem 2.5.10. By Lemma 2.5.11, using also the assumption that $\nabla^2\varphi \geqslant tI_n$, we have that

$$t\lambda_\mu = t\int_{\mathbb{R}^n} |\nabla w_1|^2\, d\mu \leqslant \int_{\mathbb{R}^n} \langle (\nabla^2\varphi)\nabla w_1, \nabla w_1\rangle\, d\mu \leqslant \lambda_\mu^3 \|\mathrm{Cov}(\mu)\|_{\mathrm{op}},$$

which implies that

$$\frac{1}{\lambda_\mu^2} \leqslant \frac{\|\mathrm{Cov}(\mu)\|_{\mathrm{op}}}{t}. \tag{2.62}$$

This is the claim of the theorem. □

2.5.3 Stochastic localization

Eldan's stochastic localization has been the key for the recent developments in Bourgain's slicing problem and the Kannan–Lovász–Simonovits conjecture that we shall discuss in the next section. Eldan introduced in [157] a variant of the stochastic localization scheme that we shall use. Then, his idea was used in a number of subsequent works [364, 131, 293].

First, we recall some preliminaries from stochastic calculus. We refer to [445, 482], and [154] for definitions and background on semimartingales and stochastic integration. Let x_t and y_t be real-valued stochastic processes. The quadratic variations $[x]_t$ and $[x,y]_t$ are real-valued stochastic processes defined by

$$[x]_t = \lim_{|P|\to 0} \sum_{i=1}^{\infty} (x_{\tau_n} - x_{\tau_{n-1}})^2 \quad \text{and} \quad [x,y]_t = \lim_{|P|\to 0} \sum_{n=1}^{\infty} (x_{\tau_n} - x_{\tau_{n-1}})(y_{\tau_n} - y_{\tau_{n-1}}),$$

where $P = \{0 = \tau_0 \leqslant \tau_1 \leqslant \tau_2 \cdots \uparrow t\}$ is a stochastic partition of the non-negative real numbers, $|P| = \max_n(\tau_n - \tau_{n-1})$ is the mesh of P, and the limit is defined using convergence in probability. Note that $[x]_t$ is non-decreasing in t and $[x,y]_t$ satisfies the equation

$$[x,y]_t = \frac{1}{4}([x+y]_t - [x-y]_t).$$

Assume that the processes x_t and y_t satisfy the differential equations $dx_t = a(x_t)\,dt + \sigma(x_t)\,dW_t$ and $dy_t = b(y_t)\,dt + \eta(y_t)\,dW_t$, where (W_t) is a standard Brownian motion. Then, we have

$$[x]_t = \int_0^t \sigma^2(x_s)\,ds \quad \text{and} \quad [x,y]_t = \int_0^t \sigma(x_s)\eta(y_s)\,ds.$$

Also, $d[x,y]_t = \sigma(x_t)\eta(y_t)\,dt$.

Analogously, for vector valued differential equations $dx_t = a(x_t)\,dt + \Sigma(x_t)\,dW_t$ and $dy_t = b(y_t)\,dt + H(y_t)\,dW_t$ we have that

$$[x^i, x^j]_t = \int_0^t (\Sigma(x_s)\Sigma^T(x_s))_{ij}\,ds \quad \text{and} \quad d[x^i, y^j]_t = (\Sigma(x_t)H^T(y_t))_{ij}\,dt.$$

We shall use Itô's formula: If x is a semimartingale and f is a twice continuously differentiable function, then

$$df(x_t) = \sum_i \frac{df(x_t)}{dx^i}\, dx^i + \frac{1}{2}\sum_{i,j} \frac{d^2 f(x_t)}{dx^i\, dx^j}\, d[x^i, x^j]_t. \tag{2.63}$$

Let μ be an isotropic log-concave probability measure on $\mathbb{R}^n$ with a log-concave density $\varrho = e^{-\varphi}$. Set $\varrho_0 = \varrho$, $\mu_0 = \mu$, $a_0 = \mathrm{bar}(\mu)$, $A_0 = \mathrm{Cov}(\mu) = I_n$, and $\lambda_0 = \lambda_\mu$. Our goal is to show that

$$\lambda_0 \geqslant \frac{c}{\|A_0\|_{\mathrm{op}}} = c$$

for some absolute constant $c > 0$. For $t \geqslant 0$ and $\xi \in \mathbb{R}^n$ consider the measure $\mu_{t,\xi}$ with density

$$\varrho_{t,\xi}(x) = \frac{1}{Z_{t,\xi}} e^{\langle x,\xi\rangle - t|x|^2/2} \varrho_0(x),$$

where $Z_{t,\xi}$ is a suitable normalizing constant. Note that $\mu_{t,\xi}$ is t-uniformly log-concave and if we set $\lambda_{t,\xi} = \lambda_{\mu_{t,\xi}}$, $A_{t,\xi} = \mathrm{Cov}(\mu_{t,\xi})$, etc., then

$$\lambda_{t,\xi} \geqslant \sqrt{\frac{t}{\|A_{t,\xi}\|_{\mathrm{op}}}} \geqslant t$$

by Theorem 2.5.10. We choose ξ_t in a random way, by the stochastic differential equation

$$\xi_0 = 0, \quad d\xi_t = dW_t + a_{t,\xi_t}\, dt,$$

where $(W_t)_{t\geqslant 0}$ is a standard Brownian motion in $\mathbb{R}^n$ with $W_0 = 0$. One can check that $(\xi_t)_{t\geqslant 0}$ coincides in law with the process $(tX + W_t)_{t\geqslant 0}$, where $X \sim \mu$ is independent of W_t. We set

$$\mu_t = \mu_{t,\xi_t}, \quad \varrho_t = \varrho_{t,\xi_t}, \quad a_t = \mathrm{bar}(\mu_t), \quad A_t = \mathrm{Cov}(\mu_t), \quad \lambda_t = \lambda_{\mu_t}.$$

From Itô's formula we see that

$$d\varrho_t(x) = \varrho_t(x)\langle x - a_t, dW_t\rangle, \quad x \in \mathbb{R}^n.$$

The process $(\varrho_t(x))_{t\geqslant 0}$ is a martingale with respect to the filtration induced by the Brownian motion, and in particular

$$\mathbb{E}(\varrho_t(x)) = \varrho_0(x) = \varrho(x) \tag{2.64}$$

for all $t \geqslant 0$ and $x \in \mathbb{R}^n$. Roughly speaking, for every $t > 0$ we have expressed the log-concave measure μ as a mixture of t-uniformly log-concave measures. Moreover, this mixture is constructed as a solution of a stochastic differential equation, and using Itô's formula we can try to control its behavior over time.

Recall that $A_0 = I_n$ because μ is isotropic, therefore $\|A_0\|_{\mathrm{op}} = 1$. One of the crucial properties of the covariance process $(A_t)_{t\geqslant 0}$ is the following fact.

Theorem 2.5.13. *For every* $0 < t \leqslant c/(\psi_n^2 \log n)$ *we have*

$$\mathbb{E}\,\|A_t\|_{\mathrm{op}} \leqslant C, \tag{2.65}$$

where $C > 0$ *is an absolute constant.*

2.5.4 Upper bound for the KLS-constant

In this subsection, we show that $\psi_n = O(\sqrt{\log n})$. We start with the next lemma.

Lemma 2.5.14. *Let* $f \in L^2(\mu)$. *For every* $t > 0$ *we have that*

$$\mathbb{E}\,(\mathrm{Var}_{\varrho_t}(f)) \leqslant \mathrm{Var}_{\varrho_0}(f) \leqslant \left(2 + \frac{t}{\lambda_0}\right)\mathbb{E}\,(\mathrm{Var}_{\varrho_t}(f)).$$

Proof. For $f \in L^2(\mu)$ we set

$$M_{t,\xi}(f) = \int_{\mathbb{R}^n} f(x)\varrho_{t,\xi}(x)\,dx.$$

This is a smooth function of ξ, and differentiation under the integral sign shows that

$$\begin{aligned}\nabla_\xi M_{t,\xi}(f) &= \int_{\mathbb{R}^n} (x - a_{t,\xi})f(x)\varrho_{t,\xi}(x)\,dx \\ &= \int_{\mathbb{R}^n} (x - a_{t,\xi})(f(x) - M_{t,\xi}(f))\varrho_{t,\xi}(x)\,dx.\end{aligned} \tag{2.66}$$

From (2.64) and Fubini's theorem we see that

$$\begin{aligned}\mathrm{Var}_{\varrho_0}(f) &= \int_{\mathbb{R}^n} (f - M_0)^2\varrho_0 = \mathbb{E}\Big(\int_{\mathbb{R}^n} (f - M_0)^2\varrho_t\Big) \\ &= \mathbb{E}\Big(\int_{\mathbb{R}^n} (f - M_t)^2\varrho_t\Big) + \mathbb{E}(M_t - M_0)^2.\end{aligned} \tag{2.67}$$

Note that

$$\mathbb{E}\Big(\int_{\mathbb{R}^n} (f - M_t)^2\varrho_t\Big) = \mathbb{E}\,(\mathrm{Var}_{\varrho_t}(f)) \leqslant \mathrm{Var}_{\varrho_0}(f).$$

Next, we observe that

$$\mathbb{E}(M_t - M_0)^2 = \mathrm{Var}(M_t) = \mathrm{Var}(M_{t,\xi_t}(f)),$$

where $\xi_t \sim tX + W_t$, where X has distribution μ and W_t is a gaussian random vector in $\mathbb{R}^n$, independent of X, with mean zero and covariance matrix tI_n. From the subadditivity property of the Poincaré constant we get

$$\vartheta_{\xi_t}^2 = \vartheta_{tX+W_t}^2 \leqslant \vartheta_{tX}^2 + \vartheta_{W_t}^2 = \frac{t^2}{\lambda_0} + t. \tag{2.68}$$

From (2.66) and the Cauchy–Schwartz inequality we see that

$$\begin{aligned}|\nabla_\xi M_{t,\xi}(f)| &= \sup_{u\in S^{n-1}} \int_{\mathbb{R}^n} \langle x - a_{t,\xi}, u\rangle (f(x) - M_{t,\xi}(f))\varrho_{t,\xi}(x)\,dx \\ &\leqslant \sqrt{\|A_{t,\xi}\|_{\mathrm{op}}}\,\sqrt{\mathrm{Var}_{\varrho_{t,\xi}}(f)}\end{aligned}$$

for any $\xi \in \mathbb{R}^n$. Using the Poincaré inequality for the random vector ξ_t and (2.68), we get

$$\begin{aligned}\mathrm{Var}(M_{t,\xi_t}(f)) &\leqslant \left(1 + \frac{t}{\lambda_0}\right) t\mathbb{E}|\nabla_{\xi_t} M_{t,\xi_t}(f)|^2 \\ &\leqslant \left(1 + \frac{t}{\lambda_0}\right) t\mathbb{E}(\|A_t\|_{\mathrm{op}}\mathrm{Var}_{\varrho_t}(f)) \\ &\leqslant \left(1 + \frac{t}{\lambda_0}\right)\mathbb{E}\,(\mathrm{Var}_{\varrho_t}(f)),\end{aligned} \tag{2.69}$$

where we have also used the fact that $\|A_t\|_{\mathrm{op}} \leqslant 1/t$. Combining (2.67) and (2.69) we see that

$$\mathrm{Var}_{\varrho_0}(f) \leqslant \mathbb{E}\,(\mathrm{Var}_{\varrho_t}(f)) + \left(1 + \frac{t}{\lambda_0}\right)\mathbb{E}\,(\mathrm{Var}_{\varrho_t}(f)) = \left(2 + \frac{t}{\lambda_0}\right)\mathbb{E}\,(\mathrm{Var}_{\varrho_t}(f)),$$

as claimed. □

A consequence of Lemma 2.5.14 is the next proposition.

Proposition 2.5.15. *For any $a > 0$ and $0 < t \leqslant a\lambda_0$*

$$\frac{1}{\lambda_0} \leqslant c_1(a+2)\mathbb{E}(1/\lambda_t) \leqslant c_1(a+2)\frac{\mathbb{E}(\sqrt{\|A_t\|_{\mathrm{op}}})}{\sqrt{t}}, \tag{2.70}$$

where $c_1 > 0$ is an absolute constant.

Proof. Recall E. Milman's Theorem 2.5.2 that states that

$$c\lambda_0^{-1} \leqslant \sup_{\varphi} \mathrm{Var}_\mu(f), \tag{2.71}$$

where the supremum is over all 1-Lipschitz functions $f : \mathbb{R}^n \to \mathbb{R}$ and $c > 0$ is an absolute constant. Therefore, we may choose a 1-Lipschitz function $f : \mathbb{R}^n \to \mathbb{R}$ such that

$$\mathrm{Var}_\mu(f) \geqslant c\lambda_0^{-1}/2. \tag{2.72}$$

Lemma 2.5.14 and the Poincaré inequality imply that

$$\begin{aligned}\mathrm{Var}_\mu(f) = \mathrm{Var}_{\varrho_0}(f) &\leqslant (2+\alpha)\mathbb{E}\left(\mathrm{Var}_{\varrho_t}(f)\right) \leqslant (2+\alpha)\mathbb{E}\Big(\frac{1}{\lambda_t}\int_{\mathbb{R}^n} |\nabla f|^2 \varrho_t\Big)\\ &\leqslant (2+\alpha)\mathbb{E}(\lambda_t^{-1}).\end{aligned} \tag{2.73}$$

Now, the left-hand side inequality in (2.70) is a consequence of (2.72) and (2.73), while the right-hand side inequality follows from (2.58). □

Theorem 2.5.16. *For any $n \geqslant 2$ we have that*

$$\psi_n \leqslant C\sqrt{\log n},$$

where $C > 0$ is an absolute constant.

Proof. We may choose an isotropic log-concave probability measure μ on $\mathbb{R}^n$ such that

$$\psi_\mu \geqslant \frac{\psi_n}{2}. \tag{2.74}$$

From (2.65) we know that, for $t = c\psi_n^{-2}/\log n$

$$\mathbb{E}(\sqrt{\|A_t\|_{\mathrm{op}}}) \leqslant \sqrt{\mathbb{E}\|A_t\|_{\mathrm{op}}} \leqslant C_1, \tag{2.75}$$

where $C_1 > 0$ is an absolute constant. By the definition of t and the facts that $\psi_n \approx \psi_\mu \approx \vartheta_\mu$ and $\lambda_0 = \vartheta_\mu^{-2}$ we get

$$t \leqslant c_1\psi_n^{-2} \leqslant c_2\psi_\mu^{-2} \leqslant c_3\lambda_0. \tag{2.76}$$

Then, we can apply Proposition 2.5.15 with $\alpha = c_3$. Combining (2.70) with (2.75) and the choice of t gives

$$\lambda_0^{-1} \leqslant C_2\frac{\mathbb{E}(\sqrt{\|A_t\|_{\mathrm{op}}})}{\sqrt{t}} \leqslant \frac{C_3}{\sqrt{t}} \leqslant C_4\psi_n\sqrt{\log n}. \tag{2.77}$$

Then, from (2.74), (2.77), and the equivalence $\lambda_0^{-1} = \vartheta_\mu^2 \approx \psi_\mu^2$ we get

$$\psi_n^2 \leqslant 4\psi_\mu^2 \leqslant C_5\lambda_0^{-1} \leqslant C_6\psi_n\sqrt{\log n},$$

which shows that $\psi_n \leqslant C\sqrt{\log n}$ for an absolute constant $C > 0$. □

2.6 The slicing theorem

In this section, we provide a proof of the hyperplane conjecture.

Theorem 2.6.1. *There exists an absolute constant $C > 0$ such that $L_n \leqslant C$ for all $n \geqslant 1$.*

We shall present the proof of P. Bizeul [76] that makes use of the reduction to small ball estimates from Section 2.4. In what follows, we say that a random vector X in $\mathbb{R}^n$ is b-subgaussian if for every $p \geqslant 1$ and any $\xi \in S^{n-1}$

$$\left(\mathbb{E}\left|\langle X - \mathbb{E}(X), \xi\rangle\right|^p\right)^{1/p} \leqslant b\sqrt{p}.$$

The small ball estimate of Theorem 2.4.12 can be also stated for a not necessarily isotropic log-concave random vector. Then, it takes the following form (see Paouris [453]).

Theorem 2.6.2. *Let X be a b-subgaussian log-concave random vector in $\mathbb{R}^n$ with covariance matrix A. For any $0 < \varepsilon \leqslant c_0$ and any $y \in \mathbb{R}^n$*

$$\mathbb{P}(|X - y|^2 \leqslant \varepsilon\,\mathrm{tr}(A)) \leqslant \varepsilon^{\frac{c_0\,\mathrm{tr}(A)}{b^2\|A\|_{\mathrm{op}}\|A^{-1}\|_{\mathrm{op}}}},$$

where $c_0 > 0$ is an absolute constant.

One can check that the theorem is tight, up to absolute constants, by considering the case where X is a standard gaussian random vector. However, for general isotropic log-concave random vectors, it provides a suboptimal exponent of order $\sqrt{n}$. Lee and Vempala [364] obtained an improved estimate using stochastic localization.

Theorem 2.6.3. *Let X be an isotropic log-concave random vector in $\mathbb{R}^n$. For any $y \in \mathbb{R}^n$ and any $0 < \varepsilon \leqslant c_0$*

$$\mathbb{P}(|X - y|^2 \leqslant \varepsilon n) \leqslant \varepsilon^{\frac{c_0 n}{\psi_n^2\log(n)}},$$

where $c_0 > 0$ is an absolute constant.

In Section 2.5, we discussed the best known bound for the KLS constant, due to Klartag [290]: we saw that

$$\psi_n^2 \leqslant C\log n$$

for some absolute constant $C > 0$. Inserting this estimate into Theorem 2.6.3 we get an exponent that is of the order of n up to a factor $(\log n)^2$. However, even if we knew that the KLS conjecture has an affirmative answer, the estimate of Lee and Vempala would still include a suboptimal factor $\log n$.

Following Bizeul [76] we shall see how one can remove the extra logarithmic factor that appears in Theorem 2.6.3 and establish an optimal small ball estimate.

Theorem 2.6.4. *Let X be an isotropic log-concave random vector in $\mathbb{R}^n$. For any $0 < \varepsilon \leqslant c_0$ and any $y \in \mathbb{R}^n$*

$$\mathbb{P}(|X-y|^2 \leqslant \varepsilon n) \leqslant \varepsilon^{c_0 n},$$

where $c_0 > 0$ is an absolute constant.

Then, we will show that Theorem 2.6.4 implies the isotropic constant conjecture. In Section 2.4, we described in detail the work of Dafnis and Paouris and the reduction of the hyperplane conjecture to the optimal small ball estimate. Theorem 2.6.4 is actually equivalent to Theorem 2.6.1. For the inverse direction note that if L_n is bounded above by some constant $C_1 > 0$, then we have seen that the same holds true for $\tilde{L}_n$. This means that for any isotropic log-concave random vector X with density f and any $\varepsilon > 0$ and $y \in \mathbb{R}^n$, we have that

$$\mathbb{P}(|X-y|^2 \leqslant \varepsilon n) = \int_{B(y,\sqrt{\varepsilon n})} f(x)\,dx \leqslant C_1^n |\sqrt{\varepsilon n} B_2^n| \leqslant (C_2\varepsilon)^{n/2},$$

which implies Theorem 2.6.4.

We start with some preliminary facts that will be used in the proof. The first one is a variant of Theorem 2.6.2.

Lemma 2.6.5. *Let X be a b-subgaussian log-concave random vector with covariance matrix A. Then, for any $y \in \mathbb{R}^n$ and any $0 < \varepsilon \leqslant c_0$ we have that*

$$\mathbb{P}(|X-y|^2 \leqslant \varepsilon\,\mathrm{tr}(A)) \leqslant \left(\frac{4n\|A\|_{\mathrm{op}}}{\mathrm{tr}(A)}\varepsilon\right)^{\frac{c_0\,\mathrm{tr}(A)^3}{8n^2b^2\|A\|_{\mathrm{op}}^2}},$$

where $c_0 > 0$ is the constant from Theorem 2.6.2.

Proof. Let $\lambda_1 \geqslant \cdots \geqslant \lambda_n$ be the eigenvalues of the matrix A. For any $2 \leqslant k \leqslant n$ we have that

$$n\lambda_k \geqslant \sum_{i=k}^{n} \lambda_i = \mathrm{tr}(A) - \sum_{i=1}^{k-1} \lambda_i \geqslant \mathrm{tr}(A) - k\lambda_1. \tag{2.78}$$

We choose

$$k = \max\left\{\left\lfloor \frac{\mathrm{tr}(A)}{2\lambda_1} \right\rfloor, 1\right\}.$$

If $k \geqslant 2$, then (2.78) implies that

$$\lambda_k \geqslant \frac{\mathrm{tr}(A)}{2n},$$

which remains true when $k = 1$. Let E be the k-dimensional subspace associated to the eigenvalues $\lambda_1, \dots, \lambda_k$, and let $X_E = P_E X$ be the projection of X onto E, and $A_E = P_E A P_E$ the covariance matrix of X_E. Then, X_E is b-subgaussian and its covariance matrix satisfies

$$\|A_E\|_{\mathrm{op}} \leqslant \|A\|_{\mathrm{op}}, \quad \|A_E^{-1}\|_{\mathrm{op}} \leqslant \frac{2n}{\mathrm{tr}(A)}, \quad \mathrm{tr}(A_E) \geqslant \frac{k}{n}\mathrm{tr}(A) \geqslant \frac{\mathrm{tr}(A)^2}{4n\|A\|_{\mathrm{op}}}.$$

Let $y \in \mathbb{R}^n$ and define $y_E = P_E y$. If $0 < \varepsilon \leqslant c_0$, then

$$\begin{aligned}
\mathbb{P}(|X-y|^2 \leqslant \varepsilon\,\mathrm{tr}(A)) &\leqslant \mathbb{P}(|X_E - y_E|^2 \leqslant \varepsilon \mathrm{tr}(A)) \\
&\leqslant \left(\frac{\varepsilon\,\mathrm{tr}(A)}{\mathrm{tr}(A_E)}\right)^{\frac{c_0 \mathrm{tr}(A_E)}{b^2\|A_E\|_{\mathrm{op}}\|A_E^{-1}\|_{\mathrm{op}}}} \\
&\leqslant \left(\frac{4n\varepsilon\|A\|_{\mathrm{op}}}{\mathrm{tr}(A)}\right)^{\frac{c_0 \mathrm{tr}(A)^3}{8b^2n^2\|A\|_{\mathrm{op}}^2}},
\end{aligned}$$

where the second inequality follows by applying Theorem 2.6.2 to X_E. □

We shall also use some classical facts about log-concave vectors. The next lemma is a special case of (2.12).

Lemma 2.6.6. *There exists an absolute constant $\kappa_0 > 1$ such that for any log-concave real random variable Y and any $p \geqslant 2$*

$$(\mathbb{E}|Y|^p)^{1/p} \leqslant \kappa_0 p\, \mathbb{E}(|Y|^2)^{1/2}.$$

Recall also that a random vector, or its density, is called t-uniformly log-concave for some $t > 0$ if it is log-concave with respect to a gaussian of variance $\frac{1}{t}$. The next lemma follows from a log-concave/convex correlation inequality of Hargé [252].

Lemma 2.6.7. *Let X be a t-uniformly log-concave random vector. Then, it is $\frac{1}{\sqrt{t}}$-subgaussian.*

Proof. Without loss of generality we may assume that $t = 1$ so that X is log-concave with respect to the standard gaussian measure denoted by γ_n. We may also assume that $\mathbb{E}(X) = 0$. Hargé's result states that if f is log-concave and g is convex, then

$$\mathrm{Cov}_{\gamma_n}(f, g) \leqslant 0. \tag{2.79}$$

Let f be the relative density of X with respect to γ_n. Applying (2.79) with $g(x) = |\langle x, \xi\rangle|^p$ for all $p \geqslant 1$ and all $\xi \in S^{n-1}$ we obtain the result. □

2.6.1 Reduction to small diameter

A first observation, which has been also used in previous approaches to the problem, is that in order to prove Theorem 2.6.4 we may assume that X is supported in a ball of radius of the order of $\sqrt{n}$ and that $y = 0$. In this subsection we explain this reduction.

Lemma 2.6.8. *Let X be an isotropic log-concave random vector in $\mathbb{R}^n$. There exists an isotropic log-concave random vector Y that has support inside the ball $B(0, \sqrt{c_1 n})$, where $c_1 > 0$ is an absolute constant, and satisfies*

$$\mathbb{P}(|X - y| \leqslant \varepsilon\sqrt{n})^2 \leqslant \mathbb{P}(|Y| \leqslant 2\varepsilon\sqrt{n})$$

for every $0 < \varepsilon < 1$ and any $y \in \mathbb{R}^n$.

Proof. We consider the symmetrized version $X_1 = \frac{X-\tilde{X}}{\sqrt{2}}$ of X, where $\tilde{X}$ is an independent copy of X. Then, X_1 is an isotropic symmetric log-concave random vector with the property that, for any $r > 0$

$$\begin{aligned}\mathbb{P}(|X - y| \leqslant r) &= [\mathbb{P}(|X - y| \leqslant r,\ |\tilde{X} - y| \leqslant r)]^{1/2} \\ &\leqslant \left[\mathbb{P}\left(\left|\frac{1}{\sqrt{2}}(X - \tilde{X})\right| \leqslant \sqrt{2}r\right)\right]^{1/2} \\ &= [\mathbb{P}(|X_1| \leqslant \sqrt{2}r)]^{1/2}.\end{aligned} \tag{2.80}$$

We set $K = B(0, 8\kappa_0^2\sqrt{n}) = \{x \in \mathbb{R}^n : |x| \leqslant 8\kappa_0^2\sqrt{n}\}$, where $\kappa_0 \geqslant 1$ is the constant from Lemma 2.6.6, and define the random vector

$$X_2 = X_1 \cdot \mathbb{1}_{X_1 \in K}.$$

If f_1 is the density of X_1, then X_2 has density

$$f_2 = \frac{f_1 \mathbb{1}_K}{\int_K f_1} \leqslant 2 f_1 \mathbb{1}_K, \tag{2.81}$$

where the inequality above holds because $\int_K f_1 \geqslant \frac{1}{2}$. To see this, we first use the fact that X_1 is isotropic and Markov's inequality to write

$$\mathbb{P}(X_1 \in K^c) = \mathbb{P}(|X_1|^2 > 64\kappa_0^4 n) \leqslant \frac{1}{64\kappa_0^4 n}\mathbb{E}|X_1^2| = \frac{1}{64\kappa_0^4} < \frac{1}{2} \tag{2.82}$$

and this implies that

$$\int_K f_1 = 1 - \mathbb{P}(X_1 \in K^c) \geqslant 1 - \frac{1}{2} = \frac{1}{2}.$$

We know that X_2 is symmetric, and hence, from (2.81) we get

$$\mathrm{Cov}(X_2) \leqslant 2I_n.$$

In fact, we claim that

$$\frac{1}{2}I_n \leqslant \mathrm{Cov}(X_2) \leqslant 2I_n. \tag{2.83}$$

To see the left-hand side inequality, for any $\xi \in S^{n-1}$ we write

$$\begin{aligned}
\int_{\mathbb{R}^n} \langle x, \xi\rangle^2 f_2(x)\,dx &\geqslant \int_K \langle x, \xi\rangle^2 f_1(x)\,dx \\
&= 1 - \int_{K^c} \langle x, \xi\rangle^2 f_1(x)\,dx \\
&\geqslant 1 - \mathbb{P}(X_1 \in K^c)^{1/2}\Big(\int_{\mathbb{R}^n} \langle x, \xi\rangle^4 f_1(x)\,dx\Big)^{1/2} \\
&\geqslant 1 - \frac{1}{8\kappa_0^2}(2\kappa_0)^2 \int_{\mathbb{R}^n} \langle x, \xi\rangle^2 f_1(x)\,dx = \frac{1}{2},
\end{aligned}$$

where we have used the Cauchy–Schwarz inequality, Lemma 2.6.6, and (2.82). Finally, we set

$$X_3 = \mathrm{Cov}(X_2)^{-1/2}X_2.$$

Note that X_3 is an isotropic log-concave random vector by construction. Furthermore, using (2.83), we see that it has support inside the ball $B(0, 8\sqrt{2}\kappa_0^2\sqrt{n})$. For any $0 \leqslant \varepsilon < 1$, setting $r = \varepsilon\sqrt{n}$ we get

$$\begin{aligned}
\mathbb{P}(X \in B(y, r))^2 &\leqslant \mathbb{P}(X_1 \in B(0, \sqrt{2}r)) \\
&\leqslant \mathbb{P}(X_2 \in B(0, \sqrt{2}r)) \\
&= \mathbb{P}(X_3 \in \mathrm{Cov}(X_2)^{-1/2}B(0, \sqrt{2}r)) \\
&\leqslant \mathbb{P}(X_3 \in B(0, 2r)),
\end{aligned} \tag{2.84}$$

where we used successively (2.80), the fact that $B(0, \sqrt{2}r) \subset K = B(0, 8\kappa_0^2\sqrt{n})$, and in the last line the inequality (2.83). □

It is now clear from (2.84) that in order to prove Theorem 2.6.4 we may assume X has support inside a ball of diameter $c_1\sqrt{n}$, where $c_1 > 0$ is an absolute constant.

2.6.2 Bounds for the shrinkage of sets

The main idea for the proof of Theorem 2.6.4 is again to use stochastic localization as in the previous section. We recall some basic facts from Section 2.5.3. If μ is a log-concave probability measure with density ϱ with respect to the Lebesgue measure, for $t \in \mathbb{R}$ and $\xi \in \mathbb{R}^n$ we define a density

$$\varrho_{t,\xi}(x) = \frac{1}{Z_{t,\xi}} e^{\langle x,\xi\rangle - \frac{t}{2}|x|^2} \varrho(x), \quad x \in \mathbb{R}^n, \tag{2.85}$$

where $Z_{t,\xi}$ is a normalizing factor. Note that $\varrho_{t,\xi}$ is t-uniformly log-concave. We also define

$$a_{t,\xi} = \int_{\mathbb{R}^n} x\varrho_{t,\xi}(x)\,dx$$

and

$$A_{t,\xi} = \int_{\mathbb{R}^n} (x - a_{t,\xi})^{\otimes 2} \varrho_{t,\xi}(x)\,dx,$$

the barycenter and covariance matrix of $\varrho_{t,\xi}$, respectively. The tilt process $(\xi_t)_{t\geqslant 0}$ is defined as the solution of the stochastic differential equation

$$d\xi_t = a_{t,\xi_t}\,dt + dW_t, \tag{2.86}$$

where $(W_t)_{t\geqslant 0}$ is a standard Brownian motion. The process ϱ_{t,ξ_t} is the stochastic localization process starting at ϱ. We abbreviate $\varrho_t = \varrho_{t,\xi_t}$ and similarly for the barycenter and covariance, a_t and A_t. We also define μ_t to be the random measure with density ϱ_t. For any $x \in \mathbb{R}^n$, ϱ_t is an Itô process satisfying the stochastic differential equation

$$d\varrho_t(x) = \langle x - a_t, dW_t\rangle \varrho_t(x). \tag{2.87}$$

Integrating (2.87) we see that the process $(\int_{\mathbb{R}^n} \varphi\,d\mu_t)_{t\geqslant 0}$ is a martingale for any integrable function φ. In particular, for any $t \geqslant 0$

$$\mathbb{E}_\mu(\varphi) = \mathbb{E}(\mathbb{E}_{\mu_t}(\varphi)). \tag{2.88}$$

Therefore, for any $t \geqslant 0$ we have decomposed the original log-concave probability measure μ into a mixture of measures μ_t that are t-uniformly log-concave.

Our goal in this subsection is to obtain estimates for the evolution of the measure of sets along the stochastic localization process under the additional assumption that μ has bounded support of diameter D.

Proposition 2.6.9. *Let μ be an isotropic log-concave probability measure with bounded support of diameter D. For any measurable set S and any $\lambda > 1$ we have that*

$$\mu(S) \leqslant e^{\frac{D^2 t}{2}} \mu_t(S)^{\frac{1}{\lambda}},$$

with probability at least $1 - \frac{1}{\lambda}$, where $(\mu_t)_{t \geqslant 0}$ is the stochastic localization process starting at μ.

Proof. Let $S \subset \mathbb{R}^n$ be a set of positive measure and let $g_t = \mu_t(S)$. Using (2.87), we can check that

$$dg_t = \left\langle \int_S (x - a_t)\, d\mu_t(x), dW_t \right\rangle.$$

Then,

$$d[g]_t = \left| \int_S (x - a_t)\, d\mu_t(x) \right|^2 dt.$$

Note that μ_t is supported in the same set as μ, which has diameter D, and hence

$$\left| \int_S (x - a_t)\, d\mu_t(x) \right|^2 \leqslant \sup_{x \in \mathrm{supp}(\mu_t)} |x - a_t|^2 \left(\int_S d\mu_t(x) \right)^2 \leqslant D^2 \mu_t(S)^2 = D^2 g_t^2.$$

It follows that

$$d[g]_t \leqslant D^2 g_t^2\, dt. \tag{2.89}$$

Using Itô's formula (2.63) we compute $d(\log(g_t^{-1})) = -\frac{dg_t}{g_t} + \frac{1}{2}\frac{d[g]_t}{g_t^2}$, which implies that $d\mathbb{E}(\log(g_t^{-1})) \leqslant \frac{D^2 t}{2}$, and hence

$$\mathbb{E}(\log(g_t^{-1})) \leqslant \log(g_0^{-1}) + \frac{D^2 t}{2}. \tag{2.90}$$

Now, let $\lambda > 1$. From (2.90), using Markov's inequality, we see that with probability at least $1 - \frac{1}{\lambda}$

$$\log(g_t^{-1}) \leqslant \lambda \left(\log(g_0^{-1}) + \frac{D^2 t}{2} \right).$$

Equivalently,

$$\log g_0 \leqslant \frac{1}{\lambda} \log g_t + \frac{D^2 t}{2}$$

and the result follows. □

2.6.3 Proof of the slicing theorem

By the log-concave Lichnerowicz inequality (2.57) we have that $A_t \leqslant \frac{1}{t} I_n$ almost surely, in the sense of symmetric matrices. Using Klartag's improved log-concave Lichnerowicz inequality of Theorem 2.5.10, Guan [243] obtained a lower bound on the trace of A_t up to a time of order 1.

Theorem 2.6.10. *Assume that the starting measure μ is an isotropic log-concave probability measure. Then,*

$$\mathbb{E}(\mathrm{tr}(A_{\delta_1})) \geqslant \delta_1 n,$$

where $\delta_1 > 0$ is an absolute constant.

Theorem 2.6.10 plays a crucial role in the proof of the optimal small ball estimate of Theorem 2.6.4.

Proof of Theorem 2.6.4. Lemma 2.6.8 shows that it is enough to consider an isotropic log-concave random vector X distributed according to an isotropic log-concave probability measure μ with support of diameter $c_1\sqrt{n}$, where $c_1 > 0$ is an absolute constant. Let $(\mu_t)_{t\geqslant 0}$ be the stochastic localization process starting at μ and let X_t be the random vector with law μ_t. Let δ_1 be the constant in Guan's Theorem 2.6.10. We consider the event

$$E_0 = \{\mathrm{tr}(A_{\delta_1}) \geqslant \delta_1 n/2\}. \tag{2.91}$$

Lemma 2.6.11. *We have $\mathbb{P}(E_0) \geqslant \delta_1^2/2$.*

Proof. We set $p_0 = \mathbb{P}(E_0)$. By the log-concave Lichnerowicz inequality we know that $A_{\delta_1} \leqslant \frac{1}{\delta_1} I_n$ almost surely, and hence

$$\delta_1 n \leqslant \mathbb{E}\left(\mathrm{tr}(A_{\delta_1})\right) = \mathbb{E}\left(\mathrm{tr}(A_{\delta_1})\mathbb{1}_{E_0}\right) + \mathbb{E}\left(\mathrm{tr}(A_{\delta_1})\mathbb{1}_{E_0^c}\right) \leqslant \frac{np_0}{\delta_1} + \frac{\delta_1 n}{2},$$

which implies that $p_0 \geqslant \delta_1^2/2$. □

Let $\bar{c}_0 > 0$ be the constant from Lemma 2.6.5. For every $0 < \varepsilon < \bar{c}_0$ we set $S_\varepsilon = B(0, \sqrt{\varepsilon n})$. We also fix $\lambda = \frac{4}{\delta_1^2}$. Proposition 2.6.9 shows that the probability p_1 of the event

$$E_1 = \{\mu(S_\varepsilon) \leqslant e^{\frac{c_1^2\delta_1 n}{2}} \mu_{\delta_1}(S_\varepsilon)^{\frac{1}{\lambda}}\}$$

satisfies $p_1 \geqslant 1 - \frac{\delta_1^2}{4}$, therefore

$$\mathbb{P}(E_0 \cap E_1) \geqslant p_0 + p_1 - 1 \geqslant \frac{\delta_1^2}{4} > 0.$$

In the event $E_0 \cap E_1$ we have $\mathrm{tr}(A_{\delta_1}) \geqslant \delta_1 n/2$, and also $\|A_{\delta_1}\|_{\mathrm{op}} \leqslant 1/\delta_1$. Moreover, X_{δ_1} is $1/\sqrt{\delta_1}$-subgaussian by Lemma 2.6.7. Inserting all these estimates into Lemma 2.6.5 we get that

$$\mu(S_\varepsilon) \leqslant e^{\frac{c_1^2\delta_1 n}{2}} \mu_{\delta_1}(S_\varepsilon)^{\frac{4}{\delta_1^2}} \leqslant e^{\frac{c_1^2\delta_1 n}{2}} \left(\frac{16\varepsilon}{\delta_1^3}\right)^{\frac{\bar{c}_0\delta_1^8 n}{256}} \leqslant \varepsilon^{\frac{\bar{c}_0\delta_1^8 n}{512}},$$

if we assume that

$$\varepsilon < \varepsilon_0 := \min\left\{\frac{\bar{c}_0\delta_1}{2}, \frac{\delta_1^6}{256}\exp\left(-\frac{256c_1^2}{\bar{c}_0\delta_1^7}\right)\right\}.$$

This establishes Theorem 2.6.4 with $c_0 = \min\{\frac{\bar{c}_0\delta_1}{2}, \frac{\delta_1^6}{256}\exp(-\frac{256c_1^2}{\bar{c}_0\delta_1^7}), \frac{\bar{c}_0\delta_1^8}{512}\}$. □

Proof of Theorem 2.6.1. Basically, we repeat the proof of Theorem 2.4.19. From Proposition 2.4.18 we know that there exists an isotropic convex body K in $\mathbb{R}^n$ with $L_n \leqslant C_1 L_K$ and $|K \cap \sqrt{C_2 n} B_2^n| > c_2^n$ for some absolute constants $C_1, C_2, c_2 > 0$.

Theorem 2.6.4 shows that there exists an absolute constant $c_0 > 0$ such that for any $0 < \varepsilon \leqslant c_0$

$$|K \cap \varepsilon\sqrt{n} L_K B_2^n| \leqslant \varepsilon^{c_0 n}.$$

Choosing $\varepsilon_1 = \min\{c_0, c_2^{\frac{1}{c_0}}\}$, and comparing the above inequalities, we get $L_K \leqslant \sqrt{C_2}/\varepsilon_1$, and hence

$$L_n^2 \leqslant C_1^2 L_K^2 \leqslant \frac{C_1^2 C_2}{\varepsilon_1^2}.$$

This shows that L_n is bounded by an absolute constant. □

2.7 Further reading

2.7.1 The thin-shell conjecture

The thin-shell conjecture has its origin in the central limit problem, the question to identify those high-dimensional distributions that have approximately gaussian marginals. Assume that μ is an isotropic Borel probability measure on $\mathbb{R}^n$, i. e., normalized so that

$$\mathbb{E}_\mu(x_j) = 0 \quad \text{and} \quad \mathbb{E}_\mu(x_i x_j) = \delta_{ij}, \quad i,j = 1,\dots,n.$$

Sudakov [518] was probably the first to observe that if μ satisfies a "thin-shell bound", then the central limit problem has an affirmative answer for μ. More precisely, if

$$\mu\left(\{x \in \mathbb{R}^n : \left||x| - \sqrt{n}\right| \geqslant \varepsilon\sqrt{n}\}\right) \leqslant \varepsilon$$

for some $\varepsilon \in (0,1/2)$, then, for all directions ξ in a subset A of S^{n-1} of measure $\sigma(A) \geqslant 1 - \exp(-c_1\sqrt{n})$, we have

$$\left|\mathbb{P}\left(\langle X,\xi\rangle \leqslant t\right) - \Phi(t)\right| \leqslant c_2(\varepsilon + n^{-\alpha}) \quad \text{for all } t \in \mathbb{R},$$

where $\Phi(t)$ is the standard gaussian distribution function and $c_1, c_2, \alpha > 0$ are absolute constants. Other early related articles are the ones by Diaconis and Freedman [149], and by von Weizsäker [536]. Anttila, Ball, and Perissinaki [20] studied the case where $\mu = \mu_K$ is the measure with density $L_K^n \mathbb{1}_{\frac{K}{L_K}}$ for an isotropic origin-symmetric convex body K in $\mathbb{R}^n$ and showed that a thin-shell estimate implies an affirmative answer to the central limit problem for μ_K. This work made the central limit problem widely known among people working in convex geometry. A clear exposition of both the general and the log-concave case can be found in Bobkov's article [80].

Klartag noted in [289] that there exist isotropic distributions for which a thin-shell estimate is not possible. For example, if we write σ_t for the uniform probability measure on the sphere tS^{n-1}, then, for $t_1 = \sqrt{n}/2$ and $t_2 = \sqrt{7n}/2$, the isotropic probability measure $\mu = \frac{1}{2}(\sigma_{t_1} + \sigma_{t_2})$ does not have gaussian marginals, and hence it does not satisfy a thin-shell bound. However, it was conjectured that the assumption of log-concavity guarantees a thin-shell bound. The authors of [83] defined the parameter

$$\sigma_\mu^2 = \frac{1}{n}\mathrm{Var}_\mu(|x|^2)$$

and asked if

$$\sigma_n = \sup\{\sigma_\mu : \mu \text{ is an isotropic log-concave probability measure on } \mathbb{R}^n\} \leqslant C$$

for some absolute constant $C > 0$. It is easily checked that this is equivalent to the question if there exists an absolute constant $C > 0$ such that, for any $n \geqslant 1$ and any isotropic log-concave probability measure μ on $\mathbb{R}^n$, one has

$$\mathbb{E}_\mu(|x| - \sqrt{n})^2 \leqslant C^2.$$

Moreover, applying the Poincaré inequality for the function $f(x) = |x|^2$ one can check that

$$\sigma_n \leqslant C\psi_n$$

for all n, where $C > 0$ is an absolute constant. In other words, the KLS conjecture implies the thin-shell conjecture. An important fact is that the thin-shell conjecture is also related to the isotropic constant conjecture. Eldan and Klartag [158] proved that there exists an absolute constant $C > 0$ such that

$$L_n \leqslant C\sigma_n$$

for every $n \geqslant 1$.

The first non-trivial upper bound for σ_n was given by Klartag in [286] in his proof of the central limit theorem for convex bodies. He proved that $\sigma_n \leqslant C\sqrt{n}/\log n$. This estimate was then improved to $\sigma_n \leqslant Cn^{2/5+o(1)}$ by Klartag [287], to $\sigma_n \leqslant Cn^{3/8}$ by Fleury [175], to $\sigma_n \leqslant Cn^{1/3}$ by Gué don and E. Milman [244], and to $\sigma_n \leqslant Cn^{1/4}$ by Lee and Vempala [363]. A consequence of the more recent developments on the KLS conjecture is that $\sigma_n \leqslant C\sqrt{\log n}$ by the corresponding bound $\psi_n = O(\sqrt{\log n})$ of Klartag in [290]. An even more recent breakthrough of Guan [243] showed that

$$\sigma_n \leqslant C \log\log n.$$

Finally, Klartag and Lehec [296] announced an affirmative answer to the thin-shell conjecture.

Theorem 2.7.1. *There exists an absolute constant $C > 0$ such that*

$$\mathrm{Var}_\mu(|x|^2) \leqslant Cn$$

for every $n \geqslant 1$ and every isotropic log-concave probability measure μ on $\mathbb{R}^n$. Equivalently, $\sup_n \sigma_n \leqslant C$.

Using reverse Hölder inequalities for polynomials of a random vector distributed uniformly in a convex body (see [102, 440]) one can deduce from Theorem 2.7.1 that, for any isotropic log-concave probability measure μ on $\mathbb{R}^n$ and for any $t > 0$

$$\mu(\{x \in \mathbb{R}^n : |\,|x| - \sqrt{n}\,| \geqslant t\}) \leqslant c_1 \exp(-c_2\sqrt{t}).$$

The proof of Theorem 2.7.1 starts from ideas that Klartag used in [288], where he established an optimal thin-shell bound for unconditional log-concave measures. This approach was extended by Barthe and Cordero-Erausquin in [61] who adapted Klartag's techniques to provide spectral gap estimates for log-concave measures with many symmetries, and by Barthe and Klartag in [62]. Given a log-concave probability measure μ on $\mathbb{R}^n$, let $H^1(\mu)$ be the space of all functions $f \in L^2(\mu)$ with weak partial derivatives in $L^2(\mu)$, equipped with the norm

$$\|f\|^2_{H^1(\mu)} = \left(\int_{\mathbb{R}^n} |f|^2\, d\mu + \int_{\mathbb{R}^n} |\nabla f|^2\, d\mu\right)^{1/2}.$$

Barthe and Klartag showed in [62] that $C_c^\infty(\mathbb{R}^n)$ is dense in $H^1(\mu)$. Now, for every $f \in L^2(\mu)$ with $\int f\, d\mu = 0$, define

$$\|f\|_{H^{-1}(\mu)} = \sup\Big\{\int_{\mathbb{R}^n} fg\,d\mu : g \in H^1(\mu),\ \int_{\mathbb{R}^n} |\nabla g|^2\,d\mu \leqslant 1\Big\}$$
$$= \sup\Big\{\int_{\mathbb{R}^n} fg\,d\mu : g \in C_c^\infty(\mathbb{R}^n),\ \int_{\mathbb{R}^n} |\nabla g|^2\,d\mu \leqslant 1\Big\}.$$

In [62] and [288] it is shown, by using Bochner's formula, that

$$\|f\|_{L^2(\mu)}^2 \leqslant \|\nabla f\|_{H^{-1}(\mu)}^2 := \sum_{i=1}^n \|\partial_i f\|_{H^{-1}(\mu)}^2 \tag{2.92}$$

for every smooth function $f \in H^1(\mu)$ that satisfies $\int f\,d\mu = 0$ and $\int \nabla f\,d\mu = 0$. Applying (2.92) for an isotropic log-concave probability measure μ and the function $f(x) = |x|^2 - n$, we see that

$$\mathrm{Var}_\mu(|x|^2) \leqslant 4\sum_{i=1}^n \|x_i\|_{H^{-1}(\mu)}^2.$$

Klartag and Lehec prove the next inequality, which immediately implies Theorem 2.7.1.

Theorem 2.7.2. *There exists an absolute constant $C > 0$ such that*

$$\sum_{i=1}^n \|x_i\|_{H^{-1}(\mu)}^2 \leqslant Cn$$

for every $n \geqslant 1$ and every isotropic log-concave probability measure μ on $\mathbb{R}^n$.

The proof is based again on stochastic localization. Klartag and Lehec consider the family of exponential tilts of log-affine perturbations of the original measure μ and construct suitable couplings between these tilts. Then, they use these couplings to bound the $H^{-1}(\mu)$-norm by the growth of the covariance process $(A_t)_{t\geqslant 0}$ using a variant of Guan's technique from [243].

2.7.2 The MM^*-estimate for isotropic convex bodies

In Section 1.2.4, we considered the parameters

$$M(K) = \int_{S^{n-1}} p_K(\xi)\,d\sigma(\xi) \quad \text{and} \quad M^*(K) = \int_{S^{n-1}} h_K(\xi)\,d\sigma(\xi),$$

where p_K is the Minkowski functional and h_K is the support function of a convex body K in $\mathbb{R}^n$ with $0 \in \mathrm{int}(K)$.

Recall that $\mathrm{vrad}(K) = (|K|/|B_2^n|)^{1/n}$ denotes the volume radius of K. We saw that

$$M(K)^{-1} \leqslant \mathrm{vrad}(K) \leqslant M^*(K) = M(K^\circ)$$

and, on the other hand, a fundamental fact in the other direction, following from results of Figiel–Tomczak-Jaegermann [169], Lewis [371], and Pisier [462] states that for any symmetric convex body K in $\mathbb{R}^n$ there exists $T \in GL_n$ such that

$$M(T(K))\,M^*(T(K)) \leqslant c\log n,$$

where $c > 0$ is an absolute constant. In the general case, without the symmetry assumption for K, one considers the parameter

$$E(K) = \inf M(T(K))\,M^*(T(K)),$$

where the infimum is taken over all invertible affine transformations of $\mathbb{R}^n$ for which $0 \in \operatorname{int}(T(K))$. The study of the question to obtain a sharp upper bound for $\max E(K)$ started with a work of Banaszczyk, Litvak, Pajor, and Szarek, who showed in [50] that if K is a convex body in $\mathbb{R}^n$ that is in John's position, then

$$M^*(K) \leqslant c\sqrt{n\log n}.$$

When K is in John's position, from the inclusion $K \supseteq B_2^n$ we also obtain the trivial upper bound $M(K) \leqslant M(B_2^n) = 1$, and this implies that $E(K) \leqslant c\sqrt{n\log n}$. This estimate was improved by Rudelson in [490]: he showed that if K is a convex body in $\mathbb{R}^n$, then

$$E(K) \leqslant c\sqrt[3]{n}(\log n)^b,$$

where $c, b > 0$ are absolute constants.

The question to obtain upper bounds for the parameter $M^*(K)$, when K is in the isotropic position, was essentially answered by E. Milman who proved in [417] that if K is an isotropic convex body in $\mathbb{R}^n$, then

$$M^*(K) \leqslant c_1\sqrt{n}(\log n)^2 L_K \leqslant c_2\sqrt{n}(\log n)^2,$$

where the second inequality takes into account the fact that $L_n \leqslant C$. Urysohn's inequality shows that the dependence on n is optimal up to the logarithmic term.

Some first non-trivial results about the dual problem, to estimate $M(K)$ in the isotropic position, were obtained in [217] and later in [210], where it was shown that $M(K) \leqslant C(n\log n)^{1/3}/\sqrt{n}$ (see also [511]). Very recently, Bizeul and Klartag [77] obtained the optimal (up to the $\log n$-term) upper bound

$$M(K) \leqslant \frac{C\log n}{\sqrt{n}}.$$

The main estimate behind this result is a comparison inequality between log-concave measures and gaussian measures, going back to an inequality due to Eldan and Lehec [159]. They considered the parameter

$$\tau_n^2 = \sup_{\mu} \sup_{\xi \in S^{n-1}} \sum_{i,j=1}^{n} \mathbb{E}_\mu \big(x_i x_j \langle x, \xi\rangle\big)^2,$$

where the first supremum is over all isotropic log-concave probability measures μ on $\mathbb{R}^n$, and proved that if μ is an isotropic log-concave probability measure on $\mathbb{R}^n$, then, for any symmetric convex body K in $\mathbb{R}^n$, we have that

$$\int_{\mathbb{R}^n} \|x\|_K \, d\mu(x) \leqslant c_1 \sqrt{\log n}\, \tau_n \int_{\mathbb{R}^n} \|x\|_K \, d\gamma_n(x),$$

where γ_n is the standard gaussian measure on $\mathbb{R}^n$ and $c_1 > 0$ is an absolute constant.

A fundamental result of Eldan [157] asserts that τ_n can be controlled in terms of the thin-shell constant: he proved that

$$\tau_n^2 \leqslant c_2 \sum_{k=1}^{n} \frac{\sigma_k^2}{k},$$

where $c_2 > 0$ is an absolute constant. This showed that progress on the thin-shell conjecture has direct implications for comparison inequalities between log-concave measures and gaussian measures.

Bizeul and Klartag observed in [77] that these comparison results extend naturally from norms to gauge functions, that is, non-negative, convex, 1-homogeneous functions on $\mathbb{R}^n$. Moreover, they obtained two-sided estimates in this broader setting.

Theorem 2.7.3. *Let X be an isotropic, log-concave random vector in $\mathbb{R}^n$ and let G be a standard gaussian random vector in $\mathbb{R}^n$. Then, for any gauge function $\|\cdot\|$ on $\mathbb{R}^n$*

$$\frac{c_1}{\psi_n \sqrt{\log n}} \mathbb{E}\|G\| \leqslant \mathbb{E}\|X\| \leqslant c_2 \psi_n \sqrt{\log n}\, \mathbb{E}\|G\|, \tag{2.93}$$

where $c_1, c_2 > 0$ are absolute constants.

Theorem 2.7.3 shows that any isotropic log-concave measure is sandwiched, in the gauge order, between two centered gaussian measures whose covariances differ by logarithmic factors. In particular, such a measure may be compared to gaussian measures with covariance matrices of order $(\log n)^2 I_n$ and $(\log n)^{-2} I_n$. The right-hand inequality in (2.93) is essentially due to Eldan and Lehec, while the left-hand inequality relies on new ideas and, once again, exploits Eldan's stochastic localization scheme.

Once Theorem 2.7.3 is established, the proof of the M-estimate follows readily. Applying the two-sided comparison (2.93) to the gauge function $x \mapsto p_K(x)$ and to the uniform measure on a convex body K in an isotropic position, we obtain

$$\sqrt{n} M(K) = \sqrt{n} \int_{S^{n-1}} p_K(x) \, d\sigma(x) \approx \mathbb{E}\,(p_K(G)) \leqslant c\psi_n \sqrt{\log n}\, \mathbb{E}\,(p_K(X)) \leqslant c\psi_n \sqrt{\log n},$$

since $p_K(x) \leqslant 1$ for all $x \in K$. Under the KLS conjecture, the logarithmic factors in this estimate would improve to their conjecturally optimal order.

2.7.3 Diameter of the Minkowski compactum

The definition of the Banach–Mazur distance can be extended to the setting of not necessarily symmetric convex bodies as follows. If K and D are convex bodies in $\mathbb{R}^n$, their Banach–Mazur distance is defined by

$$d_{\mathrm{BM}}(K,D) = \min\{t > 0 : T(K) + y \subseteq D + x \subseteq t(T(K) + y)\},$$

where the minimum is taken over all $x, y \in \mathbb{R}^n$ and all linear isomorphisms $T \in GL_n$.

A fundamental open problem for many years was to determine a sharp upper bound for the diameter

$$d_n = \max_{K,D} d_{\mathrm{BM}}(K,D)$$

of the Minkowski compactum with respect to the Banach–Mazur distance. This question was finally resolved in the recent work [77] of Bizeul and Klartag.

The connection between the diameter problem and MM^*-estimates has been known for a long time. Recall that for any convex body $K \subset \mathbb{R}^n$ we define

$$E(K) := \min M(TK)M^*(TK),$$

where the minimum is taken over all invertible affine transformations T of $\mathbb{R}^n$ such that $0 \in \mathrm{int}(TK)$. The following theorem reduces the problem of bounding the diameter of the Minkowski compactum to that of minimizing the product MM^* in the non-symmetric setting.

Theorem 2.7.4. *Let K and D be two convex bodies in $\mathbb{R}^n$. Then,*

$$d_{\mathrm{BM}}(K,D) \leqslant cn\sqrt{\log n}(E(K) + E(D)),$$

where $c > 0$ is an absolute constant.

A proof of Theorem 2.7.4 may be found in [28, Theorem 6.3.1]. It relies on the method of random orthogonal factorizations, introduced by Benyamini and Gordon [70].

In view of Theorem 2.7.4, obtaining sharp upper bounds for d_n reduces to finding a position of a convex body K for which the quantity $M(K)M^*(K)$ is as small as possible. Recall that in the centrally symmetric case one always has

$$E(K) := \min_{T \in GL_n} M(T(K))M^*(T(K)) \leqslant C \log n, \tag{2.94}$$

where $C > 0$ is an absolute constant. A position $T(K)$ satisfying (2.94) is called an *ℓ-position* of K. The almost optimal M^*- and M-estimates obtained by E. Milman and by Bizeul–Klartag show that the isotropic position already provides a canonical ℓ-position for arbitrary convex bodies. Indeed, if K is an isotropic convex body in $\mathbb{R}^n$, then

$$M(K)M^*(K) \leqslant c(\log n)^3$$

for an absolute constant. Combining this estimate with Theorem 2.7.4 yields the following essentially optimal bound for the diameter of the Minkowski compactum.

Theorem 2.7.5. *Let K and D be two convex bodies in $\mathbb{R}^n$. Then,*

$$d_{\mathrm{BM}}(K,D) \leqslant cn(\log n)^{7/2},$$

where $c > 0$ is an absolute constant. In particular, $d_n = O(n(\log n)^{7/2})$.

Theorem 2.7.5 reveals the striking fact that the diameters of the symmetric and nonsymmetric Banach–Mazur compacta are of essentially the same order.

2.7.4 Isotropic position is a regular *M*-position

Another important consequence of the MM^*-estimate for isotropic convex bodies is an almost optimal nonsymmetric analogue of Pisier's Theorem 1.2.7 on the existence of α-regular M-positions for centrally symmetric convex bodies.

Tziotziou observed in [527] that the M^*-estimate of E. Milman, together with the M-estimate of Bizeul and Klartag, implies that every isotropic convex body $K \subset \mathbb{R}^n$ is (almost) in a 2-regular M-position. The proof relies crucially on the Sudakov and dual Sudakov inequalities (see Section 1.2.5). These inequalities assert that if K is a symmetric convex body in $\mathbb{R}^n$, then for every $t > 0$

$$N(K, tB_2^n) \leqslant \exp\big(cnM^*(K)^2/t^2\big) \quad \text{and} \quad N(B_2^n, tK) \leqslant \exp\big(cnM(K)^2/t^2\big),$$

where $c > 0$ is an absolute constant. In the next theorem we set $r_n = \omega_n^{-1/n}$; recall that $r_n \approx \sqrt{n}$.

Theorem 2.7.6. *Let K be an isotropic convex body in $\mathbb{R}^n$. Then, for every $t > 0$*

$$\max\{N(K, tr_nB_2^n),\ N(B_2^n, tr_nK^\circ)\} \leqslant \exp\left(\frac{\gamma_n^2 n}{t^2}\right), \tag{2.95}$$

$$\max\{N(r_nB_2^n, tK),\ N(r_nK^\circ, tB_2^n)\} \leqslant \exp\left(\frac{\delta_n^2 n}{t^2}\right), \tag{2.96}$$

where $\gamma_n \leqslant c_1(\log n)^2$ and $\delta_n \leqslant c_2 \log n$.

Some first geometric applications of Theorem 2.7.6 were obtained in [527], including new inequalities comparing the volumes of sections and projections of a convex body whose barycenter or Santaló point lies at the origin with those of its inner and outer regularizations

$$K_{\mathrm{out}} = \mathrm{conv}(K, -K) \quad \text{and} \quad K_{\mathrm{in}} = K \cap (-K).$$

3 Comparison problems for sections and projections

This chapter focuses on the classical Busemann–Petty problem and its generalizations to arbitrary measures. The key conceptual advance for the classsical problem was Lutwak's introduction of intersection bodies, which reduced the problem to asking whether every origin-symmetric convex body is an intersection body. Using this approach, the problem was resolved completely, with an affirmative answer for $n \leqslant 4$ and a negative answer for $n \geqslant 5$. This resolution relies on geometric methods using the spherical Radon transform and perturbation techniques.

The problem was then extended to arbitrary measures. For a non-negative continuous function f on $\mathbb{R}^n$, one defines the measure $\mu(K) = \int_K f$. The generalized Busemann–Petty question asks whether inequalities for measures of central hyperplane sections

$$\mu(K \cap \xi^\perp) \leqslant \mu(L \cap \xi^\perp) \quad \text{for all } \xi \in S^{n-1}$$

imply the corresponding inequality for the total measures, $\mu(K) \leqslant \mu(L)$. The answer mirrors the classical problem: it is affirmative for $n \leqslant 4$ and negative for $n \geqslant 5$, under suitable smoothness and positivity conditions on f. The proofs employ Fourier analytic tools, the spherical Radon transform, and perturbation arguments for convex bodies.

Since the classical or measure versions fail in higher dimensions, one can ask an isomorphic version: for a measure μ with even, continuous, non-negative density and two origin-symmetric convex bodies K and L satisfying $\mu(K \cap \xi^\perp) \leqslant \mu(L \cap \xi^\perp)$ for all ξ, what is the smallest constant β_n such that $\mu(K) \leqslant \beta_n \, \mu(L)$? Using the Banach–Mazur distance from K to the class of intersection bodies, $d_{\mathcal{I}}(K)$, one can bound $\beta_n \leqslant d_{\mathcal{I}}(K)$. By John's theorem and the fact that the Euclidean ball is an intersection body, this yields $\beta_n \leqslant \sqrt{n}$ for any origin-symmetric convex body.

The lower-dimensional Busemann–Petty problem considers $1 \leqslant k \leqslant n-1$ and asks for the smallest constant $\beta_{n,k}$ such that if $|K \cap F| \leqslant |L \cap F|$ for all $(n-k)$-dimensional subspaces F, then $|K|^{(n-k)/n} \leqslant \beta_{n,k}^k |L|^{(n-k)/n}$. It is known that $\beta_{n,k} > 1$ for $k < n-3$, while uniform bounds exist: there is an absolute constant C such that $\beta_{n,k} \leqslant C$ for all n and k. The proof uses dual affine quermassintegrals and the boundedness of the isotropic constant. This framework extends to general measures, where the measure of K is bounded in terms of the measure of L and the generalized outer volume ratio distance to the class of k-intersection bodies.

The chapter also discusses stability and separation results. Stability asks whether a small discrepancy in hyperplane sections implies a controlled discrepancy in volumes, and separation asks whether a uniform gap in sections implies a lower bound on the volume difference. These phenomena extend to fractional Laplacians of the hyperplane section functions and to arbitrary even measures, with intersection bodies playing a central role in the estimates.

The Shephard problem, concerned with projections instead of sections, asks whether the inequality $|P_{\xi^\perp}(K)| \leqslant |P_{\xi^\perp}(L)|$ for all directions ξ implies $|K| \leqslant |L|$. The answer is

https://doi.org/10.1515/9783111360508-003

affirmative in two dimensions, but negative in higher dimensions, except when L is a projection body, in which case the volume inequality holds. Quantitative versions of the Shephard problem bound the ratio of volumes in terms of the maximal projection ratio and a volume ratio distance, typically of order $\sqrt{n}$. Lower-dimensional versions involve $(n-k)$-dimensional projections and constants $S_{n,k}$, whose growth depends on both n and k.

A functional-analytic viewpoint considers linear operators on continuous even functions, relating pointwise inequalities under the operator to L_p norm inequalities. For the spherical Radon transform, this reproduces the Busemann–Petty problem, while for the cosine transform, it recovers the Shephard problem and identifies associated classes of functions such as support functions of projection bodies.

Finally, convex measures in $\mathbb{R}^n$ are studied via s-concave measures, interpolating between log-concave ($s = 0$) and convex ($s = -\infty$) measures. Using K. Ball's K_p-bodies, one can associate a convex body to a convex measure, allowing one to translate measure-theoretic inequalities into classical convex geometric inequalities. This enables hyperplane section inequalities and other volume comparison results to extend from the Lebesgue measure to general convex measures.

3.1 The Busemann–Petty problem

The following problem was posed by Busemann and Petty [119] in 1956: suppose that K and L are origin-symmetric convex bodies in $\mathbb{R}^n$ such that

$$|K \cap \xi^{\perp}| \leqslant |L \cap \xi^{\perp}|, \quad \text{for all } \xi \in S^{n-1}. \tag{3.1}$$

Does it follow that

$$|K| \leqslant |L|?$$

The fact that, by the Minkowski uniqueness theorem, every convex body is uniquely determined by the volumes of its central hyperplane sections, provides some intuition in the direction of an affirmative answer. Also, in view of the Cavalieri principle one might say that the solution should be affirmative. Clearly, it is affirmative in dimension 2, because the assumption (3.1) implies in this case that the radius of K in every direction is smaller than the radius of L, which means that $K \subseteq L$. In higher dimensions, however, the situation is not so clear. When applying the Cavalieri principle one is allowed to consider all hyperplane sections perpendicular to a given direction, and then the volume of a body can be expressed as the integral of hyperplane volumes of all these sections. In the Busemann–Petty problem, one has information only about central hyperplane sections.

The first results regarding the problem all went in the positive direction. Busemann was able to give a positive answer in the case where K is an ellipsoid. He returned to the problem in [118], where he concentrated on the three-dimensional case and gave examples showing that the assumptions of convexity and symmetry are both essential. Later, Hadwiger [249] and Giertz [218] independently gave a positive answer to the problem in a special case where K and L are coaxial, convex symmetric bodies of revolution in $\mathbb{R}^3$. The first breakthrough on the Busemann–Petty problem came in 1975. Larman and Rogers [355] chose $L = B_2^n$ and proved that if $n \geqslant 12$, then there exist symmetric convex bodies K that are arbitrarily small perturbations of B_2^n such that the pair (K, B_2^n) provides a negative answer to the problem. The proof is probabilistic in nature.

Ball [38] proved that if $Q_n = [-1/2, 1/2]^n$ is the cube of volume 1 in $\mathbb{R}^n$, then $|Q_n \cap \xi^\perp| \leqslant \sqrt{2}$ for all $\xi \in S^{n-1}$. This bound leads to a concrete counterexample to the Busemann–Petty problem: Ball observed in [39] that, if $n \geqslant 10$, $K = Q_n$ and L is the Euclidean ball of volume 1, then

$$|Q_n \cap \xi^\perp| \leqslant \sqrt{2} < \omega_{n-1}\omega_n^{-\frac{n-1}{n}} = |L \cap \xi^\perp|$$

for all $\xi \in S^{n-1}$. Ball's counterexample is essentially different from the one of Larman and Rogers: the cube is far from being a perturbation of a ball. After Ball's example, Giannopoulos [207] and Bourgain [103] constructed counterexamples for $n \geqslant 7$, and Papadimitrakis [455] obtained counterexamples in dimensions $n = 6$ and $n = 5$.

The key step that led to the final solution to the Busemann–Petty problem is Lutwak's definition of an intersection body; see Section 1.7. Lutwak [392] observed that intersection bodies are closely connected with the Busemann–Petty problem. Using simple facts from the theory of dual mixed volumes one can see that if K is an intersection body, then the answer to the Busemann–Petty problem is affirmative for K and any origin-symmetric convex body L. However, if L is an origin-symmetric convex body that is not an intersection body, then K and a perturbation L of K provide a counterexample to the question (see [389, 391], and [392]). Therefore, the Busemann–Petty problem has an affirmative answer in $\mathbb{R}^n$ if and only if every origin-symmetric convex body in $\mathbb{R}^n$ is an intersection body. Using this reduction of the problem, Gardner [191, 192] and Zhang [551] gave a negative answer to the problem for $n \geqslant 5$ by providing examples of non-intersection bodies in $\mathbb{R}^5$. Around the same time, Gardner [193] proved that every origin-symmetric convex body in $\mathbb{R}^3$ is an intersection body, and hence the Busemann–Petty problem has an affirmative answer in dimension $n = 3$. For a few years it was believed that the problem has a negative answer in the remaining case $n = 4$. Through the work of Koldobsky, who established a Fourier analytic characterization of intersection bodies in [305], it was understood that the case $n = 4$ was still open. Zhang [554, 552] proved that the answer in $\mathbb{R}^4$ is affirmative; around the same time, using the Fourier analytic approach, Gardner, Koldobsky, and Schlumprecht [202] gave a unified solution to the problem in all dimensions (afterwards, an alternative proof was also given in [60]).

Thus, the answer to the Busemann–Petty problem is positive if $n \leqslant 4$ and negative for all higher dimensions (see the books [194] and [310] for a complete discussion of the problem and its history).

We start the discussion of the problem with Lutwak's reduction.

Theorem 3.1.1. *The answer to the Busemann–Petty problem in $\mathbb{R}^n$ is affirmative if and only if every origin-symmetric convex body in $\mathbb{R}^n$ is an intersection body.*

We present a geometric proof that consists of two parts.

Theorem 3.1.2. *Let K be an origin-symmetric star body in $\mathbb{R}^n$ that is an intersection body. Then, for any origin-symmetric star body L satisfying* (3.1), *we have $|K| \leqslant |L|$, i. e., the answer to the Busemann–Petty problem for K and L is affirmative.*

Proof. Using the expression (1.35) for the volume of sections in terms of the spherical Radon transform, we see that the condition (3.1) is equivalent to

$$R(\|\cdot\|_K^{-n+1})(\xi) \leqslant R(\|\cdot\|_L^{-n+1})(\xi), \quad \text{for all } \xi \in S^{n-1}. \tag{3.2}$$

Since K is an intersection body, there exists a finite measure μ on S^{n-1} such that $\|\cdot\|_K^{-1}$ and $R\mu$ are equal functionals on $C(S^{n-1})$. Applying both functionals to the continuous function $\|\cdot\|_L^{-n+1}$ we get

$$\begin{aligned}\int_{S^{n-1}} \|x\|_K^{-1}\|x\|_L^{-n+1}\,dx &= \langle \|\cdot\|_K^{-1}, \|\cdot\|_L^{-n+1}\rangle = \langle R\mu, \|\cdot\|_L^{-n+1}\rangle \\ &= \langle \mu, R(\|\cdot\|_L^{-n+1})\rangle = \int_{S^{n-1}} R(\|\cdot\|_L^{-n+1})(\xi)\,d\mu(\xi).\end{aligned}$$

Using (3.2), the positivity of the measure, and repeating the above calculation in the reverse order we get that the latter quantity is greater than or equal to

$$\int_{S^{n-1}} R(\|\cdot\|_K^{-n+1})(\xi)\,d\mu(\xi) = \int_{S^{n-1}} \|x\|_K^{-1}\|x\|_K^{-n+1}\,dx = n|K|.$$

The last equality here is the polar formula (1.3) for volume. Hence, by Hölder's inequality

$$\begin{aligned} n|K| &\leqslant \int_{S^{n-1}} \|x\|_K^{-1}\|x\|_L^{-n+1}\,dx \\ &\leqslant \Big(\int_{S^{n-1}} \|x\|_K^{-n}\,dx\Big)^{1/n}\Big(\int_{S^{n-1}} \|x\|_L^{-n}\,dx\Big)^{(n-1)/n} \\ &= n|K|^{1/n}|L|^{(n-1)/n},\end{aligned}$$

which implies $|K| \leqslant |L|$. □

For the second part of the proof of Theorem 3.1.1 we are going to use the standard perturbation procedure for convex bodies. Namely, if K is a 2-smooth convex body with strictly positive curvature, $p \in \mathbb{R}$ is a non-zero number and f is any C^2 function on the sphere, then for small enough ε the equality

$$\|x\|_L^p = \|x\|_K^p + \varepsilon f(x/|x|)|x|^p$$

defines a convex body L. This essentially follows from a simple two-dimensional argument: if h is a strictly concave function on a bounded interval so that $h'' < \delta < 0$, and g is a twice differentiable function on this interval, then for small ε the second derivative of the function $h + \varepsilon g$ is still strictly negative on the whole interval, so $f + \varepsilon g$ is also concave.

Theorem 3.1.3. *If there exists an origin-symmetric convex body in $\mathbb{R}^n$ that is not an intersection body, then one can construct two origin-symmetric convex bodies K and L in $\mathbb{R}^n$ providing a counterexample to the Busemann–Petty problem, i. e., for every $\xi \in S^{n-1}$*

$$|K \cap \xi^\perp| \leqslant |L \cap \xi^\perp|,$$

but

$$|K| > |L|.$$

Proof. First, by the approximation argument of Lemma 1.7.13, there exists an infinitely smooth, origin-symmetric convex body L in $\mathbb{R}^n$ with strictly positive curvature that is not an intersection body. The function $(1/(2^n\pi^{n-1}))\|\cdot\|_L^{-1}$ satisfies the conditions of Proposition 1.4.14, so its Fourier transform has the form $g \cdot r^{-n+1}$, where $g \in C^\infty(S^{n-1})$. Since $g \cdot r^{-n+1}$ is a homogeneous function of degree $-n+1$, by Corollary 1.5.1, the spherical Radon transform of g is equal to $\|\cdot\|_L^{-1}$, i. e., $Rg = \|\cdot\|_L^{-1}$.

The function g is sign-changing, because otherwise L would be an intersection body. Let $\Omega \subset S^{n-1}$ be an open set where $g < 0$. Consider a function $h \in C^\infty(S^{n-1})$ so that $h \geqslant 0$ everywhere on S^{n-1} and $h > 0$ only on a non-empty subset of Ω. By the same argument as before, involving Lemma 1.4.13 and Corollary 1.5.1, we see that $h = Ru$ for some $u \in C^\infty(S^{n-1})$.

Now, we are ready to construct the body K. Define K by

$$\|\xi\|_K^{-n+1} - \|\xi\|_L^{-n+1} - \varepsilon u, \quad \xi \in S^{n-1}.$$

Since L has strictly positive curvature, by the argument before this theorem, one can choose a small enough ε so that the body K is convex.

Next, for every $\xi \in S^{n-1}$, since $Ru = h \geqslant 0$

$$R(\|\cdot\|_K^{-n+1})(\xi) = R(\|\cdot\|_L^{-n+1})(\xi) - \varepsilon Ru(\xi) \leqslant R(\|\cdot\|_L^{-n+1})(\xi),$$

which means that K and L satisfy condition (3.2), so

$$|K \cap \xi^\perp| \leqslant |L \cap \xi^\perp|, \quad \text{for all } \xi \in S^{n-1}.$$

On the other hand, by the self-duality of the spherical Radon transform, Lemma 1.3.3, and the fact that the product hg is non-positive everywhere, we have

$$\begin{aligned}
\int_{S^{n-1}} \|\xi\|_K^{-n+1}\|\xi\|_L^{-1}\,d\xi &= \int_{S^{n-1}} \|\xi\|_K^{-n+1} Rg(\xi)\,d\xi = \int_{S^{n-1}} R(\|\cdot\|_K^{-n+1})(\xi)g(\xi)\,d\xi \\
&= \int_{S^{n-1}} R(\|\cdot\|_L^{-n+1})(\xi)g(\xi)\,d\xi - \varepsilon \int_{S^{n-1}} h(\xi)g(\xi)\,d\xi \\
&> \int_{S^{n-1}} R(\|\cdot\|_L^{-n+1})(\xi)g(\xi)\,d\xi = \int_{S^{n-1}} \|\xi\|_L^{-n+1} Rg(\xi)\,d\xi \\
&= \int_{S^{n-1}} \|\xi\|_L^{-n+1}\|\xi\|_L^{-1}\,d\xi = n|L|.
\end{aligned}$$

Applying Hölder's inequality in the same way as in Theorem 3.1.2, we get $|K| > |L|$. □

We are now ready to present the solution to the Busemann–Petty problem. It is an immediate consequence of Lutwak's connection (Theorems 3.1.2 and 3.1.3) and Theorem 1.7.2.

Theorem 3.1.4. *The solution to the Busemann–Petty problem is affirmative if $n \leqslant 4$ and it is negative if $n \geqslant 5$.*

Proof. From Theorem 1.7.2 we know that every origin-symmetric convex body in $\mathbb{R}^n$, $n = 2, 3, 4$ is a 1-intersection body (or simply an intersection body). Therefore, the affirmative part follows from Theorem 3.1.2.

Let $n \geqslant 5$. From Theorem 1.7.2 we also know that there exist origin-symmetric convex bodies that are not intersection bodies. Thus, the negative part follows from Theorem 3.1.3. □

In view of this negative result, one can ask for an affirmative answer to the Busemann–Petty problem up to an absolute constant. Does there exist an absolute constant $c > 0$ such that if K_1 and K_2 are centered convex bodies in $\mathbb{R}^n$ that satisfy $|K_1 \cap \xi^\perp| \leqslant |K_2 \cap \xi^\perp|$ for all $\xi \in S^{n-1}$, then $|K_1|^{\frac{n-1}{n}} \leqslant c\,|K_2|^{\frac{n-1}{n}}$? This is the so-called isomorphic Busemann–Petty problem.

3.2 The Busemann–Petty problem for arbitrary measures

An extension of the Busemann–Petty problem to arbitrary measures in place of volume was considered by Zvavitch in [555]. Let f be a continuous non-negative function on $\mathbb{R}^n$.

For any subspace H of $\mathbb{R}^n$ and any closed bounded set $B \subset H$ we set

$$\mu(B) = \int_B f(x)\,dx,$$

where integration is with respect to the Lebesgue measure on H. The "generalized Busemann–Petty problem for f", for brevity GBP(f), is the next question: Let K, L be origin-symmetric convex bodies in $\mathbb{R}^n$ such that for every $\xi \in S^{n-1}$

$$\mu(K \cap \xi^\perp) \leqslant \mu(L \cap \xi^\perp). \tag{3.3}$$

Does it necessarily follow that

$$\mu(K) \leqslant \mu(L)\,?$$

It was proved in [555] that the answer to this question is the same as in the case of volume (where $f \equiv 1$): If f is an even, continuous positive function on $\mathbb{R}^n$, then the answer to GBP(f) is affirmative if $n \leqslant 4$ and negative if $n \geqslant 5$.

For the proof, we need a formula that expresses the measure of sections of star bodies in terms of the Fourier transform. This result is the case $q = 0$ of Theorem 1.5.8, but we give here a simpler proof.

Theorem 3.2.1. *Let K be an origin-symmetric star body in $\mathbb{R}^n$. For every $\xi \in S^{n-1}$ we have that*

$$\mu(K \cap \xi^\perp) = \frac{1}{\pi}\widehat{G_K}(\xi),$$

where

$$G_K(x) = \frac{1}{|x|^{n-1}} \int_0^{|x|/\|x\|_K} t^{n-2} f\left(\frac{tx}{|x|}\right) dt$$

and the Fourier transform of G_K is a continuous, homogeneous of degree -1, function on $\mathbb{R}^n \setminus \{0\}$.

Proof. We denote by χ the indicator function of $[-1,1]$. Using polar coordinates on $\xi^\perp$ we write

$$\mu(K \cap \xi^\perp) = \int_{\xi^\perp} \chi(\|x\|_K) f(x)\,dx = \int_{S^{n-1}\cap\xi^\perp} \left(\int_0^{\|\theta\|_K^{-1}} t^{n-2} f(t\theta)\,dt \right) d\theta. \tag{3.4}$$

The function under the outer integral is the restriction to $S^{n-1} \cap \xi^\perp$ of the homogeneous of degree $-(n-1)$ continuous function G_K on $\mathbb{R}^n \setminus \{0\}$, and applying Lemma 1.5.1 we write

$$\mu(K \cap \xi^\perp) = \int_{S^{n-1}\cap\xi^\perp} G_K(\theta)\,d\theta = R(G_K)(\xi) = \frac{1}{\pi}\widehat{G_K}(\xi), \tag{3.5}$$

using the fact that the spherical Radon transform of G_K is equal, up to a constant, to the restriction to the sphere of its Fourier transform. □

We shall also use the next elementary lemma.

Lemma 3.2.2. *Let $a, b > 0$ and let $u, v : (0, \max\{a, b\}) \to \mathbb{R}$ be positive continuous functions such that $tu(t)/v(t)$ is non-decreasing. Then,*

$$\int_0^a t^{n-1}u(t)\,dt - a\frac{u(a)}{v(a)}\int_0^a t^{n-2}v(t)\,dt \leqslant \int_0^b t^{n-1}u(t)\,dt - a\frac{u(a)}{v(a)}\int_0^b t^{n-2}v(t)\,dt. \tag{3.6}$$

Proof. The claim of the lemma is equivalent to the inequality

$$a\frac{u(a)}{v(a)}\int_a^b t^{n-2}v(t)\,dt \leqslant \int_a^b t^{n-1}u(t)\,dt.$$

We observe that

$$a\frac{u(a)}{v(a)}\int_a^b t^{n-2}v(t)\,dt = \int_a^b t^{n-1}u(t)\Big(a\frac{u(a)}{v(a)}\Big)\Big(t\frac{u(t)}{v(t)}\Big)^{-1}dt \leqslant \int_a^b t^{n-1}u(t)\,dt.$$

This inequality holds true in both cases $a \leqslant b$ and $a \geqslant b$. □

The next theorem will allow us to settle the case $n \leqslant 4$.

Theorem 3.2.3. *Let K be an origin-symmetric star body in $\mathbb{R}^n$ such that $\|\cdot\|_K^{-1}$ is a positive definite distribution on $\mathbb{R}^n$. If L is an origin-symmetric star body in $\mathbb{R}^n$ such that*

$$\mu(K \cap \xi^\perp) \leqslant \mu(L \cap \xi^\perp) \tag{3.7}$$

for all $\xi \in S^{n-1}$, then

$$\mu(K) \leqslant \mu(L).$$

Proof. Theorem 3.2.1 shows that the assumption (3.7) is equivalent to the inequality

$$\widehat{G_K}(\xi) \leqslant \widehat{G_L}(\xi)$$

for all $\xi \in S^{n-1}$. Integrating this inequality over S^{n-1} with respect to the measure μ_0 that corresponds (see Proposition 1.4.7) to the positive definite homogeneous of degree -1 distribution $\|\cdot\|_K^{-1}$, we get

$$\int_{S^{n-1}} \widehat{G_K}(\xi)\,d\mu_0(\xi) \leqslant \int_{S^{n-1}} \widehat{G_L}(\xi)\,d\mu_0(\xi),$$

which, by the spherical Parseval's formula (see Proposition 1.6.5), gives

$$\int_{S^{n-1}} \|x\|_K^{-1}\Big(\int_0^{\|x\|_K^{-1}} t^{n-2}f(tx)\,dt\Big)dx \leqslant \int_{S^{n-1}} \|x\|_K^{-1}\Big(\int_0^{\|x\|_L^{-1}} t^{n-2}f(tx)\,dt\Big)dx. \tag{3.8}$$

Now, we apply Lemma 3.2.2 with $a = \|x\|_K^{-1}$, $b = \|x\|_L^{-1}$ and $u(t) = f(tx)$ to get

$$\int_0^{\|x\|_K^{-1}} t^{n-1}f(tx)\,dt - \|x\|_K^{-1}\int_0^{\|x\|_K^{-1}} t^{n-2}f(tx)\,dt \leqslant \int_0^{\|x\|_L^{-1}} t^{n-1}f(tx)\,dt - \|x\|_K^{-1}\int_0^{\|x\|_L^{-1}} t^{n-2}f(tx)\,dt$$

for all $x \in S^{n-1}$. Integrating over S^{n-1} we see that

$$\int_{S^{n-1}}\Big(\int_0^{\|x\|_K^{-1}} t^{n-1}f(tx)\,dt\Big)dx - \int_{S^{n-1}} \|x\|_K^{-1}\Big(\int_0^{\|x\|_K^{-1}} t^{n-2}f(tx)\,dt\Big)dx \leqslant \int_{S^{n-1}}\Big(\int_0^{\|x\|_L^{-1}} t^{n-1}f(tx)\,dt\Big)dx - \int_{S^{n-1}} \|x\|_K^{-1}\Big(\int_0^{\|x\|_L^{-1}} t^{n-2}f(tx)\,dt\Big)dx. \tag{3.9}$$

Adding (3.8) and (3.9), we conclude that

$$\int_{S^{n-1}}\Big(\int_0^{\|x\|_K^{-1}} t^{n-1}f(tx)\,dt\Big)dx \leqslant \int_{S^{n-1}}\Big(\int_0^{\|x\|_L^{-1}} t^{n-1}f(tx)\,dt\Big)dx.$$

Finally, note that the left- and right-hand sides of the last inequality are equal to $\mu(K)$ and $\mu(L)$, respectively, if we express them in polar coordinates. This proves the theorem. □

The next theorem will allow us to settle the case $n \geqslant 5$.

Theorem 3.2.4. *Let $f \in C^\infty(\mathbb{R}^n)$ be an even, continuous positive function. If L is an infinitely smooth, origin-symmetric convex body in $\mathbb{R}^n$ with positive curvature and the func-*

tion $\|\cdot\|_L^{-1}$ does not represent a positive definite distribution, then there exists an origin-symmetric convex body D in $\mathbb{R}^n$ such that

$$\mu(D \cap \xi^\perp) \leqslant \mu(L \cap \xi^\perp)$$

for all $\xi \in S^{n-1}$, but

$$\mu(D) > \mu(L).$$

For the proof, we need the next perturbation lemma.

Lemma 3.2.5. *Let L be an infinitely smooth origin-symmetric convex body L with positive curvature and let $f, g \in C^2(\mathbb{R}^n)$ be two even functions such that f is strictly positive on $\mathbb{R}^n$. If $\varepsilon > 0$ is small enough, then the star body D whose radial function $\|x\|_D^{-1}$ is defined by the equation*

$$\int_0^{\|x\|_D^{-1}} t^{n-2} f(tx)\,dt = \int_0^{\|x\|_L^{-1}} t^{n-2} f(tx)\,dt - \varepsilon g(x)$$

for all $x \in S^{n-1}$, is convex.

Proof. If $\varepsilon > 0$ is small enough, then we may define a function α_ε on S^{n-1} such that

$$\int_0^{\|x\|_L^{-1}} t^{n-2} f(tx)\,dt - \varepsilon g(x) = \int_0^{\|x\|_L^{-1} - \alpha_\varepsilon(x)} t^{n-2} f(tx)\,dt \tag{3.10}$$

for all $x \in S^{n-1}$. Since $f(tx) > 0$ for $tx \in \mathbb{R}^n \setminus \{0\}$, we have that $y \mapsto \int_0^y t^{n-2} f(tx)\,dt$ is strictly increasing on $[0, +\infty)$, and hence

$$\|x\|_D^{-1} = \|x\|_L^{-1} - \alpha_\varepsilon(x), \quad x \in S^{n-1}. \tag{3.11}$$

Moreover, since f and its first order partial derivatives are bounded for finite values of x, from (3.10) we see that α_ε and its first and second order partial derivatives converge uniformly to 0 on S^{n-1} as $\varepsilon \to 0$. Since L is convex and has positive curvature, we can choose $\varepsilon > 0$ small enough so that the body D is convex and has positive curvature. □

Proof of Theorem 3.2.4. Since L is infinitely smooth, Proposition 1.4.14 shows that $(\|\cdot\|_L^{-1})^\wedge$ is continuous on S^{n-1}. Also, by the assumption of the theorem, this function is negative on some open symmetric subset Ω of S^{n-1}. We consider a non-negative even function $h \in C^\infty(S^{n-1})$ supported on Ω and extend it to a continuous function $h \cdot r^{-1}$ that is homogeneous of degree -1 on $\mathbb{R}^n$. From Lemma 1.4.13 we know that the Fourier transform of $h \cdot r^{-1}$ is a function of the form $g \cdot r^{-n+1}$, where $g \in C^\infty(S^{n-1})$.

Given $\varepsilon > 0$ small enough, we consider the star body D whose radial function $\|x\|_D^{-1}$ is defined by the equation

$$\frac{1}{|x|^{n-1}} \int_0^{\frac{|x|}{\|x\|_D}} t^{n-2} f\left(\frac{tx}{|x|}\right) dt = \frac{1}{|x|^{n-1}} \int_0^{\frac{|x|}{\|x\|_L}} t^{n-2} f\left(\frac{tx}{|x|}\right) dt - \varepsilon g\left(\frac{x}{|x|}\right) \frac{1}{|x|^{n-1}}.$$

Lemma 3.2.5 shows that we can choose ε so that D is convex.

Since $h \geqslant 0$, by Theorem 3.2.1 we have that

$$\mu(D \cap \xi^\perp) = \frac{1}{\pi}\widehat{G_D}(\xi) = \frac{1}{\pi}\widehat{G_L}(\xi) - \frac{(2\pi)^n \varepsilon h(\xi)}{\pi} \leqslant \mu(L \cap \xi^\perp)$$

for all $\xi \in S^{n-1}$. On the other hand, by construction, h is positive only where $(\|\cdot\|_L^{-1})^\wedge$ is negative, therefore

$$\begin{aligned}(\|\cdot\|_L^{-1})^\wedge(\xi)\widehat{G_D}(\xi) &= (\|\cdot\|_L^{-1})^\wedge(\xi)\widehat{G_L}(\xi) - (2\pi)^n(\|\cdot\|_L^{-1})^\wedge(\xi)\varepsilon h(\xi)\\ &\geqslant (\|\cdot\|_L^{-1})^\wedge(\xi)\widehat{G_L}(\xi)\end{aligned}$$

for all $\xi \in S^{n-1}$, and this inequality is strict for all ξ on some non-empty open set.

Integrating the last inequality over S^{n-1} and applying the spherical Parseval identity of Proposition 1.6.4 and Lemma 3.2.2, we check that $\mu(D) > \mu(L)$. □

We are now able to prove the result of Zvavitch [555], under some additional assumptions on f.

Theorem 3.2.6. *If $n \leqslant 4$, then, for any even, continuous positive function on $\mathbb{R}^n$, the answer to* GBP(f) *is affirmative. If $n \geqslant 5$, then, for any even positive function $f \in C^\infty(\mathbb{R}^n)$, the answer to* GBP($f$) *is negative.*

Proof. Let $n \leqslant 4$. Then, by Theorems 1.7.1 and 1.7.2, for every origin-symmetric convex body K in $\mathbb{R}^n$, the function $\|\cdot\|_K^{-1}$ represents a positive definite distribution. Then, Theorem 3.2.3 shows that the answer to GBP(f) is affirmative.

Let $n \geqslant 5$. Then, for any $q > 2$ the unit ball of ℓ_q^n is not an intersection body; see Proposition 1.7.12. By Lemma 1.7.13 we may find an infinitely smooth origin-symmetric convex body L in $\mathbb{R}^n$, with positive curvature, which is not an intersection body. Then, by Theorem 1.7.1, $\|\cdot\|_L^{-1}$ is not a positive definite distribution, and the result follows from Theorem 3.2.4. □

Note. One can extend Theorem 3.2.6 to a slightly more general setting where measures applied to sections and to the whole body are different [555, 190]. Another application is the solution of the Busemann–Petty problem in hyperbolic and spherical spaces [545].

3.3 Isomorphic version of the Busemann–Petty problem

Since the answer to the Busemann–Petty problem for measures is negative in dimensions $n \geqslant 5$, it is natural to ask the corresponding isomorphic Busemann–Petty question: Determine the best constant β_n so that for any measure μ with continuous non-negative even density f and any two origin-symmetric convex bodies K and M in $\mathbb{R}^n$ that satisfy

$$\mu(K \cap \xi^\perp) \leqslant \mu(M \cap \xi^\perp)$$

for every $\xi \in S^{n-1}$, one has that

$$\mu(K) \leqslant \beta_n \mu(M).$$

One can give an answer to this question with a constant not depending on the measure or bodies, but dependent on the dimension, namely, we will see that $\beta_n \leqslant \sqrt{n}$. Very recently it was announced in [325] that $\beta_n \geq c\sqrt{n}$, where c is an absolute constant. As we will see in Section 3.8, it is possible to improve the bound $\sqrt{n}$ to $O(L_n)$ for convex measures.

Recall that the Banach–Mazur distance between two origin-symmetric convex bodies K and L in $\mathbb{R}^n$ is defined by

$$d_{\mathrm{BM}}(K, L) = \inf\{d > 0 \mid \text{there exists } T \in GL_n : K \subset TL \subset dK\}$$

and let

$$d_{\mathcal{I}}(K) = \min\{d_{\mathrm{BM}}(K, L) : L \text{ is an intersection body in } \mathbb{R}^n\}.$$

The following result was proved in [324].

Theorem 3.3.1. *For any measure μ with continuous, non-negative even density f on $\mathbb{R}^n$ and any two origin-symmetric convex bodies $K, M \subset \mathbb{R}^n$ such that*

$$\mu(K \cap \xi^\perp) \leqslant \mu(M \cap \xi^\perp) \tag{3.12}$$

for all $\xi \in S^{n-1}$, we have

$$\mu(K) \leqslant d_{\mathcal{I}}(K)\mu(M).$$

Proof. First, we use the polar formula for measures of sections (3.4) to write the condition (3.12) in terms of the spherical Radon transform

$$R\left(\int_0^{\|\cdot\|_K^{-1}} r^{n-2} f(r\,\cdot)\,dr\right)(\xi) \leqslant R\left(\int_0^{\|\cdot\|_M^{-1}} r^{n-2} f(r\,\cdot)\,dr\right)(\xi). \tag{3.13}$$

Next, we consider an intersection body L such that $L \subset K \subset d_{\mathcal{I}}(K)L$ (note that a linear image of an intersection body is again an intersection body; see, for example, [194, Theorem 8.1.6]) and integrate (3.13) over S^{n-1} with respect to the measure ν corresponding to the intersection body L. Using the fact that

$$\int_{S^{n-1}} \|x\|_L^{-1} f(x)\, dx = \int_{S^{n-1}} Rf(x)\, d\nu(x)$$

we get

$$\int_{S^{n-1}} \|x\|_L^{-1} \int_0^{\|x\|_K^{-1}} t^{n-2} f(tx)\, dt\, dx \leqslant \int_{S^{n-1}} \|x\|_L^{-1} \int_0^{\|x\|_M^{-1}} t^{n-2} f(tx)\, dt\, dx. \tag{3.14}$$

Now, we apply Lemma 3.2.2 to get

$$\begin{aligned} &\frac{\|x\|_L^{-1}}{\|x\|_K^{-1}} \int_0^{\|x\|_K^{-1}} t^{n-1} f(tx)\, dt - \|x\|_L^{-1} \int_0^{\|x\|_K^{-1}} t^{n-2} f(tx)\, dt \\ &\quad \leqslant \frac{\|x\|_L^{-1}}{\|x\|_K^{-1}} \int_0^{\|x\|_M^{-1}} t^{n-1} f(tx)\, dt - \|x\|_L^{-1} \int_0^{\|x\|_M^{-1}} t^{n-2} f(tx)\, dt. \end{aligned} \tag{3.15}$$

Integrating (3.15) over S^{n-1}, adding it to (3.14), and using $L \subset K \subset d_{\mathcal{I}}(K)L$ we get

$$\int_{S^{n-1}} \frac{\|x\|_L^{-1}}{\|x\|_K^{-1}} \int_0^{\|x\|_K^{-1}} t^{n-1} f(tx)\, dt\, dx \leqslant \int_{S^{n-1}} \frac{\|x\|_L^{-1}}{\|x\|_K^{-1}} \int_0^{\|x\|_M^{-1}} t^{n-1} f(tx)\, dt\, dx \tag{3.16}$$

and

$$\frac{1}{d_{\mathcal{I}}(K)} \int_{S^{n-1}} \int_0^{\|x\|_K^{-1}} t^{n-1} f(tx)\, dt\, dx \leqslant \int_{S^{n-1}} \int_0^{\|x\|_M^{-1}} t^{n-1} f(tx)\, dt\, dx.$$

By the polar formula for the measure of a star body, the latter inequality reads as

$$\mu(K) \leqslant d_{\mathcal{I}}(K)\mu(M)$$

and the theorem follows. □

Since the Euclidean ball B_2^n is an intersection body, by John's theorem we have that $d_{\mathrm{BM}}(K, B_2^n) \leqslant \sqrt{n}$ for all origin-symmetric convex bodies $K \subset \mathbb{R}^n$. This immediately shows that $d_{\mathcal{I}}(K) \leqslant \sqrt{n}$ for all origin-symmetric convex bodies $K \subset \mathbb{R}^n$. This fact, together with Theorem 3.3.1, implies the following.

Corollary 3.3.2. *For any measure μ with continuous non-negative even density on $\mathbb{R}^n$ and any two origin-symmetric convex bodies $K, M \subset \mathbb{R}^n$ such that*

$$\mu(K \cap \xi^\perp) \leqslant \mu(M \cap \xi^\perp) \tag{3.17}$$

for all $\xi \in S^{n-1}$, we have

$$\mu(K) \leqslant \sqrt{n}\, \mu(M).$$

If the body K in Theorem 3.3.1 is an intersection body, the theorem holds with constant $\beta = 1$; see also Theorem 3.2.3). There are other special classes of bodies for which the constant β does not depend on the dimension.

For the class of k-intersection bodies the claim of the theorem is true with a constant β depending only on k. To see this, we estimate the Banach–Mazur distance from k-intersection bodies to intersection bodies.

Corollary 3.3.3. *Let $k \in \mathbb{N}$. There exists a constant $C(k)$ such that for any $n > k$, any measure μ with continuous non-negative even density on $\mathbb{R}^n$, any convex k-intersection body K in $\mathbb{R}^n$ and any origin-symmetric convex body $M \subset \mathbb{R}^n$, the inequality*

$$\mu(K \cap \xi^\perp) \leqslant \mu(M \cap \xi^\perp)$$

for all $\xi \in S^{n-1}$ implies

$$\mu(K) \leqslant C(k)\, \mu(M).$$

Proof. Let $q = 1/2$. Propositions 1.7.10 and 1.7.11 imply that

$$d_{\mathcal{I}}(K) \leqslant c(k, 1/2) =: C(k).$$

The result follows from Theorem 3.3.1. □

The constant $\sqrt{n}$ in Corollary 3.3.2 can also be improved if K is the unit ball of a subspace of L_p, $p > 2$. For such K, by a result of Lewis [370] (see also [501] for a different proof), we have $d_{\mathrm{BM}}(K, B_2^n) \leqslant n^{1/2-1/p}$. Since B_2^n is an intersection body, Theorem 3.3.1 implies the following.

Corollary 3.3.4. *Let $p > 2$, let K be the unit ball of an n-dimensional subspace of L_p, and let μ be a measure with even, continuous density on $\mathbb{R}^n$. Suppose that M is an origin-symmetric convex body in $\mathbb{R}^n$ so that*

$$\mu(K \cap \xi^\perp) \leqslant \mu(M \cap \xi^\perp)$$

for all $\xi \in S^{n-1}$. Then,

$$\mu(K) \leqslant n^{1/2-1/p} \mu(M).$$

Note. The statements of Theorem 3.3.1 and Corollaries 3.3.2, 3.3.3, and 3.3.4 are still true if M is any star body.

3.4 Lower-dimensional Busemann–Petty problem

A natural generalization of the Busemann–Petty problem, for lower-dimensional sections, is the following question: Let $1 \leqslant k \leqslant n-1$ and let $\beta_{n,k}$ be the smallest constant $\beta > 0$ with the following property: For every pair of centered convex bodies K and L in $\mathbb{R}^n$ that satisfy

$$|K \cap F| \leqslant |L \cap F|$$

for all $F \in G_{n,n-k}$, one has

$$|K|^{\frac{n-k}{n}} \leqslant \beta^k |L|^{\frac{n-k}{n}}.$$

It is known that $\beta_{n,k} > 1$ if $k < n-3$. This was first shown in [105] (with a correction in [489]). A different proof was suggested in [309]. Yet another argument is based on Theorem 1.7.16 and can be found in [310, p. 112]. For $k = n-2$ and $k = n-3$ (two- and three-dimensional sections) it is not known whether $\beta_{n,k}$ has to be strictly greater than 1 (an affirmative answer has just been announced in [375]). The asymptotic lower-dimensional Busemann–Petty problem asks if the constants $\beta_{n,k}$ are uniformly bounded. The next theorem provides an affirmative answer.

Theorem 3.4.1. *There exists an absolute constant $C > 0$ such that $\beta_{n,k} \leqslant C$ for all n and k.*

For the proof, we introduce the dual affine quermassintegrals of a convex body. Let $1 \leqslant k \leqslant n-1$. For every convex body K, or more generally for any bounded Borel set, in $\mathbb{R}^n$, the k-th dual affine quermassintegral of K is defined by

$$\Psi_k(K) = |K|^{-\frac{n-k}{kn}} \left(\int_{G_{n,k}} |K \cap F^{\perp}|^n \, d\nu_{n,k}(F) \right)^{\frac{1}{kn}}.$$

Grinberg proved in [234] that this is an affinely invariant quantity and that

$$\Psi_k(K) \leqslant \Psi_k(B_2^n) \leqslant \sqrt{e} \tag{3.18}$$

for every bounded Borel set K of positive volume in $\mathbb{R}^n$, as a consequence of the inequality

$$\int_{G_{n,k}} |K \cap F|^n \, d\nu_{n,k}(F) \leqslant \frac{\omega_k^n}{\omega_n^k} |K|^k. \tag{3.19}$$

We shall discuss the proof of (3.19) in Section 5.2. Here, we shall obtain an asymptotic version of (3.18), which suffices for our purposes.

Let K be a centered convex body in $\mathbb{R}^n$. Since $\Psi_k(K) = \Psi_k(T(K))$ for every $T \in GL_n$, we may assume that K is isotropic. From Theorem 2.3.4 we know that $|K \cap F^\perp|^{1/k} \approx 1$ for every $F \in G_{n,k}$, and hence

$$\Psi_k(K) = \left(\int_{G_{n,k}} |K \cap F^\perp|^n \, d\nu_{n,k}(F) \right)^{\frac{1}{kn}} \approx 1.$$

In other words, for every centered convex body K in $\mathbb{R}^n$ and any $1 \leqslant k \leqslant n-1$ we have that

$$c_1 \leqslant \Psi_k(K) \leqslant c_2, \tag{3.20}$$

where $c_1, c_2 > 0$ are absolute constants.

Proof of Theorem 3.4.1. Let K, L be two centered convex bodies in $\mathbb{R}^n$. Assume that for some $1 \leqslant k \leqslant n-1$ we have that

$$|K \cap F| \leqslant |L \cap F|$$

for all $F \in G_{n,n-k}$. Then,

$$\begin{aligned} \Psi_k(K)\,|K|^{\frac{n-k}{kn}} &= \left(\int_{G_{n,k}} |K \cap F^\perp|^n \, d\nu_{n,k}(F) \right)^{\frac{1}{kn}} \\ &\leqslant \left(\int_{G_{n,k}} |L \cap F^\perp|^n \, d\nu_{n,k}(F) \right)^{\frac{1}{kn}} = \Psi_k(L)\,|L|^{\frac{n-k}{kn}}. \end{aligned}$$

Taking into account (3.20) we get

$$|K|^{\frac{n-k}{n}} \leqslant \left(\frac{\Psi_k(L)}{\Psi_k(K)} \right)^k |L|^{\frac{n-k}{n}} \leqslant \left(\frac{c_2}{c_1} \right)^k |L|^{\frac{n-k}{n}}$$

and the result follows with $C = c_2/c_1$. □

In particular, the case $k = 1$ of Theorem 3.4.1 shows that the isomorphic Busemann–Petty problem has an affirmative answer for the class of centered convex bodies.

Theorem 3.4.2 (isomorphic Busemann–Petty problem). *There exists an absolute constant $c > 0$ such that if K_1 and K_2 are centered convex bodies in $\mathbb{R}^n$ that satisfy $|K_1 \cap \xi^\perp| \leqslant |K_2 \cap \xi^\perp|$ for all $\xi \in S^{n-1}$, then $|K_1|^{\frac{n-1}{n}} \leqslant c\,|K_2|^{\frac{n-1}{n}}$.*

The affine quermassintegrals of a convex body K in $\mathbb{R}^n$ were introduced by Lutwak in [390]. We shall discuss an appropriately normalized variant that was considered by Dafnis and Paouris in [144]. For every convex body K in $\mathbb{R}^n$ and every $1 \leqslant k \leqslant n-1$, we define the *normalized k-th affine quermassintegral* of K by

$$\Phi_k(K) := |K|^{-\frac{1}{n}} \Big(\int_{G_{n,k}} |P_F(K)|^{-n} \, d\nu_{n,k}(F) \Big)^{-\frac{1}{kn}}.$$

Grinberg proved in [234] that these quantities are invariant under invertible affine transformations. In this language, Lutwak conjectured in [392] that the affine quermassintegrals satisfy the inequalities

$$\Phi_k(K) \geqslant \Phi_k(B_2^n). \tag{3.21}$$

Dafnis and Paouris studied in [144] an isomorphic variant of Lutwak's conjecture, which asks if there exist absolute constants $c_1, c_2 > 0$ such that for every convex body K in $\mathbb{R}^n$ and any $1 \leqslant k \leqslant n-1$

$$c_1 \sqrt{n/k} \leqslant \Phi_k(K) \leqslant c_2 \sqrt{n/k}. \tag{3.22}$$

Note that in the case $k = 1$, (3.22) follows by the Blaschke–Santaló and the Bourgain–Milman inequalities, while in the case $k = n-1$ the conjectured rate of growth for $\Phi_{n-1}(K)$ is again true, by the Petty projection inequality and its reverse, proved by Zhang [553].

The left-hand side of (3.22) was proved by Paouris and Pivovarov in [454]; it confirms Lutwak's conjecture in an isomorphic sense. The proof relies on a duality argument, that employs the Blaschke–Santaló inequality and the Bourgain–Milman inequality, combined with Grinberg's inequality (3.19).

Theorem 3.4.3. *Let K be a convex body in $\mathbb{R}^n$. For every $1 \leqslant k \leqslant n-1$ we have that*

$$\Phi_k(K) \geqslant c\sqrt{n/k},$$

where $c > 0$ is an absolute constant.

Proof. We consider first the case where K is origin-symmetric. Let $K^\circ = \{x \in \mathbb{R}^n : \langle x, y\rangle \leqslant 1 \text{ for all } y \in K\}$ be the polar body of K. We have that $P_H(K)$ is an origin-symmetric convex body in H for any $H \in G_{n,k}$, $1 \leqslant k \leqslant n-1$, and $(P_H(K))^\circ = K^\circ \cap H$. Applying the Blaschke–Santaló and the Bourgain–Milman inequalities for $K^\circ \cap H$ we see that

$$c_1^k \omega_k^2 \leqslant |P_H(K)|\, |K^\circ \cap H| \leqslant \omega_k^2, \tag{3.23}$$

where $c > 0$ is an absolute constant. Then, using (3.19) we write

$$c_1^{kn}\omega_k^{2n}\int_{G_{n,k}}|P_E(K)|^{-n}\,d\nu_{n,k}(E)\leqslant\int_{G_{n,k}}|K^\circ\cap E|^n\,d\nu_{n,k}(E)$$
$$\leqslant\frac{\omega_k^n}{\omega_n^k}|K^\circ|^k\leqslant\frac{\omega_k^n}{\omega_n^k}\omega_n^{2k}|K|^{-k}=\omega_k^n\omega_n^k|K|^{-k}. \tag{3.24}$$

This shows that

$$\Phi_k(K)=|K|^{-\frac{1}{n}}\Big(\int_{G_{n,k}}|P_F(K)|^{-n}\,d\nu_{n,k}(F)\Big)^{-\frac{1}{kn}}$$
$$\geqslant c_1\frac{\omega_k^{2/k}}{\omega_k^{1/k}\omega_n^{1/n}}=c_1\frac{\omega_k^{1/k}}{\omega_n^{1/n}}\geqslant c\sqrt{n/k}$$

for an absolute constant $c>0$. If K is not necessarily symmetric, consider the difference body $K-K$ of K. By the Rogers–Shephard inequality, for every $E\in G_{n,k}$ we have $|P_E(K-K)|=|P_E(K)-P_E(K)|\leqslant\binom{2k}{k}|P_E(K)|$, and hence $|P_E(K-K)|^{1/k}\leqslant 4|P_E(K)|^{1/n}$. Since we also have $|K-K|\geqslant|K|$, we get

$$\Phi_k(K-K)=|K-K|^{-\frac{1}{n}}\Big(\int_{G_{n,k}}|P_E(K)-P_E(K)|^{-n}\,d\nu_{n,k}(E)\Big)^{-\frac{1}{kn}}$$
$$\leqslant|K|^{-\frac{1}{n}}\Big(\int_{G_{n,k}}|P_E(K)-P_E(K)|^{-n}\,d\nu_{n,k}(E)\Big)^{-\frac{1}{kn}}$$
$$\leqslant|K|^{-\frac{1}{n}}\Big(\int_{G_{n,k}}4^{-kn}|P_E(K)|^{-n}\,d\nu_{n,k}(E)\Big)^{-\frac{1}{kn}}$$
$$=4\Phi_k(K).$$

Since $K-K$ is origin-symmetric, we know that $\Phi_k(K-K)\geqslant c\sqrt{n/k}$, and the general case follows. □

Subsequently, E. Milman and Yehudayoff [418] established the sharp lower bound $\Phi_k(B_2^n)\leqslant\Phi_k(K)$ and verified Lutwak's conjecture, including a characterization of the equality cases, for all $1\leqslant k\leqslant n-1$: ellipsoids are the only local minimizers with respect to the Hausdorff metric.

Regarding the upper bound in (3.22), an almost optimal estimate (up to a $\log n$-term) was given by Dafnis and Paouris in [144]. Let us briefly recall their argument. It is a classical fact that if K is a convex body in $\mathbb{R}^n$, then the sequence

$$Q_k(K)=\Big(\frac{1}{\omega_k}\int_{G_{n,k}}|P_F(K)|\,d\nu_{n,k}(F)\Big)^{1/k} \tag{3.25}$$

is decreasing in k. This is a consequence of the Aleksandrov–Fenchel inequality (see [116, Sections 20.1 and 20.2] or [505, Section 6.4]). In particular, for any $1 \leqslant k \leqslant n-1$ we have $Q_k(K) \leqslant Q_1(K)$, which may be written in the equivalent form

$$\left(\frac{1}{\omega_k} \int_{G_{n,k}} |P_F(K)|\, d\nu_{n,k}(F)\right)^{\frac{1}{k}} \leqslant w(K), \tag{3.26}$$

where $w(K)$ is the mean width of K. Then, by Hölder's inequality

$$\begin{aligned} &|K|^{-1/n}\left(\int_{G_{n,k}} |P_F(K)|^{-n}\, d\nu_{n,k}(F)\right)^{-\frac{1}{kn}} \\ &\qquad \leqslant |K|^{-1/n}\left(\int_{G_{n,k}} |P_F(K)|\, d\nu_{n,k}(F)\right)^{\frac{1}{k}} \leqslant \omega_k^{1/k} |K|^{-1/n} w(K). \end{aligned}$$

Since the term on the left-hand side of this inequality is invariant under volume preserving affine transformations, we may assume that K has minimal mean width, and it is known that in this case we have $w(K) \leqslant c\sqrt{n}\log n\, |K|^{1/n}$ for some absolute constant $c > 0$ (see [27, Chapter 6]). Combining the above we get

$$\Phi_k(K) \leqslant c_2 \sqrt{n/k}\, \log n. \tag{3.27}$$

It was also shown in [144] that

$$\Phi_k(K) \leqslant c_3 (n/k)^{3/2} \sqrt{\log (en/k)}.$$

In other words, if k is proportional to n, then the upper bound for $\Phi_k(K)$ is of the order of 1.

An interesting question that remains open is whether the $\log n$-term in (3.27) can actually be dropped.

The lower-dimensional Busemann–Petty problem can be posed for a general measure in place of volume. For any $1 \leqslant k \leqslant n-1$ and any measure μ on $\mathbb{R}^n$ with a locally integrable, non-negative density g one may define $\beta_{n,k}(\mu)$ as the smallest constant $\beta > 0$ with the following property: For every pair of centered convex bodies K and L in $\mathbb{R}^n$ that satisfy $\mu(K \cap F) \leqslant \mu(L \cap F)$ for every $F \in G_{n,n-k}$, one has

$$\mu(K) \leqslant \beta^k \mu(L).$$

Similarly, one may define the "symmetric" constant $\beta_{n,k}^{(s)}(\mu)$. In the previous section, we saw that $\beta_{n,1}^{(s)}(\mu) \leqslant \sqrt{n}$ for every measure μ with an even, continuous non-negative density.

We start the discussion with a result from [331] where the authors studied the lower-dimensional isomorphic Busemann–Petty problem for measures: in fact they studied a more general version with two different measures μ and ν.

Theorem 3.4.4. *Let K, L be star bodies in $\mathbb{R}^n$, let $0 < k < n$, and let μ and ν be measures on $\mathbb{R}^n$ with non-negative locally integrable densities f, g such that $\|g\|_\infty = g(0) = 1$ and*

$$\mu(K \cap E) \leqslant \nu(L \cap E) \tag{3.28}$$

for all $E \in G_{n,n-k}$. Then,

$$\mu(K) \leqslant (d_{\rm ovr}(K, \mathcal{BP}_k^n))^k \frac{n}{n-k} |K|^{k/n} (\nu(L))^{\frac{n-k}{n}}.$$

Recall that, by John's theorem, if K is an origin-symmetric convex body in $\mathbb{R}^n$, then $d_{\rm ovr}(K, \mathcal{BP}_k^n) \leqslant \sqrt{n}$. Also, as proved in [329] (see Theorem 1.7.19) for any origin-symmetric convex body K in $\mathbb{R}^n$

$$d_{\rm ovr}(K, \mathcal{BP}_k^n) \leqslant C\sqrt{n/k}(\log(en/k))^{3/2}.$$

The distance $d_{\rm ovr}(K, \mathcal{BP}_k^n)$ is bounded by an absolute constant for many classes of convex bodies K. For example, it is proved in [317] that this distance is bounded by e for unconditional convex bodies (see Proposition 4.3.3). Also, as proved by E. Milman [413] this distance is bounded by $\sqrt{p}$ if K is the unit ball of a subspace of L_p, $p > 2$ (see Proposition 4.3.4). Recall that when K is the unit ball of a subspace of L_p with $0 < p \leqslant 2$, then K belongs to $\mathcal{BP}_k^n$ for every k (as follows from Propositions 1.7.10 and 1.7.15), so the distance is 1.

For the proof of Theorem 3.4.4 we need the next simple lemmas from [423].

Lemma 3.4.5. *Let K be an origin-symmetric convex body in $\mathbb{R}^n$. For every $p > -n$ we have that*

$$\int_K \|x\|_K^p \, dx = \frac{n}{n+p}|K|.$$

Proof. We compute the integral in polar coordinates

$$\begin{aligned}\int_K \|x\|_K^p \, dx &= \int_{S^{n-1}} \int_0^{\|\theta\|_K^{-1}} r^{n+p-1} \|\theta\|_K^p \, dr \, d\theta = \int_{S^{n-1}} \frac{1}{n+p} \|\theta\|_K^{-n-p} \|\theta\|_K^p \, d\theta \\ &= \frac{1}{n+p} \int_{S^{n-1}} \|\theta\|_K^{-n} \, d\theta = \frac{n}{n+p}|K|\end{aligned}$$

by (1.4). □

Lemma 3.4.6. *Let $f : \mathbb{R}^n \to [0, +\infty)$ be a measurable function such that $\|f\|_\infty = 1$ and let K be an origin-symmetric convex body in $\mathbb{R}^n$. The function*

$$F_K(p) = \left(\frac{\int_{\mathbb{R}^n} \|x\|_K^p f(x)\, dx}{\int_K \|x\|_K^p\, dx} \right)^{1/(n+p)}$$

is increasing on $(-n, +\infty)$,

Proof. Given $p > q > -n$, for every $t > 0$ we write

$$\begin{aligned}\int_{\mathbb{R}^n} \|x\|_K^p f(x)\, dx &\geqslant t^{p-q} \int_{\mathbb{R}^n} \|x\|_K^q f(x)\, dx - \int_{tK} (t^{p-q}\|x\|_K^q - \|x\|_K^p) f(x)\, dx \\ &\geqslant t^{p-q} \int_{\mathbb{R}^n} \|x\|_K^q f(x)\, dx - \int_{\mathbb{R}^n} t^{p+n}(\|x\|_K^q - \|x\|_K^p)\, dx.\end{aligned}$$

We optimize this inequality choosing $t = F_K(q)$. Using also Lemma 3.4.5 we check that this choice gives $F_K(p) \geqslant F_K(q)$. □

Proof of Theorem 3.4.4. Fix a small $\delta > 0$ and let $D \in \mathcal{BP}_k^n$ be a body such that $K \subseteq D$ and

$$|D|^{1/n} \leqslant (1 + \delta)\, d_{\mathrm{ovr}}(K, \mathcal{BP}_k^n)\, |K|^{1/n}. \tag{3.29}$$

Now, let μ_D be the positive functional corresponding to D by the definition of generalized k-intersection bodies; see (1.92). Then, for every $g \in C(S^{n-1})$ we have that

$$\int_{S^{n-1}} \|x\|_D^{-k} g(x)\, dx = \mu_D(R_{n-k} g). \tag{3.30}$$

Since

$$|K \cap E| = \frac{1}{n-k} R_{n-k}(\|\cdot\|^{-n+k})(E)$$

and

$$\mu(K \cap H) = \int_{K \cap E} f = R_{n-k}\left(\int_0^{\|\cdot\|_K^{-1}} r^{n-k-1} f(r\,\cdot)\, dr \right)(E), \tag{3.31}$$

the assumption (3.28) of the theorem can be written as

$$R_{n-k}\left(\int_0^{\|\cdot\|_K^{-1}} r^{n-k-1} f(r\,\cdot)\, dr \right)(E) \leqslant R_{n-k}\left(\int_0^{\|\cdot\|_L^{-1}} r^{n-k-1} g(r\,\cdot)\, dr \right)(E)$$

for every $E \in G_{n,n-k}$. Applying the functional μ_D to both sides of this inequality and using (1.92) and (3.30), we get

$$\int_{S^{n-1}} \|x\|_D^{-k} \left(\int_0^{\|x\|_K^{-1}} r^{n-k-1} f(rx)\,dr \right) dx$$
$$\leqslant \int_{S^{n-1}} \|x\|_D^{-k} \left(\int_0^{\|x\|_L^{-1}} r^{n-k-1} g(rx)\,dr \right) dx, \tag{3.32}$$

which is equivalent to

$$\int_K \|x\|_D^{-k} f(x)\,dx \leqslant \int_L \|x\|_D^{-k} g(x)\,dx. \tag{3.33}$$

Since $K \subset D$, we have $1 \geqslant \|x\|_K \geqslant \|x\|_D$ for every $x \in K$. Therefore,

$$\int_K \|x\|_D^{-k} f(x)\,dx \geqslant \int_K \|x\|_K^{-k} f(x)\,dx \geqslant \int_K f(x)\,dx = \mu(K).$$

On the other hand, from Lemma 3.4.6 we know that

$$\left(\frac{\int_L \|x\|_D^{-k} g(x)\,dx}{\int_D \|x\|_D^{-k}\,dx} \right)^{1/(n-k)} \leqslant \left(\frac{\int_L g(x)\,dx}{\int_D dx} \right)^{1/n}.$$

Lemma 3.4.5 shows that $\int_D \|x\|_D^{-k}\,dx = \frac{n}{n-k}|D|$, therefore we can estimate the right-hand side of (3.33) by

$$\int_L \|x\|_D^{-k} g(x)\,dx \leqslant \frac{n}{n-k} (\nu(L))^{\frac{n-k}{n}} |D|^{\frac{k}{n}}.$$

Then, we apply (3.32) and let $\delta \to 0$ to get the result. □

An extension of Theorem 3.4.4 was obtained in [215] (see also [216]).

Theorem 3.4.7. *Let K and L be star bodies in $\mathbb{R}^n$, let $0 < k < n$ be an integer, and let f, g be non-negative, continuous functions on K and L, respectively, such that $\|g\|_\infty = g(0) = 1$. Then,*

$$\frac{\int_K f}{(\int_L g)^{\frac{n-k}{n}} |K|^{\frac{k}{n}}} \leqslant \frac{n}{n-k} (d_{\mathrm{ovr}}(K, \mathcal{BP}_k^n))^k \max_{H \in G_{n,n-k}} \frac{\int_{K\cap H} f}{\int_{L\cap H} g}. \tag{3.34}$$

Proof. For a small $\delta > 0$, let $D \in \mathcal{BP}_k^n$ be a body such that $K \subset D$ and

$$|D|^{\frac{1}{n}} \leqslant (1+\delta)\, d_{\mathrm{ovr}}(K, \mathcal{BP}_k^n)\, |K|^{\frac{1}{n}}. \tag{3.35}$$

Let μ_D be the functional corresponding to D by (1.92) and choose ε such that

$$\int_{K\cap H} f \leqslant \varepsilon \int_{L\cap H} g$$

for all $H \in G_{n,n-k}$. By (3.31), we have

$$R_{n-k}\left(\int_0^{\|\cdot\|_K^{-1}} r^{n-k-1} f(r\,\cdot)\,dr\right)(H) \leqslant \varepsilon\, R_{n-k}\left(\int_0^{\|\cdot\|_L^{-1}} r^{n-k-1} g(r\,\cdot)\,dr\right)(H)$$

for every $H \in G_{n,n-k}$. Applying the positive functional μ_D to both sides, we get

$$\begin{aligned}\int_{S^{n-1}} \|x\|_D^{-k}\left(\int_0^{\|x\|_K^{-1}} r^{n-k-1} f(rx)\,dr\right)dx \\ \leqslant \varepsilon \int_{S^{n-1}} \|x\|_D^{-k}\left(\int_0^{\|x\|_L^{-1}} r^{n-k-1} g(rx)\,dr\right)dx,\end{aligned} \tag{3.36}$$

which is equivalent to

$$\int_K \|x\|_D^{-k} f(x)\,dx \leqslant \varepsilon \int_L \|x\|_D^{-k} g(x)\,dx. \tag{3.37}$$

Since $K \subset D$, we have $1 \geqslant \|x\|_K \geqslant \|x\|_D$ for every $x \in K$. Therefore,

$$\int_K \|x\|_D^{-k} f(x)\,dx \geqslant \int_K \|x\|_K^{-k} f(x)\,dx \geqslant \int_K f.$$

On the other hand, by Lemma 3.4.6v

$$\left(\frac{\int_L \|x\|_D^{-k} g(x)\,dx}{\int_D \|x\|_D^{-k}\,dx}\right)^{1/(n-k)} \leqslant \left(\frac{\int_L g(x)\,dx}{\int_D dx}\right)^{1/n}.$$

Since $\int_D \|x\|_D^{-k}\,dx = \frac{n}{n-k}|D|$, we can estimate the right-hand side of (3.37) by

$$\int_L \|x\|_D^{-k} g(x)\,dx \leqslant \varepsilon \frac{n}{n-k}\left(\int_L g\right)^{\frac{n-k}{n}} |D|^{\frac{k}{n}}.$$

Applying (3.35) and sending δ to zero, we see that the latter inequality in conjunction with (3.37) implies

$$\int_K f \leqslant \varepsilon \frac{n}{n-k}(d_{\mathrm{ovr}}(K, \mathcal{BP}_k^n))^k |K|^{\frac{k}{n}}.$$

Choosing $\varepsilon = \max_{H\in G_{n,n-k}} \frac{\int_{K\cap H} f}{\int_{L\cap H} g}$ we conclude the proof. □

From Theorem 3.4.7, we can recover the isomorphic Busemann–Petty theorem for the Radon transform established in Theorem 3.4.4, as follows. If, in addition to the conditions of Theorem 3.4.7, we assume that

$$\int_{K\cap H} f \leqslant \int_{L\cap H} g$$

for all $H \in G_{n,n-k}$, then we get

$$\int_K f \leqslant \frac{n}{n-k}(d_{\mathrm{ovr}}(K, \mathcal{BP}_k^n))^k |K|^{\frac{k}{n}} \left(\int_L g\right)^{\frac{n-k}{n}}.$$

Another consequence of Theorem 3.4.7 is a mean value inequality for the Radon transform. If we choose $K = L$ and $g \equiv 1$, then we obtain the following inequality.

Theorem 3.4.8. *Let K be a star body in $\mathbb{R}^n$, let $0 < k < n$ be an integer, and let f be a non-negative continuous function on K. Then,*

$$\frac{\int_K f}{|K|} \leqslant \frac{n}{n-k}(d_{\mathrm{ovr}}(K, \mathcal{BP}_k^n))^k \max_H \frac{\int_{K\cap H} f}{|K\cap H|}. \tag{3.38}$$

If $f \equiv 1$, $g \equiv 1$, we can get a slightly sharper (without the factor $\frac{n}{n-k}$) inequality than what Theorem 3.4.7 gives in this case.

Theorem 3.4.9. *Let K, L be star bodies in $\mathbb{R}^n$ and $0 < k < n$, then*

$$\left(\frac{|K|}{|L|}\right)^{\frac{n-k}{n}} \leqslant (d_{\mathrm{ovr}}(K, \mathcal{BP}_k^n))^k \max_{H\in G_{n,n-k}} \frac{|K\cap H|}{|L\cap H|}.$$

Proof. Let ε be such that $|K\cap H| \leqslant \varepsilon\, |L\cap H|$ for all $H \in G_{n,n-k}$, and let D be as in the proof of Theorem 3.4.7. By (3.31), for all H

$$R_{n-k}(\|\cdot\|_K^{-n+k})(H) \leqslant \varepsilon\, R_{n-k}(\|\cdot\|_L^{-n+k})(H).$$

Applying the positive functional μ_D to both sides, we get by (1.92) that

$$\int_{S^{n-1}} \|x\|_D^{-k}\|x\|_K^{-n+k}\, dx \leqslant \varepsilon \int_{S^{n-1}} \|x\|_D^{-k}\|x\|_L^{-n+k}\, dx.$$

Since $K \subset D$, we have $1 \geqslant \|x\|_K \geqslant \|x\|_D$, and the left-hand side is greater than $n|K|$. Using this and Hölder's inequality

$$n|K| \leqslant \varepsilon \int_{S^{n-1}} \|x\|_D^{-k} \|x\|_L^{-n+k}\, dx \leqslant \varepsilon \Big(\int_{S^{n-1}} \|x\|_D^{-n}\, dx \Big)^{\frac{k}{n}} \Big(\int_{S^{n-1}} \|x\|_L^{-n}\, dx \Big)^{\frac{n-k}{n}}$$
$$= \varepsilon n |D|^{\frac{k}{n}} |L|^{\frac{n-k}{n}} \leqslant \varepsilon n\, (1+\delta)^k \,(d_{\mathrm{ovr}}(K, \mathcal{BP}_k^n))^k\, |K|^{\frac{k}{n}} |L|^{\frac{n-k}{n}}.$$

Sending δ to zero and setting

$$\varepsilon = \max_{H \in G_{n,n-k}} \frac{|K \cap H|}{|L \cap H|},$$

we get the result. □

Theorem 3.4.7 also implies a lower estimate for the sup-norm of the Radon transform. Indeed, applying Theorem 3.4.7 with $L = B_2^n$ and $g \equiv 1$ we obtain the following slicing inequality from [317]

$$\int_K f \leqslant \frac{n}{n-k} \frac{\omega_n^{\frac{n-k}{n}}}{\omega_{n-k}} (d_{\mathrm{ovr}}(K, \mathcal{BP}_k^n))^k |K|^{\frac{k}{n}} \max_H \int_{K \cap H} f.$$

In Chapter 4, we shall discuss slicing inequalities in more detail.

3.5 Stability in comparison problems

Generalizing the Busemann–Petty problem we may say that a typical volume comparison problem asks whether, for any K, L from a certain class of origin-symmetric convex bodies in $\mathbb{R}^n$, knowing that

$$g_K(\xi) \leqslant g_L(\xi)$$

for all $\xi \in S^{n-1}$ we can conclude that $|K| \leqslant |L|$, where g_K is a geometric characteristic of K. As a primary example we can have in mind the hyperplane section function $g_K(\xi) = |K \cap \xi^\perp|$.

In the case where the answer to a volume comparison problem is affirmative, one can ask a stronger stability question. Suppose that $\varepsilon > 0$ and

$$g_K(\xi) \leqslant g_L(\xi) + \varepsilon \tag{3.39}$$

for all $\xi \in S^{n-1}$. Does there exist a constant c not dependent on ε and such that for every $\varepsilon > 0$

$$|K|^{\frac{n-1}{n}} \leqslant |L|^{\frac{n-1}{n}} + c\varepsilon? \tag{3.40}$$

Stability results are related to hyperplane inequalities as follows. Suppose that stability holds for both pairs K, L and L, K with the same constant c. If

$$\varepsilon = \max_{\xi \in S^{n-1}} |g_K(\xi) - g_L(\xi)|,$$

then one can switch K and L in (3.39) and, correspondingly, in (3.40). The resulting inequality for volumes will be called a volume difference inequality

$$\left||K|^{\frac{n-1}{n}} - |L|^{\frac{n-1}{n}}\right| \leqslant c\varepsilon = c \max_{\xi \in S^{n-1}} |g_K(\xi) - g_L(\xi)|. \tag{3.41}$$

Suppose now that the function g_L converges to zero uniformly with respect to ξ when $L = \beta B_2^n$ is a multiple of the Euclidean unit ball and $\beta \to 0$. Then, when $L = \beta B_2^n$ and $\beta \to 0$, the inequality (3.41) turns into what we call a hyperplane inequality

$$|K|^{\frac{n-1}{n}} \leqslant c \max_{\xi \in S^{n-1}} g_K(\xi). \tag{3.42}$$

One can also consider a separation problem. Suppose that $\varepsilon > 0$ and

$$g_K(\xi) \leqslant g_L(\xi) - \varepsilon \tag{3.43}$$

for all $\xi \in S^{n-1}$. Does there exist a constant c not dependent on ε and such that for every $\varepsilon > 0$

$$|K|^{\frac{n-1}{n}} \leqslant |L|^{\frac{n-1}{n}} - c\varepsilon? \tag{3.44}$$

In the case where the answer is affirmative, assuming that

$$\varepsilon = \min_{\xi \in S^{n-1}} (g_L(\xi) - g_K(\xi)) > 0,$$

we get another kind of volume difference inequality

$$|L|^{\frac{n-1}{n}} - |K|^{\frac{n-1}{n}} \geqslant c\varepsilon = c \min_{\xi \in S^{n-1}} (g_L(\xi) - g_K(\xi)). \tag{3.45}$$

Again, if $g_{\beta B_2^n}$ converges to zero uniformly in ξ when $\beta \to 0$, we get the following version of a hyperplane inequality:

$$|L|^{\frac{n-1}{n}} \geqslant c \min_{\xi \in S^{n-1}} g_L(\xi). \tag{3.46}$$

This strategy was first applied in [311] to several functions g_K including the hyperplane section function and the hyperplane projection function. In [313] similar inequalities were proved for a general measure with continuous density in place of volume.

Sections of lower dimensions were considered in [321], and stability and hyperplane inequalities for complex convex bodies were proved in [312, 330].

In this section, we review stability and separation results and prove some of them with the best possible constants. We also prove several hyperplane inequalities for intersection and projection bodies.

Suppose that

$$g_K(\xi) = A_K(\xi) = |K \cap \xi^\perp|, \quad \xi \in S^{n-1}$$

is the hyperplane section function. Then, the volume comparison question is the classical Busemann–Petty problem. The corresponding stability result was proved in [311, Theorem 1]. Let us note that the theorem is stated in [311] with

$$\tau_n := \omega_n^{\frac{n-1}{n}} / \omega_{n-1}$$

replaced by 1, although the proof there actually establishes the result with the constant τ_n, which is the best possible. Also, the proof in [311] is geometric, while here we use methods of Fourier analysis.

Theorem 3.5.1. *Suppose that $\varepsilon > 0$, K and L are origin-symmetric star bodies in $\mathbb{R}^n$, and K is an intersection body. If for every $\xi \in S^{n-1}$*

$$|K \cap \xi^\perp| \leqslant |L \cap \xi^\perp| + \varepsilon, \tag{3.47}$$

then

$$|K|^{\frac{n-1}{n}} \leqslant |L|^{\frac{n-1}{n}} + \tau_n \varepsilon.$$

Proof. By approximation (see, for example, Theorem 1.2.4), we can assume that the bodies K and L are infinitely smooth. Since

$$|K \cap \xi^\perp| = \frac{1}{\pi(n-1)} \left(\|x\|_K^{-n+1}\right)^\wedge(\xi) \tag{3.48}$$

for all $\xi \in S^{n-1}$, we may write (3.47) as

$$\left(\|x\|_K^{-n+1}\right)^\wedge(\xi) \leqslant \left(\|x\|_L^{-n+1}\right)^\wedge(\xi) + \pi(n-1)\varepsilon \tag{3.49}$$

for all $\xi \in S^{n-1}$. Note that, by Lemma 1.4.13, $(\|x\|_K^{-1})^\wedge$ is an infinitely smooth non-negative function on the sphere. By (3.49), the polar formula for volume and Parseval's formula on the sphere, Proposition 1.6.4

$$(2\pi)^n n|K| = (2\pi)^n \int_{S^{n-1}} \|x\|_K^{-n+1} \|x\|_K^{-1}\, dx = \int_{S^{n-1}} (\|x\|_K^{-1})^\wedge(\theta) (\|x\|_K^{-n+1})^\wedge(\theta)\, d\theta$$

$$\leqslant \int_{S^{n-1}} (\|x\|_K^{-1})^\wedge(\theta)(\|x\|_L^{-n+1})^\wedge(\theta)\,d\theta + \pi(n-1)\varepsilon \int_{S^{n-1}} (\|x\|_K^{-1})^\wedge(\theta)\,d\theta. \tag{3.50}$$

By Parseval's formula and Hölder's inequality

$$\int_{S^{n-1}} (\|x\|_K^{-1})^\wedge(\theta)(\|x\|_L^{-n+1})^\wedge(\theta)\,d\theta$$
$$= (2\pi)^n \int_{S^{n-1}} \|x\|_L^{-n+1}\|x\|_K^{-1}\,dx \leqslant (2\pi)^n n|K|^{\frac{1}{n}}|L|^{\frac{n-1}{n}}. \tag{3.51}$$

To estimate the second summand in (3.50), we use the formula from Lemma 1.4.3

$$(|x|^{-n+1})^\wedge(\theta) = \frac{2\pi^{\frac{n+1}{2}}}{\Gamma(\frac{n-1}{2})}|\theta|^{-1}.$$

Again, using Parseval's formula and then Hölder's inequality, we see that

$$\begin{aligned}
\int_{S^{n-1}} (\|x\|_K^{-1})^\wedge(\theta)\,d\theta &= \frac{\Gamma(\frac{n-1}{2})}{2\pi^{\frac{n+1}{2}}} \int_{S^{n-1}} (\|x\|_K^{-1})^\wedge(\theta)(|x|^{-n+1})^\wedge(\theta)\,d\theta \\
&= \frac{(2\pi)^n\Gamma(\frac{n-1}{2})}{2\pi^{\frac{n+1}{2}}} \int_{S^{n-1}} \|x\|_K^{-1}\,dx \\
&\leqslant \frac{(2\pi)^n\Gamma(\frac{n-1}{2})\kappa_n^{\frac{n-1}{n}}}{2\pi^{\frac{n+1}{2}}} \left(\int_{S^{n-1}} \|x\|_K^{-n}\,dx\right)^{\frac{1}{n}} \\
&= \frac{(2\pi)^n\Gamma(\frac{n-1}{2})\kappa_n^{\frac{n-1}{n}}}{2\pi^{\frac{n+1}{2}}}(n|K|)^{\frac{1}{n}}.
\end{aligned}$$

Combining this with (3.50) and (3.51), we get

$$(2\pi)^n n|K| \leqslant (2\pi)^n n|K|^{\frac{1}{n}}|L|^{\frac{n-1}{n}} + \frac{(2\pi)^n\varepsilon\pi(n-1)n^{\frac{1}{n}}\Gamma(\frac{n-1}{2})\kappa_n^{\frac{n-1}{n}}}{2\pi^{\frac{n+1}{2}}}|K|^{1/n}.$$

Finally, to represent the coefficient in the required form, recall that $\kappa_n = n\omega_n = 2\pi^{\frac{n}{2}}/\Gamma(\frac{n}{2})$. □

Interchanging K and L in Theorem 3.5.1, we get the corresponding volume difference inequality.

Corollary 3.5.2. *If K and L are intersection bodies in $\mathbb{R}^n$, then*

$$\left||K|^{\frac{n-1}{n}} - |L|^{\frac{n-1}{n}}\right| \leqslant \tau_n \max_{\xi\in S^{n-1}} \left||K\cap\xi^\perp| - |L\cap\xi^\perp|\right|.$$

If $L = \delta B_2^n$ in the latter inequality, then sending δ to zero we get that for any intersection body K in $\mathbb{R}^n$

$$|K|^{\frac{n-1}{n}} \leqslant \tau_n \max_{\xi \in S^{n-1}} |K \cap \xi^{\perp}|. \tag{3.52}$$

Note that (3.52) immediately follows from the affirmative part of the Busemann–Petty problem; see [194, Theorem 9.4.11].

A separation result for hyperplane sections was proved in [311, Theorem 2]. The constant c in this result does not depend on ε, but depends on the dimension and on the normalized inradius of K

$$r(K) = \frac{\min_{\xi \in S^{n-1}} \rho_K(\xi)}{|K|^{1/n}}.$$

Theorem 3.5.3. *Let K and L be origin-symmetric star bodies in $\mathbb{R}^n$ and $\varepsilon > 0$. Assume that K is an intersection body. If*

$$|K \cap \xi^{\perp}| \leqslant |L \cap \xi^{\perp}| - \varepsilon \tag{3.53}$$

for every $\xi \in S^{n-1}$, then

$$|K|^{\frac{n-1}{n}} \leqslant |L|^{\frac{n-1}{n}} - \sqrt{\frac{2\pi}{n+1}}\, r(K)\varepsilon.$$

Since the answer to the Busemann–Petty problem is negative in most dimensions, one may ask what information about the hyperplane section function A_K does allow us to compare the volumes in all dimensions. An answer to this question was given in [335]: for two origin-symmetric infinitely smooth bodies K, L in $\mathbb{R}^n$ and $\alpha \in [n-4, n-1)$ the assumption that

$$(-\Delta)^{\alpha/2} A_K(\xi) \leqslant (-\Delta)^{\alpha/2} A_L(\xi) \tag{3.54}$$

for all $\xi \in S^{n-1}$ implies that $|K| \leqslant |L|$, while for $\alpha < n-4$ this is not necessarily true. Here, Δ is the Laplace operator on $\mathbb{R}^n$, and the fractional powers of the Laplacian are defined by

$$(-\Delta)^{\alpha/2} f = \frac{1}{(2\pi)^n} \left(|x|^{\alpha} \hat{f}(x) \right)^{\wedge},$$

where the Fourier transform is considered in the sense of distributions and the function A_K is extended in (3.54) to a homogeneous function of degree -1 on $\mathbb{R}^n$. The corresponding stability result was proved in [311, Theorem 3].

Theorem 3.5.4. *Let $\varepsilon > 0$, $\alpha \in [n-4, n-1)$, and let K and L be origin-symmetric, infinitely smooth, convex bodies in $\mathbb{R}^n$, $n \geqslant 4$, such that for every $\xi \in S^{n-1}$*

$$(-\Delta)^{\alpha/2}A_K(\xi) \leqslant (-\Delta)^{\alpha/2}A_L(\xi) + \varepsilon. \tag{3.55}$$

Then,

$$|K|^{\frac{n-1}{n}} \leqslant |L|^{\frac{n-1}{n}} + c\varepsilon,$$

where

$$c = c(\alpha, n) = \frac{\sqrt{\pi}(n-1)\Gamma(\frac{n-\alpha-1}{2})}{2^{\alpha+\frac{1}{n}} n^{\frac{n-1}{n}} \Gamma(\frac{\alpha+1}{2})(\Gamma(\frac{n}{2}))^{\frac{n-1}{n}}}.$$

A separation result was proved in [311, Theorem 4].

Theorem 3.5.5. *Let $\varepsilon > 0$, $\alpha \in [n-4, n-1)$, K and L be origin-symmetric, infinitely smooth, convex bodies in $\mathbb{R}^n$, $n \geqslant 4$, such that for every $\xi \in S^{n-1}$*

$$(-\Delta)^{\alpha/2}A_K(\xi) \leqslant (-\Delta)^{\alpha/2}A_L(\xi) - \varepsilon. \tag{3.56}$$

Then,

$$|K|^{\frac{n-1}{n}} \leqslant |L|^{\frac{n-1}{n}} - c\varepsilon,$$

where

$$c = r(K)\frac{\pi(n-1)\Gamma(\frac{n-\alpha-1}{2})}{n 2^{\alpha}\Gamma(\frac{\alpha+1}{2})\Gamma(\frac{n}{2})}.$$

In the more general case of arbitrary measures, a stability version of Theorem 3.2.3 was established in [313, Theorem 2]. Note that in the case of volume (when $f \equiv 1$), the result of the next theorem is weaker than that of Theorem 3.5.1.

Theorem 3.5.6. *Let f be an even, non-negative, continuous function on $\mathbb{R}^n$, let μ be the measure with density f, let K and L be origin-symmetric star bodies in $\mathbb{R}^n$, and let $\varepsilon > 0$. Suppose that K is an intersection body and that for every $\xi \in S^{n-1}$*

$$\mu(K \cap \xi^{\perp}) \leqslant \mu(L \cap \xi^{\perp}) + \varepsilon. \tag{3.57}$$

Then,

$$\mu(K) \leqslant \mu(L) + \frac{n}{n-1}\tau_n |K|^{1/n}\varepsilon. \tag{3.58}$$

Interchanging K and L, we get the following volume difference inequality.

Corollary 3.5.7. *If K and L are intersection bodies in $\mathbb{R}^n$, then*

$$|\mu(K) - \mu(L)| \leqslant \frac{n\tau_n}{n-1} \max_{\xi \in S^{n-1}} |\mu(K \cap \xi^{\perp}) - \mu(L \cap \xi^{\perp})| \max\{|K|^{\frac{1}{n}}, |L|^{\frac{1}{n}}\}. \tag{3.59}$$

Sending L to the empty set, we obtain a hyperplane inequality for arbitrary measures on intersection bodies.

Corollary 3.5.8. *If K is an intersection body in $\mathbb{R}^n$, then*

$$\mu(K) \leqslant \frac{n}{n-1}\tau_n \max_{\xi\in S^{n-1}} \mu(K \cap \xi^\perp)\, |K|^{1/n}. \tag{3.60}$$

The constant in (3.60) is sharp; it is achieved asymptotically when $K = B_2^n$ and μ converges weakly to the uniform measure on the sphere S^{n-1} (see [313]).

Definition 3.5.9. We introduce here the parameters $\tau_{n,k}$ that will be used frequently in subsequent sections

$$\tau_{n,k} = \omega_n^{\frac{n-k}{n}} / \omega_{n-k}, \quad 1 \leqslant k \leqslant n-1. \tag{3.61}$$

The constant τ_n used in this section earlier is $\tau_{n,1}$ here.

The following estimates are well known. We include the proof from [320, Lemma 2.1].

Proposition 3.5.10. *For any $n \geqslant 2$ and $1 \leqslant k \leqslant n-1$ we have that*

$$e^{-k/2} \leqslant \tau_{n,k} \leqslant 1.$$

Proof. Recall that the volume ω_n of the unit Euclidean ball in $\mathbb{R}^n$ is equal to

$$\omega_n = |B_2^n| = \frac{\pi^{n/2}}{\Gamma(1+\frac{n}{2})}.$$

Therefore, we need to show that

$$1 \leqslant \frac{\Gamma(\frac{n}{2}+1)^{\frac{n-k}{n}}}{\Gamma(\frac{n-k}{2}+1)} \leqslant e^{k/2}. \tag{3.62}$$

For the lower bound, we use the convexity of $\log\Gamma$ (see, for example, [310, Lemma 2.14]) to write

$$\frac{\log\Gamma(\frac{n}{2}+1) - \log\Gamma(1)}{n/2} \geqslant \frac{\log\Gamma(\frac{n-k}{2}+1) - \log\Gamma(1)}{(n-k)/2},$$

which implies our claim.

For the upper bound, we write (3.62) in the equivalent form

$$\frac{\Gamma(\frac{n}{2}+1)}{\Gamma(\frac{n-k}{2}+1)} \frac{1}{(\Gamma(\frac{n}{2}+1))^{\frac{k}{n}} e^{k/2}} \leqslant 1 \tag{3.63}$$

and using again the convexity of $\log\Gamma$ we get

$$\Gamma^2\left(\frac{n}{2}+1\right) \leqslant \Gamma\left(\frac{n}{2}+\frac{3}{2}\right)\Gamma\left(\frac{n}{2}+\frac{1}{2}\right) = \left(\frac{n}{2}+\frac{1}{2}\right)\Gamma^2\left(\frac{n}{2}+\frac{1}{2}\right).$$

It follows that

$$\frac{\Gamma(\frac{n}{2}+1)}{\Gamma(\frac{n}{2}+\frac{1}{2})} \leqslant \left(\frac{n}{2}+\frac{1}{2}\right)^{1/2}. \tag{3.64}$$

We immediately see now that for $k > 1$

$$\frac{\Gamma(\frac{n}{2}+1)}{\Gamma(\frac{n-k}{2}+1)} \leqslant \left(\frac{n}{2}\right)^{k/2}. \tag{3.65}$$

For the second fraction in (3.63) we use the fact that, from Stirling's formula, for every $x \geqslant 1/2$ we have

$$\begin{aligned}\Gamma(x+1) &\geqslant \sqrt{2\pi}(x+1)^{x+1/2}e^{-x-1} = \frac{\sqrt{2\pi}}{e}(x/e)^x(x+1/x)^x\sqrt{x+1} \\ &\geqslant (x/e)^x\sqrt{x+1}.\end{aligned} \tag{3.66}$$

Setting $x = n/2$ in (3.66) we see that the second fraction in (3.63) is bounded by

$$(n/2)^{-k/2}\left(\frac{n}{2}+1\right)^{-\frac{k}{2n}}. \tag{3.67}$$

Combining this bound with (3.65) we obtain (3.63) for $k > 1$. In the case $k = 1$, after combining (3.67) and (3.64) we see that it remains to prove that

$$(1+2/n)^{1/2}(n/2+1)^{-1/(2n)} < 1.$$

This inequality reduces to

$$(1+2/n)^{n/2} < e < (n/2+1)^{1/2},$$

which is true for $n \geqslant 16$. For $k = 1$ and $n < 16$ we can also check the upper bound in (3.62) directly. □

3.6 The Shephard problem

In this section, we discuss the Shephard problem concerning projections instead of sections.

Problem 3.6.1. Let K and L be two origin-symmetric convex bodies in $\mathbb{R}^n$. Suppose that $|P_{\theta^\perp}(K)| < |P_{\theta^\perp}(L)|$ for every $\theta \in S^{n-1}$. Does it follow that $|K| < |L|$?

The answer is affirmative if $n = 2$: we have $|P_{\theta^\perp}(K)| = 2h_K(\theta + \frac{\pi}{2})$, where h_K is the support function of K, thus the hypothesis implies that $h_K < h_L$, i. e., $K \subset L$. Shortly after it was posed in 1964 in [508], the Shephard problem was answered in the negative for all $n \geqslant 3$. This was done independently by Petty [458] who gave an explicit counterexample in $\mathbb{R}^3$, and by Schneider [503] for all $n \geqslant 3$. Both authors employ Minkowski's theory of mixed volumes and show that the class of projection bodies plays a central role in the problem.

It will be useful to recall a few facts from the theory of mixed volumes; see Section 1.2.3. First, Minkowski's inequality states that

$$V_{n-1}(K, L) \geqslant |K|^{\frac{n-1}{n}} |L|^{\frac{1}{n}}$$

for all convex bodies $K, L \in \mathbb{R}^n$, where $V_{n-1}(K, L) = V(K, \ldots, K, L)$. We also have

$$V_{n-1}(K, L) = \frac{1}{n} \int_{S^{n-1}} h_L(u)\, dS_K(u),$$

where S_K is the surface area measure of K. For a polytope K with facets $F_1, \ldots, F_m$ having exterior normals $u_1, \ldots, u_m$, respectively, S_K is the measure supported on $\{u_1, \ldots, u_m\}$ with $S_K(\{u_j\}) = |F_j|, j = 1, \ldots, m$. By the monotonicity of mixed volumes (or by the integral representation of V_{n-1}) one also has that $L_1 \subseteq L_2$ implies $V_{n-1}(K, L_1) \leqslant V_{n-1}(K, L_2)$.

In this section, we will also use Minkowski's existence theorem in an essential way. Recall that it states that if $u_1, \ldots, u_m$ are distinct unit vectors with 0 in the interior of their convex hull, and $\gamma_1, \ldots, \gamma_m > 0$ are such that $\sum_{j=1}^m \gamma_j u_j = 0$, then there exists a polytope K having $u_1, \ldots, u_m$ as its (only) normal vectors and satisfying $|F_j| = \gamma_j, j = 1, \ldots, m$, where F_j is the facet of K corresponding to u_j. By approximation, one obtains that a finite Borel measure μ on S^{n-1} is the surface area measure of some $K \in \mathcal{K}^n$ if and only if μ is not concentrated on any great subsphere of S^{n-1}, and

$$\int_{S^{n-1}} u\, d\mu(u) = 0.$$

Note that the second condition is always satisfied if μ is an even measure.

Definition 3.6.2. Let $\mathcal{K}^n$ denote the class of non-empty, compact, convex sets in $\mathbb{R}^n$. If $K \in \mathcal{K}^n$, the projection body ΠK of K is the origin-symmetric convex body whose support function is defined by

$$h_{\Pi K}(\theta) = \big|P_{\theta^\perp}(K)\big|, \quad \theta \in S^{n-1}.$$

One has the integral representation (Cauchy formula)

$$|P_{\theta^\perp}(K)| = \frac{1}{2}\int_{S^{n-1}} |\langle u,\theta\rangle|\, dS_K(u),$$

which is easily verified in the case of a polytope, and extends to any $K \in \mathcal{K}^n$ by approximation. It follows that the support function of a projection body is given by

$$h_{\Pi K}(\theta) = \int_{S^{n-1}} |\langle u,\theta\rangle|\, d\mu(u)$$

for some finite even Borel measure on S^{n-1} (which also shows that all projection bodies are indeed convex). This integral representation of $h_{\Pi K}$ also shows that ΠK is a zonoid: a Hausdorff limit of a sequence of finite Minkowski sums of line segments. Minkowski's existence theorem implies that, conversely, a convex body is a zonoid only if it is the projection body of some origin-symmetric convex body in $\mathbb{R}^n$. Moreover, if we denote by $\mathcal{Z}$ the class of zonoids, Alexandrov's uniqueness theorem shows that the Minkowski map $\Pi : \mathcal{C}_n \to \mathcal{Z}$ with $K \mapsto \Pi K$, is injective. Note also that $\mathcal{Z}$ is invariant under invertible linear transformations and closed in the Hausdorff metric. We refer the reader to the books of Schneider [505], Gardner [194], and the survey article of Bolker [87] for more information.

The starting point for the answer to the Shephard problem is an identity that combines these classical results. If $K, L \in \mathcal{K}^n$, then

$$V_{n-1}(K,\Pi L) = V_{n-1}(L,\Pi K). \tag{3.68}$$

The proof is straightforward

$$V_{n-1}(K,\Pi L) = \frac{1}{n}\int_{S^{n-1}} h_{\Pi L}(u)\, dS_K(u) = \frac{1}{2n}\int_{S^{n-1}}\int_{S^{n-1}} |\langle u,\theta\rangle|\, dS_L(\theta)\, dS_K(u)$$

and the identity follows by a change in the order of integration.

This observation is crucial: suppose that $K, L \in \mathcal{K}^n$ with $|P_{\theta^\perp}(K)| < |P_{\theta^\perp}(L)|$ for all $\theta \in S^{n-1}$, and also assume that $L = \Pi L_0$ is the projection body of some $L_0 \in \mathcal{K}^n$. The hypothesis implies that $\Pi(\lambda K) \subseteq \Pi L$ for some $\lambda > 1$, and using (3.68) and the monotonicity of $V_{n-1}(L_0, \cdot)$ we see that

$$\begin{aligned} |L| = V_{n-1}(L,L) &= V_{n-1}(L,\Pi L_0) = V_{n-1}(L_0,\Pi L) \\ &\geqslant V_{n-1}(L_0,\Pi(\lambda K)) = V_{n-1}(\lambda K, L). \end{aligned}$$

Then, Minkowski's inequality gives

$$|L| \geqslant |\lambda K|^{\frac{n-1}{n}} |L|^{\frac{1}{n}}$$

and since $\lambda > 1$ this means that $|L| > |K|$. So, if the body L with larger projections is a zonoid, then the answer to the Shephard problem is affirmative, even if K is non-symmetric (this was proved by both Petty and Schneider).

This argument shows that when L is a projection body, the Shephard problem has a positive answer. In general, for all $n \geqslant 3$, one can produce counterexamples.

Theorem 3.6.3. *For any $n \geqslant 3$ there exists a pair of origin-symmetric convex bodies K and L with $|K| > |L|$ and such that for all $\theta \in S^{n-1}$, we have $|P_{\theta^\perp}(K)| \leqslant |P_{\theta^\perp}(L)|$.*

Projection bodies again hold the key for the proof, as given by Schneider. In fact, he shows that for any smooth enough origin-symmetric convex body K in $\mathbb{R}^n$ that is not a projection body, one can produce a corresponding L satisfying the statement in the theorem.

Proof. Suppose that the origin-symmetric convex body K is not a projection body. Assuming that K is sufficiently smooth one can show that there exists a smooth enough even function $g \in C(S^{n-1})$ such that

$$h_K(u) = \frac{1}{2} \int_{S^{n-1}} |\langle u, \theta\rangle| g(\theta)\, d\sigma(\theta), \quad u \in S^{n-1}.$$

This follows from the fact that the cosine transform $T : C_e^\infty(S^{n-1}) \to C_e^\infty(S^{n-1})$ defined by $(Tf)(u) = \int_{S^{n-1}} |\langle u, \theta\rangle| f(\theta)\, d\sigma(\theta)$ is a continuous bijection; see [194, p. 429] or Corollary 1.4.12. Since K is not a zonoid, g has to be negative on some Borel subset of S^{n-1} of positive measure. Therefore, we can define a non-negative function $F \in C_e^\infty(S^{n-1})$ that is not identically zero and satisfies $\{\theta : F(\theta) > 0\} \subset \{\theta : g(\theta) < 0\}$. This function F is the cosine transform of some function $f \in C_e^\infty(S^{n-1})$, i. e.

$$F(\theta) = \int_{S^{n-1}} |\langle u, \theta\rangle| f(u)\, d\sigma(u), \quad \theta \in S^{n-1}.$$

Let us denote by f_K the density function of the surface area measure S_K. Multiplying F by a positive constant if necessary, we have that $f_K + f$ is the density of a positive Borel measure μ on S^{n-1}. Since f is even, Minkowski's existence theorem shows that there exists an origin-symmetric convex body L for which $\mu = S_L$. Then, L is a perturbation of K that gives a negative answer to the Shephard problem: observe that for every $\theta \in S^{n-1}$

$$|P_{\theta^\perp}(L)| - |P_{\theta^\perp}(K)| = \frac{1}{2} \int_{S^{n-1}} |\langle u, \theta\rangle| f(u)\, d\sigma(u) = \frac{1}{2} F(\theta) \geqslant 0,$$

while

$$n(|K| - V_{n-1}(L, K)) = \int_{S^{n-1}} h_K\, dS_K(u) - \int_{S^{n-1}} h_K\, dS_L(u)$$

$$= -\int_{S^{n-1}} h_K(u) f(u)\, d\sigma(u)$$
$$= -\frac{1}{2}\int_{S^{n-1}}\int_{S^{n-1}} |\langle u,\theta\rangle| g(\theta) f(u)\, d\sigma(\theta)\, d\sigma(u)$$
$$= -\frac{1}{2}\int_{S^{n-1}} F(\theta) g(\theta)\, d\sigma(\theta) > 0$$

and hence Minkowski's inequality shows that $|K| > |L|$.

The above argument shows that if, for some $n \geqslant 3$, we can find a smooth enough origin-symmetric convex body in $\mathbb{R}^n$ that is not a projection body, then the Shephard problem has a negative answer in $\mathbb{R}^n$. The existence of such bodies is easily verified. The unit ball of ℓ_1^n, $n \geqslant 3$, has non-symmetric facets and this guarantees that it is not a zonoid (see Bolker [87]). Since this body can be approximated by smooth origin-symmetric convex bodies, and since the class of zonoids is closed under Hausdorff limits, there do exist smooth origin-symmetric convex bodies in $\mathbb{R}^n$, $n \geqslant 3$, which are not projection bodies. □

After the Petty–Schneider counterexamples to the Shephard problem, one might try to relax the question, asking for the smallest constant S_n (or the order of growth of this constant S_n as $n \to \infty$) for which: if K, L are origin-symmetric convex bodies in $\mathbb{R}^n$ and $|P_{\theta^\perp}(K)| < |P_{\theta^\perp}(L)|$ for all $\theta \in S^{n-1}$, then $|K| \leqslant S_n|L|$.

Such a constant S_n does exist, and a first estimate on its size can be established in a fairly simple way after what we have already seen: Let K, L be two origin-symmetric convex bodies in $\mathbb{R}^n$ and suppose that $|P_{\theta^\perp}(K)| < |P_{\theta^\perp}(L)|$ for all $\theta \in S^{n-1}$. Let $\mathcal{E}$ be the ellipsoid of maximal volume contained in L. From John's theorem, $L \subseteq \sqrt{n}\mathcal{E}$, and hence $|L|^{\frac{1}{n}} \leqslant \sqrt{n}|\mathcal{E}|^{\frac{1}{n}}$. A ball is obviously a zonoid and the class of zonoids is linearly invariant, so there exists another body $\mathcal{E}_0$, which is actually an ellipsoid, such that $\Pi\mathcal{E}_0 = \mathcal{E}$. Our assumption is that $\Pi K \subset \Pi L$, and our standard use of Minkowski's inequality yields

$$\begin{aligned}|L| &= V_{n-1}(L,L) \geqslant V_{n-1}(L,\mathcal{E}) = V_{n-1}(\mathcal{E}_0,\Pi L) \geqslant V_{n-1}(\mathcal{E}_0,\Pi K)\\ &= V_{n-1}(K,\mathcal{E}) \geqslant |K|^{\frac{n-1}{n}}|\mathcal{E}|^{\frac{1}{n}},\end{aligned}$$

that is,

$$|K|^{\frac{n-1}{n}} \leqslant \sqrt{n}|L|^{\frac{n-1}{n}}. \tag{3.69}$$

So, under the hypothesis $|P_{\theta^\perp}(K)| < |P_{\theta^\perp}(L)|$, $\theta \in S^{n-1}$, one has

$$|K| \leqslant c\sqrt{n}|L|,$$

where $c > 0$ is an absolute constant.

It turns out that this simple estimate cannot be essentially improved, due to a result of K. Ball [42].

Theorem 3.6.4 (Ball). *There exists an absolute constant $c > 0$ such that the constant S_n described above satisfies $S_n \geqslant c\sqrt{n}$.*

Proof. If S_K is the surface area measure of some origin-symmetric convex body K in $\mathbb{R}^n$, then by Minkowski's first inequality, for any other $D \in \mathcal{K}^n$ we must have

$$\frac{1}{n}\int_{S^{n-1}} h_D(u)\,dS_K(u) = V_{n-1}(K,D) \geqslant |K|^{\frac{n-1}{n}}|D|^{\frac{1}{n}}.$$

In other words, K minimizes $\int_{S^{n-1}} h_D\,dS_K$ among all $D \in \mathcal{K}^n$ that have the same volume as K.

We shall choose m and $u_j \in S^{n-1}$, γ_j, $j = 1,\dots,m$, in such a way that the origin-symmetric polytope K with surface area measure $\sum \gamma_j(\delta_{u_j} + \delta_{-u_j})$ has a very small volume when compared to its projections. First, we choose $\gamma_1 = \dots = \gamma_m = \gamma > 0$. All the facets of K will have the same area. Then, for every $\theta \in S^{n-1}$

$$|P_{\theta^\perp}(K)| = \frac{1}{2}\int_{S^{n-1}} |\langle u,\theta\rangle|\,dS_K(u) = \gamma\sum_{j=1}^m |\langle u_j,\theta\rangle|.$$

Define $L = \{x \in \mathbb{R}^n : |\langle x,u_j\rangle| \leqslant 1,\ j = 1,\dots,m\}$. Then, $h_L(u_j) \leqslant 1$, $j = 1,\dots,m$, and hence

$$\frac{2\gamma m}{n} \geqslant \frac{1}{n}\int_{S^{n-1}} h_L(u)\,dS_K(u) \geqslant |K|^{\frac{n-1}{n}}|L|^{\frac{1}{n}}.$$

It follows that for every $\theta \in S^{n-1}$

$$|P_{\theta^\perp}(K)| \geqslant \frac{n}{2m}|K|^{\frac{n-1}{n}}|L|^{\frac{1}{n}}\sum_{j=1}^m |\langle u_j,\theta\rangle|. \tag{3.70}$$

Here, we use a theorem of Vaaler [530].

Theorem 3.6.5 (Vaaler). *Let $u_1,\dots,u_m$ be vectors spanning $\mathbb{R}^n$. Consider the centrally symmetric convex body*

$$L = \{x \in \mathbb{R}^n : |\langle x,u_i\rangle| \leqslant 1,\ i = 1,\dots,m\}.$$

Then,

$$|L|^{1/n} \geqslant 2\sqrt{\frac{n}{\sum_{i=1}^m |u_i|^2}}.$$

In our case, we have $|u_i| = 1$ for all $i = 1, \dots, m$, and hence

$$|L|^{\frac{1}{n}} \geqslant 2\sqrt{n/m}.$$

Inserting this estimate into (3.70) we see that

$$|P_{\theta^\perp}(K)| \geqslant \left(\frac{n}{m}\right)^{\frac{3}{2}} \sum_{j=1}^{m} |\langle u_j, \theta\rangle|\, |K|^{\frac{n-1}{n}}, \quad \theta \in S^{n-1} \tag{3.71}$$

and we are left with the choice of m and $u_j, j = 1, \dots, m$.

The final step depends on the following consequence of Kashin's theorem on Euclidean subspaces of ℓ_1^N [277]: there exist vectors $u_1, \dots, u_{2n} \in S^{n-1}$ such that

$$c\sqrt{n}|x| \leqslant \sum_{j=1}^{2n} |\langle x, u_j\rangle| \leqslant 2\sqrt{n}|x| \tag{3.72}$$

for every $x \in \mathbb{R}^n$, where $c > 0$ is an absolute constant. Combining (3.71) and (3.72) we summarize as follows: there exists an origin-symmetric polytope K (corresponding to the choice of $m = 2n$, γ_j and $u_j, j = 1, \dots, m$ described above) with the following property: every projection $P_{\theta^\perp}(K)$, $\theta \in S^{n-1}$ of K has volume

$$|P_{\theta^\perp}(K)| \geqslant c_1\sqrt{n}|K|^{\frac{n-1}{n}}$$

for some absolute constant $c_1 > 0$. Let now λ be any positive number for which $|\lambda B_2^n| > S_n|K|$. There exists $\theta \in S^{n-1}$ such that $|P_{\theta^\perp}(\lambda B_2^n)| \geqslant |P_{\theta^\perp}(K)|$, which means that

$$\lambda^{n-1}\omega_{n-1} \geqslant c_1\sqrt{n}|K|^{\frac{n-1}{n}}.$$

It follows that

$$\left(\frac{S_n|K|}{\omega_n}\right)^{\frac{n-1}{n}} \omega_{n-1} \geqslant c_1\sqrt{n}|K|^{\frac{n-1}{n}},$$

that is,

$$S_n \geqslant c_1\sqrt{n}\,\tau_n \geqslant c_2\sqrt{n}.$$

Therefore, S_n is of the order of $\sqrt{n}$ as $n \to \infty$. □

We conclude this section with a quotient estimate from [215]. For an origin-symmetric convex body L in $\mathbb{R}^n$, we denote by

$$d_{\mathrm{vr}}(L,\Pi_n)=\inf\left\{\left(\frac{|L|}{|D|}\right)^{1/n}: D\in\Pi_n,\ D\subset L\right\}$$

the volume ratio distance from L to the class Π_n of projection bodies in $\mathbb{R}^n$.

Theorem 3.6.6. *Let K, L be origin-symmetric convex bodies in $\mathbb{R}^n$. Then,*

$$\left(\frac{|K|}{|L|}\right)^{\frac{n-1}{n}}\leqslant d_{\mathrm{vr}}(L,\Pi_n)\max_{\theta\in S^{n-1}}\frac{|P_{\theta^\perp}(K)|}{|P_{\theta^\perp}(L)|}.$$

Proof. Let $\varepsilon>0$ be a constant such that

$$|P_{\theta^\perp}(K)|\leqslant\varepsilon|P_{\theta^\perp}(L)|$$

for all $\theta\in S^{n-1}$. By the Cauchy formula

$$\int_{S^{n-1}}|\langle\xi,\theta\rangle|\,dS_K(\xi)\leqslant\varepsilon\int_{S^{n-1}}|\langle\xi,\theta\rangle|\,dS_L(\xi),\quad\forall\theta\in S^{n-1}.\tag{3.73}$$

Let $\delta>0$, and let D be a projection body in $\mathbb{R}^n$ such that

$$(1+\delta)d_{\mathrm{vr}}(L,\Pi_n)|D|^{1/n}\geqslant|L|^{1/n}.$$

Since D is a projection body, there exists a measure μ_D on S^{n-1} such that

$$h_D(\xi)=\int_{S^{n-1}}|\langle\xi,\theta\rangle|\,d\mu_D(\theta)$$

for all $\xi\in S^{n-1}$. Integrating both sides of (3.73) by θ over the sphere with respect to the measure μ_D, we get

$$\int_{S^{n-1}}h_D(\xi)\,dS_K(\xi)\leqslant\varepsilon\int_{S^{n-1}}h_D(\xi)\,dS_L(\xi).\tag{3.74}$$

Since $D\subset L$, we have $h_D\leqslant h_L$ in every direction, so by (1.22)

$$\int_{S^{n-1}}h_D(\xi)\,dS_L(\xi)\leqslant\int_{S^{n-1}}h_L(\xi)\,dS_L(\xi)=n|L|.$$

On the other hand, by Minkowski's first inequality and by the choice of D

$$\begin{aligned}V_{n-1}(K,D)&=\int_{S^{n-1}}h_D(\xi)\,dS_K(\xi)\geqslant n|D|^{\frac{1}{n}}|K|^{\frac{n-1}{n}}\\&\geqslant\frac{n}{(1+\delta)d_{\mathrm{vr}}(L,\Pi_n)}|L|^{\frac{1}{n}}|K|^{\frac{n-1}{n}}.\end{aligned}$$

Combining the last two inequalities with (3.74) we get

$$\frac{n}{(1+\delta)d_{\mathrm{vr}}(L,\Pi_n)}|L|^{\frac{1}{n}}|K|^{\frac{n-1}{n}} \leqslant n\varepsilon|L|.$$

Now, put

$$\varepsilon = \max_{\theta\in S^{n-1}} \frac{|P_{\theta^\perp}(K)|}{|P_{\theta^\perp}(L)|}$$

and send δ to 0. □

Since ellipsoids are projection bodies, by John's theorem, $d_{\mathrm{vr}}(L,\Pi_n) \leqslant \sqrt{n}$. Also, $S_n \leqslant \sup_L d_{\mathrm{vr}}(L,\Pi_n)$. Combining this with Theorem 3.6.4, we get the following.

Corollary 3.6.7. *There exists an absolute constant $c > 0$ such that*

$$c\sqrt{n} \leqslant \sup_L d_{\mathrm{vr}}(L,\Pi_n) \leqslant \sqrt{n},$$

where the supremum is taken over all origin-symmetric convex bodies in $\mathbb{R}^n$.

3.7 Lower-dimensional Shephard problem

Let $1 \leqslant k \leqslant n-1$ and let $S_{n,k}$ be the smallest constant $S > 0$ with the following property: For every pair of convex bodies K and L in $\mathbb{R}^n$ that satisfy $|P_F(K)| \leqslant |P_F(L)|$ for all $F \in G_{n,n-k}$, one has that

$$|K|^{\frac{n-k}{n}} \leqslant S\,|L|^{\frac{n-k}{n}}.$$

The isomorphic lower-dimensional Shephard problem can be formulated as follows.

Question 3.7.1. Provide sharp estimates for $S_{n,k}$ for all n and k.

Goodey and Zhang [226] proved that $S_{n,k} > 1$ if $n-k > 1$. In this section, we collect some known upper bounds for $S_{n,k}$. Our first observation is that, since any projection of a zonoid is a zonoid, using an inductive argument we can prove the following (for a detailed account on this topic, see [194, Chapter 4]).

Theorem 3.7.2. *Let K be a convex body and let Z be a zonoid in $\mathbb{R}^n$ such that*

$$|P_F(K)| \leqslant |P_F(Z)|$$

for every $F \in G_{n,n-k}$. Then,

$$|K| \leqslant |Z|. \tag{3.75}$$

Using Theorem 3.7.2 and the fact that every ellipsoid is a zonoid, we can give a simple bound for the constants $S_{n,k}$.

Proposition 3.7.3. *For all n and $1 \leqslant k \leqslant n-1$ we have $S_{n,k} \leqslant (c\sqrt{n})^{n-k}$, where $c > 0$ is an absolute constant.*

Proof. Let K and L be two convex bodies in $\mathbb{R}^n$ such that $|P_F(K)| \leqslant |P_F(L)|$ for every $F \in G_{n,n-k}$. There exists an ellipsoid $\mathcal{E}$ in $\mathbb{R}^n$ such that $L \subseteq \mathcal{E}$ and $|\mathcal{E}|^{\frac{1}{n}} \leqslant c\sqrt{n}\,|L|^{\frac{1}{n}}$, where $c > 0$ is an absolute constant (for example, see [44], where a sharp estimate for c is also given). Since $L \subseteq \mathcal{E}$, we have

$$|P_F(K)| \leqslant |P_F(L)| \leqslant |P_F(\mathcal{E})|$$

for all $F \in G_{n,n-k}$. Since $\mathcal{E}$ is a zonoid, Theorem 3.7.2 implies that

$$|K|^{\frac{1}{n}} \leqslant |\mathcal{E}|^{\frac{1}{n}} \leqslant c\sqrt{n}\,|L|^{\frac{1}{n}}.$$

This shows that $S_{n,k} \leqslant (c\sqrt{n})^{n-k}$. □

We can elaborate on this argument if we use Pisier's theorem on the existence of α-regular M-ellipsoids for origin-symmetric convex bodies in $\mathbb{R}^n$ (see Theorem 1.2.7).

Theorem 3.7.4. *Let $1 \leqslant k \leqslant n-1$ and let K and L be two convex bodies in $\mathbb{R}^n$ such that*

$$|P_F(K)| \leqslant |P_F(L)|$$

for every $F \in G_{n,n-k}$. Then,

$$|K|^{\frac{1}{n}} \leqslant c\sqrt{n/(n-k)}\log(en/(n-k))\,|L|^{\frac{1}{n}},$$

where $c > 0$ is an absolute constant. Therefore,

$$S_{n,k} \leqslant \left(c\sqrt{n/(n-k)}\log(en/(n-k))\right)^{n-k}.$$

Proof. Consider the difference body $L-L$ of L, and the ellipsoid $\mathcal{E}_\alpha$ from Theorem 1.2.7, where $\alpha \in (0,2)$ will be chosen in the end, that corresponds to $L-L$. Note that

$$N(\mathcal{E}_\alpha, c(\alpha)^{1/\alpha}(L-L)) \leqslant e^n,$$

therefore

$$|\mathcal{E}_\alpha|^{\frac{1}{n}} \leqslant ec(\alpha)^{1/\alpha}|L-L|^{\frac{1}{n}}. \tag{3.76}$$

Since

$$N(P_F(L-L), P_F(t\mathcal{E}_\alpha)) \leqslant N(L-L, t\mathcal{E}_\alpha) \leqslant \exp\left(\frac{c(\alpha)n}{t^\alpha}\right)$$

for every $F \in G_{n,n-k}$, we have

$$|P_F(L-L)| \leqslant \exp\left(\frac{c(\alpha)n}{t^\alpha_{n,n-k,\alpha}}\right)|P_F(t_{n,n-k,\alpha}\mathcal{E}_\alpha)| = |P_F(et_{n,n-k,\alpha}\mathcal{E}_\alpha)|,$$

if we choose

$$t_{n,n-k,\alpha} = \left(\frac{c(\alpha)n}{n-k}\right)^{\frac{1}{\alpha}}.$$

Now, if we set $\mathcal{E} := et_{n,n-k,\alpha}\mathcal{E}_\alpha$, we have

$$|P_F(K)| \leqslant |P_F(L)| \leqslant |P_F(L-L)| \leqslant |P_F(\mathcal{E})|$$

for every $F \in G_{n,n-k}$, and since $\mathcal{E}$ is a zonoid, Theorem 3.7.2 shows that $|K| \leqslant |\mathcal{E}|$. Using also (3.76) and the fact that $c(\alpha) = O((2-\alpha)^{-\alpha/2})$, we get

$$\begin{aligned}|K|^{\frac{1}{n}} &\leqslant et_{n,n-k,\alpha}|\mathcal{E}_\alpha|^{\frac{1}{n}} \leqslant e^2 t_{n,n-k,\alpha}c(\alpha)^{1/\alpha}|L-L|^{\frac{1}{n}}\\ &\leqslant \frac{c_1}{2-\alpha}\left(\frac{n}{n-k}\right)^{\frac{1}{\alpha}}|L|^{\frac{1}{n}},\end{aligned}$$

where $c_1 > 0$ is an absolute constant (we have also used the fact that $|L-L|^{\frac{1}{n}} \leqslant 4|L|^{\frac{1}{n}}$ by the Rogers–Shephard inequality). Choosing $\alpha = 2 - \frac{1}{\log(\frac{en}{n-k})}$ we get the result. □

We can also give an upper bound for $S_{n,k}$ using the known bounds for the affine quermassintegrals of a convex body from Section 3.4.

Proposition 3.7.5. *Let $1 \leqslant k \leqslant n-1$ and let K and L be two convex bodies in $\mathbb{R}^n$ such that*

$$|P_F(K)| \leqslant |P_F(L)|$$

for every $F \in G_{n,n-k}$. Then,

$$|K|^{\frac{1}{n}} \leqslant c\min\{\log n, (n/(n-k))^{3/2}\sqrt{\log(en/(n-k))}\}\,|L|^{\frac{1}{n}},$$

where $c > 0$ is an absolute constant. Therefore,

$$S_{n,k} \leqslant (c\min\{\log n, (n/(n-k))^{3/2}\sqrt{\log(en/(n-k))}\})^{n-k}.$$

Proof. Our assumption implies that

$$\Phi_{n-k}(K) = \Bigg(\int\limits_{G_{n,n-k}} |P_F(K)|^{-n} \, d\nu_{n,n-k}(F) \Bigg)^{-\frac{1}{(n-k)n}}$$

$$\leqslant \Bigg(\int\limits_{G_{n,n-k}} |P_F(L)|^{-n} \, d\nu_{n,n-k}(F) \Bigg)^{-\frac{1}{(n-k)n}} = \Phi_{n-k}(L).$$

Taking into account the known bounds for Φ_{n-k} (Theorems (3.4.3) and (3.27)) we get

$$c_1 \sqrt{n/k}\, |K|^{\frac{1}{n}} \leqslant \Phi_{n-k}(K) \leqslant \Phi_{n-k}(L) \leqslant c_2 \sqrt{n/k}\, \phi_{n,n-k}\, |L|^{\frac{1}{n}},$$

where $\phi_{n,k} = c \min\{\log n, (n/(n-k))^{3/2} \sqrt{\log(en/(n-k))}\}$. □

The next lemma from [526] provides a third estimate for $S_{n,k}$, which is reasonably good if k is small.

Lemma 3.7.6. *For every* $1 \leqslant k \leqslant n-1$ *we have that* $S_{n,k} \leqslant n^{k/2}$.

Proof. Let $1 \leqslant k \leqslant n-1$ and let K and L be two convex bodies in $\mathbb{R}^n$ such that $|P_F(K)| \leqslant |P_F(L)|$ for every $F \in G_{n,n-k}$. We use inductively the following claim. For any $1 \leqslant s \leqslant n-1$ let a_{n-s} be the best positive constant with the property that

$$|P_E(K)| \leqslant a_{n-s} |P_E(L)| = |P_F(a_{n-s}^{\frac{1}{n-s}} L)|$$

for all $E \in G_{n,n-s}$. Then,

$$a_{n-s+1} \leqslant a_{n-s} s^{\frac{n-s+1}{n-s}} (n-s+1)^{\frac{n-s+1}{2(n-s)}}. \tag{3.77}$$

To see this, consider any $F \in G_{n,n-s+1}$. Since the assumption is, in particular, true for all 1-codimensional subspaces of F, from (3.69) we see that

$$|P_F(K)|^{\frac{n-s}{n-s+1}} \leqslant (n-s+1)^{\frac{1}{2}} |P_F(a_{n-s}^{\frac{1}{n-s}} L)|^{\frac{n-s}{n-s+1}}$$

$$= (n-s+1)^{\frac{1}{2}} a_{n-s} |P_F(L)|^{\frac{n-s}{n-s+1}},$$

therefore $|P_F(K)| \leqslant a_s^{\frac{n-s+1}{n-s}} (n-s+1)^{\frac{n-s+1}{2(n-s)}} |P_F(L)|$ for all $F \in G_{n,n-s+1}$. Assume that $a_{n-k} = 1$, i. e., $|P_E(K)| \leqslant |P_E(L)$ for all $E \in G_{n,n-k}$. From (3.77) we see that $a_{n-k+s} \leqslant \prod_{j=1}^{s} (n-k+j)^{\frac{n-k+s}{2(n-k+j)}}$, and in particular

$$a_n \leqslant \prod_{j=1}^{k} (n-k+j)^{\frac{n}{2(n-k+j)}} \leqslant n^{\frac{n}{2} \sum_{j=1}^{k} \frac{1}{n-k+j}} \leqslant n^{\frac{nk}{2(n-k)}}.$$

This means that $|K| \leqslant n^{\frac{nk}{2(n-k)}} |L|$, therefore

$$|K|^{\frac{n-k}{n}} \leqslant n^{\frac{k}{2}} |L|^{\frac{n-k}{n}},$$

which shows that $S_{n,k} \leqslant \sqrt{n}$. □

Note. It is not hard to check that in the range $k \ll \frac{n}{\log n}$ the upper bound of Lemma 3.7.6 for $S_{n,k}$ is better than the ones provided by Theorem 3.7.4 and Proposition 3.7.5.

3.8 A comparison problem for integral transforms

In this section, we consider the following problem introduced in [319]. In the case where $T = R$ is the spherical Radon transform this problem was solved in [334].

Problem 3.8.1. Let $G \subset \mathbb{R}^n$ be a compact origin-symmetric set, and let $T : C_e(G) \to C_e(G)$ be a linear operator, where $C_e(G)$ is the space of even real-valued continuous functions on G. Suppose that $f, g \in C_e(G)$ are positive functions, and $Tf(x) \leqslant Tg(x)$ for every $x \in G$. Does this condition alone allow us to compare the L_p-norms of the functions f and g for a given $p > 1$?

It appears that one can make a meaningful observation in the general situation, for an arbitrary linear operator T. Following [319], we define a class of functions $\mathrm{Pos}(T)$ that in some sense controls Problem 3.8.1. Namely, if $f^{p-1} \in \mathrm{Pos}(T)$, then the L_p-norm of f is smaller than the L_p-norm of g provided that $Tf \leqslant Tg$ pointwise. On the other hand, if $g^{p-1} \notin \mathrm{Pos}(T)$, then one can construct a function f giving, together with g, a counterexample.

Definition 3.8.2. We say that a function $h \in C_e(G)$ is positively associated with the operator T if for any $f \in C_e(G)$ satisfying $Tf(x) \geqslant 0$ for every $x \in G$ we have

$$\int_G f(x)h(x)\,dx \geqslant 0.$$

We denote the set of such functions h by $\mathrm{Pos}(T)$.

The set G can be a subset of lower dimension in $\mathbb{R}^n$, and then dx is the Lebesgue measure of the corresponding dimension in G. Below, we consider the case where $G = S^{n-1}$ is the unit sphere in $\mathbb{R}^n$, and dx is the uniform measure on the sphere with density 1.

Let us prove the claims made above. We say that a function f is positive on a set D if $f(x) \geqslant 0$ for every $x \in D$. If $f(x) > 0$ for every $x \in D$, we say that f is strictly positive on D.

Lemma 3.8.3. *Let $p > 1$. Suppose that a positive function $f \in C_e(G)$ has the property that $f^{p-1} \in \mathrm{Pos}(T)$. If $g \in C_e(G)$ is any positive function satisfying $Tf(x) \leqslant Tg(x)$ for every*

$x \in G$, then

$$\|f\|_{L_p(G)} \leqslant \|g\|_{L_p(G)}.$$

Proof. Since $f^{p-1} \in \mathrm{Pos}(T)$ and $T(g-f) \geqslant 0$ pointwise in G, we have

$$\int_G (g(x) - f(x)) f^{p-1}(x)\,dx \geqslant 0.$$

Rearranging the latter inequality and applying Hölder's inequality, we get

$$\begin{aligned}\|f\|^p_{L_p(G)} &= \int_G f^p(x)\,dx \leqslant \int_G f^{p-1}(x) g(x)\,dx \\ &\leqslant \Big(\int_G f^p(x)\,dx\Big)^{\frac{p-1}{p}} \Big(\int_G g^p(x)\,dx\Big)^{\frac{1}{p}} \\ &= \|f\|^{p-1}_{L_p(G)} \|g\|_{L_p(G)}.\end{aligned}$$

□

Lemma 3.8.4. *Let $p > 1$, and suppose that a function $g \in C_e(G)$ is strictly positive and $g^{p-1} \notin \mathrm{Pos}(T)$. Then, there exists a strictly positive function $f \in C_e(G)$ such that $Tf(x) \leqslant Tg(x)$ for every $x \in G$, but*

$$\|f\|_{L_p(G)} > \|g\|_{L_p(G)}.$$

Proof. Since $g^{p-1} \notin \mathrm{Pos}(T)$, there exists a function $h \in C_e(G)$ such that $Th \geqslant 0$ pointwise and

$$\int_G g^{p-1}(x) h(x)\,dx < 0. \tag{3.78}$$

Since g is strictly positive, both g and h are continuous, and also G is compact, there exists $\varepsilon > 0$ such that the function $f = g - \varepsilon h$ is strictly positive on G.

Since $Th \geqslant 0$ pointwise in G, we have $Tf = Tg - \varepsilon Th \leqslant Tg$ pointwise in G. On the other hand, by the definition of f and by inequality (3.78)

$$\int_G g^{p-1}(x) f(x)\,dx = \int_G g^p(x)\,dx - \varepsilon \int_G g^{p-1}(x) h(x)\,dx > \|g\|^p_{L_p(G)}.$$

By Hölder's inequality

$$\|g\|^{p-1}_{L_p(G)} \|f\|_{L_p(G)} \geqslant \int_G g^{p-1}(x) f(x)\,dx > \|g\|^p_{L_p(G)}.$$

□

Let $\mathbb{1}_G$ be the function identically equal to 1 on G. Lemma 3.8.3 implies a lower estimate for the operator T. We call it the slicing inequality for the operator T.

Corollary 3.8.5. *Let $p > 1$, and suppose that the operator T satisfies $T\mathbb{1}_G = c_T\mathbb{1}_G$, where $c_T > 0$ is a constant. Let f be a positive function on G such that $f^{p-1} \in \mathrm{Pos}(T)$. Then,*

$$\|f\|_{L_p(G)} \leqslant \frac{|G|^{\frac{1}{p}}}{c_T} \max_{z\in G} Tf(z), \tag{3.79}$$

where $|G|$ is the Lebesgue measure of G.

Proof. First, $\max_{z\in G} Tf(z) > 0$ unless $f \equiv 0$. In fact, if $Tf(z) \leqslant 0$ for every $z \in G$, then, since $f^{p-1} \in \mathrm{Pos}(T)$, we have $\int_G f^p(x)\,dx \leqslant 0$. But f is a positive continuous function on G, so $f \equiv 0$.

Let g be a constant function on G defined by

$$g = \frac{1}{c_T} \max_{z\in G} Tf(z)\mathbb{1}_G.$$

Then, g is positive on G and $Tg(x) = \max_{z\in G} Tf(z) \geqslant Tf(x)$ for every $x \in G$. By Lemma 3.8.3

$$\|f\|_{L_p(G)} \leqslant \|g\|_{L_p(G)} = \frac{|G|^{\frac{1}{p}}}{c_T} \max_{z\in G} Tf(z). \qquad \square$$

We see that the classes $\mathrm{Pos}(T)$ play the same role for Problem 3.8.1 as intersection bodies for the Busemann–Petty problem. It is easy to see that in the case where $T = R$ is the spherical Radon transform, $p = \frac{n}{n-1}$, $f = \|\cdot\|_K^{-n+1}$, $g = \|\cdot\|_L^{-n+1}$ Problem 3.8.1 turns into the statement of the Busemann–Petty problem for star bodies K and L. In fact, by (1.2)

$$Rf(\theta) = \int\limits_{S^{n-1}\cap\theta^\perp} \|\xi\|_K^{-n+1}\,d\xi = (n-1)|K\cap\theta^\perp|,$$

so the condition $Rf(\theta) \leqslant Rg(\theta)$ is equivalent to $|K\cap\theta^\perp| \leqslant |L\cap\theta^\perp|$. On the other hand, by (1.4)

$$\|f\|_{L_{\frac{n}{n-1}}(S^{n-1})} = \Big(\int\limits_{S^{n-1}} \|\xi\|_K^{-n}\,d\xi\Big)^{\frac{n-1}{n}} = (n|K|)^{\frac{n-1}{n}},$$

so the inequality for $L_{\frac{n}{n-1}}$-norms is equivalent to the inequality for volumes of K and L.

Positively associated functions appeared in the papers [506, 542, 543, 222, 191, 551, 553]. In particular, Goodey, Lutwak, and Weil [222] proved that the radial function ρ_K of an origin-symmetric star body K belongs to the class $\mathrm{Pos}(R)$ if and only if K is an intersection body (note that any strictly positive continuous function on S^{n-1} is the radial function of a star body). They also proved that a convex body K in $\mathbb{R}^n$ is a projection body if and only if the support function h_K of K satisfies the inequality $\int_{S^{n-1}} h_K(x)\,d\mu(x) \geqslant 0$ for every even signed measure μ on the sphere such that the cosine transform C of μ is positive, i. e.

$$C\mu(x) = \int\limits_{S^{n-1}} |(x,\xi)|\,d\mu(\xi) \geqslant 0$$

for all $x \in S^{n-1}$. The authors of [222] then applied their results to characterize the classes of intersection and projection bodies in terms of dual mixed volumes and mixed volumes.

In the general setting of Problem 3.8.1, let us consider the case where $G = S^{n-1}$ is the unit sphere in $\mathbb{R}^n$, and the operator T is the spherical Radon transform R. In the next two lemmas we establish an analytic characterization of the class $\operatorname{Pos}(R)$. We prove that a function $f \in C_e^\infty(S^{n-1})$ belongs to $\operatorname{Pos}(R)$ if and only if the extension $f \cdot r^{-1}$ of f to an even homogeneous function of degree -1 on $\mathbb{R}^n$ represents a positive definite distribution.

Lemma 3.8.6. *Suppose that $h \in C_e(S^{n-1})$ and the distribution $h \cdot r^{-1}$ is positive definite in $\mathbb{R}^n$. Then, $h \in \operatorname{Pos}(R)$.*

Proof. Let $f \in C_e(S^{n-1})$ be such that $Rf(\theta) \geqslant 0$ for every $\theta \in S^{n-1}$. By Lemma 1.5.1

$$(f \cdot r^{-n+1})^\wedge = \pi Rf \cdot r^{-1},$$

so $(f \cdot r^{-n+1})^\wedge$ is a positive continuous function on the sphere.

Let μ_0 be the measure on S^{n-1} corresponding to the positive definite distribution $h \cdot r^{-1}$ by Proposition 1.4.7. By Parseval's formula on the sphere with $k = 1$, we get

$$\int\limits_{S^{n-1}} f(\theta)h(\theta)\,d\theta = \int\limits_{S^{n-1}} (f \cdot r^{-n+1})^\wedge(\xi)\,d\mu_0(\xi) \geqslant 0,$$

so $h \in \operatorname{Pos}(R)$. □

Lemma 3.8.7. *Let $h \in C_e^\infty(S^{n-1})$ be such that the distribution $h \cdot r^{-1}$ is not positive definite. Then, $h \notin \operatorname{Pos}(R)$.*

Proof. By Lemma 1.4.13, the Fourier transform of the distribution $h \cdot r^{-1}$ is a C^∞-function on the sphere extended to an even homogeneous of degree $-n+1$ function on $\mathbb{R}^n$. Since this function cannot be positive, there exists an open symmetric set $\Omega \subset S^{n-1}$ on which the function $(h \cdot r^{-1})^\wedge$ is strictly negative. Consider a function $\phi \in C_e^\infty(S^{n-1})$ such that $\phi \geqslant 0$ on the whole sphere and is not identically zero, and such that ϕ is supported in Ω. Again, by Lemma 1.4.13, the Fourier transform $(\phi \cdot r^{-1})^\wedge = \psi \cdot r^{-n+1}$, where $\psi \in C_e^\infty(S^{n-1})$. Since ϕ is an even function, we have $(\psi \cdot r^{-n+1})^\wedge = (2\pi)^n \phi \cdot r^{-1}$.

By Lemma 1.5.1 and the definition of the function ϕ, for every $\theta \in S^{n-1}$

$$R\psi(\theta) = \frac{1}{\pi}(\psi \cdot r^{-n+1})^\wedge(\theta) = 2(2\pi)^{n-1}(\phi \cdot r^{-1})(\theta) = 2(2\pi)^{n-1}\phi(\theta) \geqslant 0.$$

On the other hand, by Parseval's formula for infinitely smooth functions on the sphere (Proposition 1.6.4)

$$\begin{aligned}(2\pi)^n \int_{S^{n-1}} \psi(\theta)h(\theta)\,d\theta &= \int_{S^{n-1}} (h \cdot r^{-1})^\wedge(\xi)(\psi \cdot r^{-n+1})^\wedge(\xi)\,d\xi \\ &= (2\pi)^n \int_{S^{n-1}} (h \cdot r^{-1})^\wedge(\xi)\phi(\xi)\,d\xi < 0,\end{aligned}$$

because the positive function ϕ can be strictly positive only in Ω, where $(h \cdot r^{-1})^\wedge$ is strictly negative. So, $h \notin \mathrm{Pos}(R)$. □

Using Lemmas 3.8.3, 3.8.4, 3.8.6, and 3.8.7, we immediately get the solution of Problem 3.8.1 for $T = R$. This result was first proved in [334].

Theorem 3.8.8. (i) *Let $p > 1$. Suppose that $f \in C_e(S^{n-1})$ is a positive function such that $f^{p-1} \cdot r^{-1}$ is a positive definite distribution. If $g \in C_e(S^{n-1})$ is a positive function such that $Rf \leqslant Rg$ pointwise on S^{n-1}, then*

$$\|f\|_{L_p(S^{n-1})} \leqslant \|g\|_{L_p(S^{n-1})}.$$

(ii) *Let $p > 1$. If $g \in C_e^\infty(S^{n-1})$ is a strictly positive function such that $g^{p-1} \cdot r^{-1}$ is not a positive definite distribution, then there exists a strictly positive function $f \in C_e^\infty(S^{n-1})$ such that $Rf \leqslant Rg$ pointwise on S^{n-1}, but*

$$\|f\|_{L_p(S^{n-1})} > \|g\|_{L_p(S^{n-1})}.$$

Note that in the case $p = 1$ the answer to the comparison problem for $T = R$ is affirmative in the sense that the inequality for L_1-norms follows from $Rf \leqslant Rg$ pointwise. One can see this using the well-known Fubini-type formula on the sphere

$$\int_{S^{n-1}} Rf(\theta)\,d\sigma(\theta) = \int_{S^{n-1}} f(\theta)\,d\sigma(\theta),$$

where σ is the uniform probability measure on S^{n-1}; see, for example, Lemma 1.3.2.

Also, by the argument of Lemma 1.7.13, any function $h \in C_e(S^{n-1})$ such that $h \cdot r^{-1}$ is not a positive definite distribution can be approximated in $C_e(S^{n-1})$ by infinitely smooth functions h_m for which $h_m \cdot r^{-1}$ are not positive definite distributions.

Since in our case $\mu(G) = \kappa_n$ and $c_T = \kappa_{n-1}$, the slicing inequality of Corollary 3.8.5 takes the following form.

Corollary 3.8.9. *Let $p > 1$. Suppose that $f \in C_e(S^{n-1})$ is a positive function such that $f^{p-1} \cdot r^{-1}$ represents a positive definite distribution on $\mathbb{R}^n$. Then,*

$$\|f\|_{L_p(S^{n-1})} \leqslant \frac{\kappa_n^{1/p}}{\kappa_{n-1}} \max_{\xi \in S^{n-1}} Rf(\xi). \tag{3.80}$$

Note that this inequality is in the spirit of estimates for the Radon transform proved in [443, 69].

As follows from Proposition 1.7.1, a strictly positive continuous function f on the sphere has the property that the distribution $f^{p-1} \cdot r^{-1}$ is positive definite if and only if $f = \|\cdot\|_K^{-\frac{1}{p-1}}$ for some intersection body K. Combining this observation with (3.80), we get that for any intersection body K in $\mathbb{R}^n$ and any $p > 1$

$$\left(\int_{S^{n-1}} \|x\|_K^{-\frac{p}{p-1}} \, dx \right)^{\frac{1}{p}} \leqslant \frac{\kappa_n^{1/p}}{\kappa_{n-1}} \max_{\xi \in S^{n-1}} \left(\int_{S^{n-1} \cap \xi^\perp} \|x\|_K^{-\frac{1}{p-1}} \, dx \right). \tag{3.81}$$

When $p = \frac{n}{n-1}$, we get Bourgain's slicing inequality in the case where K is an intersection body

$$|K|^{\frac{n-1}{n}} \leqslant \tau_n \max_{\xi \in S^{n-1}} |K \cap \xi^\perp|,$$

where

$$\tau_n = \frac{\omega_n^{\frac{n-1}{n}}}{\omega_{n-1}} < 1.$$

Note that we can also immediately get the latter inequality from the affirmative part of the Busemann–Petty problem; see [194, Theorem 9.4.11]. This inequality is sharp in the class of intersection bodies.

The fact that the distribution $f^{p-1} \cdot r^{-1}$ is positive definite if and only if $f = \|\cdot\|_K^{-\frac{1}{p-1}}$ for some intersection body K allows us to give concrete examples in the affirmative and negative directions of Theorem 3.8.8.

In [319], Problem 3.8.1 was also solved for the cosine transform

$$Cf(x) = \int_{S^{n-1}} |\langle x, \xi \rangle| f(\xi) \, d\xi.$$

It was proved that $f \in C_e^\infty(S^{n-1})$ belongs to $\mathrm{Pos}(C)$ if and only if the Fourier transform of the extension $f \cdot r$ of f to an even homogeneous function of degree 1 on $\mathbb{R}^n$ is a negative distribution on $\mathbb{R}^n \setminus \{0\}$. This means that f is the support function of a projection body. The answer to Problem 3.8.1 in this case is as follows.

Theorem 3.8.10. (i) *Let $p > 1$. Suppose that $f \in C_e(S^{n-1})$ is a positive function such that the Fourier transform of $f^{p-1} \cdot r$ is a negative distribution on $\mathbb{R}^n \setminus \{0\}$. If $g \in C_e(S^{n-1})$ is a positive function such that $Cf \leqslant Cg$ pointwise on S^{n-1}, then*

$$\|f\|_{L_p(S^{n-1})} \leqslant \|g\|_{L_p(S^{n-1})}.$$

(ii) *Let $p > 1$. If $g \in C_e^\infty(S^{n-1})$ is a strictly positive function such that the Fourier transform of $g^{p-1} \cdot r$ is not a negative distribution on $\mathbb{R}^n \setminus \{0\}$, then there exists a strictly positive function $f \in C_e^\infty(S^{n-1})$ such that $Cf \leqslant Cg$ pointwise on S^{n-1}, but*

$$\|f\|_{L_p(S^{n-1})} > \|g\|_{L_p(S^{n-1})}.$$

3.9 Convex measures

Following the works of Borell [92, 93], we define the classes of s-concave measures. Let $s \in [-\infty, 1]$. A measure μ on $\mathbb{R}^n$ is called s-concave if for any compact $A, B \subset \mathbb{R}^n$, with $\mu(A)\mu(B) > 0$ and $0 < \lambda < 1$, we have

$$\mu(\lambda A + (1-\lambda)B) \geqslant \left(\lambda\mu(A)^s + (1-\lambda)\mu(B)^s\right)^{1/s}.$$

The case where $s = 0$ corresponds to log-concave measures that satisfy

$$\mu(\lambda A + (1-\lambda)B) \geqslant \mu(A)^\lambda \mu(B)^{(1-\lambda)}$$

and the case $s = -\infty$ corresponds to convex measures that satisfy

$$\mu(\lambda A + (1-\lambda)B) \geqslant \min\{\mu(A), \mu(B)\}.$$

We also note that the class of convex measures is the largest class in this group in the sense that it contains all other s-concave measures. Due to this fact, we concentrate our attention on convex measures.

Borell [92, 93] has shown that a measure μ on $\mathbb{R}^n$ whose support is not contained in any affine hyperplane is a convex measure if and only if it is absolutely continuous with respect to the Lebesgue measure, and its density f is a $-1/n$-concave function on its support, i. e.

$$f(\lambda x + (1-\lambda)y) \geqslant \left(\lambda f(x)^{-1/n} + (1-\lambda)f(y)^{-1/n}\right)^{-n}$$

for all x, y such that $f(x), f(y) > 0$ and $\lambda \in [0, 1]$. From the latter definition it follows that if $f(x)$ is a $-1/n$-concave function, then $f^{-1/n}$ is a convex function on its support.

The main tool for the study of convex measures is the family of K_p-bodies of a function, introduced by K. Ball in [40]. Given a measurable function $f : \mathbb{R}^n \to [0, \infty)$ with $f(0) > 0$, for any $p > 0$ we define the set

$$K_p(f) = \left\{ x \in \mathbb{R}^n : \int_0^\infty f(rx) r^{p-1}\, dr \geqslant \frac{f(0)}{p} \right\}. \tag{3.82}$$

From the definition, it follows that the radial function of $K_p(f)$ is given by

$$\rho_{K_p(f)}(x) = \left(\frac{1}{f(0)} \int_0^\infty p r^{p-1} f(rx)\, dr \right)^{1/p} \tag{3.83}$$

for $x \neq 0$. It was proved by K. Ball that if f is also log-concave and integrable, then, for every $p > 0$, $K_p(f)$ is a convex set (see also [139] for a simple proof). This fact was generalized by Bobkov [81, Theorem 2.1].

Theorem 3.9.1. *Let $f : \mathbb{R}^n \to [0, \infty)$ be an even $-1/n$-concave function on its support, satisfying $0 < \int_{\mathbb{R}^n} f < \infty$. Then, the map*

$$x \to \left(\int_0^\infty f(rx) r^{n-2}\, dr \right)^{-\frac{1}{n-1}}$$

defines a norm on $\mathbb{R}^n$.

An immediate consequence of the Ball–Bobkov theorem is a very useful technique for connecting a convex measure of a convex body with the volume of another convex body. This technique allows us to generalize a number of classical results on the Lebesque measure to the case of convex measures (see [40, 81, 282], and [139]). Namely, given the density f of a convex measure μ and a convex symmetric body K we define a body K_f by

$$\|x\|_{K_f} = \left((n-1) \int_0^\infty (\mathbb{1}_K f)(rx) r^{n-2}\, dr \right)^{-\frac{1}{n-1}},$$

where $\mathbb{1}_K$ is the indicator function of K. Theorem 3.9.1 guarantees that K_f is convex. Moreover,

$$\begin{aligned} |K_f \cap \xi^\perp| &= \frac{1}{n-1} \int_{S^{n-1}\cap\xi^\perp} \|x\|_{K_f}^{-(n-1)}\, dx \\ &= \int_{S^{n-1}\cap\xi^\perp} \int_0^\infty (\mathbb{1}_K f)(rx) r^{n-2}\, dr\, dx = \mu(K \cap \xi^\perp). \end{aligned} \tag{3.84}$$

Our next goal is to estimate $|K_f|$. We start with a lemma on the behavior of $-1/n$-concave functions, the proof of which may be found in [284, Lemma 2.4] and [81, Lemma 4.2].

Lemma 3.9.2. *Let $n \geqslant 1$ be an integer, and let $g : [0,\infty) \to [0,\infty)$ be a $-1/n$-concave, non-increasing function with $g(0) = 1$, $0 < \int_0^\infty g(t)t^{n-2}\,dt < \infty$. Then,*

$$c_1 \leqslant \frac{\int_0^\infty t^{n-1} g(t)\,dt}{(\int_0^\infty t^{n-2} g(t)\,dt)^{\frac{n}{n-1}}} \leqslant c_2,$$

where $c_1, c_2 > 0$ are absolute constants.

Note. We need the $-1/n$-concavity assumption only to prove the right-hand side inequality in the above lemma. The left-hand side does not require this assumption, but does require $g \leqslant e^n$.

Now, assume that $f(0) = 1$, f is even and $-1/n$-concave, then $f(tx)$ is non-increasing for $t \geqslant 0$ and the function $g(t) = (\mathbb{1}_K f)(tx)$ satisfies the conditions of Lemma 3.9.2. We write

$$|K_f| = \frac{1}{n} \int_{S^{n-1}} \|x\|_{K_f}^{-n}\,dx = \frac{1}{n} \int_{S^{n-1}} \left((n-1) \int_0^\infty (\mathbb{1}_K f)(rx) r^{n-2}\,dr \right)^{\frac{n}{n-1}} dx$$

and applying Lemma 3.9.2 we get

$$c_1 \mu(K) \leqslant |K_f| \leqslant c_2 \mu(K), \tag{3.85}$$

where $c_1, c_2 > 0$ are absolute constants (and the right-hand side inequality does not require $-1/n$-concavity, but does require the boundedness of f).

Recall that if

$$|K \cap \xi^\perp| \leqslant |L \cap \xi^\perp|$$

for all $\xi \in S^{n-1}$, then $|K| \leqslant cL_n|L|$. Applying this fact to bodies K_f and L_f we immediately get the following theorem.

Theorem 3.9.3. *For any measure μ with continuous, non-negative, even $-1/n$-concave density f on $\mathbb{R}^n$ and any two origin-symmetric convex bodies $K, L \subset \mathbb{R}^n$ such that*

$$\mu(K \cap \xi^\perp) \leqslant \mu(L \cap \xi^\perp) \tag{3.86}$$

for all $\xi \in S^{n-1}$, we have

$$\mu(K) \leqslant cL_n \mu(L).$$

Theorem 3.9.3 was independently obtained in [324, Theorem 4], and [526]. The paper [526] contains several stronger results that are mentioned in Section 5.6.1.

3.10 Further reading

3.10.1 The complex Busemann–Petty problem

The complex version of the Busemann–Petty problem was considered in [327]. For $\xi \in \mathbb{C}^n$, $|\xi| = 1$, denote by

$$H_\xi = \left\{ z \in \mathbb{C}^n : (z, \xi) = \sum_{k=1}^{n} z_k \overline{\xi_k} = 0 \right\}$$

the complex hyperplane through the origin perpendicular to ξ.

Origin-symmetric convex bodies in $\mathbb{C}^n$ are the unit balls of norms on $\mathbb{C}^n$. We denote by $\|\cdot\|_K$ the norm on $\mathbb{C}^n$ corresponding to the body K

$$K = \{z \in \mathbb{C}^n : \|z\|_K \leqslant 1\}.$$

In order to define volume, we identify $\mathbb{C}^n$ with $\mathbb{R}^{2n}$ using the mapping

$$\xi = (\xi_1, \dots, \xi_n) = (\xi_{11} + i\xi_{12}, \dots, \xi_{n1} + i\xi_{n2}) \mapsto (\xi_{11}, \xi_{12}, \dots, \xi_{n1}, \xi_{n2}).$$

Under this mapping, the hyperplane H_ξ turns into a $(2n-2)$-dimensional subspace of $\mathbb{R}^{2n}$ orthogonal to the vectors

$$\xi = (\xi_{11}, \xi_{12}, \dots, \xi_{n1}, \xi_{n2}) \quad \text{and} \quad \xi^\perp = (-\xi_{12}, \xi_{11}, \dots, -\xi_{n2}, \xi_{n1}).$$

Since norms on $\mathbb{C}^n$ satisfy the equality

$$\|\lambda z\| = |\lambda| \|z\|, \quad \forall z \in \mathbb{C}^n, \ \forall \lambda \in \mathbb{C},$$

origin-symmetric complex convex bodies correspond to those origin-symmetric convex bodies K in $\mathbb{R}^{2n}$ that are invariant with respect to any coordinate-wise two-dimensional rotation, namely for each $\theta \in [0, 2\pi]$ and each $\xi = (\xi_{11}, \xi_{12}, \dots, \xi_{n1}, \xi_{n2}) \in \mathbb{R}^{2n}$

$$\|\xi\|_K = \|R_\theta(\xi_{11}, \xi_{12}), \dots, R_\theta(\xi_{n1}, \xi_{n2})\|_K, \tag{3.87}$$

where R_θ stands for the counterclockwise rotation of $\mathbb{R}^2$ by the angle θ with respect to the origin. We shall simply say that K is invariant with respect to all R_θ if it satisfies equation (3.87).

Now, the complex Busemann–Petty problem can be formulated as follows: suppose K and L are origin-symmetric invariant with respect to all R_θ convex bodies in $\mathbb{R}^{2n}$ such that

$$|K \cap H_\xi|_{2n-2} \leqslant |L \cap H_\xi|_{2n-2}$$

for each ξ from the unit sphere S^{2n-1} of $\mathbb{R}^{2n}$. Does it follow that

$$|K|_{2n} \leqslant |L|_{2n}?$$

It was proved in [327] that the answer is affirmative if $n \leqslant 3$ and negative if $n \geqslant 4$. Zymonopoulou [557] showed that the answer remains the same if volume is replaced by a measure with an even, continuous, positive density invariant with respect to all rotations R_θ. The proof is based on Lemma 1.6.7 with $H = H_\xi$ and the observation that the Fourier transform of an infinitely smooth complex norm is a constant function on $S^{2n-1} \cap H_\xi^\perp$. More generalizations to bodies with certain symmetries can be found in [235, 145, 487, 488, 160].

3.10.2 The Busemann–Petty problem with dilations

Recall that the Busemann–Petty problem for arbitrary measures has a positive answer if the dimension $n \leqslant 4$, and a negative answer if $n \geqslant 5$ for an arbitrary strictly positive, even, continuous density; see Theorem 3.2.1. Lafi and Zvavitch [353] consider this problem with an additional assumption that the inequality for measures of sections extends to the dilates of the bodies. Namely, suppose that μ is a measure on $\mathbb{R}^n$ with even, continuous positive density f, K, L are origin-symmetric convex bodies in $\mathbb{R}^n$ so that

$$\int\limits_{tK\cap\xi^\perp} f(x)\,dx \leqslant \int\limits_{tL\cap\xi^\perp} f(x)\,dx$$

for every $t > 0$ and every $\xi \in S^{n-1}$. Does it necessarily follow that

$$\int\limits_K f \leqslant \int\limits_L f\ ?$$

Lafi and Zvavitch show that, in general, the answer is still negative in dimensions $n \geqslant 5$, even if one considers only log-concave densities. This study was initiated by a question of V. Milman concerning the case where f is the standard gaussian density. In the gaussian case, the answer is still negative if $n \geqslant 7$, and the question is open if $n = 5, 6$.

3.10.3 Variants of the Shephard problem for measures

A functional approach to the Shephard problem was suggested by Livshyts [381]. Let μ be a measure on $\mathbb{R}^n$ with density g continuous on its support, and let K be a convex body in $\mathbb{R}^n$. The surface area measure $\sigma_{\mu,K}$ of the body K with respect to the measure μ is defined as a measure on S^{n-1} such that for any Borel set $\Omega \in S^{n-1}$

$$\sigma_{\mu,K}(\Omega) = \int\limits_{\eta_K^{-1}(\Omega)} g(x)\,d\lambda_K(x),$$

where $\eta_K : \partial K \to S^{n-1}$ is the Gauss map of K assigning to every $x \in \partial K$ the unit normal vector to ∂K at x, and λ_K is the Lebesgue measure on ∂K. Now, for a fixed $\theta \in S^{n-1}$ define the μ-projection of the body K by

$$P_{\mu,K}(\theta) = \int\limits_0^1 p_{\mu,K}(\theta,t)\,dt,$$

where

$$p_{\mu,K}(\theta,t) = \frac{n}{2}\int\limits_{S^{n-1}} |\langle\theta,u\rangle|\,d\sigma_{\mu,tK}(u).$$

It is easy to see that if μ is the Lebesgue measure on $\mathbb{R}^n$ we get $P_{\mu,K}(\theta)$ equal to the $(n-1)$-dimensional volume of the orthogonal projection of K to the hyperplane $\theta^\perp$.

Livshyts [381] proves the following version of the Shephard problem. Let K, L be symmetric convex bodies, and suppose that L is a projection body. Also, suppose that g is even r-homogeneous for some $r > 0$, and it is also p-concave on some half-space with $p > 0$. Then, the inequalities

$$P_{\mu,K}(\theta) \leqslant P_{\mu,L}(\theta), \quad \forall\theta \in S^{n-1}$$

imply $\mu(K) \leqslant \mu(L)$. If the support of μ is the whole space, and L is not a projection body, it is possible to find K giving a counterexample to this version of the Shephard problem. This version works well with the Fourier techniques for projections developed in [333].

Langharst, Roysdon, and Zvavitch [354] introduced a variant of the projection body. Using the same notations as above, they define the projection body $\Pi_{\mu,K}$ corresponding to a convex body K and a measure μ with density g on R^n by

$$h_{\Pi_{\mu,K}}(\theta) = \frac{1}{2}\int\limits_{S^{n-1}} |\langle\theta,u\rangle|\,d\sigma_{\mu,K}(u)\,du, \quad \forall\theta \in S^{n-1}.$$

Kryvonos and Langharst [350] proved an isomorphic version of the Shephard problem for measures with homogeneous densities. Suppose that μ is a measure with continuous density in $\mathbb{R}^n$, and μ is p-concave and α-homogeneous with either $\alpha \geqslant n$, $p > 0$ or $\alpha < n$, $p = 1/\alpha$. If K and L are origin-symmetric convex bodies in $\mathbb{R}^n$ such that

$$\Pi_{\mu,K} \subset \Pi_{\mu,L},$$

then

$$\mu(K) \leqslant d_{BM}(L,\Pi)^{\frac{\alpha}{\alpha-1}}\nu(L),$$

where $d_{BM}(L,\Pi)$ is the Banach–Mazur distance from L to the class of projection bodies. Note that by John's theorem this distance is less than or equal to $\sqrt{n}$.

A different approach was suggested by Tziotziou [526]. Suppose that f, g are log-concave integrable functions on $\mathbb{R}^n$ such that $\|f\|_\infty = f(0) = \|g\|_\infty = g(0) = 1$. Let E be an $(n-k)$-dimensional subspace of $\mathbb{R}^n$, $1 \leqslant k \leqslant n-1$. The projection of f to E is defined by

$$P_E f(x) = \max\{f(y) : y \in x + E^\perp\}, \quad x \in E.$$

For $t > 0$, denote by $R_t(f)$ the set $\{x : f(x) \geqslant t\}$. It is easy to see that if $f = \chi_K$ is the indicator function of a convex body K, then $R_t(P_E\chi_K)$, with $0 \leqslant t < 1$, is the orthogonal projection of K to E. Tziotziou proves the following generalization of the isomorphic lower-dimensional Shephard problem. If

$$|R_t(P_E f)| \leqslant |R_t(P_E g)|$$

for every $0 \leqslant t \leqslant 1$ and every $(n-k)$-dimensional subspace E of $\mathbb{R}^n$, then

$$\|f\|_{L_1(\mathbb{R}^n)}^{\frac{n-k}{n}} \leqslant S_{n,k}^k \|g\|_{L_1(\mathbb{R}^n)}^{\frac{n-k}{n}},$$

where the constant $S_{n,k}$ is the same as in the case of volume; see Section 3.7. If $f = \chi_K$ and $g = \chi_L$ the result is exactly the same as in the isomorphic Shephard problem for the volume of convex bodies.

4 Slicing inequalities for measures

This chapter explores the classical slicing problem and its extensions to lower-dimensional sections and general measures. It begins with the foundational question of whether the volume of a centered convex body $K \subset \mathbb{R}^n$ can be controlled by the maximal volume of its hyperplane sections, which is equivalent to the boundedness of isotropic constants. The discussion then generalizes to $(n-k)$-dimensional sections, establishing uniform bounds for constants relating the volume of a convex body to the maximal volume of its intersections with subspaces of arbitrary codimension. We extend these results to general measures with even, continuous densities. Analogous inequalities relate the measure of a convex body to the measures of its lower-dimensional sections. Key findings include sharp bounds for intersection bodies and general symmetric convex bodies, with estimates expressed via the outer volume ratio distance to classes of intersection bodies. These results unify classical slicing inequalities with measure-theoretic versions, highlighting the roles of convexity, symmetry, and functional-analytic techniques.

We also address the sharpness of the slicing inequality. For sufficiently large dimensions, a probabilistic construction shows that there exist origin-symmetric convex bodies and even, continuous densities whose hyperplane sections are small relative to the volume.

The study of p-th absolute moments of measures on origin-symmetric convex bodies further generalizes slicing inequalities. For a body K of volume 1 and an even non-negative function f, the minimal directional p-th moment is controlled by a constant $\gamma(p,n)$, which is bounded above and below by multiples of $\sqrt{pn}$ and $\sqrt{n}$. These results yield nearly sharp inequalities for $p > 2$ and a Busemann–Petty-type comparison, connecting directional moment inequalities to global integrals over the body.

Finally, we extend these ideas to fractional derivatives of the Radon transform. For an origin-symmetric, infinitely smooth convex body K and a smooth function f, the fractional derivative of Rf at zero along a direction ξ provides directional estimates for the integral of f over K, with constants controlled by the outer volume ratio distance to L^n_{-1-q}-bodies. Generalized Busemann–Petty inequalities show that if fractional derivatives of f are dominated by those of a comparison function g, the integral of f is bounded in terms of g, the volume of K, and the outer volume ratio distance. Corresponding slicing inequalities guarantee the existence of directions with lower bounds on fractional derivatives, with special cases recovering classical derivatives for integer orders. Examples demonstrate that these derivatives can be uniformly small, confirming that the outer volume ratio distance governs sharpness.

https://doi.org/10.1515/9783111360508-004

4.1 The slicing problem

Recall that the classical slicing problem asks if there exists an absolute constant $C_1 > 0$ such that for every $n \geqslant 1$ and every centered convex body K in $\mathbb{R}^n$ one has

$$|K|^{\frac{n-1}{n}} \leqslant C_1 \max_{\xi \in S^{n-1}} |K \cap \xi^{\perp}|. \tag{4.1}$$

In Chapter 2, we saw that this problem is equivalent to the question of whether there exists an absolute constant $C_2 > 0$ such that

$$L_n := \max\{L_K : K \text{ is isotropic in } \mathbb{R}^n\} \leqslant C_2 \tag{4.2}$$

for all $n \geqslant 1$. From the equivalence of the two questions and the affirmative answer to the isotropic constant problem it follows that

$$|K|^{\frac{n-1}{n}} \leqslant c_1 L_n \max_{\xi \in S^{n-1}} |K \cap \xi^{\perp}| \leqslant c_2 \max_{\xi \in S^{n-1}} |K \cap \xi^{\perp}| \tag{4.3}$$

for every centered convex body K in $\mathbb{R}^n$. Thus, the slicing problem has an affirmative answer.

The natural generalization, the lower-dimensional slicing problem, is the following question: Let $1 \leqslant k \leqslant n-1$ and let $\alpha_{n,k}$ be the smallest positive constant $\alpha > 0$ with the following property: For every centered convex body K in $\mathbb{R}^n$ one has

$$|K|^{\frac{n-k}{n}} \leqslant \alpha^k \max_{F \in G_{n,n-k}} |K \cap F|. \tag{4.4}$$

The question is whether the constants $\alpha_{n,k}$ are uniformly bounded. This is now known to be true.

Theorem 4.1.1. *There exists an absolute constant $C > 0$ such that $\alpha_{n,k} \leqslant C$ for all n and k.*

Proof. We observe that any upper bound for $\beta_{n,k}$ implies an upper bound for the lower-dimensional slicing problem. To see this, consider a centered convex body K in $\mathbb{R}^n$, fix $1 \leqslant k \leqslant n-1$ and choose $r > 0$ such that

$$\max_{F \in G_{n,n-k}} |K \cap F| = \omega_{n-k} r^{n-k}. \tag{4.5}$$

If we set $B(r) = rB_2^n$, then we have $|K \cap F| \leqslant |B(r) \cap F|$ for all $F \in G_{n,n-k}$, therefore

$$|K|^{\frac{n-k}{n}} \leqslant (\beta_{n,k})^k |B(r)|^{\frac{n-k}{n}} = (\beta_{n,k})^k \omega_n^{\frac{n-k}{n}} r^{n-k}. \tag{4.6}$$

It follows that

$$|K|^{\frac{n-k}{n}} \leqslant \tau_{n,k} (\beta_{n,k})^k \max_{F \in G_{n,n-k}} |K \cap F|. \tag{4.7}$$

From Theorem 3.4.1 we know that there exists an absolute constant $C > 0$ such that $\beta_{n,k} \leqslant C$ for all n and k. Since we also have $\tau_{n,k} \leqslant 1$ (see Proposition 3.5.10) the proof is complete. □

The slicing problem can be posed for a general measure in place of volume. Let g be a locally integrable non-negative function on $\mathbb{R}^n$. For every Borel subset $B \subseteq \mathbb{R}^n$ we define

$$\mu(B) = \int_B g(x)\,dx, \tag{4.8}$$

where, if $B \subseteq F$ for some subspace $F \in G_{n,s}$, $1 \leqslant s \leqslant n-1$, integration is understood with respect to the s-dimensional Lebesgue measure on F. Then, for any $1 \leqslant k \leqslant n-1$ one may define $\alpha_{n,k}(\mu)$ as the smallest constant $\alpha > 0$ with the following property: For every centered convex body K in $\mathbb{R}^n$ one has

$$\mu(K) \leqslant \alpha^k \max_{F \in G_{n,n-k}} \mu(K \cap F)\,|K|^{\frac{k}{n}}. \tag{4.9}$$

The question was first considered in [313], where it was proved that if K is an intersection body in $\mathbb{R}^n$ and μ is a measure on $\mathbb{R}^n$ with an even, continuous density f, then

$$\mu(K) \leqslant \frac{n}{n-1}\tau_{n,1} \max_{\xi \in S^{n-1}} \mu(K \cap \xi^\perp)|K|^{1/n}. \tag{4.10}$$

The constant $\frac{n}{n-1}\tau_{n,1}$ in (4.10) is the best possible. Note that the classical slicing problem has an affirmative answer for intersection bodies K; in fact, one has that

$$|K|^{\frac{n-1}{n}} \leqslant \tau_{n,1} \max_{\xi \in S^{n-1}} |K \cap \xi^\perp|. \tag{4.11}$$

A proof of (4.11) can be found in [194, p. 374].

The question of whether it is possible to remove the assumption that K is an intersection body and establish the inequality (4.10) for any origin-symmetric convex body K in $\mathbb{R}^n$, maybe with a larger constant in the right-hand side, was answered in the affirmative in [315]. Recall that $\tau_{n,k} = \omega_n^{\frac{n-k}{n}}/\omega_{n-k} < 1$ (see Proposition 3.5.10) for all $1 \leqslant k \leqslant n-1$, and $\tau_n = \tau_{n,1}$.

Theorem 4.1.2. *Let K be an origin-symmetric convex body in $\mathbb{R}^n$. If μ is a measure with even, continuous, non-negative density on K, then*

$$\mu(K) \leqslant \frac{n}{n-1}\tau_{n,1}\sqrt{n} \max_{\xi \in S^{n-1}} \mu(K \cap \xi^\perp)|K|^{1/n}. \tag{4.12}$$

In [316], Theorem 4.1.2 was extended to lower-dimensional sections.

Theorem 4.1.3. *If K is an origin-symmetric convex body in $\mathbb{R}^n$ and if μ is a measure on K with even and continuous density, then*

$$\mu(K) \leqslant \tau_{n,k}\frac{n}{n-k}(\sqrt{n})^k \max_{F\in G_{n,n-k}} \mu(K\cap F)\,|K|^{\frac{k}{n}} \tag{4.13}$$

for every $1 \leqslant k \leqslant n-1$.

In other words, for the symmetric analogue $\alpha_{n,k}^{(s)}$ of $\alpha_{n,k}$ one has

$$\sup_{\mu} \alpha_{n,k}^{(s)}(\mu) \leqslant c\sqrt{n}. \tag{4.14}$$

A more general result was proved in [317]. It shows that the slicing problem for measures is controlled by the outer volume ratio distance from a body to the classes of intersection bodies; recall the definition in Section 1.7.2.

Theorem 4.1.4. *Let K be an origin-symmetric star body in $\mathbb{R}^n$. Then, for any measure μ with even, continuous density on K we have*

$$\mu(K) \leqslant d_{\mathrm{ovr}}(K,\mathcal{I}_n)\,\frac{n}{n-1}\tau_{n,1}\max_{\xi\in S^{n-1}} \mu(K\cap\xi^{\perp})\,|K|^{1/n},$$

where $\mathcal{I}_n$ is the class of intersection bodies in $\mathbb{R}^n$.

A lower-dimensional version of Theorem 4.1.4 was also proved in [317] and reads as follows.

Theorem 4.1.5. *Let K be an origin-symmetric star body in $\mathbb{R}^n$ and $1 \leqslant k \leqslant n-1$. Then, for any measure μ with even, continuous density on K we have*

$$\mu(K) \leqslant (d_{\mathrm{ovr}}(K,\mathcal{BP}_k^n))^k\,\frac{n}{n-k}\tau_{n,k}\max_{F\in G_{n,n-k}} \mu(K\cap F)\,|K|^{k/n},$$

where $\mathcal{BP}_k^n$ is the class of generalized k-intersection bodies.

We prove these results in the next two sections. Note that convexity of K is not required in Theorems 4.1.4 and 4.1.5. If, in addition, K is origin-symmetric and convex, then by John's theorem $d_{\mathrm{ovr}}(K,\mathcal{I}_n) \leqslant \sqrt{n}$ and $d_{\mathrm{ovr}}(K,\mathcal{BP}_k^n) \leqslant \sqrt{n}$, since the classes $\mathcal{I}_n$ and $\mathcal{BP}_k^n$ contain origin-symmetric ellipsoids. Thus, Theorems 4.1.4 and 4.1.5 imply Theorems 4.1.2 and 4.1.3, respectively.

4.2 Proof of the hyperplane case

For the proof of Theorem 4.1.2 we use the following stability theorem.

Theorem 4.2.1. *Let K be an intersection body in $\mathbb{R}^n$ and f be an even, continuous function on K such that $f \geqslant 1$ everywhere on K. Assume that for some $\varepsilon > 0$ and for all $\xi \in S^{n-1}$ we have that*

$$\int_{K\cap\xi^\perp} f \leqslant |K\cap\xi^\perp| + \varepsilon. \tag{4.15}$$

Then,

$$\int_K f \leqslant |K| + \frac{n}{n-1}\tau_n |K|^{1/n}\varepsilon. \tag{4.16}$$

Proof. Recall that

$$\mu(K\cap\xi^\perp) = R\left(\int_0^{\|\cdot\|_K^{-1}} r^{n-2} f(r\cdot)\,dr\right)(\xi)$$

and

$$|K\cap\xi^\perp| = \frac{1}{n-1} R(\|\cdot\|_K^{-n+1})(\xi),$$

therefore (4.15) can be expressed in terms of the Radon transform as follows:

$$R\left(\int_0^{\|\cdot\|_K^{-1}} r^{n-2} f(r\cdot)\,dr\right)(\xi) \leqslant \frac{1}{n-1} R(\|\cdot\|_K^{-n+1})(\xi) + \varepsilon. \tag{4.17}$$

Since K is an intersection body, there exists a finite Borel measure μ on S^{n-1} such that, for every continuous function g on S^{n-1} we have that

$$\int_{S^{n-1}} \|x\|_K^{-1} g(x)\,dx = \int_{S^{n-1}} Rg(x)\,d\mu(x). \tag{4.18}$$

Integrating (4.17) over S^{n-1} with respect to μ and using (4.18) we get

$$\begin{aligned}\int_{S^{n-1}} \|\theta\|_K^{-1}\left(\int_0^{\|\theta\|_K^{-1}} r^{n-2} f(r\theta)\,dr\right)d\theta &\leqslant \frac{1}{n-1}\int_{S^{n-1}} \|\theta\|_K^{-n}\,d\theta + \varepsilon\int_{S^{n-1}} d\mu(\xi)\\ &= \frac{1}{n-1}|K| + \varepsilon\int_{S^{n-1}} d\mu(\xi).\end{aligned} \tag{4.19}$$

Using the assumption that $f \geqslant 1$ and the formula

$$\mu(K) = \int_K f(x)\,dx = \int_{S^{n-1}}\left(\int_0^{\|\theta\|_K^{-1}} r^{n-1} f(r\theta)\,dr\right)d\theta$$

we see that the integral on the left-hand side is equal to

$$\begin{aligned}
\int_{S^{n-1}} \|\theta\|_K^{-1}\Bigg(\int_0^{\|\theta\|_K^{-1}} r^{n-2}f(r\theta)\,dr\Bigg)\,d\theta
&= \int_{S^{n-1}}\Bigg(\int_0^{\|\theta\|_K^{-1}} r^{n-1}f(r\theta)\,dr\Bigg)\,d\theta \\
&\quad + \int_{S^{n-1}}\Bigg(\int_0^{\|\theta\|_K^{-1}} (\|\theta\|_K^{-1}-r)r^{n-2}f(r\theta)\,dr\Bigg)\,d\theta \\
&\geqslant \int_K f + \int_{S^{n-1}}\Bigg(\int_0^{\|\theta\|_K^{-1}} (\|\theta\|_K^{-1}-r)r^{n-2}\,dr\Bigg)\,d\theta \\
&= \int_K f + \frac{1}{(n-1)n}\int_{S^{n-1}} \|\theta\|_K^{-n}\,d\theta \\
&= \int_K f + \frac{1}{n-1}|K|. \qquad (4.20)
\end{aligned}$$

Next, using the fact that $\|\cdot\|_K^{-1} = R\mu$ and (4.18), we compute

$$\begin{aligned}
\int_{S^{n-1}} d\mu(\xi) &= \frac{1}{\kappa_{n-1}}\int_{S^{n-1}} R\mathbb{1}(\xi)\,d\mu(\xi) \\
&= \frac{1}{\kappa_{n-1}}\int_{S^{n-1}} \|\theta\|_K^{-1}\,d\theta \\
&= \frac{\kappa_n^{\frac{n-1}{n}}}{\kappa_{n-1}}\Bigg(\int_{S^{n-1}} \|\theta\|_K^{-n}\,d\theta\Bigg)^{1/n} \\
&= \frac{\kappa_n^{\frac{n-1}{n}}}{\kappa_{n-1}} n^{1/n}|K|^{1/n} = \frac{n}{n-1}\tau_n|K|^{1/n}, \qquad (4.21)
\end{aligned}$$

where in the last step recall that $\kappa_m = |S^{m-1}| = m\omega_m$ for all $m \geqslant 1$. Combining (4.19), (4.20), and (4.21) we conclude that

$$\int_K f + \frac{1}{n-1}|K| \leqslant \frac{n}{n-1}|K| + \frac{n}{n-1}\tau_n|K|^{1/n}\varepsilon$$

and the result follows. □

Remark 4.2.2. Using Theorem 4.2.1 we can give a proof of (4.10) as follows. Let g be the density of μ; then, g is an even, continuous function on K. Set $f = (1+g)\mathbb{1}_K$, where $\mathbb{1}_K$ is the indicator function of K. Note that f is an even, continuous function on K and $f \geqslant 1$. We set

$$\varepsilon = \max_{\xi \in S^{n-1}} \Big(\int_{K \cap \xi^\perp} f - |K \cap \xi^\perp| \Big) = \max_{\xi \in S^{n-1}} \int_{K \cap \xi^\perp} g.$$

Then,

$$\int_{K \cap \xi^\perp} f \leqslant |K \cap \xi^\perp| + \varepsilon$$

for all $\xi \in S^{n-1}$, therefore we may apply Theorem 4.2.1 to write

$$\begin{aligned} \mu(K) = \int_K g = \int_K f - |K| &\leqslant \frac{n}{n-1} \tau_n |K|^{1/n} \max_{\xi \in S^{n-1}} \int_{K \cap \xi^\perp} g \\ &= \frac{n}{n-1} \tau_n |K|^{1/n} \max_{\xi \in S^{n-1}} \mu(K \cap \xi^\perp). \end{aligned}$$

A variant of this argument, also making use of John's theorem, provides a proof of Theorem 4.1.2.

Proof of Theorem 4.1.2. Let g be the density of μ; then, g is an even, continuous function on K. By John's theorem we can find an origin-symmetric ellipsoid $\mathcal{E}$ such that

$$\frac{1}{\sqrt{n}} \mathcal{E} \subseteq K \subseteq \mathcal{E}.$$

Recall that $\mathcal{E}$ is an intersection body. Let $f = \mathbb{1}_{\mathcal{E}} + g \mathbb{1}_K$, where $\mathbb{1}_{\mathcal{E}}, \mathbb{1}_K$ are the indicator functions of $\mathcal{E}$ and K. We clearly have $f \geqslant 1$ on $\mathcal{E}$. The function f is not continuous but we may approximate it by even, continuous functions. We set

$$\varepsilon = \max_{\xi \in S^{n-1}} \Big(\int_{\mathcal{E} \cap \xi^\perp} f - |\mathcal{E} \cap \xi^\perp| \Big) = \max_{\xi \in S^{n-1}} \int_{K \cap \xi^\perp} g.$$

Then,

$$\int_{\mathcal{E} \cap \xi^\perp} f \leqslant |\mathcal{E} \cap \xi^\perp| + \varepsilon$$

for all $\xi \in S^{n-1}$, therefore we may apply Theorem 4.2.1 to write

$$\mu(K) = \int_K g = \int_{\mathcal{E}} f - |\mathcal{E}| \leqslant \frac{n}{n-1} \tau_n |\mathcal{E}|^{1/n} \max_{\xi \in S^{n-1}} \int_{K \cap \xi^\perp} g.$$

Since $|\mathcal{E}|^{1/n} \leqslant \sqrt{n}|K|^{1/n}$, it follows that

$$\mu(K) \leqslant \frac{n}{n-1}\tau_n \sqrt{n}|K|^{1/n} \max_{\xi\in S^{n-1}} \int_{K\cap\xi^\perp} g$$
$$= \frac{n}{n-1}\tau_n \sqrt{n}|K|^{1/n} \max_{\xi\in S^{n-1}} \mu(K\cap\xi^\perp). \qquad \square$$

We now prove Theorem 4.1.4. Instead of the original proof from [317], we present here a slightly shorter proof that appears in [291].

Proof of Theorem 4.1.4. For every $\xi \in S^{n-1}$, we have

$$\mu(K\cap\xi^\perp) \leqslant \max_{\theta\in S^{n-1}} \mu(K\cap\theta^\perp). \tag{4.22}$$

Let f be the continuous density of the measure μ. Writing the integral in spherical coordinates, we see that for every $\xi \in S^{n-1}$

$$\mu(K\cap\xi^\perp) = \int_{K\cap\xi^\perp} f = \int_{S^{n-1}\cap\xi^\perp} \left(\int_0^{\rho_K(\theta)} r^{n-2} f(r\theta)\,dr \right) d\theta = \int_{S^{n-1}\cap\xi^\perp} F(\theta)\,d\theta,$$

where

$$F(\theta) = \int_0^{\rho_K(\theta)} r^{n-2} f(r\theta)\,dr, \quad \theta\in S^{n-1}.$$

Therefore, inequality (4.22) can be written in terms of the spherical Radon transform as

$$RF(\xi) \leqslant \max_{\theta\in S^{n-1}} \mu(K\cap\theta^\perp) \tag{4.23}$$

for all $\xi \in S^{n-1}$. Note that the right-hand side of (4.23) does not depend on ξ.

Let D be an intersection body such that the distance $d_{\mathrm{ovr}}(K, \mathcal{I}_n)$ is almost realized, i. e., $K \subset D$ and for some small $\delta > 0$

$$|D|^{1/n} \leqslant (1+\delta) d_{\mathrm{ovr}}(K, \mathcal{I}_n)|K|^{1/n}. \tag{4.24}$$

Integrating both sides of (4.23) by ξ over the sphere with respect to the measure μ_D corresponding to D, we get

$$\int_{S^{n-1}} \rho_D(\theta) \left(\int_0^{\rho_K(\theta)} r^{n-2} f(r\theta)\,dr \right) d\theta \leqslant \mu_D(S^{n-1}) \max_{\theta\in S^{n-1}} \mu(K\cap\theta^\perp). \tag{4.25}$$

The left-hand side of (4.25) is equal to

$$\int_{S^{n-1}}\left(\int_0^{\rho_K(\theta)}(\rho_D(\theta)-r)r^{n-2}f(r\theta)\,dr\right)d\theta+\int_{S^{n-1}}\left(\int_0^{\rho_K(\theta)}r^{n-1}f(r\theta)\,dr\right)d\theta$$
$$\geqslant\int_{S^{n-1}}\left(\int_0^{\rho_K(\theta)}r^{n-1}f(r\theta)\,dr\right)d\theta=\int_K f=\mu(K),\tag{4.26}$$

because $K\subset D$ implies $\rho_D(\theta)\geqslant\rho_K(\theta)$ for every θ.

Now, we estimate the right-hand side of (4.25) from above. Let $\mathbf{1}$ be the constant function with value 1 on S^{n-1}. We use the fact that $R\mathbf{1}(\xi)=\kappa_{n-1}$ for every $\xi\in S^{n-1}$, Hölder's inequality, and the standard formula for volume

$$\mu_D(S^{n-1})=\frac{1}{\kappa_{n-1}}\int_{S^{n-1}}R\mathbf{1}(\xi)\,d\mu_D(\xi)=\frac{1}{\kappa_{n-1}}\int_{S^{n-1}}\rho_D(\xi)\,d\xi$$
$$\leqslant\frac{\kappa_n^{\frac{n-1}{n}}}{\kappa_{n-1}}\left(\int_{S^{n-1}}\rho_D^n(\xi)\,d\xi\right)^{\frac{1}{n}}=\frac{n^{1-1/n}}{n-1}\tau_n\,|D|^{1/n}\leqslant\frac{n}{n-1}\tau_n|D|^{\frac{1}{n}}.$$

By using (4.24), sending $\delta\to 0$, and combining the estimates above, we obtain the conclusion of the theorem. □

4.3 Proof of the lower-dimensional case

For the proof of Theorem 4.1.3 we use the next theorem, which generalizes Theorem 4.2.1.

Theorem 4.3.1. *Let $1\leqslant k\leqslant n-1$. Let K be a generalized k-intersection body in $\mathbb{R}^n$ and f be an even, continuous function on K such that $f\geqslant 1$ everywhere on K. Assume that for some $\varepsilon>0$ and for all $H\in G_{n,n-k}$ we have that*

$$\int_{K\cap H}f\leqslant|K\cap H|+\varepsilon.\tag{4.27}$$

Then,

$$\int_K f\leqslant|K|+\frac{n}{n-k}\tau_{n,k}|K|^{k/n}\varepsilon.\tag{4.28}$$

Proof. Let $R_{n-k}:C(S^{n-1})\to C(G_{n,n-k})$ denote the $(n-k)$-dimensional spherical Radon transform, defined by

$$R_{n-k}g(H)=\int_{S^{n-1}\cap H}g(x)\,dx,\quad H\in G_{n,n-k}$$

for every function $g \in C(S^{n-1})$. Recall that

$$\mu(K \cap H) = R_{n-k}\left(\int_0^{\|\cdot\|_K^{-1}} r^{n-k-1} f(r\cdot)\, dr \right)(H)$$

and

$$|K \cap H| = \frac{1}{n-k} R_{n-k}(\| \cdot \|_K^{-n+k})(H),$$

therefore (4.27) can be expressed in terms of R_{n-k} as follows:

$$R_{n-k}\left(\int_0^{\|\cdot\|_K^{-1}} r^{n-k-1} f(r\cdot)\, dr \right)(H) \leqslant \frac{1}{n-k} R_{n-k}(\| \cdot \|_K^{-n+k})(H) + \varepsilon. \tag{4.29}$$

Set $X = R_{n-k}(C(S^{n-1}))$. Since K is a generalized k-intersection body, there exists a functional $\nu \in M^+(X)$ such that

$$\int_{S^{n-1}} \|x\|_K^{-k} g(x)\, dx = \nu(R_{n-k} g) \tag{4.30}$$

for every $g \in S^{n-1}$. Applying ν to both sides of (4.29) and using (4.30) we get

$$\int_{S^{n-1}} \|\theta\|_K^{-k} \left(\int_0^{\|\theta\|_K^{-1}} r^{n-k-1} f(r\theta)\, dr \right) d\theta \leqslant \frac{1}{n-k} \int_{S^{n-1}} \|\theta\|_K^{-n}\, d\theta + \varepsilon\nu(\mathbf{1}). \tag{4.31}$$

Note that the inequality is preserved because ν is a positive functional. Using the assumption that $f \geqslant 1$ we write the integral on the left-hand side as

$$\begin{aligned}
&\int_{S^{n-1}} \|\theta\|_K^{-k} \left(\int_0^{\|\theta\|_K^{-1}} r^{n-k-1} f(r\theta)\, dr \right) d\theta \\
&= \int_{S^{n-1}} \left(\int_0^{\|\theta\|_K^{-1}} r^{n-1} f(r\theta)\, dr \right) d\theta \\
&+ \int_{S^{n-1}} \left(\int_0^{\|\theta\|_K^{-1}} (\|\theta\|_K^{-k} - r^k) r^{n-k-1} f(r\theta)\, dr \right) d\theta \\
&\geqslant \int_K f + \int_{S^{n-1}} \left(\int_0^{\|\theta\|_K^{-1}} (\|\theta\|_K^{-k} - r^k) r^{n-k-1} f(r\theta)\, dr \right) d\theta
\end{aligned}$$

$$= \int_K f + \frac{k}{(n-k)n} \int_{S^{n-1}} \|\theta\|_K^{-n}\, d\theta$$

$$= \int_K f + \frac{k}{n-k}|K|. \tag{4.32}$$

Next, writing $\mathbf{1} = R_{n-k}\mathbf{1}/\kappa_{n-k}$ and using (4.30), we compute

$$\begin{aligned} \nu(\mathbf{1}) &= \frac{1}{\kappa_{n-k}}\nu(R_{n-k}\mathbf{1}) = \frac{1}{\kappa_{n-k-1}} \int_{S^{n-1}} \|\theta\|_K^{-k}\, d\theta \\ &= \frac{\kappa_n^{\frac{n-k}{n}}}{\kappa_{n-k}} \left(\int_{S^{n-1}} \|\theta\|_K^{-n}\, d\theta \right)^{1/n} \\ &= \frac{\kappa_n^{\frac{n-k}{n}}}{\kappa_{n-k}} n^{k/n}|K|^{k/n} = \frac{n}{n-k}\tau_{n,k}|K|^{k/n}. \end{aligned} \tag{4.33}$$

Combining (4.31), (4.32), and (4.33) we conclude that

$$\int_K f + \frac{k}{n-k}|K| \leqslant \frac{n}{n-k}|K| + \frac{n}{n-k}\tau_{n,k}|K|^{k/n}\varepsilon$$

and the result follows. □

Remark 4.3.2. Using Theorem 4.3.1 in a way analogous to the one described in Remark 4.2.2, Koldobsky and Ma proved in [321] that if K is a generalized k-intersection body in $\mathbb{R}^n$ and if μ is a measure with even, continuous non-negative density g on K and $1 \leqslant k \leqslant n-1$, then

$$\mu(K) \leqslant \frac{n}{n-k}\tau_{n,k} \max_{H \in G_{n,n-k}} \mu(K \cap H)|K|^{k/n}. \tag{4.34}$$

With a similar argument, also making use of John's theorem, we can give a proof of Theorem 4.1.3.

Proof of Theorem 4.1.3. We can find an origin-symmetric ellipsoid $\mathcal{E}$ such that

$$\frac{1}{\sqrt{n}}\mathcal{E} \subseteq K \subseteq \mathcal{E}.$$

Recall that $\mathcal{E}$ is an intersection body, and hence $\mathcal{E}$ is a generalized k-intersection body. Let $f = \mathbb{1}_{\mathcal{E}} + g\mathbb{1}_K$, where $\mathbb{1}_{\mathcal{E}}$, $\mathbb{1}_K$ are the indicator functions of $\mathcal{E}$ and K. We clearly have $f \geqslant 1$ on $\mathcal{E}$. The function f is not continuous but we may approximate it by even, continuous functions. We set

$$\varepsilon = \max_{H \in G_{n,n-k}} \left(\int_{\mathcal{E}\cap H} f - |\mathcal{E} \cap H| \right) = \max_{H \in G_{n,n-k}} \int_{K \cap H} g.$$

Then,

$$\int_{\mathcal{E}\cap H} f \leqslant |\mathcal{E}\cap H| + \varepsilon$$

for all $H \in G_{n,n-k}$, therefore we may apply Theorem 4.3.1 to write

$$\mu(K) = \int_K g = \int_{\mathcal{E}} f - |\mathcal{E}| \leqslant \frac{n}{n-k}\tau_{n,k}|\mathcal{E}|^{k/n} \max_{H\in G_{n,n-k}} \int_{K\cap H} g.$$

Since $|\mathcal{E}|^{1/n} \leqslant \sqrt{n}|K|^{1/n}$, it follows that

$$\begin{aligned}\mu(K) &\leqslant \frac{n}{n-k}\tau_{n,k}n^{k/2}|K|^{k/n} \max_{H\in G_{n,n-k}} \int_{K\cap H} g \\ &= \frac{n}{n-k}\tau_{n,k}n^{k/2}|K|^{k/n} \max_{H\in G_{n,n-k}} \mu(K\cap H).\end{aligned}$$

□

Next, we prove Theorem 4.1.5.

Proof of Theorem 4.1.5. For every $H \in G_{n,n-k}$, we have

$$\mu(K\cap H) \leqslant \max_{F\in G_{n,n-k}} \mu(K\cap F). \tag{4.35}$$

Let f be the continuous density of the measure μ. Writing the integral in spherical coordinates, we see that for every $H \in G_{n,n-k}$

$$\mu(K\cap H) = \int_{K\cap H} f = \int_{S^{n-1}\cap H} \left(\int_0^{\rho_K(\theta)} r^{n-k-1} f(r\theta)\,dr \right) d\theta = \int_{S^{n-1}\cap H} u(\theta)\,d\theta,$$

where

$$u(\theta) = \int_0^{\rho_K(\theta)} r^{n-k-1} f(r\theta)\,dr, \quad \theta \in S^{n-1}.$$

Therefore, inequality (4.35) can be written in terms of the $(n-k)$-dimensional spherical Radon transform

$$R_{n-k}u(H) \leqslant \max_{F\in G_{n,n-k}} \mu(K\cap F) \tag{4.36}$$

for all $H \in G_{n,n-k}$. Note that the right-hand side of (4.36) does not depend on H.

Let D be a generalized k-intersection body such that the distance $d_{\mathrm{ovr}}(K, \mathcal{BP}_n)$ is almost realized, i. e., $K \subset D$ and for some small $\delta > 0$

$$|D|^{1/n} \leqslant (1+\delta) d_{\mathrm{ovr}}(K, \mathcal{BP}_n)|K|^{1/n}. \tag{4.37}$$

Set $X = R_{n-k}(C(S^{n-1}))$. Since D is a generalized k-intersection body, there exists a functional $\nu \in M^+(X)$ such that

$$\int_{S^{n-1}} \|x\|_D^{-k} g(x)\,dx = \nu(R_{n-k}g) \tag{4.38}$$

for every $g \in C(S^{n-1})$. Applying ν to both sides of (4.36) we get

$$\int_{S^{n-1}} \|\theta\|_D^{-k} \left(\int_0^{\|\theta\|_K^{-1}} r^{n-k-1} f(r\theta)\,dr \right) d\theta \leqslant \nu(\mathbf{1}) \max_{F \in G_{n,n-k}} \mu(K \cap F). \tag{4.39}$$

Note that the inequality is preserved because ν is a positive functional.

The left-hand side of (4.39) is equal to

$$\begin{aligned}
&\int_{S^{n-1}} \left(\int_0^{\|\theta\|_K^{-1}} (\|\theta\|_D^{-k} - r^k) r^{n-k-1} f(r\theta)\,dr \right) d\theta \\
&+ \int_{S^{n-1}} \left(\int_0^{\rho_K(\theta)} r^{n-1} f(r\theta)\,dr \right) d\theta \\
&\geqslant \int_{S^{n-1}} \left(\int_0^{\rho_K(\theta)} r^{n-1} f(r\theta)\,dr \right) d\theta \\
&= \int_K f = \mu(K),
\end{aligned} \tag{4.40}$$

because $K \subset D$ implies $\|\theta\|_D^{-1} \geqslant \|\theta\|_K^{-1}$ for every θ, and also $f \geqslant 0$.

Now, we estimate the right-hand side of (4.39) from above. We use (4.38), the fact that $R_{n-k}\mathbf{1}(\xi) = \kappa_{n-k}$ for every $\xi \in S^{n-1}$, Hölder's inequality, and the polar formula for volume

$$\begin{aligned}
\nu(S^{n-1}) &= \frac{1}{\kappa_{n-k}} \int_{S^{n-1}} R\mathbf{1}(\xi)\,d\nu(\xi) = \frac{1}{\kappa_{n-k}} \int_{S^{n-1}} \|\xi\|_D^{-k}\,d\xi \\
&\leqslant \frac{\kappa_n^{\frac{n-k}{n}}}{\kappa_{n-k}} \left(\int_{S^{n-1}} \|\theta\|_D^{-n}\,d\xi \right)^{\frac{n-k}{n}} = \frac{n^{1-k/n}}{n-k} \tau_{n,k} \,(n|D|)^{k/n} \\
&= \frac{n}{n-k} \tau_{n,k} |D|^{\frac{k}{n}}.
\end{aligned}$$

We finish the proof by using (4.37), sending $\delta \to 0$, and combining the estimates above.

□

The distances $d_{\mathrm{ovr}}(K, L_{-k}^n)$ and $d_{\mathrm{ovr}}(K, \mathcal{BP}_k^n)$ are bounded by absolute constants for certain classes of bodies K. By Propositions 1.7.15 and 1.7.16, for every $k = 1, \dots, n-1$ we have

$$\mathcal{I}_n = \mathcal{BP}_1^n \subset \mathcal{BP}_k^n \subset L_{-k}^n,$$

so

$$d_{\mathrm{ovr}}(K, L_{-k}^n) \leqslant d_{\mathrm{ovr}}(K, \mathcal{BP}_k^n) \leqslant d_{\mathrm{ovr}}(K, \mathcal{I}_n)$$

and it is enough to get an upper bound for the distance from K to the class of intersection bodies $\mathcal{I}_n$.

Recall that a convex body K in $\mathbb{R}^n$ is called unconditional if for every vector $(x_1, \dots, x_n) \in K$ the vectors $(\pm x_1, \dots, \pm x_n)$ belong to K for all choices of signs.

Proposition 4.3.3. *If K is an unconditional convex body in $\mathbb{R}^n$, then for every $k = 1, \dots, n-1$ we have*

$$d_{\mathrm{ovr}}(K, L_{-k}^n) \leqslant d_{\mathrm{ovr}}(K, \mathcal{BP}_k^n) \leqslant d_{\mathrm{ovr}}(K, \mathcal{I}_n) \leqslant e.$$

Proof. By Proposition 1.7.10, the ℓ_1^n-ball

$$B_1^n = \{x \in \mathbb{R}^n : |x_1| + \cdots + |x_n| \leqslant 1\}$$

is an intersection body. Also, the class $\mathcal{I}_n$ is invariant with respect to linear transformations of $\mathbb{R}^n$, so $T(B_1^n) \in \mathcal{I}_n$ for every linear operator T on $\mathbb{R}^n$ with $\det(T) \neq 0$.

By a result of Lozanovskii [388] (for a proof, see [464, Corollary 3.4]), there exists a linear operator T on $\mathbb{R}^n$ such that $T(B_\infty^n) \subset K \subset nT(B_1^n)$, where B_∞^n is the cube with side-length 2 in $\mathbb{R}^n$. Let $D = nT(B_1^n) \in \mathcal{I}_n$. Since $|B_1^n| = 2^n/n!$, we have $|D|^{1/n} \leqslant 2e|\det T|^{1/n}$. On the other hand, $|T(B_\infty^n)| = 2^n|\det T|$, and $T(B_\infty^n) \subset K$, so $|D|^{1/n} \leqslant e\,|K|^{1/n}$. □

Our next result estimates the distance from the unit balls of subspaces of L_p, $p > 2$ to the class of intersection bodies. This estimate was first proved by E. Milman [413] and also follows from the results of [147]. We present here the proof from [322]. Note that the unit balls of subspaces of L_p with $0 < p \leqslant 2$ are intersection bodies, so the distance to $\mathcal{I}_n$ is 1.

Proposition 4.3.4. *If K is the unit ball of an n-dimensional subspace of L_p, $p > 2$, then for every $k = 1, \dots, n-1$*

$$d_{\mathrm{ovr}}(K, L_{-k}^n) \leqslant d_{\mathrm{ovr}}(K, \mathcal{BP}_k^n) \leqslant d_{\mathrm{ovr}}(K, \mathcal{I}_n) \leqslant C\sqrt{p},$$

where C is an absolute constant.

Proof. Since the classes $\mathcal{I}_n$ are invariant under linear transformations, we can assume that K is in the Lewis position. By a result of Lewis in the form of [396, Theorem 8.2], this means that there exists a measure ν on the sphere such that for every $x \in \mathbb{R}^n$

$$\|x\|_K^p = \int_{S^{n-1}} |\langle x,u\rangle|^p \, d\nu(u)$$

and

$$|x|^2 = \int_{S^{n-1}} \langle x,u\rangle^2 \, d\nu(u).$$

Also, by the same result of Lewis [370], $K \subset n^{1/2-1/p} B_2^n$.

Let us estimate the volume of K from below. By the Fubini theorem, formula (1.4.9), and Stirling's formula, we get

$$\begin{aligned}\int_{S^{n-1}} \|x\|_K^p d\sigma(x) &= \int_{S^{n-1}}\int_{S^{n-1}} |\langle x,u\rangle|^p \, d\sigma(x)\, d\nu(u)\\ &= \int_{S^{n-1}} |x_1|^p \, d\sigma(x) \int_{S^{n-1}} d\nu(u) \leqslant \left(\frac{Cp}{n+p}\right)^{p/2} \int_{S^{n-1}} d\nu(u).\end{aligned}$$

Here, σ is the uniform probability measure on the sphere. Now,

$$\begin{aligned}\frac{Cp}{n+p}\left(\int_{S^{n-1}} d\nu(u)\right)^{2/p} &\geqslant \left(\int_{S^{n-1}} \|x\|_K^p \, d\sigma(x)\right)^{2/p}\\ &\geqslant \left(\int_{S^{n-1}} \|x\|_K^{-n} \, d\sigma(x)\right)^{-2/n}\\ &= \left(\frac{|K|}{|B_2^n|}\right)^{-2/n} \approx \frac{1}{n}|K|^{-2/n},\end{aligned}$$

because $|B_2^n|^{1/n} \approx n^{-1/2}$. On the other hand

$$\begin{aligned}1 &= \int_{S^{n-1}} |x|^2 d\sigma(x) = \int_{S^{n-1}}\int_{S^{n-1}} \langle x,u\rangle^2 \, d\nu(u)\, d\sigma(x)\\ &= \int_{S^{n-1}}\int_{S^{n-1}} |x_1|^2 \, d\sigma(x)\, d\nu(u) = \frac{1}{n}\int_{S^{n-1}} d\nu(u),\end{aligned}$$

so

$$\frac{Cp}{n+p} n^{2/p} \geqslant \frac{1}{n}|K|^{-2/n}$$

and

$$|K|^{1/n} \geqslant cn^{-1/p}\sqrt{\frac{n+p}{np}} \geqslant \frac{cn^{1/2-1/p}}{\sqrt{p}}|B_2^n|^{1/n}.$$

Finally, since $K \subset n^{1/2-1/p}B_2^n$, and $B_2^n \in \mathcal{I}_n$ for every k, we have

$$d_{\mathrm{ovr}}(K, \mathcal{I}_n) \leqslant \left(\frac{|n^{1/2-1/p}B_2^n|}{|K|}\right)^{1/n} \leqslant C\sqrt{p},$$

where C is an absolute constant. □

4.4 Lower bound for the slicing inequality

The question of whether the constant $\sqrt{n}$ in the estimate of Theorem 4.1.2 is optimal was studied in [291, 297]. The answer turns out to be affirmative.

Theorem 4.4.1. *For any $n \geqslant 3$ there exists an origin-symmetric convex body $T \subseteq \mathbb{R}^n$ and an even, continuous probability density $f : T \to [0, \infty)$ such that for any affine hyperplane $H \subseteq \mathbb{R}^n$*

$$\int_{T\cap H} f \leqslant \frac{C}{\sqrt{n}}|T|^{-1/n}, \tag{4.41}$$

where $C > 0$ is an absolute constant.

More precisely, Klartag and Koldobsky originally obtained in [291] the slightly weaker estimate

$$\int_{T\cap H} f \leqslant C\frac{\sqrt{\log\log n}}{\sqrt{n}}|T|^{-1/n} \tag{4.42}$$

and the $\sqrt{\log\log n}$-term was removed by Klartag and Livshyts in [297]. In this section we present the proof of (4.42).

We may clearly assume that the dimension n is large enough. We shall use the following Bernstein-type inequality.

Lemma 4.4.2. *Let $Y_1, \ldots, Y_N$ be independent, identically distributed random variables with values in $[0, 1]$. If $p \in [0, 1]$ satisfies $p \geqslant \mathbb{E}(Y_1)$, then*

$$\mathbb{P}\left(\frac{1}{N}\sum_{i=1}^{N} Y_i \geqslant 3p\right) \leqslant e^{-pN}.$$

Proof. Since $Y_1 \in [0,1]$ and $\mathbb{E}(Y_1) \leqslant p$, we have that

$$\mathbb{E}(e^{Y_1}) = 1 + \sum_{k=1}^{\infty} \frac{\mathbb{E}(Y_1^{k-1}Y_1)}{k!} \leqslant 1 + \sum_{k=1}^{\infty} \frac{\mathbb{E}(Y_1)}{k!} \leqslant 1 + p(e-1).$$

Therefore,

$$\mathbb{E}(e^{\sum_{i=1}^{N} Y_i}) = (\mathbb{E}(e^{Y_1}))^N \leqslant (1 + p(e-1))^N \leqslant e^{p(e-1)N}.$$

Then, the Markov–Chebyshev inequality implies that

$$\mathbb{P}\left(\frac{1}{N}\sum_{i=1}^{N} Y_i \geqslant 3p\right) \leqslant e^{-3pN}\,\mathbb{E}(e^{\sum_{i=1}^{N} Y_i}) \leqslant e^{p(e-4)N} < e^{-pN},$$

because $4 - e > 1$. □

We shall apply Lemma 4.4.2 for the random variables $Y_i = \varphi(t + r\langle \xi, U_i\rangle)$, where ξ is a fixed unit vector, U_i is a random point distributed uniformly in S^{n-1}, and $\varphi(t) = e^{-t^2/2}$, $t \in \mathbb{R}$.

Lemma 4.4.3. *Let $n \geqslant 4$ and let U be a random point distributed uniformly in S^{n-1}. Let also $\xi \in S^{n-1}$ be a fixed unit vector. Then, for any $t \in \mathbb{R}$ and $r > \sqrt{n}$*

$$\mathbb{E}(\varphi(t + r\langle U, \xi\rangle)) \leqslant C\frac{\sqrt{n}}{r}\,\varphi\left(\frac{c\sqrt{n}}{r}t\right),$$

where $C > 0$ and $0 < c < 1$ are absolute constants.

Proof. The random variable $Z = \langle U, \xi\rangle$ is supported in the interval $[-1,1]$ and its density is proportional to the function $h(s) = (1-s^2)^{(n-3)/2}$. Setting $k = n-3$ we see that we need to prove that

$$a_k \int_{-1}^{1} \varphi(rs + t)\,(1-s^2)^{k/2}\,ds \leqslant C\frac{\sqrt{k}}{r}\,e^{-\frac{kt^2}{2r^2}}, \tag{4.43}$$

where

$$a_k^{-1} = \int_{-1}^{1} (1-s^2)^{k/2}\,ds \geqslant \int_{-1/\sqrt{k}}^{1/\sqrt{k}} (1-s^2)^{k/2}\,ds \geqslant \frac{c}{\sqrt{k}}. \tag{4.44}$$

Using (4.44) and the inequality $(1-x)^m \leqslant \exp(-xm)$, which is true for all $0 < x < 1$ and $m > 0$, we see that

$$a_k \int_{-1}^{1} \varphi(rs+t)\,(1-s^2)^{k/2}\,ds \leqslant C\sqrt{k} \int_{-\infty}^{\infty} e^{-(rs+t)^2/2-ks^2/2}\,ds$$
$$= C\sqrt{\frac{2\pi k}{r^2+k}} \cdot e^{-\frac{kt^2}{2(r^2+k)}}.$$

This proves the assertion of the lemma. □

Combining Lemma 4.4.2 and Lemma 4.4.3 we obtain the next corollary.

Corollary 4.4.4. *Let $N \geqslant n \geqslant 4$ and let $U_1, \dots, U_N \in S^{n-1}$ be independent random vectors distributed uniformly in S^{n-1}. Fix $\xi \in S^{n-1}$, $t \in \mathbb{R}$, $r > \sqrt{n}$, and $\alpha > 0$. Then,*

$$\mathbb{P}\left(\frac{1}{N}\sum_{i=1}^{N} \varphi(t + r\langle \xi, U_i\rangle) \geqslant C \max\left\{\alpha, \frac{\sqrt{n}}{r}\varphi\left(\frac{c\sqrt{n}}{r}t\right)\right\}\right) \leqslant e^{-\alpha N},$$

where $C > 0$ and $0 < c < 1$ are absolute constants.

Proof. Set $Y_i = \varphi(t + r\langle \xi, U_i\rangle)$. Then, $Y_1, \dots, Y_N$ are independent, identically distributed random variables with values in $[0,1]$. We define

$$p = \max\left\{\alpha, C\frac{\sqrt{n}}{r} \cdot \varphi\left(\frac{c\sqrt{n}}{r}t\right)\right\},$$

where $c, C > 0$ are the constants from Lemma 4.4.3. Then, Lemma 4.4.3 shows that $\mathbb{E}(Y_1) \leqslant p$, and hence we may apply Lemma 4.4.2 to get

$$\mathbb{E}\left(\frac{1}{N}\sum_{i=1}^{N} Y_i \geqslant 3p\right) \leqslant \exp(-pN) \leqslant \exp(-\alpha N),$$

also using the fact that $p \geqslant \alpha$. □

Note that the function $\varphi(t) = e^{-t^2/2}$ is 1-Lipschitz because $|\varphi'(t)| = |t|e^{-t^2/2} \leqslant 1$ for all $t \in \mathbb{R}$. This allows us to make the estimate of Corollary 4.4.4 uniform in $\xi \in S^{n-1}$.

Proposition 4.4.5. *Let $n \geqslant 5$, $N \geqslant 10n\log n$ and $\sqrt{n} \leqslant r \leqslant n$. Let $U_1, \dots, U_N$ be independent random vectors distributed uniformly in S^{n-1}. Then, with probability greater than $1 - n^{-n}$ we have that for all $\xi \in S^{n-1}$ and $t \in \mathbb{R}$*

$$\frac{1}{N}\sum_{i=1}^{N} \varphi(t + r\langle \xi, U_i\rangle) \leqslant \frac{Cn\log n}{\min\{N, n^3\}} + C\frac{\sqrt{n}}{r}\varphi\left(\frac{c\sqrt{n}}{r}t\right), \tag{4.45}$$

where $c, C > 0$ are absolute constants.

Proof. Let $\theta_1,\dots,\theta_N \in S^{n-1}$. The function

$$G_{\theta_1,\dots,\theta_N}(t,\xi) = \frac{1}{N}\sum_{i=1}^{N}\varphi(t + r\langle \xi,\theta_i\rangle), \qquad t \in \mathbb{R},\ \xi \in S^{n-1}$$

is Lipschitz on $\mathbb{R}\times S^{n-1}$ with Lipschitz constant bounded by $r+1 \leqslant n+1$. We set $\delta = n^{-3}$ and consider a δ-net $\mathcal{F}$ of S^{n-1} with cardinality

$$|\mathcal{F}| \leqslant \left(\frac{5}{\delta}\right)^n \leqslant e^{6n\log n}. \tag{4.46}$$

Let $I \subseteq \mathbb{R}$ be the set of all integer multiples of n^{-3} that lie in the interval $[-n^3, n^3]$. Then, for any $\xi \in S^{n-1}$ and $t \in [-n^3, n^3]$ we may find $\eta \in \mathcal{F}$ and $s \in I$ such that

$$G_{\theta_1,\dots,\theta_N}(t,\xi) \leqslant (n+1)\frac{2}{n^3} + G_{\theta_1,\dots,\theta_N}(s,\eta), \tag{4.47}$$

also using the Lipschitz property of $G_{\theta_1,\dots,\theta_N}$. Applying Corollary 4.4.4 with $\alpha = 10(n\log n)/N$, we see that

$$\begin{aligned}&\mathbb{P}\left(G_{U_1,\dots,U_N}(t,\xi) \leqslant C\max\left\{\alpha, \frac{\sqrt{n}}{r}\cdot\varphi\left(\frac{c\sqrt{n}}{r}t\right)\right\} \text{ for all } \xi \in \mathcal{F} \text{ and } t \in I\right)\\ &\quad\geqslant 1 - |\mathcal{F}|\,|I|\,e^{-\alpha N} \geqslant 1 - n^7\cdot e^{6n\log n}e^{-10n\log n} \geqslant 1 - n^{-n}. \end{aligned}\tag{4.48}$$

The constant c in (4.48) is at most 1. Hence, for any $t, s \in [-n^3, n^3]$ with $|t-s| \leqslant n^{-3}$ we have that $\varphi(c\sqrt{n}\cdot s/r) \leqslant 5\varphi(c\sqrt{n}\cdot t/r)$. It follows from (4.47) and (4.48) that

$$\begin{aligned}&\mathbb{P}\left(G_{U_1,\dots,U_N}(t,\xi) \leqslant \frac{Cn\log n}{N} + \frac{\tilde{C}\sqrt{n}}{r}\cdot\varphi\left(\frac{c\sqrt{n}}{r}t\right) \text{ for all } \xi \in S^{n-1} \text{ and } |t| \leqslant n^3\right)\\ &\quad\geqslant 1 - n^{-n}.\end{aligned}$$

This shows that with probability greater than $1-n^{-n}$, the inequality (4.45) holds true for all $\xi \in S^{n-1}$ and $|t| \leqslant n^3$. We can also check that (4.45) is true when $|t| > n^3$ because, in this case

$$\begin{aligned}\frac{1}{N}\sum_{i=1}^{N}\varphi(t + r\langle\xi,U_i\rangle) &\leqslant \frac{1}{N}\sum_{i=1}^{N}\varphi(|t| - n)\\ &\leqslant \varphi(n^3 - n) \leqslant e^{-10n} \leqslant \frac{n\log n}{\min\{N, n^3\}},\end{aligned}$$

with probability 1. This completes the proof. □

By iterating Proposition 4.4.5 twice we obtain the next lemma.

Lemma 4.4.6. *Assume that $n \geqslant C_1$ and set $N_1 = n^3$, $N_2 = \lceil n\log^3 n\rceil$, $r_1 = n/\sqrt{\log n}$ and $r_2 = n/\sqrt{\log\log n}$. Then, there exist $\theta_1,\dots,\theta_{N_1},\eta_1,\dots,\eta_{N_2} \in S^{n-1}$ such that for all $\xi \in S^{n-1}$ and $t \in \mathbb{R}$*

$$\frac{1}{N_1N_2}\sum_{i=1}^{N_1}\sum_{j=1}^{N_2}\varphi(t + r_1\langle\xi,\theta_i\rangle + r_2\langle\xi,\eta_j\rangle) \leqslant \frac{C}{\sqrt{n}\log n} + C\frac{\sqrt{n}}{r_2}\cdot\varphi\left(c\frac{\sqrt{n}}{r_2}t\right),$$

where $c, C, C_1 > 0$ are absolute constants.

Proof. We may assume that $n \geqslant 5$ is sufficiently large so that $cr_2/r_1 > 1$, where, everywhere in this proof, $c > 0$ is the constant from Proposition 4.4.5. By Proposition 4.4.5, we may fix $\theta_1,\dots,\theta_{N_1} \in S^{n-1}$ such that for all $\xi \in S^{n-1}$ and $s \in \mathbb{R}$

$$\frac{1}{N_1}\sum_{i=1}^{N_1}\varphi(s + r_1\langle\xi,\theta_i\rangle) \leqslant \frac{Cn\log n}{N_1} + C\frac{\sqrt{n}}{r_1}\varphi\left(\frac{c\sqrt{n}}{r_1}s\right). \tag{4.49}$$

Note that $Cn(\log n)/N_1 \leqslant C/n$. It follows that, for any $\eta_1,\dots,\eta_{N_2} \in S^{n-1}$ and for any $\xi \in S^{n-1}$ and $t \in \mathbb{R}$

$$\begin{aligned}&\frac{1}{N_1N_2}\sum_{i=1}^{N_1}\sum_{j=1}^{N_2}\varphi(t + r_1\langle\xi,\theta_i\rangle + r_2\langle\xi,\eta_j\rangle)\\ &\quad\leqslant \frac{C}{n} + C\frac{\sqrt{n}}{r_1}\frac{1}{N_2}\sum_{j=1}^{N_2}\varphi\left(\frac{c\sqrt{n}}{r_1}t + \frac{cr_2\sqrt{n}}{r_1}\langle\xi,\eta_j\rangle\right),\end{aligned}$$

where we used (4.49) with $s = t + r_2\langle\xi,\eta_j\rangle$. Applying Proposition 4.4.5 with

$$r = cr_2\sqrt{n}/r_1 > \sqrt{n} \quad\text{and}\quad N = N_2 \geqslant n\log^3 n \geqslant 10n\log n,$$

we see that there exist $\eta_1,\dots,\eta_{N_2} \in S^{n-1}$ such that for all $\xi \in S^{n-1}$ and $t \in \mathbb{R}$

$$\begin{aligned}\frac{1}{N_2}\sum_{j=1}^{N_2}\varphi\left(\frac{c\sqrt{n}}{r_1}t + \frac{cr_2\sqrt{n}}{r_1}\langle\xi,\eta_j\rangle\right) &\leqslant \frac{Cn\log n}{N_2} + C\frac{\sqrt{n}}{r}\varphi\left(\frac{c\sqrt{n}}{r}\frac{c\sqrt{n}}{r_1}t\right)\\ &\leqslant \frac{\tilde{C}}{\log^2 n} + C\frac{\sqrt{n}}{r}\varphi\left(\frac{c\sqrt{n}}{r_2}t\right).\end{aligned}$$

Combining the two inequalities above we see that for all $\xi \in S^{n-1}$ and $t \in \mathbb{R}$

$$\frac{1}{N_1N_2}\sum_{i=1}^{N_1}\sum_{j=1}^{N_2}\varphi(t + r_1\langle\xi,\theta_i\rangle + r_2\langle\xi,\eta_j\rangle) \leqslant \frac{\tilde{C}}{\sqrt{n}\log^{3/2} n} + C\frac{\sqrt{n}}{r_2}\varphi\left(\frac{c\sqrt{n}}{r_2}t\right).$$ □

Let $\{e_1, \dots, e_n\}$ denote the standard basis in $\mathbb{R}^n$ and consider unit vectors $\theta_1, \dots, \theta_{N_1}$ and $\eta_1, \dots, \eta_{N_2}$ that satisfy the conclusion of Lemma 4.4.6. We define an origin-symmetric convex body $K \subseteq \mathbb{R}^n$ as the convex hull of the $2(N_1 + N_2 + n)$ vectors

$$\pm r_1\theta_1, \dots, \pm r_1\theta_{N_1}, \quad \pm r_2\eta_1, \dots, \pm r_2\eta_{N_2}, \quad \pm ne_1, \dots, \pm ne_n.$$

We write γ_n for the standard gaussian measure in $\mathbb{R}^n$, whose density is

$$x \mapsto (2\pi)^{-n/2} \exp(-|x|^2/2).$$

We will use below the standard bound $(2\pi)^{-1/2} \int_t^\infty \varphi(s)\, ds \leqslant \varphi(t)/2$ for all $t \geqslant 0$.

Lemma 4.4.7. *We have that $|K| \leqslant C^n$, where $C > 0$ is an absolute constant.*

Proof. Consider the polar body K° of K. If Z is a standard gaussian random vector in $\mathbb{R}^n$, then by the Khatri–Sidák lemma [279, 510] we have that

$$\begin{aligned}
&\mathbb{P}(Z \in 5nK^\circ) \\
&\quad = \mathbb{P}(|r_1\langle Z, \theta_i\rangle| \leqslant 5n,\ |r_2\langle Z, \eta_j\rangle| \leqslant 5n \text{ and } |n\langle Z, e_k\rangle| \leqslant 5 \text{ for all } i, j, k) \\
&\quad \geqslant \prod_{i=1}^{N_1} \mathbb{P}(|r_1\langle Z, \theta_i\rangle| \leqslant 5n) \cdot \prod_{j=1}^{N_2} \mathbb{P}(|r_2\langle Z, \eta_j\rangle| \leqslant 5n) \cdot \prod_{k=1}^{n} \mathbb{P}(|n\langle Z, e_k\rangle| \leqslant 5n).
\end{aligned}$$

Since $\langle Z, \theta_i\rangle$ is a standard gaussian random variable, we know that

$$\mathbb{P}(|r_1\langle Z, \theta_i\rangle| \leqslant 5n) = 1 - \frac{2}{\sqrt{2\pi}} \int_{5n/r_1}^{\infty} \varphi(s)\, ds \geqslant 1 - \varphi(5n/r_1).$$

Therefore,

$$\gamma_n(5nK^\circ) = \mathbb{P}(Z \in 5nK^\circ) \geqslant (1 - \varphi(5n/r_1))^{N_1} (1 - \varphi(5n/r_2))^{N_2} (1 - \varphi(5))^n.$$

By the choice of the values of the parameters in Lemma 4.4.6 we obtain

$$\gamma_n(5nK^\circ) \geqslant \left(1 - \frac{1}{n^{10}}\right)^{n^3} \left(1 - \frac{1}{(\log n)^{10}}\right)^{1+n\log^3 n} c^n \geqslant e^{-\tilde{c}n}.$$

Since the density of γ_n is bounded by $(2\pi)^{-n/2}$, we conclude that

$$|5nK^\circ| > c^n.$$

Now, the lemma follows from the Blaschke–Santaló inequality. □

Since $e_1, \dots, e_n \in K/n$, while $K \subseteq \mathbb{R}^n$ is convex and origin-symmetric, we know that

$$K \supseteq \sqrt{n}B_2^n. \tag{4.50}$$

Recall that $\int_{\mathbb{R}^n} |x|^2 d\gamma_n(x) = n$. By the Markov–Chebyshev inequality

$$\gamma_n(2\sqrt{n}B_2^n) \geqslant 1/2. \tag{4.51}$$

Let us now define the probability measure μ to be the convolution

$$\mu = \gamma_n * \left(\frac{1}{N_1 N_2} \sum_{i=1}^{N_1} \sum_{j=1}^{N_2} \frac{\delta_{r_1\theta_i + r_2\eta_j} + \delta_{-r_1\theta_i - r_2\eta_j}}{2} \right),$$

where δ_y is the Dirac measure at the point $y \in \mathbb{R}^n$. Then, μ is a probability measure in $\mathbb{R}^n$.

Lemma 4.4.8. $\mu(4K) \geqslant 1/2$.

Proof. From (4.50) we have that, for any i and j, the convex body $4K = 2K + 2K$ contains the set $\pm r_1\theta_i \pm r_2\eta_j + 2\sqrt{n}B_2^n$. The measure μ is an average of translates of γ_n, each centered at a point of the form $\pm r_1\theta_i \pm r_2\eta_j$. It follows that

$$\mu(4K) \geqslant \gamma_n(2\sqrt{n}B_2^n) \geqslant \frac{1}{2},$$

as claimed. □

Write g for the continuous density of the measure μ. If $\varphi_n(x) = \exp(-|x|^2/2)$ for $x \in \mathbb{R}^n$, we have

$$g(x) = \frac{1}{N_1 N_2} \sum_{i=1}^{N_1} \sum_{j=1}^{N_2} \frac{\varphi_n(x + r_1\theta_i + r_2\eta_j) + \varphi_n(x - r_1\theta_i - r_2\eta_j)}{2 \cdot (2\pi)^{n/2}}.$$

Note that for any $\xi \in S^{n-1}$, $z \in \mathbb{R}^n$ and $t \geqslant 0$

$$\int_{t\xi + \xi^\perp} \frac{\varphi_n(x+z)}{(2\pi)^{n/2}} \, dx = \frac{\varphi(t + \langle z, \xi \rangle)}{\sqrt{2\pi}}. \tag{4.52}$$

Lemma 4.4.9. *For any* $\xi \in S^{n-1}$ *and* $t \geqslant 0$

$$\int_{t\xi + \xi^\perp} g \leqslant C \frac{\sqrt{\log \log n}}{\sqrt{n}}.$$

Proof. By (4.52) we have that

$$\int_{t\xi + \xi^\perp} g = \frac{1}{N_1 N_2} \sum_{i=1}^{N_1} \sum_{j=2}^{N_2} \frac{1}{2\sqrt{2\pi}} (\varphi(t + r_1\langle \xi, \theta_i \rangle + r_2 \langle \xi, \eta_j \rangle)$$
$$+ \varphi(-t + r_1\langle \xi, \theta_i \rangle + r_2 \langle \xi, \eta_j \rangle)).$$

Therefore, according to Lemma 4.4.6 and using the fact that $\varphi \leqslant 1$, we get

$$\int_{t\xi+\xi^\perp} g \leqslant \frac{C}{\sqrt{n}\log n} + C\frac{\sqrt{n}}{r_2} \leqslant \tilde{C}\frac{\sqrt{\log\log n}}{\sqrt{n}},$$

completing the proof. □

Proof of Theorem 4.4.1. We present the proof of the theorem with an extra double logarithmic term. We set

$$f(x) = g(x) \cdot \mathbb{1}_{4K}(x)/\mu(4K),$$

where $\mathbb{1}_{4K}$ is the indicator function of $4K$. Then, f is an even, continuous probability density supported on $T = 4K$. From Lemma 4.4.8 we have that

$$f(x) \leqslant 2g(x) \quad \text{for } x \in \mathbb{R}^n.$$

Therefore, from Lemma 4.4.9, for any $\xi \in S^{n-1}$ and $t \geqslant 0$

$$\int_{T\cap(\xi^\perp+t\xi)} f(y)\,dy \leqslant 2\int_{\xi^\perp+t\xi} g(y)\,dy \leqslant \frac{C\sqrt{\log\log n}}{\sqrt{n}}. \tag{4.53}$$

We also know that $|T|^{1/n} \leqslant C$, by Lemma 4.4.7. Therefore, the desired estimate (4.41) follows from (4.53). □

As noted in [291], the inequalities of Theorems 4.1.4 and 4.4.1 immediately imply a lower estimate for the maximal distance from an origin-symmetric convex body to the class of intersection bodies, matching the upper estimate provided by John's theorem.

Corollary 4.4.10. *There exists an absolute constant $c > 0$ such that*

$$c\sqrt{n} \leqslant \sup_K d_{\text{ovr}}(K, \mathcal{I}_n) \leqslant \sqrt{n},$$

where the supremum is taken over all origin-symmetric convex bodies in $\mathbb{R}^n$.

4.5 Moments of measures on convex bodies

In this section, we discuss moments of measures on convex bodies. The following problem was considered in [85]. For any origin-symmetric convex set K of volume 1 in $\mathbb{R}^n$ and any even, continuous probability density $f : K \to [0, \infty)$ and $p \geqslant 1$, we consider the p-th absolute moment

$$M_{K,f,p}(\xi) = \int_K |\langle x, \xi\rangle|^p f(x)\,dx \tag{4.54}$$

in the direction of $\xi \in S^{n-1}$. The question is to estimate the constant $\gamma(p,n)$ that is defined as the smallest number $\gamma > 0$ satisfying

$$\min_{\xi \in S^{n-1}} M_{K,f,p}(\xi) \leqslant \gamma^p \, |K|^{p/n} \int_K f(x)\,dx \tag{4.55}$$

for all origin-symmetric convex bodies $K \subseteq \mathbb{R}^n$ and all even, continuous functions $f \geqslant 0$ on K. The continuity property of f in the definition (4.54) is irrelevant and may easily be replaced by measurability. It was shown in [85] that there is a two-sided bound for $\gamma(p,n)$.

Theorem 4.5.1. *There exist two absolute positive constants c and C such that, for any $p \geqslant 1$*

$$c\sqrt{n} \leqslant \gamma(p,n) \leqslant C\sqrt{pn}.$$

To describe the way the upper bound is obtained, denote by L_p^n the class of the unit balls of n-dimensional subspaces of L_p. Equivalently (see [310, p. 117]), L_p^n is the class of all origin-symmetric convex bodies D in $\mathbb{R}^n$ such that there exists a finite Borel measure μ_D on S^{n-1} satisfying

$$\|x\|_D^p = \int_{S^{n-1}} |\langle x, \theta\rangle|^p \, d\mu_D(\theta) \tag{4.56}$$

for all $x \in \mathbb{R}^n$. Note that $L_1^n = \Pi_n^*$ is the class of polar projection bodies that, in particular, contains the cross-polytopes; see [310, Chapter 8] for details.

For a (bounded) set K in $\mathbb{R}^n$, define the quantity

$$V(K, L_p^n) = \inf\{|D|^{1/n} : K \subseteq D,\ D \in L_p^n\}.$$

If K is measurable and has positive volume, we have the relation

$$V(K, L_p^n) = d_{\mathrm{ovr}}(K, L_p^n)\, |K|^{1/n},$$

where

$$d_{\mathrm{ovr}}(K, L_p^n) = \inf\left\{\left(\frac{|D|}{|K|}\right)^{1/n} : K \subseteq D,\ D \in L_p^n\right\}. \tag{4.57}$$

For convex K, the latter may be interpreted as the outer volume ratio distance from K to the class of unit balls of n-dimensional subspaces of L_p. The next body-wise estimates refine the upper bound of Theorem 4.5.1 in terms of the d_{ovr}-distance.

Theorem 4.5.2. *Given a probability measure μ on $\mathbb{R}^n$ with a compact support K, for every $p \geqslant 1$*

$$\min_{\xi\in S^{n-1}}\Big(\int |\langle x,\xi\rangle|^p\,d\mu(x)\Big)^{1/p} \leqslant C\sqrt{p}\,V(K,L_p^n),$$

where C is an absolute constant. In particular, if f is a non-negative, continuous function on a compact set $K \subseteq \mathbb{R}^n$ of positive volume, then

$$\min_{\xi\in S^{n-1}} M_{K,f,p}(\xi) \leqslant (Cp)^{p/2}\, d^p_{\mathrm{ovr}}(K,L_p^n)\,|K|^{p/n}\int_K f(x)\,dx.$$

Proof. Let $D \subseteq \mathbb{R}^n$ be the unit ball of an n-dimensional subspace of L_p, so that (4.56) holds for some measure μ_D on S^{n-1}. Then, integrating the inequality

$$\min_{\theta\in S^{n-1}}\int_K |\langle x,\theta\rangle|^p\,d\mu(x) \leqslant \int_K |\langle x,\xi\rangle|^p\,d\mu(x), \quad \xi\in S^{n-1}$$

over the variable ξ with respect to μ_D, we get

$$\mu_D(S^{n-1})\min_{\theta\in S^{n-1}}\int_K |\langle x,\theta\rangle|^p\,d\mu(x) \leqslant \int_K \|x\|_D^p\,d\mu(x).$$

In the case $K \subseteq D$, we have $\|x\|_D \leqslant \|x\|_K \leqslant 1$ on K, so that the last integral does not exceed $\mu(K) = 1$, and thus

$$\mu_D(S^{n-1})\min_{\theta\in S^{n-1}}\int_K |\langle x,\theta\rangle|^p\,d\mu(x) \leqslant 1. \tag{4.58}$$

In order to estimate the left-hand side of (4.58) from below, we represent the value $\mu_D(S^{n-1})$ as the integral $\int_{S^{n-1}} |x|^p\,d\mu_D(x)$ and apply the well-known formula

$$|x|^p = \frac{\Gamma(\frac{p+n}{2})}{2\pi^{\frac{n-1}{2}}\,\Gamma(\frac{p+1}{2})}\int_{S^{n-1}} |\langle x,\theta\rangle|^p\,d\theta, \quad x\in\mathbb{R}^n$$

(see Lemma 1.4.9). Using (4.56), we obtain

$$\begin{aligned}\mu_D(S^{n-1}) &= \frac{\Gamma(\frac{p+n}{2})}{2\pi^{\frac{n-1}{2}}\Gamma(\frac{p+1}{2})}\int_{S^{n-1}}\int_{S^{n-1}} |\langle x,\theta\rangle|^p\,d\theta\,d\mu_D(x)\\ &= \frac{\Gamma(\frac{p+n}{2})}{2\pi^{\frac{n-1}{2}}\Gamma(\frac{p+1}{2})}\int_{S^{n-1}} \|\theta\|_D^p\,d\theta.\end{aligned}$$

The last integral may be related to the volume of D, by using the polar formula for the volume of D

$$n\,|D| = \int_{S^{n-1}} \|\theta\|_D^{-n}\,d\theta = \kappa_{n-1}\int_{S^{n-1}} \|\theta\|_D^{-n}\,d\sigma(\theta),$$

where σ denotes the normalized Lebesgue measure on S^{n-1} and $\kappa_{n-1} = \frac{2\pi^{\frac{n}{2}}}{\Gamma(\frac{n}{2})}$ is its $(n-1)$-dimensional volume. Namely, by Jensen's inequality, we have

$$\int_{S^{n-1}} \|\theta\|_D^{-n}\, d\sigma_{n-1}(\theta) \geqslant \Big(\int_{S^{n-1}} \|\theta\|_D^{p}\, d\sigma_{n-1}(\theta) \Big)^{-\frac{n}{p}},$$

or equivalently

$$\int_{S^{n-1}} \|\theta\|_D^{p}\, d\theta \geqslant \kappa_{n-1}^{\frac{p+n}{n}} \,(n\,|D|)^{-\frac{p}{n}}.$$

Thus,

$$\begin{aligned} \mu_D(S^{n-1}) &\geqslant \frac{\Gamma(\frac{p+n}{2})\,\kappa_{n-1}^{\frac{p+n}{n}}}{2\pi^{\frac{n-1}{2}}\,\Gamma(\frac{p+1}{2})\,n^{\frac{p}{n}}\,|D|^{\frac{p}{n}}} \\ &= \sqrt{\pi}\,\frac{\Gamma(\frac{p+n}{2})}{\Gamma(\frac{p+1}{2})\,\Gamma(\frac{n}{2})} \Big(\frac{\kappa_{n-1}}{n\,|D|}\Big)^{\frac{p}{n}} \geqslant \frac{c^p}{\Gamma(\frac{p+1}{2})\,|D|^{\frac{p}{n}}}, \end{aligned}$$

where $c > 0$ is an absolute constant. Here, we used the well-known asymptotic relation $\sqrt{n}\,\kappa_{n-1}^{\frac{1}{n}} \to c_0$ as $n \to \infty$, for some absolute $c_0 > 0$, as well as the estimate $\Gamma(\frac{p+n}{2})/\Gamma(\frac{n}{2}) \geqslant (cn)^{p/2}$.

Applying this lower estimate on the left-hand side of (4.58), we get

$$\min_{\theta\in S^{n-1}} \int_K |\langle x,\theta\rangle|^p\, d\mu(x) \leqslant C^p\,\Gamma\Big(\frac{p+1}{2}\Big)\,|D|^{\frac{p}{n}}.$$

It remains to take the minimum over all admissible D and note that $\Gamma(\frac{p+1}{2})^{1/p} \leqslant c\sqrt{p}$ for $p \geqslant 1$. □

In the class of origin-symmetric convex bodies K in $\mathbb{R}^n$, there is a dimensional bound $d_{\mathrm{ovr}}(K, L_p^n) \leqslant \sqrt{n}$, which follows from John's theorem and the fact that ellipsoids belong to L_p^n for all $p \geqslant 1$ (see [310, Lemma 3.12]). Hence, the second upper bound of Theorem 4.5.2 is more accurate in comparison with the universal bound of Theorem 4.5.1.

In order to prove the lower estimate of Theorem 4.5.1, we first establish the connection between our question and the slicing inequality for arbitrary measures. In Theorem 4.1.2, we saw that

$$\int_K f(x)\,dx \leqslant 2\,d_{\mathrm{ovr}}(K,\mathcal{I}_n)\, \max_{\xi\in S^{n-1}} \int_{K\cap\xi^\perp} f(x)\,dx\, |K|^{1/n}$$

for any origin-symmetric convex body $K \subseteq \mathbb{R}^n$ and any even continuous non-negative function f on K, where $d_{\mathrm{ovr}}(K,\mathcal{I}_n)$ is the outer volume ratio distance from K to the class

$\mathcal{I}_n$ of intersection bodies in $\mathbb{R}^n$. The slicing problem for arbitrary measures can be modified to include non-central sections. For an origin-symmetric convex body $K \subseteq \mathbb{R}^n$, let $T_{n,K}$ be the smallest positive number T satisfying

$$\int_K f(x)\,dx \leqslant T\,|K|^{\frac{1}{n}} \sup_H \int_{K\cap H} f(x)\,dx, \tag{4.59}$$

where the supremum is taken over all affine hyperplanes H in $\mathbb{R}^n$, and then let

$$T_n = \sup_{K\subseteq\mathbb{R}^n} T_{n,K}.$$

We have $T_n \leqslant 2d_{\mathrm{ovr}}(K,\mathcal{I}_n) \leqslant 2\sqrt{n}$. On the other hand, in the previous section we saw that for any $n \geqslant 3$, there exists an origin-symmetric convex body $M \subseteq \mathbb{R}^n$ and an even, continuous probability density $f : M \to [0,\infty)$ such that, for any affine hyperplane $H \subseteq \mathbb{R}^n$

$$\int_{M\cap H} f(x)\,dx \leqslant \frac{C}{\sqrt{n}}\,|M|^{-1/n}. \tag{4.60}$$

This implies

$$T_n \geqslant c\sqrt{n}.$$

The connection between (4.55) and the slicing inequality for arbitrary measures (4.59) is given by the next lemma.

Lemma 4.5.3. *Given a Borel measurable function $f \geqslant 0$ on $\mathbb{R}^n$, for any $\xi \in S^{n-1}$ and $p > 0$*

$$2^p\,(p+1)\Big(\sup_{s\in\mathbb{R}} \int_{(x,\xi)=s} f(x)\,dx\Big)^p \int |(x,\xi)|^p f(x)\,dx \geqslant \Big(\int f(x)\,dx\Big)^{p+1}.$$

To prove Lemma 4.5.3, we need the following simple assertion.

Lemma 4.5.4. *Given a measurable function $g : \mathbb{R} \to [0,1]$, the function*

$$q \mapsto \left(\frac{q+1}{2}\int_{-\infty}^{\infty} |t|^q g(t)\,dt\right)^{\frac{1}{q+1}}$$

is non-decreasing on $(-1,\infty)$.

Proof. We use a standard argument, similar to the one in [283, Lemma 2.4]. Given $-1 < q < p$, let $A > 0$ be defined by the equation

$$\int_{-\infty}^{\infty} |t|^q g(t)\,dt = \int_{-A}^{A} |t|^q\,dt = \frac{2}{q+1} A^{q+1}.$$

Using

$$|t|^p \leqslant A^{p-q}\,|t|^q\,(|t| \leqslant A) \quad \text{and} \quad |t|^p \geqslant A^{p-q}\,|t|^q\,(|t| \geqslant A),$$

together with the assumption $0 \leqslant g \leqslant 1$, we then have

$$\begin{aligned}\int_{|t|\leqslant A} (1-g(t))\,|t|^p\,dt - \int_{|t|>A} g(t)\,|t|^p\,dt \\ \leqslant A^{p-q}\Big(\int_{|t|\leqslant A} (1-g(t))\,|t|^q\,dt - \int_{|t|>A} g(t)\,|t|^q\,dt\Big) = 0.\end{aligned}$$

It follows that

$$\int_{-\infty}^{\infty} g(t)\,|t|^p\,dt \geqslant \int_{-A}^{A} |t|^p\,dt = \frac{2}{p+1} A^{p+1},$$

that is,

$$\Big(\frac{p+1}{2}\int_{-\infty}^{\infty} g(t)\,|t|^p\,dt\Big)^{\frac{1}{p+1}} \geqslant A = \Big(\frac{q+1}{2}\int_{-\infty}^{\infty} g(t)\,|t|^q\,dt\Big)^{\frac{1}{q+1}}.$$ □

Proof of Lemma 4.5.3. We may assume that f is integrable. For $t \in \mathbb{R}$, introduce the hyperplanes $H_t = \{x : \langle x, \xi\rangle = t\}$. Since f is Borel measurable on $\mathbb{R}^n$, the function

$$g(t) = \frac{\int_{H_t} f(x)\,dx}{\sup_s \int_{H_s} f(x)\,dx}$$

is Borel measurable on the line and satisfies $\|g\|_\infty = 1$. By Fubini's theorem

$$\int_{-\infty}^{\infty} |t|^p g(t)\,dt = \frac{\int |\langle x, \xi\rangle|^p f(x)\,dx}{\sup_s \int_{H_s} f(x)\,dx}$$

and

$$\int_{-\infty}^{\infty} g(t)\,dt = \frac{\int f(x)\,dx}{\sup_s \int_{H_s} f(x)\,dx}.$$

Applying Lemma 4.5.4 to the function g with $q = 0$ and p, we get

$$\frac{1}{2}\int_{-\infty}^{\infty} g(t)\,dt \leqslant \left(\frac{p+1}{2}\int_{-\infty}^{\infty} |t|^p g(t)\,dt\right)^{\frac{1}{p+1}},$$

which in our case becomes

$$\left(\int_{\mathbb{R}^n} f(x)\,dx\right)^{p+1} \leqslant (p+1)\left(2\sup_s \int_{H_s} f(x)\,dx\right)^p \int_{\mathbb{R}^n} |\langle x,\xi\rangle|^p f(x)\,dx. \qquad \square$$

If f is defined on a set K in $\mathbb{R}^n$, we then have

$$2^p\,(p+1)\left(\sup_{s\in\mathbb{R}} \int_{K\cap\{x:\langle x,\xi\rangle=s\}} f(x)\,dx\right)^p M_{K,f,p}(\xi) \geqslant \left(\int_K f(x)\,dx\right)^{p+1}.$$

The lower bound in Theorem 4.5.1 then follows, by combining the above inequality with (4.55) and Theorem 4.4.1.

Corollary 4.5.5. *There exist absolute positive constants c and C such that, for every $p \geqslant 1$*

$$c\sqrt{n} \leqslant T_n \leqslant C\gamma(p,n).$$

Combining Lemma 4.5.3 with Theorem 4.5.2 we obtain a new slicing inequality. In the case of volume, where $f \equiv 1$, this inequality was established earlier by Ball [43] for $p = 1$ and by E. Milman [413] for arbitrary p. The next theorem comes from [85].

Theorem 4.5.6. *Let $f \geqslant 0$ be a Borel measurable function on a compact set $K \subseteq \mathbb{R}^n$ of positive volume. Then, for any $p > 2$*

$$\int_K f(x)\,dx \leqslant C\sqrt{p}\,d_{\mathrm{ovr}}(K,L_p^n)\,|K|^{1/n}\sup_H \int_{K\cap H} f(x)\,dx,$$

where the supremum is taken over all affine hyperplanes H in $\mathbb{R}^n$, and C is an absolute constant.

Theorem 4.5.6 also holds for $1 \leqslant p \leqslant 2$, but in this case it is weaker than Theorem 4.1.2, because the unit ball of every finite dimensional subspace of L_p, $0 < p \leqslant 2$, is an intersection body; see [305]. However, for $p > 2$ the unit balls of subspaces of L_p are not necessarily intersection bodies. For example, the unit balls of ℓ_p^n are not intersection bodies if $p > 2$, $n \geqslant 5$; see [310, Theorem 4.13]. So, the result of Theorem 4.5.6 is new for $p > 2$, and generalizes the estimate from [322] in the case where K itself belongs to the class L_p^n.

Theorem 4.5.6 gives another reason to estimate the outer volume ratio distance $d_{\mathrm{ovr}}(K,L_p^n)$ from an arbitrary symmetric convex body to the class of unit balls of subspaces of L_p. Combining Theorems 4.5.6 and 4.4.1, we get a lower estimate.

Corollary 4.5.7. *There exists an origin-symmetric convex body $M \subseteq \mathbb{R}^n$ such that*

$$d_{\mathrm{ovr}}(M, L_p^n) \geqslant c\sqrt{n/p}$$

for every $p \geqslant 1$, where $c > 0$ is an absolute constant.

A matching upper estimate was proved in [331]. Let $p \geqslant 1$ and let C be an origin-symmetric convex body in $\mathbb{R}^n$ of volume 1. Recall the definition of the L_p-centroid body $Z_p(C)$ via its support function

$$h_{Z_p(C)}(\theta) := \left(\int_C |\langle x, \theta\rangle|^p \, dx\right)^{\frac{1}{p}}, \quad \theta \in S^{n-1}. \tag{4.61}$$

We define $\mathcal{M}_p^n$ as the set of symmetric, convex bodies in $\mathbb{R}^n$ that are polar to some $Z_p(C)$, i. e.

$$\mathcal{M}_p^n := \{L \in \mathcal{K}_n : \exists C \in \mathcal{K}_n,\ |C| = 1,\ L = Z_p^\circ(C)\}.$$

We have that

$$\mathcal{M}_p^n \subseteq L_p^n.$$

Indeed, writing (4.61) in polar coordinates we get

$$\begin{aligned} \|\theta\|_L^p = h_{Z_p(C)}^p(\theta) &= \int_{S^{n-1}} |\langle x, \theta\rangle|^p \frac{\delta_n}{\|x\|_C^{n+p}} \, d\sigma(x) \\ &= \int_{S^{n-1}} |\langle x, \theta\rangle|^p d\mu(x), \end{aligned}$$

where $d\mu(x) := \frac{\delta_n}{\|x\|_C^{n+p}} \, d\sigma(x)$ is a finite Borel measure on S^{n-1} and δ_n is a constant depending only on n. It follows that $(\mathbb{R}^n, \|\cdot\|_L)$ can be embedded isometrically into L_p.

Proposition 4.5.8. *There exists $c > 0$ such that for every origin-symmetric convex body K in $\mathbb{R}^n$ and every $p \geqslant 1$*

$$d_{\mathrm{ovr}}(K, L_p^n) \leqslant d_{\mathrm{ovr}}(K, \mathcal{M}_p^n) \leqslant c\sqrt{\frac{n+p}{p}}. \tag{4.62}$$

Proof. Let K be an origin-symmetric convex body in $\mathbb{R}^n$, and let $C := K^\circ/|K^\circ|^{\frac{1}{n}}$. Note that

$$|C| = 1 \text{ and } C^\circ = |K^\circ|^{\frac{1}{n}} K. \tag{4.63}$$

Define $L \in \mathcal{M}_p^n$ by

$$L := |K^\circ|^{-\frac{1}{n}} Z_p^\circ(C) \quad \text{and} \quad L^\circ = |K^\circ|^{\frac{1}{n}} Z_p(C). \tag{4.64}$$

Note that, since $|C| = 1$, (4.61) implies that $Z_p(C) \subseteq C$. So, by (4.63) and (4.64)

$$C^\circ \subseteq Z_p^\circ(C) \quad \text{or} \quad \frac{C^\circ}{|K^\circ|^{\frac{1}{n}}} \subseteq \frac{Z_p^\circ(C)}{|K^\circ|^{\frac{1}{n}}} \quad \text{or} \quad K \subseteq L.$$

By the Bourgain–V. Milman inequality (1.12), for every symmetric convex body $C \subseteq \mathbb{R}^n$ of volume 1

$$|C^\circ|^{\frac{1}{n}} \geqslant \frac{c_0}{n}, \tag{4.65}$$

where $c_0 > 0$ is an absolute constant. The last ingredient of the proof is an inequality of Lutwak and Zhang [395] stating that for every star body C of volume 1 and every $p > 0$

$$|Z_p^\circ(C)|^{\frac{1}{n}} \leqslant |Z_p^\circ(D_n)|^{\frac{1}{n}} \leqslant c_1 \frac{1}{n} \sqrt{\frac{n+p}{p}}, \tag{4.66}$$

where $D_n = B_2^n / |B_2^n|$ is the Euclidean ball of volume 1 and $c_1 > 0$ is an absolute constant. Using (4.63), (4.64), (4.65), and (4.66), we conclude that

$$\left(\frac{|L|}{|K|}\right)^{\frac{1}{n}} = \left(\frac{|Z_p^\circ(C)|}{|C^\circ|}\right)^{\frac{1}{n}} \leqslant c \sqrt{\frac{n+p}{p}},$$

where $c := \frac{c_1}{c_0} > 0$ is an absolute constant. □

Combining Corollary 4.5.7 and Proposition 4.5.8, we obtain the next fact.

Corollary 4.5.9. *There exist absolute constants $c_1, c_2 > 0$ such that for every $p \geqslant 1$*

$$c_1 \sqrt{\frac{n}{p}} \leqslant \sup_K d_{\mathrm{ovr}}(K, L_p^n) \leqslant c_2 \sqrt{\frac{n+p}{p}},$$

where the supremum is taken over all origin-symmetric convex bodies in $\mathbb{R}^n$.

We continue with a comparison result for the moments, in the spirit of the Busemann–Petty problem for arbitrary measures. For $p \geqslant 1$, introduce the Banach–Mazur distance

$$d_{\mathrm{BM}}(M, L_p^n) = \inf\{a \geqslant 1 : \text{there exists } D \in L_p^n \text{ such that } D \subset M \subset aD\}$$

from a star body M in $\mathbb{R}^n$ to the class L_p^n. Recall that L_p^n is invariant with respect to linear transformations. By John's theorem, if M is origin-symmetric and convex, then $d_{\mathrm{BM}}(M, L_p^n) \leqslant \sqrt{n}$. The following theorem was proved in [85].

Theorem 4.5.10. *Let K and M be origin-symmetric star bodies in $\mathbb{R}^n$, and let $f \geqslant 0$, be an even, continuous function on $\mathbb{R}^n$. Given $p \geqslant 1$, suppose that for every $\xi \in S^{n-1}$*

$$\int_K |\langle x,\xi\rangle|^p f(x)\,dx \leqslant \int_M |\langle x,\xi\rangle|^p f(x)\,dx. \tag{4.67}$$

Then,

$$\int_K f(x)\,dx \leqslant d^p_{BM}(M, L_p^n) \int_M f(x)\,dx.$$

Proof. Let $D \in L_p^n$ be such that the distance $d_{\mathrm{BM}}(M, L_p^n)$ is almost realized, i. e., for small $\delta > 0$, suppose that $D \subseteq M \subseteq (1+\delta)\, d_{\mathrm{BM}}(M, L_p^n)\, D$. Integrating both sides of (4.67) over $\xi \in S^{n-1}$ with respect to the measure μ_D from (4.56), we get

$$\int_K \|x\|_D^p f(x)\,dx \leqslant \int_M \|x\|_D^p f(x)\,dx.$$

Equivalently, writing the integrals in spherical coordinates, we have

$$0 \leqslant \int_{S^{n-1}} \|\theta\|_D^p \left(\int_{\|\theta\|_K^{-1}}^{\|\theta\|_M^{-1}} r^{n+p-1} f(r\theta)\,dr \right) d\theta = \int_{S^{n-1}} \frac{\|\theta\|_D^p}{\|\theta\|_M^p} I(\theta)\,d\theta,$$

where

$$I(\theta) = \|\theta\|_M^p \int_{\|\theta\|_K^{-1}}^{\|\theta\|_M^{-1}} r^{n+p-1} f(r\theta)\,dr.$$

For $\theta \in S^{n-1}$ such that $\|\theta\|_K \geqslant \|\theta\|_M$, the latter quantity is non-negative, and we may write

$$\begin{aligned} I(\theta) &= \int_{\|\theta\|_K^{-1}}^{\|\theta\|_M^{-1}} (\|\theta\|_M^p - r^{-p})\, r^{n+p-1} f(r\theta)\,dr + \int_{\|\theta\|_K^{-1}}^{\|\theta\|_M^{-1}} r^{n-1} f(r\theta)\,dr \\ &\leqslant \int_{\|\theta\|_K^{-1}}^{\|\theta\|_M^{-1}} r^{n-1} f(r\theta)\,dr. \end{aligned}$$

If we assume that $\|\theta\|_K \leqslant \|\theta\|_M$, then we have

$$-I(\theta) = \|\theta\|_M^p \int_{\|\theta\|_M^{-1}}^{\|\theta\|_K^{-1}} r^p r^{n-1} f(r\theta)\, dr \geqslant \int_{\|\theta\|_M^{-1}}^{\|\theta\|_K^{-1}} r^{n-1} f(r\theta)\, dr,$$

which is the same upper bound on $I(\theta)$ as before. Thus,

$$0 \leqslant \int_{S^{n-1}} \frac{\|\theta\|_D^p}{\|\theta\|_M^p} \left(\int_{\|\theta\|_K^{-1}}^{\|\theta\|_M^{-1}} r^{n-1} f(r\theta)\, dr \right) d\theta,$$

that is,

$$\int_{S^{n-1}} \frac{\|\theta\|_D^p}{\|\theta\|_M^p} \left(\int_0^{\|\theta\|_K^{-1}} r^{n-1} f(r\theta)\, dr \right) d\theta \leqslant \int_{S^{n-1}} \frac{\|\theta\|_D^p}{\|\theta\|_M^p} \left(\int_0^{\|\theta\|_M^{-1}} r^{n-1} f(r\theta)\, dr \right) d\theta.$$

Now, by the choice of D

$$\|\theta\|_M \leqslant \|\theta\|_D \leqslant (1+\delta)\, d_{\mathrm{BM}}(M, L_p^n) \|\theta\|_M$$

for every $\theta \in S^{n-1}$. Hence,

$$\begin{aligned}
\int_K f(x)\, dx &= \int_{S^{n-1}} \left(\int_0^{\|\theta\|_K^{-1}} r^{n-1} f(r\theta)\, dr \right) d\theta \\
&\leqslant \int_{S^{n-1}} \frac{\|\theta\|_D^p}{\|\theta\|_M^p} \left(\int_0^{\|\theta\|_K^{-1}} r^{n-1} f(r\theta)\, dr \right) d\theta \\
&\leqslant \int_{S^{n-1}} \frac{\|\theta\|_D^p}{\|\theta\|_M^p} \left(\int_0^{\|\theta\|_M^{-1}} r^{n-1} f(r\theta)\, dr \right) d\theta \\
&\leqslant (1+\delta)\, d_{BM}^p(M, L_p^n) \int_{S^{n-1}} \left(\int_0^{\|\theta\|_M^{-1}} r^{n-1} f(r\theta)\, dr \right) d\theta \\
&= (1+\delta)\, d_{BM}^p(M, L_p^n) \int_M f(x)\, dx.
\end{aligned}$$

Sending δ to zero, we get the result. □

4.6 Derivatives of the Radon transform

In this section, we generalize the isomorphic Busemann–Petty problem and the slicing inequality for measures to inequalities for the fractional derivatives of the Radon transform. The distance involved here is the outer volume ratio distance from an arbitrary star body to the unit balls of spaces that embed in L_{-p}; see Section 1.7.1. Recall that for an origin-symmetric infinitely smooth, convex body K in $\mathbb{R}^n$, and for an infinitely differentiable function f on K, a fixed $\xi \in S^{n-1}$, and $q \in \mathbb{C}$, $\Re q > -1$, we denote by $(Rf(\xi,t))_t^{(q)}(0)$ the fractional derivative of order q at zero of the function $t \to Rf(\xi,t)$, $t \in \mathbb{R}$; see Section 1.5.

We start with a generalization of the isomorphic Busemann–Petty problem from [233].

Theorem 4.6.1. *Let K, L be infinitely smooth, origin-symmetric, convex bodies in $\mathbb{R}^n$, f, g non-negative, infinitely differentiable functions on K and L, respectively, and assume that $\|g\|_\infty = g(0) = 1$, and $q \in (-1, n-1)$ is not an odd integer. If for every $\xi \in S^{n-1}$*

$$\frac{1}{\cos(\frac{\pi q}{2})}(\mathcal{R}f(\xi,t))_t^{(q)}(0) \leqslant \frac{1}{\cos(\frac{\pi q}{2})}(\mathcal{R}g(\xi,t))_t^{(q)}(0), \tag{4.68}$$

then

$$\int_K f(x)\,dx \leqslant \frac{n}{n-q-1}(d_{\mathrm{ovr}}(K, L_{-1-q}^n))^{q+1}\Big(\int_L g(x)\,dx\Big)^{\frac{n-q-1}{n}} |K|^{\frac{q+1}{n}}.$$

Proof. By Theorem 1.5.8 the inequality (4.68) can be written as

$$\begin{aligned}&\left(|x|^{-n+q+1}\left(\int_0^{\frac{|x|}{\|x\|_K}} r^{n-q-2} f\left(r\frac{x}{|x|}\right)dr\right)\right)_x^\wedge(\xi)\\ &\leqslant \left(|x|^{-n+q+1}\left(\int_0^{\frac{|x|}{\|x\|_L}} r^{n-q-2} g\left(r\frac{x}{|x|}\right)dr\right)\right)_x^\wedge(\xi).\end{aligned} \tag{4.69}$$

Let $\delta > 0$, and let $D \in L_{-1-q}^n$ be such that $K \subset D$ and

$$|D|^{1/n} \leqslant (1+\delta) d_{\mathrm{ovr}}(K, L_{-1-q}^n)|K|^{1/n}. \tag{4.70}$$

We can integrate both sides of (4.6) over the sphere with respect to the measure μ_0 corresponding to the positive definite distribution $\|x\|_D^{-1-q}$ by Proposition 1.4.7. Then, we apply Parseval's formula on the sphere in the form of Proposition 1.6.5. We get

$$\int_{S^{n-1}} \|\theta\|_D^{-1-q} \left(\int_0^{\|\theta\|_K^{-1}} r^{n-q-2} f(r\theta)\, dr \right) d\theta$$
$$\leqslant \int_{S^{n-1}} \|\theta\|_D^{-1-q} \left(\int_0^{\|\theta\|_L^{-1}} r^{n-q-2} g(r\theta)\, dr \right) d\theta,$$

which is equivalent to

$$\int_K \|x\|_D^{-1-q} f(x)\, dx \leqslant \int_L \|x\|_D^{-1-q} g(x)\, dx. \tag{4.71}$$

Since $K \subset D$, we have $\|x\|_D \leqslant 1$ for every $x \in K$, and thus

$$\int_K \|x\|_D^{-1-q} f(x)\, dx \geqslant \int_K f(x)\, dx. \tag{4.72}$$

On the other hand, by Lemma 3.4.6

$$\left(\frac{\int_L \|x\|_D^{-1-q} g(x)\, dx}{\int_D \|x\|_D^{-1-q}\, dx} \right)^{1/(n-q-1)} \leqslant \left(\frac{\int_L g(x)\, dx}{\int_D dx} \right)^{1/n}. \tag{4.73}$$

Writing the integral in spherical coordinates we get

$$\int_D \|x\|_D^{-1-q}\, dx = \int_{\mathbb{R}^n} \|x\|_D^{-1-q} \mathbb{1}_D(x)\, dx$$
$$= \int_{S^{n-1}} \|\theta\|_D^{-1-q} \left(\int_0^{\|\theta\|_D^{-1}} r^{n-q-2}\, dr \right) d\theta$$
$$= \frac{1}{n-q-1} \int_{S^{n-1}} \|\theta\|_D^{-n}\, d\theta = \frac{n}{n-q-1} |D|.$$

Now, we can rewrite (4.73) as

$$\int_L \|x\|_D^{-1-q} g(x)\, dx \leqslant \frac{n}{n-q-1} \left(\int_L g(x)\, dx \right)^{\frac{n-q-1}{n}} |D|^{\frac{q+1}{n}}. \tag{4.74}$$

Combining estimates (4.71), (4.72), and (4.74) with the definition of D, (4.70), and sending δ to zero, we get

$$\int_K f(x)\,dx \leqslant \frac{n}{n-q-1}\Big(\int_L g(x)\,dx\Big)^{\frac{n-q-1}{n}} (d_{\mathrm{ovr}}(K, L_{-1-q}^n))^{q+1}|K|^{\frac{q+1}{n}},$$

which proves the theorem. □

We now move on to the corresponding slicing inequality, also proved in [233].

Theorem 4.6.2. *There exist absolute constants $c_1, c_2 > 0$ such that for any infinitely smooth, origin-symmetric convex body K of volume 1 in $\mathbb{R}^n$, any even infinitely smooth probability density f on K, and any $q \in \mathbb{R}$, $0 \leqslant q \leqslant n-2$, which is not an odd integer, there exists a direction $\xi \in S^{n-1}$ such that*

$$\max\Big\{\Big(\frac{c_1(q+1)}{\sqrt{n \log^2(\frac{ne}{q+1})}}\Big)^{q+1}, \Big(\frac{c_2(q+1)}{n}\Big)^{\frac{q+1}{2}}\Big\} \leqslant \frac{1}{\cos(\frac{\pi q}{2})}(Rf(\xi, t))_t^{(q)}(0). \tag{4.75}$$

If $q = 2k$, $k \in \mathbb{N} \cup \{0\}$, then $(Rf(\xi, t))_t^{(2k)}(0)$ is the usual derivative and there exists ξ such that

$$\Big(\frac{c(2k+1)}{\sqrt{n \log^2(\frac{ne}{2k+1})}}\Big)^{2k+1} \leqslant (-1)^k (Rf(\xi, t))_t^{(2k)}(0).$$

If $q = 2k-1$, $k \in \mathbb{N}$, then computing the limit as $q \to 2k-1$ we get that there exists a direction $\xi \in S^{n-1}$ such that

$$\begin{aligned}&\Big(\frac{2kc}{\sqrt{n \log^2(\frac{ne}{2k})}}\Big)^{2k}\\ &\leqslant (-1)^k (2k-1)! \int_0^\infty t^{-2k}\Big(Rf(\xi, t) - \sum_{j=0}^{k-1} \frac{t^{2j}}{(2j)!}(Rf(\xi, t))_t^{(2j)}(0)\Big)dt.\end{aligned}$$

We deduce Theorem 4.6.2 from a more general result. Recall that by L_{-1-q}^n we denote the class of star bodies D in $\mathbb{R}^n$ for which the space $(\mathbb{R}^n, \|\cdot\|_D)$ embeds in L_{-1-q}, i. e., the function $\|\cdot\|_D^{-1-q}$ represents a positive definite distribution.

Before we start, let us compute the fractional derivatives of the Radon transform in the case where $f \equiv 1$ and $K = B_2^n$, the Euclidean unit ball.

Corollary 4.6.3. *For $-1 < q < n-1$ and every $\xi \in S^{n-1}$*

$$(R(\mathbb{1}_{B_2^n})(\xi, t))_t^{(q)}(0) = \frac{2^{q+1}\pi^{\frac{n-2}{2}}\Gamma(\frac{q+1}{2})\cos(\frac{\pi q}{2})}{(n-q-1)\Gamma(\frac{n-q-1}{2})}.$$

Proof. First, we use Lemma 1.5.8 with $f \equiv 1$ and $K = B_2^n$

$$(R(\mathbb{1}_{B_2^n})(\xi,t))_t^{(q)}(0) = \frac{\cos(\frac{\pi q}{2})}{\pi(n-q-1)}(|x|^{-n+q+1})^\wedge(\xi).$$

Next, we apply the formula for the Fourier transform of powers of the Euclidean norm, Lemma 1.4.3

$$(|x|^{-n+q+1})^\wedge(\xi) = \frac{2^{q+1}\pi^{\frac{n}{2}}\Gamma(\frac{q+1}{2})}{\Gamma(\frac{n-q-1}{2})}|\xi|^{-1-q} \tag{4.76}$$

for all $\xi \in \mathbb{R}^n \setminus \{0\}$. □

The following theorem was also proved in [233]. Note that the proof initially follows the lines of the proof of Theorem 4.6.1, but then we are able to use Hölder's inequality, which leads to a slightly stronger result.

Theorem 4.6.4. *Suppose K is an infinitely smooth, origin-symmetric convex body in $\mathbb{R}^n$, f is a non-negative, even, infinitely smooth function on K and $-1 < q < n-1$ is not an odd integer. Then,*

$$\int_K f(x)\,dx \leqslant \frac{n}{(n-q-1)\,2^q\pi^{\frac{q-1}{2}}\Gamma(\frac{q+1}{2})}|K|^{\frac{q+1}{n}}(d_{\mathrm{ovr}}(K,L_{-1-q}^n))^{q+1}$$
$$\times \max_{\xi\in S^{n-1}} \frac{1}{\cos(\frac{\pi q}{2})}(Rf(\xi,t))_t^{(q)}(0), \tag{4.77}$$

the fractional derivative of order q at zero of the function $t \to Rf(\xi,t)$, $t \in \mathbb{R}$,

Proof. Consider a number $\varepsilon > 0$ such that, for every $\xi \in S^{n-1}$

$$\frac{1}{\cos(\pi q/2)}(Rf(\xi,t))_t^{(q)}(0) \leqslant \frac{\varepsilon}{\cos(\pi q/2)}(R(\mathbb{1}_{B_2^n})(\xi,t))_t^{(q)}(0).$$

By Lemma 1.5.8, for every $\xi \in S^{n-1}$

$$\left(|x|^{-n+q+1}\left(\int_0^{\frac{|x|}{\|x\|_K}} r^{n-q-2}f\left(r\frac{x}{|x|}\right)dr\right)\right)_x^\wedge(\xi) \leqslant \frac{\varepsilon}{n-q-1}(|x|^{-n+q+1})^\wedge(\xi).$$

Let $\delta > 0$, and let $D \in L_{-1-q}^n$ be such that $K \subset D$ and

$$|D|^{1/n} \leqslant (1+\delta)d_{\mathrm{ovr}}(K,L_{-1-q}^n)|K|^{\frac{1}{n}}. \tag{4.78}$$

By approximation, we can assume that D is infinitely smooth. Then, $(\|x\|_D^{-1-q})^\wedge$ is a non-negative function on the sphere. Multiplying both sides of the latter inequality by

$(\|x\|_D^{-1-q})^\wedge(\xi)$, integrating over the sphere and using Parseval's formula on the sphere we get (as in the proof of Theorem 4.5.10)

$$\int_K \|x\|_D^{-1-q} f(x)\,dx \leqslant \frac{\varepsilon}{n-q-1} \int_{S^{n-1}} \|x\|_D^{-1-q}\,dx. \tag{4.79}$$

Since $K \subset D$, we have

$$\int_K \|x\|_D^{-1-q} f(x)\,dx \geqslant \int_K \|x\|_K^{-1-q} f(x)\,dx \geqslant \int_K f(x)\,dx. \tag{4.80}$$

On the other hand, by Hölder's inequality with the exponents $\frac{n}{q+1}$ and $\frac{n}{n-q-1}$ and by the polar formula for volume

$$\int_{S^{n-1}} \|x\|_D^{-1-q}\,dx \leqslant \kappa_n^{\frac{n-q-1}{n}} n^{\frac{q+1}{n}} |D|^{\frac{q+1}{n}}. \tag{4.81}$$

Combining (4.79), (4.80), and (4.81) we get

$$\int_K f(x)\,dx \leqslant \frac{\varepsilon \kappa_n^{\frac{n-q-1}{n}} n^{\frac{q+1}{n}}}{n-q-1} |D|^{\frac{q+1}{n}}.$$

Now, replace D by K using (4.78)

$$\int_K f(x)\,dx \leqslant \frac{\varepsilon \kappa_n^{\frac{n-q-1}{n}} n^{\frac{q+1}{n}}}{n-q-1} (1+\delta)^{q+1} (d_{\text{ovr}}(K, L_{-1-q}^n))^{q+1} |K|^{\frac{q+1}{n}}$$

and put

$$\varepsilon = \max_{\xi \in S^{n-1}} \frac{\frac{1}{\cos(\pi q/2)} (Rf(\xi,t))_t^{(q)}(0)}{\frac{1}{\cos(\pi q/2)} R(\mathbb{1}_{B_2^n})(\xi,t))_t^{(q)}(0)}.$$

Replace the denominator in the expression for ε using Corollary 4.6.3, and substitute the formula for κ_n

$$\int_K f(x)\,dx \leqslant c(n,q)(1+\delta)^{q+1} (d_{\text{ovr}}(K, L_{-1-q}^n))^{q+1} |K|^{\frac{q+1}{n}}$$
$$\times \max_{\xi \in S^{n-1}} \frac{1}{\cos(\pi q/2)} (Rf(\xi,t))_t^{(q)}(0), \tag{4.82}$$

where

$$c(n,q) = \frac{\pi\Gamma(\frac{n-q-1}{2}) n^{\frac{q+1}{n}} 2^{\frac{n-q-1}{n}} \pi^{\frac{n-q-1}{2}}}{2^{q+1}\pi^{\frac{n}{2}}\Gamma(\frac{q+1}{2})(\Gamma(\frac{n}{2}))^{\frac{n-q-1}{n}}}.$$

Now, use $\Gamma(x+1) = x\Gamma(x)$ and the inequality

$$\frac{\Gamma(\frac{n-q-1}{2}+1)}{(\Gamma(\frac{n}{2}+1))^{\frac{n-q-1}{n}}} \leqslant 1,$$

which follows from the log-convexity of the Γ-function (see [310, Lemma 2.14]). We get

$$\begin{aligned} c(n,q) &= \frac{n\,\Gamma(\frac{n-q-1}{2}+1)}{2^q\pi^{\frac{q-1}{2}}\Gamma(\frac{q+1}{2})(n-q-1)(\Gamma(\frac{n}{2}+1))^{\frac{n-q-1}{n}}} \\ &\leqslant \frac{n}{2^q\pi^{\frac{q-1}{2}}\Gamma(\frac{q+1}{2})(n-q-1)} \end{aligned}$$

and (4.82) implies

$$\begin{aligned} \int_K f(x)\,dx \leqslant{}& \frac{n}{2^q\pi^{\frac{q-1}{2}}\Gamma(\frac{q+1}{2})(n-q-1)}(1+\delta)^{q+1}(d_{\mathrm{ovr}}(K, L^n_{-1-q}))^{q+1}|K|^{\frac{q+1}{n}} \\ &\times \max_{\xi\in S^{n-1}} \frac{1}{\cos(\pi q/2)}(Rf(\xi,t))_t^{(q)}(0). \end{aligned}$$

Finally, send δ to zero to get the result. □

Proof of Theorem 4.6.2. By Propositions 1.7.18 and 1.7.19 and by John's theorem, for any $0 \leqslant q < n-1$

$$d_{\mathrm{ovr}}(K, L^n_{-1-q}) \leqslant d_{\mathrm{ovr}}(K, \mathcal{BP}^n_{q+1}) \leqslant \min\left\{C\sqrt{\frac{n\log^2(\frac{ne}{q+1})}{q+1}}, \sqrt{n}\right\}. \tag{4.83}$$

Now, the first part of Theorem 4.6.2 follows from Theorem 4.6.4 combined with (4.83).

In order to prove the case of even integers in Theorem 4.6.2, put $q = 2k$ in (4.75). In the case of odd integers, we use the expression for the fractional derivative (1.59) to find the limit in (4.75) as $q \to 2k-1$

$$\begin{aligned} \lim_{q\to 2k-1} \frac{(Rf(\xi,t))_t^{(q)}(0)}{\cos(\frac{\pi q}{2})} &= \lim_{q\to 2k-1} \frac{1}{\Gamma(-q)\cos(\frac{\pi q}{2})} \\ &\times \int_0^\infty t^{-2k}\left(Rf(\xi,t) - \sum_{j=0}^{k-1}\frac{t^{2j}}{(2j)!}(Rf(\xi,t))_t^{(2j)}(0)\right)dt. \end{aligned}$$

Now, use $\Gamma(x+1) = x\Gamma(x)$ to compute

$$\lim_{q\to 2k-1}\Gamma(-q)\cos\left(\frac{\pi q}{2}\right)$$
$$= \lim_{q\to 2k-1}\frac{\Gamma(-q+2k)}{(-q)(1-q)\cdots(2k-1-q)}\sin\left(\frac{(q-2k+1)\pi}{2}\right)(-1)^k$$
$$= \frac{\pi}{2(2k-1)!}(-1)^k. \qquad \square$$

4.7 Further reading

4.7.1 Lower bound for the slicing inequality

A lower-dimensional version of Theorem 4.4.1 was proved by Haddad [246]. The argument is similar to that of Theorem 4.4.1, but the result applies only to central sections.

Theorem 4.7.1. *There exists an absolute constant $c > 0$ such that for every $n \geqslant 2$ and $1 \leqslant k \leqslant n-1$ there exist an origin-symmetric convex body K of volume 1 in $\mathbb{R}^n$ and an even, continuous probability density f on K such that for every subspace H of codimension k in $\mathbb{R}^n$*

$$\int_{K\cap H} f \leqslant \left(c\sqrt{\frac{k\log n}{n}}\right)^k.$$

Together with Theorem 4.1.5, the latter theorem implies a matching complement (up to logarithmic terms) to the estimate for the distance from origin-symmetric convex bodies to the classes $\mathcal{BP}_k^n$ given by Theorem 1.7.19.

Corollary 4.7.2. *There exist absolute constants $c_1, c_2 > 0$ such that for all $n \geqslant 2$ and $1 \leqslant k \leqslant n-1$*

$$c_1\sqrt{\frac{n}{k}}(\log n)^{-1/2} \leqslant \sup_K d_{\mathrm{ovr}}(K, \mathcal{BP}_k^n) \leqslant c_2\sqrt{\frac{n}{k}}\log\left(\frac{en}{k}\right),$$

where the supremum is taken over all origin-symmetric convex bodies in $\mathbb{R}^n$.

4.7.2 Derivatives of the Radon transform

An example showing that the derivatives in Theorem 4.6.2 can be uniformly small was constructed in [247]. The proof uses the example from Theorem 4.4.1.

Theorem 4.7.3. *Let $q \in [0, n-1)$ that is not an odd integer. There exists an infinitely smooth, symmetric, convex body $D \subseteq \mathbb{R}^n$ of volume 1, and an even, continuous, positive function $g : \mathbb{R}^n \to \mathbb{R}$ for which $\int_D g = 1$ and*

$$\max_{\xi\in S^{n-1}} \frac{1}{\cos(\frac{\pi q}{2})}(\mathcal{R}g(\xi,t))_t^{(q)}(0) \leqslant \frac{1}{\sqrt{n}}(c(q+1))^{\frac{q+1}{2}}, \quad \forall \xi \in S^{n-1},$$

where $c > 0$ is an absolute constant.

As a consequence of Theorems 4.6.4 and 4.7.3, John's theorem and inequality (4.83), we get the following distance estimate.

Corollary 4.7.4. *For every $q \in [0, n-1)$*

$$c_1 n^{\frac{1}{2(q+1)}} \leqslant \sup_K d_{\mathrm{ovr}}(K, L_{-1-q}^n) \leqslant c_2 \min\left\{\sqrt{\frac{n\log^2(\frac{ne}{q+1})}{q+1}}, \sqrt{n}\right\},$$

where c_1, c_2 are absolute constants, and the supremum is taken over all origin-symmetric convex bodies in $\mathbb{R}^n$.

Recall that when $q + 1 = k$ is an integer, L_{-k}^n is the class of k-intersection bodies in $\mathbb{R}^n$; see Section 1.7.1.

5 Integral geometric approach

This chapter provides an alternative approach to the slicing and comparison inequalities, emphasizing integral-geometric and affine isoperimetric methods. In Section 5.1, we begin with the introduction of the Blaschke–Petkantschin formulas, which allow integration over k-tuples of points in $\mathbb{R}^n$ to be expressed in terms of integrals over lower-dimensional subspaces and the Haar measures on Grassmannians. Then, in Section 5.2, we introduce the Busemann–Straus inequality and its generalization by Grinberg, which provide sharp bounds on the average volume of intersections of a bounded set with lower-dimensional subspaces. A dual version for projections of origin-symmetric convex bodies is also discussed, using polarity and classical inequalities such as Blaschke–Santaló and Bourgain–Milman, thus connecting volume estimates under intersections and projections.

Following this, Section 5.3 addresses a variant of the classical Busemann–Petty and Shephard problems, posed by V. Milman. Here, one compares the volumes of projections of a convex body with the volumes of sections of another set and asks whether inequalities for all lower-dimensional projections imply an inequality for the full volumes. Using Grinberg's inequality as a key tool, we see that such volume comparisons indeed hold under general conditions, extending to convex bodies without symmetry assumptions and even to non-convex compact sets. The framework also allows for stability and separation results, giving quantitative bounds on how close the volumes are when the inequalities hold approximately.

In Section 5.4, we provide a proof of the slicing inequality that removes symmetry and continuity assumptions, relating the measure of a convex body to the measures of its lower-dimensional sections. By combining multiple integration over the body with the Busemann–Straus/Grinberg inequality, we obtain explicit bounds depending on the codimension of the sections, generalizing classical slicing results. Section 5.5 extends these ideas to comparison inequalities, bounding integrals of functions over a set in terms of the maximal ratio of integrals over corresponding lower-dimensional sections of another set. The argument employs the outer volume ratio as a controlling parameter and, besides Blaschke–Petkantschin formulas, uses rearrangement inequalities and estimates for convex hulls within ellipsoids. Although this approach may yield less sharp estimates due to potentially large outer volume ratios, it provides a clear geometric framework connecting high-dimensional measures with their lower-dimensional sections.

5.1 Blaschke–Petkantschin formulas

In this chapter, we introduce an integral geometric approach to the problems that we discussed in the previous chapters, which is based on Blaschke–Petkantschin formulas. These are integral identities that allow us to compute the integral of a function of

https://doi.org/10.1515/9783111360508-005

k-tuples $(x_1, \dots, x_k)$ of points in $\mathbb{R}^n$ by first integrating over all such k-tuples in some $F \in G_{n,k}$ and then averaging with respect to the Haar measure $\nu_{n,k}$ on $G_{n,k}$.

In what follows, for any $F \in G_{n,k}$ we write λ_F for the k-dimensional Lebesgue measure on F, which we view as a measure on $\mathbb{R}^n$, i. e., $\lambda_F(A) = |A \cap F|$ for any Borel subset of $\mathbb{R}^n$. We also write λ_F^s for the product measure on the product $F^s := F \times F \times \cdots \times F$ of s copies of F and λ for the Lebesgue measure on $\mathbb{R}^n$.

Given $E \in G_{n,m}$ and $k < m$, we denote by $\nu_{E,k}$ the measure on the set $G_{E,k}$ of all k-dimensional subspaces of E, and given $F \in G_{n,k}$ and $k < m$, we denote by $\nu_{F,m}$ the measure on the set $G_{F,m}$ of all m-dimensional subspaces of $\mathbb{R}^n$ that contain F. For any $0 \leqslant k < m \leqslant n$ we define

$$G(n,k,m) = \{(F,E) \in G_{n,k} \times G_{n,m} : F \subset E\}.$$

The space $G(n,k,m)$ is a homogeneous $SO(n)$-space and can be equipped with a rotationally invariant probability measure $\nu_{n,k,m}$. We shall use the following fact (for a proof, see [507, Theorem 7.1.1]). If $0 \leqslant k < m \leqslant n-1$ and $g : G(n,k,m) \to \mathbb{R}$ is a non-negative, $\nu_{n,k,m}$-measurable function, then (by the uniqueness of invariant measures)

$$\begin{aligned}\int_{G(n,k,m)} g \, d\nu_{n,k,m} &= \int_{G_{n,k}} \int_{G_{F,m}} g(E,F) \, d\nu_{F,m}(E) \, d\nu_{n,k}(F) \\ &= \int_{G_{n,m}} \int_{G_{E,k}} g(E,F) \, d\nu_{E,k}(F) \, d\nu_{n,m}(E). \end{aligned} \tag{5.1}$$

For any $1 \leqslant k \leqslant n$ and $x_0, x_1, \dots, x_k \in \mathbb{R}^n$ we write

$$Q_k(x_1, \dots, x_k)$$

for the k-dimensional volume of the parallelepiped spanned by $x_1, \dots, x_k$ and

$$\Delta_k(x_0, x_1, \dots, x_k)$$

for the k-dimensional volume of the simplex $\operatorname{conv}\{x_0, x_1, \dots, x_k\}$. Note that

$$\Delta_k(x_0, x_1, \dots, x_k) = \frac{1}{k!} Q_k(x_1 - x_0, \dots, x_k - x_0). \tag{5.2}$$

With this notation, the linear Blaschke–Petkantschin formula is the next statement.

Theorem 5.1.1 (Blaschke–Petkantschin). *Let $1 \leqslant k \leqslant n$ and $f : (\mathbb{R}^n)^k \to \mathbb{R}$ be a non-negative, measurable function. Then,*

$$\int_{(\mathbb{R}^n)^k} f \, d\lambda^k = q_{n,k} \int_{G_{n,k}} \int_{F^k} f \, Q_k^{n-k} \, d\lambda_F^k \, d\nu_{n,q}(F), \tag{5.3}$$

where

$$q_{n,k} = \frac{\kappa_{n-k+1}\cdots\kappa_n}{\kappa_1\cdots\kappa_k} \tag{5.4}$$

and $\kappa_j = j\omega_j$ *is the surface area of* S^{j-1}.

We shall prove (5.3) by induction on k, using the next lemma.

Lemma 5.1.2. *Let* $0 \leqslant q \leqslant n-1$ *and* $F \in G_{n,q}$. *If* $f : \mathbb{R}^n \to \mathbb{R}^+$ *is a non-negative measurable function, then*

$$\int_{\mathbb{R}^n} f\,d\lambda = \frac{\kappa_{n-q}}{2} \int_{G_{F,q+1}} \int_E f(x)\,\mathrm{dist}(x,F)^{n-q-1}\,d\lambda_E\,d\nu_{F,q+1}(E),$$

where $\nu_{F,q+1}$ *is the invariant measure on the space of all* $(q+1)$*-dimensional subspaces of* $\mathbb{R}^n$ *containing* F, *and* $\mathrm{dist}(x,F)$ *is the distance from* x *to* F.

Proof. For any u we write F_u for the positive hull of F and u. Using spherical coordinates in $F^\perp$ and Fubini's theorem we may write

$$\begin{aligned}
\int_{\mathbb{R}^n} f(x)\,d\lambda(x) &= \int_F \int_{F^\perp} f(y+z)\,d\lambda_{F^\perp}(z)\,d\lambda_F(y) \\
&= \kappa_{n-q} \int_F \int_0^\infty \int_{S^{n-1}\cap F^\perp} f(y+tu)t^{n-q-1}\,d\sigma_{n-q-1}(u)\,dt\,d\lambda_F(y) \\
&= \kappa_{n-q} \int_{S^{n-1}\cap F^\perp} \int_F \int_0^\infty f(y+tu)t^{n-q-1}\,dt\,d\lambda_F(y)\,d\sigma_{n-q-1}(u) \\
&= \kappa_{n-q} \int_{S^{n-1}\cap F^\perp} \int_{F_u} f(x)\,\mathrm{dist}(x,F)^{n-q-1}\,d\lambda_{F_u}(x)\,d\sigma_{n-q-1}(u) \\
&= \frac{\kappa_{n-q}}{2} \int_{G_{F,q+1}} \int_E f(x)\,\mathrm{dist}(x,F)^{n-q-1}\,d\lambda_E\,d\nu_{F,q+1}(E),
\end{aligned}$$

as claimed. □

Proof of Theorem 5.1.1. The case $k = 1$ follows if we apply Lemma 5.1.2 with $q = 0$. For the inductive step we assume that (5.3) has been proved for some k and all $n \geqslant k$. We denote $(x_1,\dots,x_k)$ by $\mathbf{x}$. Using Fubini's theorem, the induction hypothesis, and Lemma 5.1.2 we write

$$\int_{(\mathbb{R}^n)^{k+1}} f\,d\lambda^{k+1} = \int_{\mathbb{R}^n} \int_{(\mathbb{R}^n)^k} f(\mathbf{x},x)\,d\lambda^k(\mathbf{x})\,d\lambda(x)$$

$$= q_{n,k} \int_{\mathbb{R}^n} \int_{G_{n,k}} \int_{F^k} f(\mathbf{x}, x) Q_k(\mathbf{x})^{n-k} \, d\lambda_F^k(\mathbf{x}) d\nu_{n,k}(F) \, d\lambda(x)$$

$$= q_{n,k} \int_{G_{n,k}} \int_{F^k} Q_k(\mathbf{x})^{n-k} \int_{\mathbb{R}^n} f(\mathbf{x}, x) \, d\lambda(x) \, d\lambda_F^k(\mathbf{x}) d\nu_{n,k}(F)$$

$$= \frac{q_{n,k}\kappa_{n-k}}{2} \int_{G_{n,k}} \int_{F^k} Q_k(\mathbf{x})^{n-k} \int_{G_{F,k+1}} \int_{E} f(\mathbf{x}, x) \operatorname{dist}(x, F)^{n-k-1}$$
$$\times \, d\lambda_E(x) \, d\nu_{F,k+1}(E) \, d\lambda_F^k(\mathbf{x}) \, d\nu_{n,k}(F).$$

Now, using (5.1) in order to change the order of integration and also using the identity

$$Q_{k+1}(x_1, \ldots, x_{k+1}) = Q_k(x_1, \ldots, x_k) \operatorname{dist}(x_{k+1}, F), \tag{5.5}$$

which holds true for any $x_1, \ldots, x_k \in F \in G_{n,k}$ and $x_{k+1} \in \mathbb{R}^n$, we write

$$\int_{(\mathbb{R}^n)^{k+1}} f \, d\lambda^{k+1} = \frac{q_{n,k}\kappa_{n-k}}{2} \int_{G_{n,k+1}} \int_{G_{E,k}} \int_{F^k} \int_{E} f(\mathbf{x}, x) Q_k(\mathbf{x})^{n-k} \operatorname{dist}(x, F)^{n-k-1}$$
$$\times \, d\lambda_E(x) \, d\lambda_F^k(\mathbf{x}) \, d\nu_{E,k}(F) \, d\nu_{n,k+1}(E)$$
$$= \frac{q_{n,k}\kappa_{n-k}}{2} \int_{G_{n,k+1}} \int_{E} \int_{G_{E,k}} \int_{F^k} f(\mathbf{x}, x) Q_{k+1}(\mathbf{x}, x)^{n-k-1} Q_k(\mathbf{x})$$
$$\times \, d\lambda_F^k(\mathbf{x}) \, d\nu_{E,k}(F) \, d\lambda_E(x) \, d\nu_{n,k+1}(E).$$

Finally, for each $E \in G_{n,k+1}$ we apply the induction hypothesis for the function $f(\cdot, x) Q_{k+1}(\cdot, x)^{n-k-1}$ on E^k. It follows that

$$\int_{(\mathbb{R}^n)^{k+1}} f \, d\lambda^{k+1} = \frac{q_{n,k}\kappa_{n-k}}{2q_{k+1,k}} \int_{G_{n,k+1}} \int_{E} \int_{E^k} f(\mathbf{x}, x) Q_{k+1}(\mathbf{x}, x)^{n-k-1}$$
$$\times \, d\lambda_E^k(\mathbf{x}) \, d\lambda_E(x) \, d\nu_{n,k+1}(E)$$
$$= q_{n,k+1} \int_{G_{n,k+1}} \int_{E^{k+1}} f Q_{k+1}^{n-k-1} \, d\lambda_E^{k+1} \, d\nu_{n,k+1}(E)$$

and (5.3) has been proved for $k + 1$. □

In view of (5.2) we may also state Theorem 5.1.1 as follows.

Theorem 5.1.3. *Let $1 \leqslant k \leqslant n - 1$. There exists a constant $p_{n,k} > 0$ such that, for every non-negative, bounded, Borel measurable function $f : (\mathbb{R}^n)^k \to \mathbb{R}$*

$$\int_{\mathbb{R}^n} \cdots \int_{\mathbb{R}^n} f(x_1, \ldots, x_k) \, dx_k \cdots dx_1$$

$$= p_{n,k} \int_{G_{n,k}} \int_F \cdots \int_F f(x_1,\ldots,x_k)\,|\mathrm{conv}\{0,x_1,\ldots,x_k\}|^{n-k}\,dx_k\ldots dx_1\,d\nu_{n,s}(F). \qquad (5.6)$$

The exact value of the constant $p_{n,k}$ is

$$p_{n,k} = (k!)^{n-k} q_{n,k} = (k!)^{n-k} \frac{\kappa_{n-k+1}\cdots\kappa_n}{\kappa_1\cdots\kappa_k}. \qquad (5.7)$$

5.1.1 Estimates for the constants $p_{n,k}$

The dependence of the constants $p_{n,k}$ on n and k will be important for the applications in this chapter. We provide two arguments leading to the next result.

Lemma 5.1.4. *For every $1 \leqslant k \leqslant n-1$ we have*

$$p_{n,n-k} \leqslant \tau_{n,k}^n (c\sqrt{n-k})^{k(n-k)}, \qquad (5.8)$$

where $c > 0$ is an absolute constant.

Proof. We start from the fact that $\omega_s = \pi^{\frac{s}{2}}/\Gamma(\frac{s}{2}+1)$ and use Stirling's approximation. Recall that

$$\begin{aligned} p_{n,n-k} &= ((n-k)!)^k \frac{(n\omega_n)\cdots((k+1)\omega_{k+1})}{((n-k)\omega_{n-k})\cdots(2\omega_2)\omega_1} \\ &= ((n-k)!)^k \binom{n}{k} \frac{\prod_{s=k+1}^n \frac{\pi^{s/2}}{\Gamma(\frac{s}{2}+1)}}{\prod_{s=1}^{n-k} \frac{\pi^{s/2}}{\Gamma(\frac{s}{2}+1)}} \\ &= ((n-k)!)^k \binom{n}{k} \pi^{\frac{k(n-k)}{2}} \frac{\prod_{s=1}^{n-k}\Gamma(\frac{s}{2}+1)}{\prod_{s=k+1}^n \Gamma(\frac{s}{2}+1)}, \end{aligned} \qquad (5.9)$$

where we have used the identity

$$\frac{1}{2}\sum_{s=k+1}^n s - \frac{1}{2}\sum_{s=1}^{n-k} s = \frac{1}{4}(n(n+1) - k(k+1) - (n-k)(n-k+1)) = \frac{1}{2}k(n-k). \qquad (5.10)$$

Using the estimate

$$\left(\frac{s}{2e}\right)^{\frac{s}{2}} \sqrt{2\pi s} \leqslant \Gamma\left(\frac{s}{2}+1\right) \leqslant \left(\frac{s}{2e}\right)^{\frac{s}{2}} \sqrt{2\pi s}\, e^{\frac{1}{6s}} \leqslant \left(\frac{s}{2e}\right)^{\frac{s}{2}} \sqrt{2\pi s}\, e^{\frac{1}{6}} \qquad (5.11)$$

we get

$$p_{n,n-k} \leqslant ((n-k)!)^k (2\pi e)^{\frac{k(n-k)}{2}} e^{\frac{n-k}{6}} \binom{n}{k}^{1/2} \frac{\prod_{s=1}^k s^{\frac{s}{2}} \prod_{s=1}^{n-k} s^{\frac{s}{2}}}{\prod_{s=1}^n s^{\frac{s}{2}}} \qquad (5.12)$$

$$= \tau_{n,k}^{n} C_{n,k} \left(\tau_{n,k}^{-n} \frac{t_k t_{n-k}}{t_n} \right), \tag{5.13}$$

where

$$C_{n,k} = ((n-k)!)^k (2\pi e)^{\frac{k(n-k)}{2}} e^{\frac{n-k}{6}} \binom{n}{k}^{1/2}$$

and

$$t_m = \prod_{r=1}^{m} r^r. \tag{5.14}$$

It is known that

$$t_m \sim A m^{\frac{m^2}{2} + \frac{m}{2} + \frac{1}{12}} e^{-\frac{m^2}{4}} \tag{5.15}$$

as $m \to \infty$, where $A > 0$ is an absolute constant (the Glaisher–Kinkelin constant, see, e. g., [173]). Note that

$$\begin{aligned} \tau_{n,k}^{-n} = \frac{\omega_{n-k}^{n}}{\omega_{n}^{n-k}} = \frac{\Gamma(\frac{n}{2}+1)^{n-k}}{\Gamma(\frac{n-k}{2}+1)^{n}} &\leqslant \left(\frac{n}{n-k} \right)^{\frac{n(n-k)}{2}} \frac{(\pi n)^{\frac{n-k}{2}} e^{\frac{n-k}{6}}}{(\pi(n-k))^{\frac{n}{2}}} \\ &\leqslant e^{\frac{n-k}{6}} \left(\frac{n}{n-k} \right)^{\frac{(n+1)(n-k)}{2}}. \end{aligned} \tag{5.16}$$

Using the fact that $n^2 = k^2 + (n-k)^2 + 2k(n-k)$ we get

$$\begin{aligned} \tau_{n,k}^{-\frac{n}{k(n-k)}} \left(\frac{t_k t_{n-k}}{t_n} \right)^{\frac{1}{2k(n-k)}} &\leqslant \frac{c_1}{\sqrt{n}} \left(\frac{k}{n} \right)^{\frac{k+1}{4(n-k)}} \left(\frac{n-k}{n} \right)^{\frac{n-k+1}{4k}} \left(\frac{n}{n-k} \right)^{\frac{n+1}{2k}} \\ &\leqslant \frac{c_1}{\sqrt{n}} \left(\frac{k}{n} \right)^{\frac{k+1}{4(n-k)}} \left(\frac{n}{n-k} \right)^{\frac{n+k+1}{4k}} \\ &\leqslant \frac{c_1}{\sqrt{n}} \left(\frac{k}{n} \right)^{\frac{k+1}{4(n-k)}} \left(\frac{n}{n-k} \right)^{\frac{n-k}{2k}} \left(\frac{n}{n-k} \right)^{\frac{2k+1}{4k}} \\ &\leqslant \frac{c_2}{\sqrt{n}} \cdot \frac{\sqrt{n}}{\sqrt{n-k}} = \frac{c_2}{\sqrt{n-k}}. \end{aligned} \tag{5.17}$$

Since

$$C_{n,k}^{\frac{1}{k(n-k)}} = \left(((n-k)!)^k (2\pi e)^{\frac{k(n-k)}{2}} e^{\frac{n-k}{6}} \binom{n}{k}^{1/2} \right)^{\frac{1}{k(n-k)}} \leqslant c_3(n-k), \tag{5.18}$$

we see that

$$(\tau_{n,k}^{-n} p_{n,n-k})^{\frac{1}{k(n-k)}} \leqslant c\sqrt{n-k} \tag{5.19}$$

for every $1 \leqslant k \leqslant n-1$, where $c > 0$ is an absolute constant. □

We can give an alternative proof of the upper bound for $p_{n,n-k}$ using some basic facts about Sylvester-type functionals. Let K be a convex body in $\mathbb{R}^m$. For every $p > 0$ we consider the normalized p-th moment of the expected volume of the random simplex $\mathrm{conv}\{0, x_1, \ldots, x_m\}$, the convex hull of the origin and m points from K, defined by

$$S_p(K) = \left(\frac{1}{|K|^{m+p}} \int_K \cdots \int_K |\mathrm{conv}\{0, x_1, \ldots, x_m\}|^p \, dx_1 \cdots dx_m\right)^{1/p}. \tag{5.20}$$

Also, for any Borel probability measure ν on $\mathbb{R}^m$ we define

$$S_p(\nu) = \left(\int_{\mathbb{R}^m} \cdots \int_{\mathbb{R}^m} |\mathrm{conv}\{0, x_1, \ldots, x_m\}|^p \, d\nu(x_1) \cdots d\nu(x_m)\right)^{1/p}. \tag{5.21}$$

Note that $S_p(K)$ is invariant under invertible linear transformations: $S_p(K) = S_p(T(K))$ for every $T \in GL_n$. The next fact is well known and goes back to Blaschke (see, e. g., [113, Proposition 3.5.5]).

Lemma 5.1.5. *Let ν be a centered Borel probability measure on $\mathbb{R}^m$. Then,*

$$m!\, S_2^2(\nu) = \det(\mathrm{Cov}(\nu)). \tag{5.22}$$

In particular, if K is centered, then

$$S_2^2(K) = \frac{L_K^{2m}}{m!}. \tag{5.23}$$

Hölder's inequality shows that the function $p \mapsto S_p(K)$ is increasing on $(0, \infty)$. We will need the next reverse Hölder inequality.

Lemma 5.1.6. *There exists an absolute constant $\delta > 0$ such that, for every log-concave probability measure ν on $\mathbb{R}^m$ and every $p > 1$*

$$S_p(\nu) \leqslant (\delta p)^m S_1(\nu). \tag{5.24}$$

In particular, for every convex body K in $\mathbb{R}^m$ and every $p > 1$

$$S_p(K) \leqslant (\delta p)^m S_1(K). \tag{5.25}$$

Proof. We know that there exists an absolute constant $\delta > 0$ with the following property: if ν is a log-concave probability measure, then, for any seminorm $u : \mathbb{R}^m \to \mathbb{R}$ and any $q > p \geqslant 1$

$$\left(\int\limits_{\mathbb{R}^m} |u(x)|^q dv(x)\right)^{1/q} \leqslant \frac{\delta q}{p}\left(\int\limits_{\mathbb{R}^m} |u(x)|^p dv(x)\right)^{1/p}. \tag{5.26}$$

This is a consequence of Borel's lemma (see Section 2.2). Next, recall that

$$\left|\mathrm{conv}\{0, x_1, \dots, x_m\}\right| = \frac{1}{m!}\left|\det(x_1, \dots, x_m)\right|. \tag{5.27}$$

The function $u_i : \mathbb{R}^m \to \mathbb{R}$ defined by $x_i \mapsto |\det(x_1, \dots, x_n)|$ for fixed x_j in $\mathbb{R}^m, j \neq i$, is a seminorm, as is the function $v_i : \mathbb{R}^m \to \mathbb{R}$ defined by

$$x_i \mapsto \int\limits_{\mathbb{R}^m} \cdots \int\limits_{\mathbb{R}^m} |\det(x_1, \dots, x_m)| dx_{i+1} \cdots dx_m \tag{5.28}$$

for fixed x_j $(1 \leqslant j < i)$ in $\mathbb{R}^m$. By consecutive applications of Fubini's theorem and of (5.26) we obtain (5.24). □

Second proof of Lemma 5.1.4. For any convex body K in $\mathbb{R}^n$, applying Theorem 5.1.3 for the function $f(x_1, \dots, x_{n-k}) = \prod_{i=1}^{n-k} \mathbb{1}_K(x_i)$ we get

$$|K|^{n-k} = p_{n,n-k} \int\limits_{G_{n,n-k}} \int\limits_{K\cap F} \cdots \int\limits_{K\cap F} \left|\mathrm{conv}\{0, x_1, \dots, x_{n-k}\}\right|^k dx_1 \dots dx_{n-k}\, d\nu_{n,n-k}(F). \tag{5.29}$$

We can rewrite (5.29) in the form

$$|K|^{n-k} = p_{n,n-k} \int\limits_{G_{n,n-k}} |K \cap F|^n [S_k(K \cap F)]^k\, d\nu_{n,n-k}(F). \tag{5.30}$$

In particular, setting $K = B_2^n$ we see that if $k \geqslant 2$, then

$$\begin{aligned} \omega_n^{n-k} &= p_{n,n-k}\omega_{n-k}^n [S_k(B_2^{n-k})]^k \\ &\geqslant p_{n,n-k}\omega_{n-k}^n [S_2(B_2^{n-k})]^k \\ &\geqslant p_{n,n-k}\omega_{n-k}^n \left(\frac{L_{B_2^{n-k}}}{\sqrt{n-k}}\right)^{k(n-k)} \\ &\geqslant p_{n,n-k}\omega_{n-k}^n \left(\frac{c_1}{\sqrt{n-k}}\right)^{k(n-k)}, \end{aligned} \tag{5.31}$$

where $c_1 > 0$ is an absolute constant, which implies that

$$p_{n,n-k} \leqslant \tau_{n,k}^n (c_0\sqrt{n-k})^{k(n-k)}, \tag{5.32}$$

where $c_0 = c_1^{-1}$. For the case $k = 1$ we can use the fact that $S_1(K \cap F) \geqslant \delta^{-(n-1)} S_2(K \cap F)$ for every $F \in G_{n,n-1}$, and then continue as above. □

5.2 Busemann–Straus/Grinberg inequality

The main result of this section is an inequality obtained by Busemann–Straus [120] and in a more general form by Grinberg [234].

Theorem 5.2.1 (Grinberg). *Let C be a bounded Borel set in $\mathbb{R}^n$ and $1 \leqslant k \leqslant n-1$. Then,*

$$\int_{G_{n,k}} |C \cap F|^n \, d\nu_{n,k}(F) \leqslant \frac{\omega_k^n}{\omega_n^k} |C|^k. \tag{5.33}$$

Recall that the *volume radius* of a bounded Borel set C in $\mathbb{R}^m$ is the quantity

$$\mathrm{vrad}(C) = \left(\frac{|C|}{\omega_m}\right)^{1/m}.$$

Therefore, the inequality of Theorem 5.2.1 may be equivalently written in the form

$$\int_{G_{n,k}} \mathrm{vrad}(C \cap F)^{kn} \, d\nu_{n,k}(F) \leqslant \mathrm{vrad}(C)^{kn}. \tag{5.34}$$

It is also useful to note that if K is a convex body, then equality in (5.33) holds for $k = 1$ if and only if K is origin-symmetric, while if $k \geqslant 2$, then equality holds in (5.33) if and only if K is an origin-symmetric ellipsoid.

For the proof of Theorem 5.2.1 we shall use the results of the previous section, in particular Theorem 5.1.3. Let C be a bounded Borel set in $\mathbb{R}^n$. For every $m \in \mathbb{N}$ we define

$$g_m(C) = \int_C \cdots \int_C |\mathrm{conv}\{0, x_1, \ldots, x_k\}|^m \, d\lambda(x_k) \cdots d\lambda(x_1).$$

More generally, if $F \in G_{n,k}$ and C is a bounded Borel set in F, then for every $m \in \mathbb{N}$ we define

$$g_m(C) = \int_C \cdots \int_C |\mathrm{conv}\{0, x_1, \ldots, x_k\}|^m \, d\lambda_F(x_k) \cdots d\lambda_F(x_1).$$

The next lemma shows that this quantity is decreasing under Steiner symmetrization.

Lemma 5.2.2. *Let C be a bounded Borel set in $\mathbb{R}^n$. For any $m \in \mathbb{N}$ and any $\xi \in S^{n-1}$ we have that*

$$g_m(K) \geqslant g_m(S_\xi(K)),$$

where $S_\xi(K)$ is the Steiner symmetral of K in the direction ξ.

A proof for the case where C is a convex body can be found in Gardner's book; see [194, Lemma 9.2.2] Pfiefer established Lemma 5.2.2 for bounded Borel sets in [460] (see also [459]) using a generalization of Anderson's theorem on integrals of symmetric unimodal functions (see [196, Theorem 11.1]).

Now, recall that starting from C we can find a sequence $\{\xi_j\}$ of directions so that applying successive Steiner symmetrizations with respect to ξ_j we obtain a sequence of convex bodies that converges to a Euclidean ball rB_2^n in the Hausdorff metric. Since Steiner symmetrization preserves volume, it is also clear that $|rB_2^n| = |C|$. From Lemma 5.2.2, and by the continuity of the functional g_m with respect to the Hausdorff metric, we obtain the next proposition.

Proposition 5.2.3. *Let C be a bounded Borel set in $\mathbb{R}^n$. For any $m \in \mathbb{N}$ we have that*

$$g_m(K) \geqslant g_m(rB_2^n),$$

where $r = \mathrm{vrad}(C)$, or equivalently, $|C| = |rB_2^n|$.

We move on now to the proof of Theorem 5.2.1. A crucial observation is that using Theorem 5.1.3 we can express the volume of a compact set $C \subset \mathbb{R}^n$ in terms of the quantities $g_{n-k}(C \cap F)$, where $F \in G_{n,k}$. This idea was already used in our second proof of Lemma 5.1.4.

Recall that for every non-negative bounded Borel measurable function $f : (\mathbb{R}^n)^k \to \mathbb{R}$ we have

$$\begin{aligned}\int_{\mathbb{R}^n} \cdots \int_{\mathbb{R}^n} f(x_1, \dots, x_k)\, dx_k \cdots dx_1 \\ = p_{n,k} \int_{G_{n,k}} \int_F \cdots \int_F f(x_1, \dots, x_k)\, |\mathrm{conv}\{0, x_1, \dots, x_k\}|^{n-k}\, dx_k \dots dx_1\, d\nu_{n,k}(F).\end{aligned}$$

Choosing $f(x_1, \dots, x_k) = \prod_{i=1}^k \mathbb{1}_C(x_i)$, we immediately see that

$$\begin{aligned}|C|^k &= p_{n,k} \int_{G_{n,k}} \int_F \cdots \int_F \prod_{i=1}^k \mathbb{1}_C(x_i)\, |\mathrm{conv}\{0, x_1, \dots, x_k\}|^{n-k}\, dx_k \dots dx_1\, d\nu_{n,k}(F) \\ &= p_{n,k} \int_{G_{n,k}} \int_{C\cap F} \cdots \int_{C\cap F} |\mathrm{conv}\{0, x_1, \dots, x_k\}|^{n-k}\, dx_k \dots dx_1\, d\nu_{n,k}(F).\end{aligned}$$

Equivalently, we have the following.

Proposition 5.2.4. *Let C be a compact set in $\mathbb{R}^n$. For every $1 \leqslant k \leqslant n-1$ we have*

$$|C|^k = p_{n,k} \int_{G_{n,k}} g_{n-k}(C \cap F)\, d\nu_{n,k}(F).$$

Using Proposition 5.2.3 and Proposition 5.2.4 we can prove the Busemann–Straus/Grinberg inequality (5.33).

Proof of Theorem 5.2.1. Proposition 5.2.3 shows that

$$g_{m-k}(C \cap F) \geqslant g_{n-k}(r_F B_2^n \cap F)$$

for every $F \in G_{n,k}$, where $r_F = (|C \cap F|/\omega_k)^{1/k}$. From Proposition 5.2.4 we get

$$|C|^k \geqslant p_{n,k} \int_{G_{n,k}} g_{n-k}(r_F B_2^n \cap F)\, d\nu_{n,k}(F). \tag{5.35}$$

Let $r > 0$ and $F \in G_{n,k}$. Applying Proposition 5.2.4 for the ball rB_2^n we see that

$$|rB_2^n|^k = p_{n,k} \int_{G_{n,k}} g_{n-k}(rB_2^n \cap E)\, d\nu_{n,k}(E) = p_{n,k} g_{n-k}(rB_2^n \cap F),$$

because $E \mapsto g_{n-k}(rB_2^n \cap E)$ is constant on $G_{n,k}$. Going back to (5.35) we can write

$$\begin{aligned} |C|^k &\geqslant p_{n,k} \int_{G_{n,k}} g_{n-k}(r_F B_2^n \cap F)\, d\nu_{n,k}(F) \\ &= \int_{G_{n,k}} |r_F B_2^n|^k d\nu_{n,k}(F) = \omega_n^k \int_{G_{n,k}} (r_F^k)^n\, d\nu_{n,k}(F) \\ &= \omega_n \int_{G_{n,k}} \frac{|C \cap F|^n}{\omega_k^n}\, d\nu_{n,k}(F) = \frac{\omega_n^k}{\omega_k^n} \int_{G_{n,k}} |C \cap F|^n\, d\nu_{n,k}. \end{aligned}$$

This proves (5.33). □

If we assume that K is an origin-symmetric convex body in $\mathbb{R}^n$, then a duality argument, based on the Blaschke–Santaló and the Bourgain–Milman inequalities leads to a related inequality about the volume of projections.

Theorem 5.2.5. *Let K be an origin-symmetric convex body in $\mathbb{R}^n$, and $1 \leqslant k < m \leqslant n$. Then,*

$$\int_{G_{n,k}} |P_E(K)|^{-n}\, d\nu_{n,k}(E) \leqslant C^{kn} \frac{\omega_n^k}{\omega_k^n} |K|^{-k}, \tag{5.36}$$

where $C > 0$ is an absolute constant.

Proof. Let $K^\circ = \{x \in \mathbb{R}^n : \langle x, y\rangle \leqslant 1 \text{ for all } y \in K\}$ be the polar body of K. Since K is origin-symmetric, $P_H(K)$ is an origin-symmetric convex body in H for any $H \in G_{n,k}$, $1 \leqslant k \leqslant n-1$, and $(P_H(K))^\circ = K^\circ \cap H$. Applying the Blaschke–Santaló and the Bourgain–Milman inequalities for $K^\circ \cap H$ we have that

$$c \leqslant \mathrm{vrad}(P_H(K))\, \mathrm{vrad}(K^\circ \cap H) \leqslant 1, \tag{5.37}$$

where $c > 0$ is an absolute constant. Then, using Theorem 5.2.1 we write

$$c^{kn} \int_{G_{n,k}} \mathrm{vrad}(P_E(K))^{-kn} \, d\nu_{n,k}(E) \leqslant \int_{G_{n,k}} \mathrm{vrad}(K^\circ \cap E)^{kn} \, d\nu_{n,k}(E)$$
$$\leqslant \mathrm{vrad}(K^\circ)^{kn} \leqslant \mathrm{vrad}(K)^{-kn}. \tag{5.38}$$

The result follows from the definition of the volume radius. □

5.3 A variant of the Busemann–Petty and Shephard problems

As a first application of Grinberg's inequality we present the positive answer to a variant of the Busemann–Petty and Shephard problems, proposed by V. Milman.

Question 5.3.1. Assume that K and C are origin-symmetric convex bodies in $\mathbb{R}^n$ that satisfy

$$|P_{\xi^\perp}(K)| \leqslant |C \cap \xi^\perp| \tag{5.39}$$

for all $\xi \in S^{n-1}$. Does it follow that $|K| \leqslant |C|$?

It was proved in [208] that the answer to this question is affirmative. In fact, the lower-dimensional analogue of the problem has an affirmative answer. Moreover, one can drop the symmetry assumptions and even the assumption of convexity for C.

Theorem 5.3.2. *Let K be a convex body in $\mathbb{R}^n$ and let C be a compact subset of $\mathbb{R}^n$ such that, for some $1 \leqslant k \leqslant n-1$*

$$|P_F(K)| \leqslant |C \cap F| \tag{5.40}$$

for all $F \in G_{n,n-k}$. Then,

$$|K| \leqslant |C|. \tag{5.41}$$

The proof of Theorem 5.3.2 is based on the Busemann–Straus/Grinberg inequality (Theorem 5.2.1) and on the fact that for every convex body K in $\mathbb{R}^n$ the sequence

$$Q_k(K) = \left(\frac{1}{\omega_k} \int_{G_{n,k}} |P_F(K)| \, d\nu_{n,k}(F) \right)^{1/k} \tag{5.42}$$

is decreasing in k, which has been also used in Section 3.4 and [505, Section 6.4]). In particular, for every $1 \leqslant k \leqslant n-1$ we have

$$\left(\frac{|K|}{\omega_n}\right)^{\frac{1}{n}} \leqslant \left(\frac{1}{\omega_k}\int_{G_{n,k}} |P_F(K)|\, d\nu_{n,k}(F)\right)^{\frac{1}{k}} \leqslant w(K). \tag{5.43}$$

Proof of Theorem 5.3.2. Let K be a convex body in $\mathbb{R}^n$ and C be a compact subset of $\mathbb{R}^n$. Assume that for some $1 \leqslant k \leqslant n-1$ we have

$$|P_F(K)| \leqslant |C \cap F| \tag{5.44}$$

for all $F \in G_{n,n-k}$. From (5.43) we get

$$\left(\frac{|K|}{\omega_n}\right)^{\frac{n-k}{n}} \leqslant \frac{1}{\omega_{n-k}}\int_{G_{n,n-k}} |P_F(K)|\, d\nu_{n,n-k}(F). \tag{5.45}$$

Our assumption, Hölder's inequality, and Grinberg's inequality give

$$\begin{aligned}\frac{1}{\omega_{n-k}}\int_{G_{n,n-k}} |P_F(K)|\, d\nu_{n,n-k}(F) &\leqslant \frac{1}{\omega_{n-k}}\int_{G_{n,n-k}} |C \cap F|\, d\nu_{n,n-k}(F)\\ &\leqslant \frac{1}{\omega_{n-k}}\left(\int_{G_{n,n-k}} |C \cap F|^n\, d\nu_{n,n-k}(F)\right)^{\frac{1}{n}}\\ &\leqslant \frac{1}{\omega_{n-k}}\frac{\omega_{n-k}}{\omega_n^{\frac{n-k}{n}}}|C|^{\frac{n-k}{n}} = \left(\frac{|C|}{\omega_n}\right)^{\frac{n-k}{n}}.\end{aligned} \tag{5.46}$$

Therefore, $|K| \leqslant |C|$. □

Remark 5.3.3. Assume that $1 \leqslant k \leqslant n-2$, and let K and C be two convex bodies in $\mathbb{R}^n$ such that $|P_F(K)| \leqslant |C \cap F|$ for all $F \in G_{n,n-k}$ and $|K| = |C|$. Then, (5.45) must be an equality, which implies that K is a ball. We must also have equality in Grinberg's inequality for C, which implies that C is an ellipsoid. Finally, we must have $|P_F(K)| = |C \cap F|$ for all $F \in G_{n,n-k}$, which means that all $(n-k)$-dimensional sections of the ellipsoid C have the same volume, and hence C is also a ball.

Remark 5.3.4. Slightly modifying the proof of Theorem 5.3.2 one can get stability and separation results, as follows. Let $\varepsilon > 0$, and let K and C be as in Theorem 5.3.2. Suppose that for every $F \in G_{n,n-k}$

$$|P_F(K)| \leqslant |C \cap F| \pm \varepsilon.$$

Then,

$$|K|^{\frac{n-k}{n}} \leqslant |C|^{\frac{n-k}{n}} \pm \tau_{n,k}\varepsilon.$$

The plus sign corresponds to stability, and the minus sign to separation. Assuming that $\varepsilon = \max_F(|P_F(K) - |C \cap F|)$ in the stability result, we get

$$|K|^{\frac{n-k}{n}} - |C|^{\frac{n-k}{n}} \leqslant \tau_{n,k} \max_F(|P_F(K) - |C \cap F|).$$

On the other hand, if $\varepsilon = \min_F(|C \cap F| - |P_F(K)|)$ in the separation result, then

$$|C|^{\frac{n-k}{n}} - |K|^{\frac{n-k}{n}} \geqslant \tau_{n,k} \min_F(|C \cap F| - |P_F(K)|).$$

5.4 A proof of the slicing inequality

In this section, we provide a different proof of the slicing inequality (Theorem 4.1.3). This approach was initiated in [126] and allows us to drop the symmetry and continuity assumptions.

Theorem 5.4.1. *Let K be a convex body in $\mathbb{R}^n$ with $0 \in \operatorname{int}(K)$. Let g be a bounded, non-negative, measurable function on $\mathbb{R}^n$ and let μ be the measure on $\mathbb{R}^n$ with density g. For every $1 \leqslant k \leqslant n-1$*

$$\mu(K) \leqslant (c\sqrt{n-k})^k \max_{F \in G_{n,n-k}} \mu(K \cap F) \cdot |K|^{\frac{k}{n}}, \tag{5.47}$$

where $c > 0$ is an absolute constant. In particular, $a_{n,k}(\mu) \leqslant c\sqrt{n-k}$.

In fact, the proof of Theorem 5.4.1 leads to the stronger estimate

$$\mu(K) \leqslant (c\sqrt{n-k})^k \Big(\int_{G_{n,n-k}} \mu(K \cap F)^n \, d\nu_{n,n-k}(F) \Big)^{\frac{1}{n}} |K|^{\frac{k}{n}}. \tag{5.48}$$

Proof of Theorem 5.4.1. Let μ be a Borel measure with a bounded locally integrable, non-negative density g on $\mathbb{R}^n$. We consider a convex body K in $\mathbb{R}^n$ with $0 \in \operatorname{int}(K)$, and fix $1 \leqslant k \leqslant n-1$. Applying Theorem 5.1.3 for the function $f(x_1, \dots, x_{n-k}) = \prod_{i=1}^{n-k} g(x_i) \mathbb{1}_K(x_i)$ we get

$$\begin{aligned}
\mu(K)^{n-k} &= \prod_{i=1}^{n-k} \int_K g(x_i)\, dx = \int_{\mathbb{R}^n} \cdots \int_{\mathbb{R}^n} f(x_1, \dots, x_{n-k})\, dx_1 \dots dx_{n-k} \\
&= p_{n,n-k} \int_{G_{n,n-k}} \int_{K \cap F} \cdots \int_{K \cap F} g(x_1) \cdots g(x_{n-k}) \\
&\quad \times \big|\operatorname{conv}\{0, x_1, \dots, x_{n-k}\}\big|^k dx_1 \dots dx_{n-k}\, d\nu_{n,n-k}(F) \\
&\leqslant p_{n,n-k} \int_{G_{n,n-k}} \int_{K \cap F} \cdots \int_{K \cap F} g(x_1) \cdots g(x_{n-k})
\end{aligned}$$

$$\times\, |K \cap F|^k dx_1 \dots dx_{n-k}\, d\nu_{n,n-k}(F)$$
$$= p_{n,n-k} \int_{G_{n,n-k}} |K \cap F|^k \mu(K \cap F)^{n-k}\, d\nu_{n,n-k}(F)$$
$$\leqslant p_{n,n-k} \Big(\int_{G_{n,n-k}} \mu(K \cap F)^n\, d\nu_{n,n-k}(F) \Big)^{\frac{n-k}{n}}$$
$$\times \Big(\int_{G_{n,n-k}} |K \cap F|^n\, d\nu_{n,n-k}(F) \Big)^{\frac{k}{n}}. \tag{5.49}$$

In order to estimate the last integral, note that if $\overline{K} = |K|^{-\frac{1}{n}} K$, then

$$\int_{G_{n,n-k}} |K \cap F|^n\, d\nu_{n,n-k}(F) = |K|^{n-k} \int_{G_{n,n-k}} |\overline{K} \cap F|^n\, d\nu_{n,n-k}(F)$$
$$\leqslant |K|^{n-k} \int_{G_{n,n-k}} |\overline{B}_2^n \cap F|^n\, d\nu_{n,n-k}(F)$$
$$= \tau_{n,k}^{-n} |K|^{n-k}, \tag{5.50}$$

by Theorem 5.2.1 and the fact that $\tau_{n,k}^{-1/k} \leqslant \sqrt{e}$. Taking into account Lemma 5.1.4 we see that

$$\mu(K)^{n-k} \leqslant (c\sqrt{n-k})^{k(n-k)} \Big(\int_{G_{n,n-k}} \mu(K \cap F)^n\, d\nu_{n-k}(F) \Big)^{\frac{n-k}{n}} |K|^{\frac{k(n-k)}{n}}. \tag{5.51}$$

This proves (5.48) and the result follows. □

5.5 A proof of the comparison inequality

In this section, we describe an alternative approach to Theorem 3.4.7 that is again based on Blaschke–Petkantschin formulas and affine isoperimetric inequalities. Below, for any bounded Borel set K in $\mathbb{R}^n$ we denote by $\mathrm{ovr}(K)$ the outer volume ratio $\mathrm{ovr}(K) = d_{\mathrm{ovr}}(K, L_2) = \inf_{\mathcal{E}} (|\mathcal{E}|/|K|)^{1/n}$, where the infimum is over all origin-symmetric ellipsoids $\mathcal{E}$ in $\mathbb{R}^n$ with $K \subseteq \mathcal{E}$.

Theorem 5.5.1. *Let K and L be two bounded Borel sets in $\mathbb{R}^n$. Let f and g be two bounded, non-negative, measurable functions on K and L, respectively, and assume that $\|g\|_1 > 0$ and $\|g\|_\infty = 1$. For every $1 \leqslant k \leqslant n-1$ we have that*

$$\frac{\int_K f}{(\int_L g)^{\frac{n-k}{n}} |K|^{\frac{k}{n}}} \leqslant (c\,\mathrm{ovr}(K))^k \max_{H \in G_{n,n-k}} \frac{\int_{K\cap H} f}{\int_{L\cap H} g}, \tag{5.52}$$

where $c > 0$ is an absolute constant.

It should be noted that even for origin-symmetric convex bodies K in $\mathbb{R}^n$ the outer volume ratio $\mathrm{ovr}(K)$ can be as large as $\sqrt{n}$. This is a disadvantage of Theorem 5.5.1 that does not provide estimates depending on k in contrast to Theorem 3.4.7.

We shall use a theorem of Dann, Paouris, and Pivovarov from [146]; the proof of this fact combines Blaschke–Petkantschin formulas with rearrangement inequalities.

Lemma 5.5.2. *Let g be a non-negative, bounded, integrable function on $\mathbb{R}^n$ with $\|g\|_1 > 0$. For every $1 \leqslant k \leqslant n-1$ we have*

$$\int_{G_{n,n-k}} \frac{1}{\|g|_H\|_\infty^k} \left(\int_H g(x)\,dx \right)^n d\nu_{n,n-k}(H) \leqslant \tau_{n,k}^{-n} \left(\int_{\mathbb{R}^n} g(x)\,dx \right)^{n-k}. \tag{5.53}$$

We also need a well-known theorem of Bárány and Füredi [53]: if $s \geqslant m+1$ and $w_j \in \mathbb{R}^m$ satisfy $|w_j| \leqslant 1$ for $j = 1, \dots, s$, then

$$|\mathrm{conv}\{w_1, \dots, w_s\}|^{1/m} \leqslant c \frac{\sqrt{\log(1+s/m)}}{m}.$$

Equivalently, this says that if $w_j \in B_2^m$, $1 \leqslant j \leqslant s$, then the volume radius of their convex hull is bounded by $c\sqrt{\log(1+s/m)/m}$. By affine invariance we obtain the following.

Lemma 5.5.3. *There exists an absolute constant $c > 0$ such that if $\mathcal{E}$ is an ellipsoid in $\mathbb{R}^m$, $s \geqslant m+1$ and $w_1, \dots, w_s \in \mathcal{E}$, then*

$$\left(\frac{|\mathrm{conv}\{w_1, \dots, w_s\}|}{|\mathcal{E}|} \right)^{1/m} \leqslant c\sqrt{\log(1+s/m)/m}.$$

Proof of Theorem 5.5.1. Let $\mathcal{E}$ be a centered ellipsoid such that $K \subseteq \mathcal{E}$ and

$$\mathrm{ovr}(K) = (|\mathcal{E}|/|K|)^{1/n}.$$

We shall now use the next fact, which is a consequence of Lemma 5.5.3: if $F \in G_{n,n-k}$ and $x_1, \dots, x_{n-k} \in K \cap H \subseteq \mathcal{E} \cap H$, then $\mathrm{conv}\{0, x_1, \dots, x_{n-k}\} \subseteq \mathcal{E} \cap H$, and since $\mathcal{E} \cap H$ is an $(n-k)$-dimensional centered ellipsoid we must have

$$\begin{aligned} |\mathrm{conv}\{0, x_1, \dots, x_{n-k}\}| &\leqslant \left(c_1 \frac{\sqrt{\log(1+(n-k+1)/(n-k))}}{\sqrt{n-k}} \right)^k |\mathcal{E} \cap H| \\ &\leqslant \left(\frac{c_1}{\sqrt{n-k}} \right)^k |\mathcal{E} \cap H|. \end{aligned} \tag{5.54}$$

Applying Theorem 5.1.1 for the function $F(x_1,\ldots,x_{n-k})=\prod_{i=1}^{n-k} f(x_i)\mathbb{1}_K(x_i)$, we get

$$\begin{aligned}\Big(\int_K f(x)\,dx\Big)^{n-k} &= \int_{\mathbb{R}^n}\cdots\int_{\mathbb{R}^n} F(x_1,\ldots,x_{n-k})\,dx_{n-k}\cdots dx_1\\ &= p_{n,n-k}\int_{G_{n,n-k}}\int_{K\cap H}\cdots\int_{K\cap H} g(x_1)\cdots g(x_{n-k})\\ &\quad\times \big|\mathrm{conv}\{0,x_1,\ldots,x_{n-k}\}\big|^k\,dx_{n-k}\cdots dx_1\,d\nu_{n,n-k}(H).\end{aligned}$$

Let

$$M:=\max_{H\in G_{n,n-k}}\frac{\int_{K\cap H} f}{\int_{L\cap H} g}.$$

Then, (5.54) shows that

$$\begin{aligned}\Big(\int_K f(x)\,dx\Big)^{n-k} &\leqslant p_{n,n-k}\Big(\frac{c_1}{\sqrt{n-k}}\Big)^{k(n-k)}\int_{G_{n,n-k}}\int_{K\cap H}\cdots\int_{K\cap H}|\mathcal{E}\cap H|^k\\ &\quad\times f(x_1)\cdots f(x_{n-k})\,dx_{n-k}\cdots dx_1\,d\nu_{n,n-k}(H)\\ &= p_{n,n-k}\Big(\frac{c_1}{\sqrt{n-k}}\Big)^{k(n-k)}\int_{G_{n,n-k}}|\mathcal{E}\cap H|^k\Big(\int_{K\cap H} f(x)\,dx\Big)^{n-k}d\nu_{n,n-k}(H)\\ &\leqslant p_{n,n-k}\Big(\frac{c_1}{\sqrt{n-k}}\Big)^{k(n-k)}M^{n-k}\int_{G_{n,n-k}}|\mathcal{E}\cap H|^k\\ &\quad\times\Big(\int_{L\cap H} g(x)\,dx\Big)^{n-k}d\nu_{n,n-k}(H).\end{aligned}$$

Now, by Hölder's inequality and Grinberg's inequality (Theorem 5.2.1) we get

$$\begin{aligned}&\int_{G_{n,n-k}}|\mathcal{E}\cap H|^k\Big(\int_{L\cap H} g(x)\,dx\Big)^{n-k}d\nu_{n,n-k}(H)\\ &\leqslant\Big(\int_{G_{n,n-k}}|\mathcal{E}\cap H|^n\,d\nu_{n,n-k}(H)\Big)^{\frac{k}{n}}\Big(\int_{G_{n,n-k}}\Big(\int_{L\cap H} g(x)\,dx\Big)^n d\nu_{n,n-k}(H)\Big)^{\frac{n-k}{n}}\\ &\leqslant \tau_{n,k}^{-n}|\mathcal{E}|^{\frac{k(n-k)}{n}}\Big(\int_{G_{n,n-k}}\Big(\int_{L\cap H} g(x)\,dx\Big)^n d\nu_{n,n-k}(H)\Big)^{\frac{n-k}{n}}\\ &= \tau_{n,k}^{-n}|K|^{\frac{k(n-k)}{n}}\mathrm{ovr}(K)^{k(n-k)}\Big(\int_{G_{n,n-k}}\Big(\int_{L\cap H} g(x)\,dx\Big)^n d\nu_{n,n-k}(H)\Big)^{\frac{n-k}{n}}.\end{aligned}$$

Finally, since $\|g|_H\|_\infty \leqslant \|g\|_\infty = 1$ for all $H \in G_{n,n-k}$, we may apply (5.53) to get

$$\int_{G_{n,n-k}} \Big(\int_{L\cap H} g(x)\,dx \Big)^n d\nu_{n,n-k}(H) \leqslant \tau_{n,k}^{-n} \Big(\int_L g(x)\,dx \Big)^{n-k}. \tag{5.55}$$

Combining the above we get

$$\begin{aligned}\Big(\int_K f(x)\,dx \Big)^{n-k} &\leqslant (\tau_{n,k}^{-n} p_{n,n-k} (c_1/\sqrt{n-k})^{k(n-k)}) \\ &\quad \times M^{n-k} |K|^{\frac{k(n-k)}{n}} \operatorname{ovr}(K)^{k(n-k)} \Big(\int_L g(x)\,dx \Big)^{\frac{(n-k)^2}{n}}. \end{aligned} \tag{5.56}$$

Note that, by Lemma 5.1.4

$$(\tau_{n,k}^{-n} p_{n,n-k})^{\frac{1}{n-k}} \Big(\frac{c_1}{\sqrt{n-k}} \Big)^k \leqslant c^k$$

for some absolute constant $c > 0$. Then, the result follows from (5.56). □

5.6 Further reading

5.6.1 Functional analogues of the Shephard, Busemann–Petty, and Milman problems

Let K, L be origin-symmetric convex bodies in $\mathbb{R}^n$. Recall that by Theorem 5.3.2 the inequalities $|P_{\xi^\perp} K| \leqslant |L \cap \xi^\perp|$ for all directions imply that $|K| \leqslant |L|$. Hosle [261] studied the reverse problem. Suppose that for every $\xi \in S^{n-1}$ we have

$$|K \cap \xi^\perp| \leqslant |P_{\xi^\perp} L|. \tag{5.57}$$

What can one say about the volumes of K and L? Since the section of a convex body always has a smaller area than the corresponding projection of the same body, the condition (5.57) is weaker than the condition of Shephard's problem. Hosle showed that the conditions (5.57) imply that $|K| \leqslant \sqrt{n}|L|$ in the case where K and L are in John's position, i. e., the ellipsoid of maximal volume contained in each of the bodies is a Euclidean ball. In his recent work [262] he shows that if K, L are convex bodies in $\mathbb{R}^n$ with K centered and satisfy (5.57) then $|K| \leqslant c\sqrt{n}|L|$, which is sharp up to the choice of the absolute constant, thus removing the John's position assumption from his earlier work, up to an absolute constant factor.

Moreover, in [261], Hosle extended the results to measures in the framework suggested by Livshyts in [381]. Namely, in the notation of Section 3.10.3, if μ is an absolutely continuous measure on $\mathbb{R}^n$, then the conditions

$$\mu(K \cap \xi^{\perp}) \leqslant P_{\mu,L}(\xi), \quad \forall \xi \in S^{n-1}$$

imply

$$\mu(K) \leqslant \frac{\sqrt{n}}{1-\frac{1}{n}} \mu(L).$$

Hosle also extended the result of Theorem 5.3.2 to p-homogeneous and $\frac{1}{p}$-concave measures, where $p > 0$. He actually proved a stability result. If for such μ

$$P_{\mu,K}(\xi) \leqslant \mu(L \cap \xi^{\perp}) + \varepsilon, \quad \forall \xi \in S^{n-1},$$

then

$$\mu(K)^{1-q} \leqslant \left(\frac{1-\frac{1}{n}}{1-q}\right) \mu(L)^{1-q} + \frac{\omega_n}{\mu(B_2^n)^q \omega_{n-1}} \varepsilon.$$

Functional analogues of geometric inequalities related to the main problems discussed in Chapters 3–5 are also provided by Gorev and Kosov [228]. Some examples are the following results:

(i) If $f \in L^1(\mathbb{R}^n)$ belongs to the class $BV(\mathbb{R}^n)$ of functions with bounded variation and $g \in L^1(\mathbb{R}^n)$ belongs to the class $LC_0(\mathbb{R}^n)$ of log-concave upper semicontinuous functions with $\|g\|_1 > 0$, and if

$$\|D_\theta f\|_{\mathrm{TV}} \leqslant \|D_\theta g\|_{\mathrm{TV}}$$

for all $\theta \in S^{n-1}$, where D_θ denotes a distributional derivative and $\|\cdot\|_{\mathrm{TV}}$ is the total variation of a Borel measure, then

$$\|f\|_{\frac{n}{n-1}} \leqslant 8\sqrt{n}\|g\|_\infty^{\frac{1}{n}} \|g\|_1^{\frac{n-1}{n}}.$$

This is a functional analogue of the Shephard-type inequality relating the volumes of two convex bodies K and L such that every $(n-1)$-dimensional projection of K has smaller volume than the corresponding projection of L. A more general inequality, which corresponds to the lower-dimensional Shephard problem, is also established.

(ii) If $f, g \in LC_0(\mathbb{R}^n)$ satisfy the Busemann–Petty condition

$$\|f|_H\|_{L^1(H)} \leqslant \|g|_H\|_{L^1(H)}$$

for all $H \in G_{n,k}$, where $f|_H$ is the restriction of f onto H, and if $f^{\frac{n}{k}}$ is centered, then

$$\|f\|_{\frac{n}{k}} \leqslant (cL)^{n-k} \|g\|_{1}^{\frac{k}{n}} \|g\|_{\infty}^{\frac{n-k}{n}},$$

where L is the isotropic constant of $f^{\frac{n}{k}}$. This is a functional analogue of the Busemann–Petty-type inequality relating the volumes of two symmetric convex bodies K and L such that every k-dimensional section of K has smaller volume than the corresponding section of L. Another version of this inequality, under the assumption that f is centered, is also established.

(iii) If $f, g \in LC_0(\mathbb{R}^n)$ satisfy the condition

$$\|P_H f\|_{L^1(H)} \leqslant \|g|_H\|_{L^1(H)}$$

for all $H \in G_{n,k}$, where $P_H f(x) = \sup_{y \in H^\perp} f(x+y)$ is the projection of f onto H, then

$$\|f\|_{\frac{n}{k}} \leqslant \|g\|_{1}^{\frac{k}{n}} \|g\|_{\infty}^{\frac{n-k}{n}}.$$

This is a functional analogue of Theorem 5.3.2.

Further extensions of some geometric inequalities related to the main problems discussed in Chapters 3–5 to log-concave integrable functions are provided by Tziotziou [527]. First, an analogue of Theorem 5.3.2 is the following: Let $f, g : \mathbb{R}^n \to [0, +\infty)$ be log-concave integrable functions with $\|f\|_\infty = f(0)$ and $\|g\|_\infty = g(0)$ such that, for some $1 \leqslant k \leqslant n-1$, we have that

$$\|P_H f\|_1 \leqslant \|g|_H\|_1 \quad \text{for all } H \in G_{n,n-k}.$$

Then,

$$\|f\|_1 \leqslant \frac{n!}{[(n-k)!]^{\frac{n}{n-k}}} \|g\|_1. \tag{5.58}$$

Using Stirling's formula we check that the constant $n!/[(n-k)!]^{\frac{n}{n-k}}$ that appears in (5.58) is of the order of $C^{\frac{kn}{n-k}}$.

Tziotziou also extends to log-concave integrable functions a number of inequalities related to the Busemann–Petty problem and the slicing problem. They all follow from the next general inequality about the Radon transform on convex sets: Let $f, g : \mathbb{R}^n \to [0, \infty)$ be non-negative integrable functions such that f is log-concave with $f(0) > 0$ and $\|g\|_\infty = g(0) = 1$. If K is a convex body in $\mathbb{R}^n$ and T is a compact subset of $\mathbb{R}^n$ with $0 \in K \cap T$, then

$$\left(\frac{f(0)^{-1} \int_K f(x)\,dx}{\int_T g(x)\,dx} \right)^{\frac{n-k}{n}} \leqslant C^k \max_{E \in G_{n,n-k}} \frac{f(0)^{-1} \int_{K\cap E} f(x)\,dx}{\int_{T\cap E} g(x)\,dx}, \tag{5.59}$$

where $C > 0$ is an absolute constant.

In particular, this implies a lower estimate for the sup-norm of the Radon transform that sharpens the slicing inequality for arbitrary functions (see [317]) under the log-concavity assumption: Let f be a log-concave integrable function with $\|f\|_\infty = f(0) = 1$. Then, for every $1 \leqslant k \leqslant n-1$ we have that

$$\int_K f(x)\,dx \leqslant C^k |K|^{\frac{k}{n}} \max_{E\in G_{n,n-k}} \int_{K\cap E} f(x)\,dx,$$

where $C > 0$ is an absolute constant.

6 Volume difference inequalities

This chapter develops a systematic theory of volume difference inequalities, which quantify how differences in the volumes of convex or star bodies can be estimated in terms of differences of their lower-dimensional sections or projections. These inequalities refine classical slicing problems by addressing stability: not only whether sectional information bounds volume, but how discrepancies in sections propagate to discrepancies in full volume.

The first part of the chapter focuses on section-based inequalities. It is shown that any general volume difference inequality would imply slicing inequalities, explaining the necessity of additional geometric assumptions. Positive results are obtained for generalized k-intersection bodies and for bodies close to this class in outer volume ratio. Using Radon transform and Fourier-analytic methods, sharp inequalities are established, including extensions to arbitrary measures with even, continuous densities. Complementary results demonstrate that bodies failing to belong to the intersection body class exhibit strong instability, highlighting the sharpness of the theory.

The second part of the chapter addresses projection-based volume difference inequalities. Motivated by the Shephard problem, the analysis identifies projection bodies as the natural setting in which projection differences can control volume differences. Quantitative inequalities are obtained using Fourier representations of curvature functions and geometric distances to the class of projection bodies, with sharp $\sqrt{n}$ dependence on the dimension. Together, these results provide a unified understanding of when and how lower-dimensional data control volume in high dimensions.

An alternative approach based on integral geometry is developed in the further reading section. By exploiting Blaschke–Petkantschin-type formulas, volume and measure differences are related to averaged sectional data for general convex bodies. This framework yields upper bounds with explicit dimensional dependence and universal lower bounds valid for arbitrary sets and measures, clarifying which aspects of volume difference phenomena are structural and which are universal.

6.1 Volume difference inequalities for sections

Volume difference inequalities are designed to estimate the error in computations of the volume of a body out of the areas of its sections and projections. In this section, we present a number of such results for the case of sections. Let $\gamma_{n,k}$ denote the smallest constant $\gamma > 0$ satisfying the inequality

$$|K|^{\frac{n-k}{n}} - |L|^{\frac{n-k}{n}} \leqslant \gamma^k \max_{F\in G_{n,n-k}} \left(|K \cap F| - |L \cap F|\right) \tag{6.1}$$

for all $1 \leqslant k \leqslant n-1$ and all origin-symmetric convex bodies K and L in $\mathbb{R}^n$ such that $L \subset K$.

https://doi.org/10.1515/9783111360508-006

Question 6.1.1. Does there exist an absolute constant C so that $\gamma_{n,k} \leqslant C$ for all n and k?

Question 6.1.1 is stronger than the slicing problem. For any origin-symmetric convex body K in $\mathbb{R}^n$ and any $1 \leqslant k \leqslant n-1$, putting $L = \beta B_2^n$ in (6.1) and then sending β to zero, we get the slicing inequality

$$|K|^{\frac{n-k}{n}} \leqslant C^k \max_{F \in G_{n,n-k}} |K \cap F|, \tag{6.2}$$

with $C = \gamma_{n,k}$.

In the case where K is a generalized k-intersection body in $\mathbb{R}^n$ and L is any origin-symmetric star body in $\mathbb{R}^n$, the inequality (6.1) was proved in [311] for $k = 1$, and in [321] for $1 < k < n$

$$|K|^{\frac{n-k}{n}} - |L|^{\frac{n-k}{n}} \leqslant c_{n,k}^k \max_{F \in G_{n,n-k}} (|K \cap F| - |L \cap F|), \tag{6.3}$$

where $c_{n,k}^k = \tau_{n,k} = \omega_n^{\frac{n-k}{n}} / \omega_{n-k}$, and ω_n is the volume of the Euclidean unit ball in $\mathbb{R}^n$. Recall that $e^{-k/2} \leqslant \tau_{n,k} \leqslant 1$, and hence $1/\sqrt{e} \leqslant c_{n,k} \leqslant 1$ for all n, k.

Note that in Question 6.1.1 we have added the extra assumption that $L \subset K$, as opposed to (6.3). Without extra assumptions on K and L, we cannot have (6.1) with any $\gamma > 0$. One may see this using counterexamples to the Busemann–Petty problem. Every counterexample provides a pair of bodies K and L that contradict inequality (6.1). If K is a generalized k-intersection body, then we know that the answer to the question of Busemann and Petty is affirmative, therefore (6.3) may be viewed as a quantified version of this fact.

The first result of this section comes from [209] and extends (6.3) to arbitrary origin-symmetric star bodies.

Theorem 6.1.2. *Let $1 \leqslant k \leqslant n-1$, and let K and L be origin-symmetric star bodies in $\mathbb{R}^n$ such that $L \subset K$. Then,*

$$|K|^{\frac{n-k}{n}} - |L|^{\frac{n-k}{n}} \leqslant c_{n,k}^k d_{\mathrm{ovr}}^k(K, \mathcal{BP}_k^n) \max_{F \in G_{n,n-k}} (|K \cap F| - |L \cap F|). \tag{6.4}$$

Proof. For every $H \in G_{n,n-k}$ we have

$$|K \cap H| - |L \cap H| \leqslant \max_{F \in G_{n,n-k}} (|K \cap F| - |L \cap F|).$$

Expressing volume in terms of the Radon transform, we get

$$\begin{aligned}
&\frac{1}{n-k}(R_{n-k}(\|\cdot\|_K^{-n+k})(H) - R_{n-k}(\|\cdot\|_L^{-n+k})(H)) \\
&\quad \leqslant \max_{F \in G_{n,n-k}} (|K \cap F| - |L \cap F|).
\end{aligned}$$

Let $D \in \mathcal{BP}_k^n$, $K \subset D$. Integrating both sides by $H \in G_{n,n-k}$ with respect to the measure μ_D that corresponds to D by (1.92), we get

$$\begin{aligned}\frac{1}{n-k} &\int\limits_{S^{n-1}} \|\theta\|_D^{-k}(\|\theta\|_K^{-n+k} - \|\theta\|_L^{-n+k})\, d\theta \\ &\leqslant \max_{F \in G_{n,n-k}} (|K \cap F| - |L \cap F|)\mu_D(G_{n,n-k}). \qquad (6.5)\end{aligned}$$

We have $\|\theta\|_D^{-1} \geqslant \|\theta\|_K^{-1} \geqslant \|\theta\|_L^{-1}$, because $L \subset K \subset D$. Using this, Hölder's inequality, and the polar formula for volume, we bound from below the left-hand side of (6.5) as follows:

$$\frac{1}{n-k} \int\limits_{S^{n-1}} \|\theta\|_K^{-k}(\|\theta\|_K^{-n+k} - \|\theta\|_L^{-n+k})\, d\theta \geqslant \frac{n}{n-k}(|K| - |K|^{\frac{k}{n}}|L|^{\frac{n-k}{n}}).$$

To estimate $\mu_D(G_{n,n-k})$ from above, we combine the fact that

$$1 = \frac{1}{\kappa_{n-k}} R_{n-k}\mathbf{1}(E)$$

for every $E \in G_{n,n-k}$ with (1.92) and Hölder's inequality to write

$$\begin{aligned}\mu_D(G_{n,n-k}) &= \frac{1}{\kappa_{n-k}} \int\limits_{G_{n,n-k}} R_{n-k}\mathbf{1}(E)\, d\mu_D(E) \\ &= \frac{1}{\kappa_{n-k}} \int\limits_{S^{n-1}} \|\theta\|_D^{-k}\, d\theta \\ &\leqslant \frac{1}{\kappa_{n-k}} \kappa_n^{\frac{n-k}{n}} \Bigg(\int\limits_{S^{n-1}} \|\theta\|_D^{-n}\, d\theta \Bigg)^{\frac{k}{n}} \\ &= \frac{1}{\kappa_{n-k}} \kappa_n^{\frac{n-k}{n}} n^{\frac{k}{n}} |D|^{\frac{k}{n}}. \qquad (6.6)\end{aligned}$$

These estimates show that

$$\begin{aligned}\frac{n}{n-k}&(|K| - |K|^{\frac{k}{n}}|L|^{\frac{n-k}{n}}) \\ &\leqslant \frac{1}{\kappa_{n-k}} \kappa_n^{\frac{n-k}{n}} n^{\frac{k}{n}} |D|^{\frac{k}{n}} \max_{F \in G_{n,n-k}} (|K \cap F| - |L \cap F|) \\ &= \frac{n}{n-k} c_{n,k}^k |D|^{\frac{k}{n}} \max_{F \in G_{n,n-k}} (|K \cap F| - |L \cap F|). \qquad (6.7)\end{aligned}$$

Finally, we choose D so that $|D|^{1/n} \leqslant (1+\delta) d_{\mathrm{ovr}}(K, \mathcal{BP}_k^n)|K|^{1/n}$, and then send δ to zero. □

Recall from Theorem 1.7.19 that the outer volume ratio distance of an origin-symmetric convex body K in $\mathbb{R}^n$ from $\mathcal{BP}_k^n$ satisfies

$$d_{\mathrm{ovr}}(K, \mathcal{BP}_k^n) \leqslant c\sqrt{n/k}\,\log(en/k), \tag{6.8}$$

where $c > 0$ is an absolute constant. In conjunction with Theorem 6.1.2, this estimate provides an affirmative answer to Question 6.1.1 for sections of proportional dimensions.

Corollary 6.1.3. *Let $1 \leqslant k \leqslant n-1$, let K be an origin-symmetric convex body in $\mathbb{R}^n$, and let L be an origin-symmetric star body in $\mathbb{R}^n$ such that $L \subset K$. Then,*

$$|K|^{\frac{n-k}{n}} - |L|^{\frac{n-k}{n}} \leqslant \left(C\sqrt{n/k}\,\log(en/k)\right)^k \max_{F \in G_{n,n-k}} (|K \cap F| - |L \cap F|), \tag{6.9}$$

where $C > 0$ is an absolute constant.

The inequality of Theorem 6.1.2 can be extended to arbitrary measures in place of volume. Let f be a bounded, non-negative, measurable function on $\mathbb{R}^n$ and let μ be the measure with density f. Writing integrals in polar coordinates, we get

$$\mu(K) = \int_K f(x)\,dx = \int_{S^{n-1}} \left(\int_0^{\rho_K(\theta)} r^{n-1} f(r\theta)\,dr \right) d\theta \tag{6.10}$$

and for $H \in G_{n,n-k}$

$$\begin{aligned} \mu(K \cap H) &= \int_{K \cap H} f(x)\,dx = \int_{S^{n-1} \cap H} \left(\int_0^{\rho_K(\theta)} r^{n-k-1} f(r\theta)\,dr \right) d\theta \\ &= R_{n-k}\left(\int_0^{\rho_K(\cdot)} r^{n-k-1} f(r\cdot)\,dr \right)(H). \end{aligned} \tag{6.11}$$

In Theorem 4.1.5 we saw that for any $1 \leqslant k \leqslant n-1$, any origin-symmetric star body K in $\mathbb{R}^n$ and any measure μ on $\mathbb{R}^n$ with even non-negative continuous density f

$$\mu(K) \leqslant \frac{n}{n-k} \tau_{n,k}\, |K|^{\frac{k}{n}}\, d_{\mathrm{ovr}}^k(K, \mathcal{BP}_k^n) \max_{F \in G_{n,n-k}} \mu(K \cap F). \tag{6.12}$$

The next measure difference inequality was proved in [209].

Theorem 6.1.4. *Let $1 \leqslant k \leqslant n-1$, let K and L be origin-symmetric star bodies in $\mathbb{R}^n$ such that $L \subset K$, and let μ be a measure on $\mathbb{R}^n$ with even non-negative continuous density. Then,*

$$\begin{aligned} &\mu(K) - \mu(L) \\ &\leqslant \frac{n}{n-k} c_{n,k}^k\, |K|^{\frac{k}{n}}\, d_{\mathrm{ovr}}^k(K, \mathcal{BP}_k^n) \max_{F \in G_{n,n-k}} (\mu(K \cap F) - \mu(L \cap F)). \end{aligned} \tag{6.13}$$

Proof. Let f be the density of the measure μ. For every $H \in G_{n,n-k}$ we have

$$\mu(K \cap H) - \mu(L \cap H) \leqslant \max_{F \in G_{n,n-k}} (\mu(K \cap F) - \mu(L \cap F)).$$

Using (6.11), we get

$$R_{n-k}\left(\int_{\rho_L(\cdot)}^{\rho_K(\cdot)} r^{n-k-1} f(r\cdot)\, dr\right)(H) \leqslant \max_{F \in G_{n,n-k}} (\mu(K \cap F) - \mu(L \cap F)).$$

Let $D \in \mathcal{BP}_k^n$, $K \subset D$. Integrating both sides by $H \in G_{n,n-k}$ with respect to the measure μ_D that corresponds to D by (1.92), we get

$$\int_{S^{n-1}} \rho_D^k(\theta)\left(\int_{\rho_L(\theta)}^{\rho_K(\theta)} r^{n-k-1} f(r\theta)\, dr\right) d\theta$$
$$\leqslant \max_{F \in G_{n,n-k}} (\mu(K \cap F) - \mu(L \cap F))\mu_D(G_{n,n-k}). \tag{6.14}$$

We have $\rho_D \geqslant \rho_K \geqslant \rho_L$, because $L \subset K \subset D$. Using this fact and (6.10), we estimate the left-hand side of (6.14) from below

$$\int_{S^{n-1}} \rho_D^k(\theta)\left(\int_{\rho_L(\theta)}^{\rho_K(\theta)} r^{n-k-1} f(r\theta)\, dr\right) d\theta$$
$$\geqslant \int_{S^{n-1}} \rho_K^k(\theta)\left(\int_{\rho_L(\theta)}^{\rho_K(\theta)} r^{n-k-1} f(r\theta)\, dr\right) d\theta$$
$$\geqslant \int_{S^{n-1}} \left(\int_{\rho_L(\theta)}^{\rho_K(\theta)} r^{n-1} f(r\theta)\, dr\right) d\theta = \mu(K) - \mu(L).$$

Then, we estimate $\mu_D(G_{n,n-k})$ and finally choose D in the same way as in the proof of Theorem 6.1.2. □

Remark 6.1.5. Note that in the case of volume ($f \equiv 1$), Theorem 6.1.4 implies that if K is an origin-symmetric convex body in $\mathbb{R}^n$, and L is an origin-symmetric star body in $\mathbb{R}^n$ such that $L \subset K$, then

$$|K|^{\frac{n-k}{n}} - |L|^{\frac{n-k}{n}} \leqslant \frac{|K| - |L|}{|K|^{\frac{k}{n}}} \leqslant \frac{n}{n-k} c_{n,k}^k\, d_{\mathrm{ovr}}^k(K, \mathcal{BP}_k^n) \max_{F \in G_{n,n-k}} (|K \cap F| - |L \cap F|).$$

This estimate differs from the one of Theorem 6.1.2 by a factor $\frac{n}{n-k}$; however, note that also $(|K| - |L|)/|K|^{\frac{k}{n}}$ is greater than $|K|^{\frac{n-k}{n}} - |L|^{\frac{n-k}{n}}$.

A different kind of volume difference inequality was proved in [208]. If K is any origin-symmetric star body in $\mathbb{R}^n$, L is an intersection body, and

$$\min_{\xi\in S^{n-1}}(|K\cap\xi^\perp|-|L\cap\xi^\perp|)>0,$$

where $\xi^\perp$ is the subspace of $\mathbb{R}^n$ perpendicular to ξ, then

$$|K|^{\frac{n-1}{n}}-|L|^{\frac{n-1}{n}}\geqslant c\frac{1}{\sqrt{n}M(\overline{L})}\min_{\xi\in S^{n-1}}(|K\cap\xi^\perp|-|L\cap\xi^\perp|), \tag{6.15}$$

where $c>0$ is an absolute constant, $\overline{L}=L/|L|^{\frac{1}{n}}$, $M(L)=\int_{S^{n-1}}\|\theta\|_L\,d\sigma(\theta)$, and σ is the normalized Lebesgue measure on the sphere. We shall prove a version of this inequality for sections of lower dimension, in terms of the next variant of $d_{\rm ovr}(K,\mathcal{BP}_k^n)$.

Definition 6.1.6. For a star body K in $\mathbb{R}^n$ and $1\leqslant k\leqslant n-1$, we define

$$d_k(K,\mathcal{BP}_k^n)=\inf\left\{\left(\frac{\int_{S^{n-1}}\|\theta\|_K^{-k}\,d\sigma(\theta)}{\int_{S^{n-1}}\|\theta\|_D^{-k}\,d\sigma(\theta)}\right)^{\frac{1}{k}}:D\subset K,\ D\in\mathcal{BP}_k^n\right\}.$$

By John's theorem, if K is origin-symmetric and convex, then $d_k(K,\mathcal{BP}_k^n)\leqslant\sqrt{n}$.

With this definition of distance, we prove the next generalization of (6.15).

Theorem 6.1.7. *Let $1\leqslant k\leqslant n-1$, and let K and L be origin-symmetric star bodies in $\mathbb{R}^n$ such that $L\subset K$. Then,*

$$\begin{aligned}d_k^k(L,\mathcal{BP}_k^n)&(|K|^{\frac{n-k}{n}}-|L|^{\frac{n-k}{n}})\\&\geqslant c^k\frac{1}{(\sqrt{n}M(\overline{L}))^k}\min_{F\in G_{n,n-k}}(|K\cap F|-|L\cap F|),\end{aligned} \tag{6.16}$$

where $c>0$ is an absolute constant.

Proof. For every $H\in G_{n,n-k}$ we have

$$|K\cap H|-|L\cap H|\geqslant\min_{F\in G_{n,n-k}}(|K\cap F|-|L\cap F|).$$

Writing volume in terms of the Radon transform, we get

$$\begin{aligned}&\frac{1}{n-k}(R_{n-k}(\|\cdot\|_K^{-n+k})(H)-R_{n-k}(\|\cdot\|_L^{-n+k})(H))\\&\quad\geqslant\min_{F\in G_{n,n-k}}(|K\cap F|-|L\cap F|).\end{aligned}$$

Let $D\in\mathcal{BP}_k^n$, $D\subset L$. Integrating both sides by $H\in G_{n,n-k}$ with respect to the measure μ_D that corresponds to D by (1.92), we get

$$\frac{1}{n-k}\int_{S^{n-1}} \|\theta\|_D^{-k}(\|\theta\|_K^{-n+k} - \|\theta\|_L^{-n+k})\,d\theta$$
$$\geqslant \min_{F\in G_{n,n-k}} (|K\cap F| - |L\cap F|)\mu_D(G_{n,n-k}). \tag{6.17}$$

We have $\|\theta\|_D^{-1} \leqslant \|\theta\|_L^{-1} \leqslant \|\theta\|_K^{-1}$, because $D \subset L \subset K$. Using this, Hölder's inequality, and the polar formula for volume, we bound from above the left-hand side of (6.17) as follows:

$$\frac{1}{n-k}\int_{S^{n-1}} \|\theta\|_L^{-k}(\|\theta\|_K^{-n+k} - \|\theta\|_L^{-n+k})\,d\theta$$
$$\leqslant \frac{n}{n-k}(|L|^{\frac{k}{n}}|K|^{\frac{n-k}{n}} - |L|).$$

We need to estimate $\mu_D(G_{n,n-k})$ from below. To do this, we combine (1.92) with the fact that $1 = \frac{1}{\kappa_{n-k}}R_{n-k}\mathbf{1}(E)$ for every $E \in G_{n,n-k}$ to write

$$\mu_D(G_{n,n-k}) = \frac{1}{\kappa_{n-k}}\int_{G_{n,n-k}} R_{n-k}\mathbf{1}(E)\,d\mu_D(E)$$
$$= \frac{\kappa_n}{\kappa_{n-k}}\int_{S^{n-1}} \|\theta\|_D^{-k}\,d\sigma(\theta). \tag{6.18}$$

These estimates show that

$$\frac{n}{n-k}(|L|^{\frac{k}{n}}|K|^{\frac{n-k}{n}} - |L|)$$
$$\geqslant \frac{\kappa_n}{\kappa_{n-k}}\int_{S^{n-1}} \|\theta\|_D^{-k}\,d\sigma(\theta) \min_{F\in G_{n,n-k}} (|K\cap F| - |L\cap F|).$$

Finally, for $\delta > 0$, we choose D so that

$$\int_{S^{n-1}} \|\theta\|_D^{-k} d\sigma(\theta) \geqslant \frac{1}{(1+\delta)d_k^k(L, \mathcal{BP}_k^n)}\int_{S^{n-1}} \|\theta\|_L^{-k}\,d\sigma(\theta)$$

and send δ to zero. Then, we use Jensen's inequality and homogeneity to write

$$\left(\int_{S^{n-1}} \|\theta\|_L^{-k}\,d\sigma(\theta)\right)^{\frac{1}{k}} \geqslant \left(\int_{S^{n-1}} \|\theta\|_L\,d\sigma(\theta)\right)^{-1} = \frac{1}{M(\overline{L})}|L|^{\frac{1}{n}} \tag{6.19}$$

and apply standard estimates for the Γ-function. □

An inequality going in the direction opposite to (6.16) was proved in [317]. Suppose that K is an infinitely smooth origin-symmetric convex body in $\mathbb{R}^n$, with strictly positive curvature, that is not an intersection body. Then, there exists an origin-symmetric convex body L in $\mathbb{R}^n$ such that $L \subset K$ and

$$|K|^{\frac{n-1}{n}} - |L|^{\frac{n-1}{n}} < c_{n,1} \min_{\xi \in S^{n-1}} (|K \cap \xi^\perp| - |L \cap \xi^\perp|). \tag{6.20}$$

Here, we prove a similar inequality going in the direction opposite to (6.4).

Theorem 6.1.8. *Suppose that L is an infinitely smooth origin-symmetric convex body in $\mathbb{R}^n$, with strictly positive curvature, that is not an intersection body. Then, there exists an origin-symmetric convex body K in $\mathbb{R}^n$ such that $L \subset K$ and*

$$|K|^{\frac{n-1}{n}} - |L|^{\frac{n-1}{n}} > c \frac{1}{\sqrt{n} M(\overline{L})} \max_{\xi \in S^{n-1}} (|K \cap \xi^\perp| - |L \cap \xi^\perp|), \tag{6.21}$$

where $c > 0$ is an absolute constant.

For the proof of Theorem 6.1.8 we use distributions (see Section 1.4.1). Recall that the Fourier transform of a distribution f is defined by $\langle \hat{f}, \phi \rangle = \langle f, \hat{\phi} \rangle$ for every test function $\phi \in \mathcal{S}(\mathbb{R}^n)$. For any even distribution f, we have $(\hat{f})^\wedge = (2\pi)^n f$.

If K is an origin-symmetric convex body and $0 < p < n$, then $\|\cdot\|_K^{-p}$ is a locally integrable function on $\mathbb{R}^n$ and represents a distribution acting by integration; see Lemma 1.1.1. Suppose that K is infinitely smooth, i. e., $\|\cdot\|_K \in C^\infty(S^{n-1})$ is an infinitely differentiable function on the sphere. Then, by Lemma 1.4.13, the Fourier transform of $\|\cdot\|_K^{-p}$ is an extension of some function $g \in C^\infty(S^{n-1})$ to a homogeneous function of degree $-n + p$ on $\mathbb{R}^n$. When we write $(\|\cdot\|_K^{-p})^\wedge(\xi)$, we mean $g(\xi)$, $\xi \in S^{n-1}$.

For $f \in C^\infty(S^{n-1})$ and $0 < p < n$, we denote by

$$(f \cdot r^{-p})(x) = f(x/|x|)|x|^{-p}$$

the extension of f to a homogeneous function of degree $-p$ on $\mathbb{R}^n$. By Lemma 1.4.13, there exists $g \in C^\infty(S^{n-1})$ such that

$$(f \cdot r^{-p})^\wedge = g \cdot r^{-n+p}.$$

If K, L are infinitely smooth origin-symmetric convex bodies, the following spherical version of Parseval's formula holds by Proposition 1.6.4: for any $p \in (-n, 0)$

$$\int_{S^{n-1}} (\|\cdot\|_K^{-p})^\wedge(\xi)(\|\cdot\|_L^{-n+p})^\wedge(\xi) = (2\pi)^n \int_{S^{n-1}} \|x\|_K^{-p} \|x\|_L^{-n+p} \, dx. \tag{6.22}$$

Theorem 1.7.1 states that an origin-symmetric convex body K in $\mathbb{R}^n$ is an intersection body if and only if the function $\|\cdot\|_K^{-1}$ represents a positive definite distribution. In the case where K is infinitely smooth, this means that the function $(\|\cdot\|_K^{-1})^\wedge$ is non-negative on the sphere.

We also need to recall Theorem 1.5.2 that expresses the volume of central hyperplane sections in terms of the Fourier transform. For any origin-symmetric star body K in $\mathbb{R}^n$, the distribution $(\|\cdot\|_K^{-n+1})^\wedge$ is a continuous function on the sphere extended to a homogeneous function of degree -1 on the whole of $\mathbb{R}^n$, and for every $\xi \in S^{n-1}$

$$|K \cap \xi^\perp| = \frac{1}{\pi(n-1)}(\|\cdot\|_K^{-n+1})^\wedge(\xi). \tag{6.23}$$

In particular, if $K = B_2^n$, then for every $\xi \in S^{n-1}$

$$(|\cdot|^{-n+1})^\wedge(\xi) = \pi(n-1)\omega_{n-1}. \tag{6.24}$$

Finally, we shall use the fact that, by Lemma 1.7.13, every non-intersection body can be approximated in the radial metric by infinitely smooth non-intersection bodies with strictly positive curvature.

Proof of Theorem 6.1.8. Since L is infinitely smooth, the Fourier transform of $\|\cdot\|_L^{-1}$ is a continuous function on the sphere S^{n-1}. Also, L is not an intersection body, so $(\|\cdot\|_L^{-1})^\wedge < 0$ on an open set $\Omega \subset S^{n-1}$. Let $\phi \in C^\infty(S^{n-1})$ be an even non-negative, not identically zero, infinitely smooth function on S^{n-1} with support in $\Omega \cup -\Omega$. Extend ϕ to an even homogeneous of degree -1 function $\phi \cdot r^{-1}$ on $\mathbb{R}^n \setminus \{0\}$. The Fourier transform of this function in the sense of distributions is $\psi \cdot r^{-n+1}$, where ψ is an infinitely smooth function on the sphere.

Let ε be a number such that $\omega_{n-1}\|\theta\|_L^{-n+1} > \varepsilon > 0$ for every $\theta \in S^{n-1}$. Define a star body K by

$$\|\theta\|_K^{-n+1} = \|\theta\|_L^{-n+1} - \delta\psi(\theta) + \frac{\varepsilon}{\omega_{n-1}}, \qquad \theta \in S^{n-1}, \tag{6.25}$$

where $\delta > 0$ is small enough so that for every θ

$$|\delta\psi(\theta)| < \min\left\{\|\theta\|_L^{-n+1} - \frac{\varepsilon}{\omega_{n-1}}, \frac{\varepsilon}{\omega_{n-1}}\right\}.$$

The latter condition implies that $L \subset K$. Since L has strictly positive curvature, by an argument from [310, p. 96], we can make ε, δ smaller (if necessary) to ensure that the body K is convex.

Now, we extend the functions in (6.25) from the sphere to $\mathbb{R}^n \setminus \{0\}$ as homogeneous functions of degree $-n+1$ and apply the Fourier transform. We get that for every $\xi \in S^{n-1}$

$$(\|\cdot\|_K^{-n+1})^\wedge(\xi) = (\|\cdot\|_L^{-n+1})^\wedge(\xi) - (2\pi)^n\delta\phi(\xi) + \pi(n-1)\varepsilon. \tag{6.26}$$

Here, we used (6.24) to compute the last term. By (6.26), (6.23), and the fact that the function ϕ is non-negative and is equal to zero at some points, we have

$$\varepsilon = \max_{\xi \in S^{n-1}} \left(|K \cap \xi^\perp| - |L \cap \xi^\perp|\right). \tag{6.27}$$

Multiplying both sides of (6.26) by $(\|\cdot\|_L^{-1})^\wedge(\xi)$, integrating over S^{n-1}, and using Parseval's formula on the sphere, we get

$$\begin{aligned}(2\pi)^n \int_{S^{n-1}} \|\theta\|_L^{-1} \|\theta\|_K^{-n+1}\, d\theta &= (2\pi)^n n|L| - (2\pi)^n \delta \int_{S^{n-1}} \phi(\theta)(\|\cdot\|_L^{-1})^\wedge(\theta)\, d\theta \\ &\quad + \pi(n-1)\varepsilon \int_{S^{n-1}} (\|\cdot\|_L^{-1})^\wedge(\theta)\, d\theta.\end{aligned}$$

Since ϕ is a non-negative function supported in Ω, where $(\|\cdot\|_L^{-1})^\wedge$ is negative, the latter equality implies

$$\begin{aligned}(2\pi)^n n|L| &+ \pi(n-1)\varepsilon \int_{S^{n-1}} (\|\cdot\|_L^{-1})^\wedge(\theta)\, d\theta \\ &< (2\pi)^n \int_{S^{n-1}} \|\theta\|_L^{-1} \|\theta\|_K^{-n+1}\, d\theta \\ &\leqslant (2\pi)^n \left(\int_{S^{n-1}} \|\theta\|_K^{-n}\, d\theta\right)^{\frac{n-1}{n}} \left(\int_{S^{n-1}} \|\theta\|_L^{-n} d\theta\right)^{\frac{1}{n}} \\ &= (2\pi)^n n|L|^{\frac{1}{n}} |K|^{\frac{n-1}{n}}.\end{aligned}$$

Finally, by (6.24), Parseval's formula, and Jensen's inequality

$$\begin{aligned}\pi(n-1) \int_{S^{n-1}} (\|\cdot\|_L^{-1})^\wedge(\theta)\, d\theta &= \frac{1}{\omega_{n-1}} \int_{S^{n-1}} (\|\cdot\|_L^{-1})^\wedge(\theta)(\|\cdot\|^{-n+1})^\wedge(\theta)\, d\theta \\ &= \frac{(2\pi)^n \kappa_n}{\omega_{n-1}} \int_{S^{n-1}} \|\theta\|_L^{-1}\, d\sigma(\theta) \\ &\geqslant \frac{(2\pi)^n \kappa_n}{\omega_{n-1}} \frac{1}{M(\overline{L})} |L|^{\frac{1}{n}} \\ &\geqslant c \frac{(2\pi)^n \sqrt{n} |L|^{\frac{1}{n}}}{M(\overline{L})}.\end{aligned}$$

Combining these estimates we get

$$(2\pi)^n n|L| + c\varepsilon \frac{(2\pi)^n \sqrt{n}|L|^{\frac{1}{n}}}{M(\bar{L})} \leqslant (2\pi)^n n|L|^{\frac{1}{n}}|K|^{\frac{n-1}{n}}.$$

The result follows after we recall (6.27). □

6.2 Volume difference inequalities for projections

In this section, we present a number of volume difference inequalities for the case of projections. In the hyperplane case, we denote by δ_n the smallest constant $\delta > 0$ with the property that

$$\delta(|L|^{\frac{n-1}{n}} - |K|^{\frac{n-1}{n}}) \geqslant \min_{\xi\in S^{n-1}} (|P_{\xi^\perp}(L)| - |P_{\xi^\perp}(K)|) \tag{6.28}$$

for all origin-symmetric convex bodies K, L in $\mathbb{R}^n$ whose curvature functions f_K and f_L exist and satisfy $f_K(\xi) \leqslant f_L(\xi)$ for all $\xi \in S^{n-1}$, where for $\xi \in S^{n-1}$ and a convex body L, we denote by $P_{\xi^\perp}(L)$ the orthogonal projection of L to $\xi^\perp$. Note that we formulate (6.28) with the condition $f_K \leqslant f_L$. The reason for this is that without an extra condition we cannot have (6.28), in general, with any $\delta > 0$. This follows from counterexamples to the Shephard problem asking whether, for any origin-symmetric convex bodies K and L, the inequalities $|P_{\xi^\perp}(K)| \leqslant |P_{\xi^\perp}(L)|$ for all $\xi \in S^{n-1}$ necessarily imply $|K| \leqslant |L|$.

It was proved in [311, 314] that if L is a projection body and K is an origin-symmetric convex body, then

$$|L|^{\frac{n-1}{n}} - |K|^{\frac{n-1}{n}} \geqslant c_{n,1} \min_{\xi\in S^{n-1}} (|P_{\xi^\perp}(L)| - |P_{\xi^\perp}(K)|). \tag{6.29}$$

Note that the condition $f_K \leqslant f_L$ is not needed for (6.29). If L is a projection body, then the answer to the question of Shephard is affirmative, therefore (6.29) is a quantified version of this fact.

For a convex body L in $\mathbb{R}^n$ denote by

$$d_{\mathrm{vr}}(L, \Pi) = \inf\left\{\left(\frac{|L|}{|D|}\right)^{1/n} : D \subset L,\ D \in \Pi\right\}$$

the volume ratio distance from L to the class of projection bodies. We extend (6.29) to arbitrary origin-symmetric convex bodies, as follows.

Theorem 6.2.1. *Suppose that K and L are origin-symmetric convex bodies in $\mathbb{R}^n$, and their curvature functions exist and satisfy $f_K(\xi) \leqslant f_L(\xi)$ for all $\xi \in S^{n-1}$. Then,*

$$d_{\mathrm{vr}}(L, \Pi)\,(|L|^{\frac{n-1}{n}} - |K|^{\frac{n-1}{n}}) \geqslant c_{n,1} \min_{\xi\in S^{n-1}} (|P_{\xi^\perp}(L)| - |P_{\xi^\perp}(K)|). \tag{6.30}$$

Proof. By approximation (see [505, Theorem 3.3.1] and [236, Section 5]), we can assume that the support functions h_K, h_L and the curvature functions f_K, f_L are infinitely smooth on the sphere. We have

$$|P_{\xi^\perp}(L)| - |P_{\xi^\perp}(K)| \geqslant \min_{\eta \in S^{n-1}} (|P_{\eta^\perp}(L)| - |P_{\eta^\perp}(K)|). \tag{6.31}$$

From Theorem 1.5.10 we know that

$$|P_{\xi^\perp}(K)| = -\frac{1}{\pi}(f_K \cdot r^{-n-1})^\wedge(\xi), \quad \xi \in S^{n-1}, \tag{6.32}$$

where f_K is extended from the sphere to a homogeneous function of degree $-n-1$ on the whole $\mathbb{R}^n$. Therefore, (6.31) can be written as

$$\begin{aligned} &-\frac{1}{\pi}(f_L \cdot r^{-n-1})^\wedge(\xi) + \frac{1}{\pi}(f_K \cdot r^{-n-1})^\wedge(\xi) \\ &\geqslant \min_{\eta \in S^{n-1}} (|P_{\eta^\perp}(L)| - |P_{\eta^\perp}(K)|), \quad \xi \in S^{n-1}. \end{aligned} \tag{6.33}$$

Let D be an infinitely smooth projection body such that $D \subset L$. Then, $h_D \leqslant h_L$ in every direction. It was proved in [333] that an infinitely smooth origin-symmetric convex body D in $\mathbb{R}^n$ is a projection body if and only if $(h_D \cdot r)^\wedge \leqslant 0$ on the sphere S^{n-1}. Integrating (6.33) with respect to this negative density, we get

$$\begin{aligned} &-\int_{S^{n-1}} (h_D \cdot r)^\wedge(\xi)(f_L \cdot r^{-n-1})^\wedge(\xi)\, d\xi + \int_{S^{n-1}} (h_D \cdot r)^\wedge(\xi)(f_K \cdot r^{-n-1})^\wedge(\xi)\, d\xi \\ &\leqslant \pi \int_{S^{n-1}} (h_D \cdot r)^\wedge(\xi)\, d\xi \min_{\eta \in S^{n-1}} (|P_{\eta^\perp}(L)| - |P_{\eta^\perp}(K)|). \end{aligned}$$

Using Parseval's formula in the form of (1.77) we see that

$$\begin{aligned} &(2\pi)^n \int_{S^{n-1}} h_D(\xi)(f_L(\xi) - f_K(\xi))\, d\xi \\ &\geqslant -\pi \int_{S^{n-1}} (h_D \cdot r)^\wedge(\xi)\, d\xi \min_{\eta \in S^{n-1}} (|P_{\eta^\perp}(L)| - |P_{\eta^\perp}(K)|). \end{aligned} \tag{6.34}$$

We estimate the left-hand side of (6.34) from above using (1.22) and (1.23) (recall that $f_K \leqslant f_L$)

$$\begin{aligned} (2\pi)^n \int_{S^{n-1}} h_D(\xi)(f_L(\xi) - f_K(\xi))\, d\xi &\leqslant (2\pi)^n \int_{S^{n-1}} h_L(\xi)(f_L(\xi) - f_K(\xi))\, d\xi \\ &\leqslant (2\pi)^n n(|L| - |K|^{\frac{n-1}{n}} |L|^{\frac{1}{n}}). \end{aligned} \tag{6.35}$$

To estimate the right-hand side of (6.35) from below, note that, by (6.32), the Fourier transform of the curvature function f_2 of the unit Euclidean ball is equal to

$$(f_2 \cdot r^{-n-1})^\wedge(\xi) = -\pi\omega_{n-1}, \quad \xi \in S^{n-1}.$$

Therefore, by Parseval's formula (1.77) and (1.23) (recall that $f_2 \equiv 1$)

$$\begin{aligned}-\pi \int_{S^{n-1}} (h_D \cdot r)^\wedge(\xi)\, d\xi &= \frac{1}{\omega_{n-1}} \int_{S^{n-1}} (h_D \cdot r)^\wedge(\xi)(f_2 \cdot r^{-n-1})^\wedge(\xi)\, d\xi \\ &= \frac{(2\pi)^n}{\omega_{n-1}} \int_{S^{n-1}} h_D(x) f_2(x)\, dx = \frac{(2\pi)^n}{\omega_{n-1}} n V_{n-1}(B_2^n, D) \\ &\geqslant \frac{(2\pi)^n n}{\omega_{n-1}} |D|^{\frac{1}{n}} |B_2^n|^{\frac{n-1}{n}} = (2\pi)^n n\, c_{n,1} |D|^{\frac{1}{n}}.\end{aligned}$$

Now, for $\delta > 0$ we choose D so that $(1+\delta)\, d_{\mathrm{vr}}(L, \Pi)\ |D|^{\frac{1}{n}} \geqslant |L|^{\frac{1}{n}}$. Combining the resulting inequality with (6.34) and (6.35) and sending δ to zero, we conclude the proof. □

Note that, by K. Ball's volume ratio estimate, for any convex body K in $\mathbb{R}^n$, $d_{\mathrm{vr}}(K, \Pi) \leqslant \sqrt{n}$. In fact, we can show that $\delta_n \approx \sqrt{n}$.

Theorem 6.2.2. *There exist absolute constants $a, b > 0$ such that $a\sqrt{n} \leqslant \delta_n \leqslant b\sqrt{n}$ for all $n \in \mathbb{N}$.*

Proof. Putting $K = \delta B_2^n$ in (6.28) and sending δ to zero, we get

$$\delta |L|^{\frac{n-1}{n}} \geqslant \min_{\xi \in S^{n-1}} |P_{\xi^\perp}(L)|.$$

In the proof of Theorem 3.6.4 we saw that for every $n \in \mathbb{N}$ there exists an origin-symmetric convex body L_n in $\mathbb{R}^n$ such that

$$\min_{\xi \in S^{n-1}} |P_{\xi^\perp}(L_n)| \geqslant c_1 \sqrt{n} |L_n|^{\frac{n-1}{n}},$$

where $c_1 > 0$ is an absolute constant. This shows that $\delta_n \geqslant c_1\sqrt{n}$. On the other hand, since ellipsoids are projection bodies, we have $d_{\mathrm{vr}}(L, \Pi) \leqslant \sqrt{n}$ for every origin-symmetric convex body L in $\mathbb{R}^n$. We can approximate L by a sequence of bodies L_k that have a curvature function (see [236]), so we can apply Theorem 6.2.1 to the bodies L_k and $K = \delta B_2^n$, $\delta \to 0$, to see that $\delta_n \leqslant (1/c_{n,1})\sqrt{n} < \sqrt{en}$. □

Remark 6.2.3. From Theorem 6.2.1 we see that the bodies L_n defined in the proof of Theorem 3.6.4 satisfy

$$d_{\mathrm{vr}}(L_n, \Pi) |L_n|^{\frac{n-1}{n}} \geqslant c_{n,1} \min_{\xi \in S^{n-1}} |P_{\xi^\perp}(L)| \geqslant c_{n,1} c_1 \sqrt{n} |L_n|^{\frac{n-1}{n}}.$$

This shows that $d_{\mathrm{vr}}(L_n, \Pi) \geqslant c_1\sqrt{n/e}$, and hence

$$\sup_L d_{\mathrm{vr}}(L, \Pi_n) \approx \sqrt{n},$$

where the supremum is over all origin-symmetric convex bodies L in $\mathbb{R}^n$.

For our next result we introduce one more notion of distance.

Definition 6.2.4. The mean width distance from a convex body K to the class of projection bodies is the quantity

$$d_w(K, \Pi) = \inf\left\{\frac{w(D)}{w(K)} : K \subset D,\ D \in \Pi\right\},$$

where $w(K)$ is the mean width of the body K.

Theorem 6.2.5. *Suppose that K and L are origin-symmetric convex bodies in $\mathbb{R}^n$, and their curvature functions exist and satisfy $f_K(\xi) \leqslant f_L(\xi)$ for all $\xi \in S^{n-1}$. Then,*

$$|L|^{\frac{n-1}{n}} - |K|^{\frac{n-1}{n}} \leqslant c\, d_w(K, \Pi)\frac{w(K)}{\sqrt{n}} \max_{\xi \in S^{n-1}} \left(|P_{\xi^\perp}(L)| - |P_{\xi^\perp}(K)|\right), \tag{6.36}$$

where $c > 0$ is an absolute constant.

Proof. Again, by approximation, we can assume that K, L are infinitely smooth. Let D be a projection body such that $K \subset D$. Then, $h_K \leqslant h_D$ in every direction. Similarly to the proof of Theorem 6.2.1

$$\begin{aligned}(2\pi)^n &\int_{S^{n-1}} h_D(\xi)(f_L(\xi) - f_K(\xi))\, d\xi \\ &\leqslant -\pi \int_{S^{n-1}} \widehat{h_D}(\xi)\, d\xi \max_{\eta \in S^{n-1}} \left(|P_{\eta^\perp}(L)| - |P_{\eta^\perp}(K)|\right).\end{aligned} \tag{6.37}$$

We estimate the left-hand side of (6.37) from below using (1.22) and (1.23) (recall that $f_K \leqslant f_L$ and $h_K \leqslant h_D$)

$$\begin{aligned}(2\pi)^n &\int_{S^{n-1}} h_D(\xi)(f_L(\xi) - f_K(\xi))\, d\xi \\ &\geqslant (2\pi)^n \int_{S^{n-1}} h_K(\xi)(f_L(\xi) - f_K(\xi))\, d\xi \\ &\geqslant (2\pi)^n n\left(|L|^{\frac{n-1}{n}}|K|^{\frac{1}{n}} - |K|\right).\end{aligned} \tag{6.38}$$

Now, for $\delta > 0$ choose D so that

$$w(D) \leqslant (1+\delta)d_{\mathrm{W}}(K,\Pi)w(\overline{K})|K|^{\frac{1}{n}}.$$

As in the proof of Theorem 6.2.1

$$-\pi\int_{S^{n-1}}\widehat{h_D}(\xi)\,d\xi = \frac{(2\pi)^n}{\omega_{n-1}}\int_{S^{n-1}}h_D(x)\,dx = \frac{(2\pi)^n\kappa_n}{\omega_{n-1}}w(D)$$
$$\leqslant (1+\delta)(2\pi)^n c\, d_{\mathrm{W}}(K,\Pi)\sqrt{n}\,w(\overline{K})|K|^{\frac{1}{n}}.$$

To complete the proof we combine the latter with (6.37) and (6.38), and send δ to zero. □

The distance d_w can be of the order $\sqrt{n}$, up to a logarithmic term. Note that if K is an origin-symmetric convex body of volume 1 in $\mathbb{R}^n$ and is in the minimal mean width position, then $w(K) \leqslant c\sqrt{n}(\log n)$. We will use the fact that projection bodies have positions with "small diameter". More precisely, we have the following statement: For every $D \in \Pi$ there exists $T \in GL_n$ such that

$$R(T(D)) \leqslant \frac{\sqrt{n}}{2}|T(D)|^{1/n}. \tag{6.39}$$

In particular, this holds true if T is chosen so that $T(D)$ in Lewis or Löwner or minimal mean width position (see, e. g., [113, Chapter 4]). Let $K = B_1^n$ be the cross-polytope, and consider a projection body D such that $B_1^n \subseteq D$. We may find T so that (6.39) is satisfied. In Chapter 5, we have already used the next result of Bárány and Füredi: if $x_1,\dots,x_N \in RB_2^n$, then

$$|\mathrm{conv}\{x_1,\dots,x_N\}|^{1/n} \leqslant \frac{c_1R\sqrt{\log(1+N/n)}}{n}.$$

Since

$$T(B_1^n) = \mathrm{conv}\{\pm Te_1,\dots,\pm Te_n\} \subseteq R(T(D))B_2^n,$$

we get

$$|T(B_1^n)|^{1/n} \leqslant \frac{c_2}{\sqrt{n}}|T(D)|^{1/n}.$$

It follows that

$$|B_1^n|^{1/n} \leqslant \frac{c_2}{\sqrt{n}}|D|^{1/n}.$$

From Urysohn's inequality we know that $w(D) \geqslant c_3\sqrt{n}\,|D|^{1/n}$, and a direct computation shows that $w(B_1^n) \leqslant c_4\sqrt{n\log n}\,|B_1^n|^{1/n}$. This shows that

$$w(D) \geqslant c_5 \sqrt{n/\log n}\, w(B_1^n).$$

Since $D \supset B_1^n$ was arbitrary, we conclude that

$$d_w(B_1^n) \geqslant c\sqrt{n/\log n}, \tag{6.40}$$

where $c > 0$ is an absolute constant.

Remark 6.2.6. Theorems 6.2.1 and 6.2.5 are complemented by the following results, going in the opposite directions, that were proved in [318]. The constant in the first result is written in a more general form than in [318].

(i) Let L be an origin-symmetric convex body in $\mathbb{R}^n$, with strictly positive curvature, that is not a projection body. Then, there exists an origin-symmetric convex body K in $\mathbb{R}^n$ so that $f_L(\xi) \geqslant f_K(\xi)$ for all $\xi \in S^{n-1}$ and

$$\max_{\xi\in S^{n-1}} \left(|P_{\xi^\perp}(L)| - |P_{\xi^\perp}(K)|\right) \leqslant \frac{1}{c_{n,1}}\left(|L|^{\frac{n-1}{n}} - |K|^{\frac{n-1}{n}}\right).$$

(ii) Let K be an origin-symmetric convex body in $\mathbb{R}^n$ that is not a projection body. Then, there exists an origin-symmetric convex body L in $\mathbb{R}^n$ so that $f_L(\xi) \geqslant f_K(\xi)$ for all $\xi \in S^{n-1}$ and

$$\min_{\xi\in S^{n-1}} \left(|P_{\xi^\perp}(L)| - |P_{\xi^\perp}(K)|\right) \geqslant \frac{c\sqrt{n}}{w(\overline{K})}\left(|L|^{\frac{n-1}{n}} - |K|^{\frac{n-1}{n}}\right),$$

where c is an absolute constant.

6.3 Further reading

6.3.1 Integral geometric approach

Using the approach developed in Chapter 5 we may prove a different version of Theorem 6.1.4, where the symmetry and continuity assumptions are dropped, but the body K is required to be convex.

Theorem 6.3.1. *Let $1 \leqslant k \leqslant n-1$, let K be a convex body with $0 \in K$ and let $L \subseteq K$ be a Borel set in $\mathbb{R}^n$. For any measure μ with a bounded measurable non-negative density, we have*

$$\begin{aligned}&\mu(K)^{n-k} - \mu(L)^{n-k}\\ &\quad\leqslant (c_0\sqrt{n-k})^{k(n-k)}|K|^{\frac{k(n-k)}{n}} \max_{F\in G_{n,n-k}} \left(\mu(K\cap F)^{n-k} - \mu(L\cap F)^{n-k}\right),\end{aligned}$$

where $c_0 > 0$ is an absolute constant.

The proof is based on the Blaschke–Petkantschin formula (in the form of Theorem 5.1.3) and on Grinberg's inequality (Theorem 5.2.1). We will also use the estimate of Lemma 5.1.4: for every $1 \leqslant k \leqslant n-1$ we have

$$[\tau_{n,k}^{-n}\, p_{n,n-k}]^{\frac{1}{k(n-k)}} \leqslant c_0\sqrt{n-k} \tag{6.41}$$

for some absolute constant $c_0 > 0$.

Proof of Theorem 6.3.1. Let g be the density of the measure μ. To simplify the notation we set $\overline{dx} := dx_{n-k}\dots dx_1$ and $dF := d\nu_{n,n-k}(F)$. Applying the Blaschke-Petkantschin formula for the functions

$$f(x_1,\dots,x_{n-k}) = \prod_{i=1}^{n-k} g(x_i)\mathbb{1}_K(x_i) \quad \text{and} \quad h(x_1,\dots,x_{n-k}) = \prod_{i=1}^{n-k} g(x_i)\mathbb{1}_L(x_i),$$

we get

$$\begin{aligned}\mu(K)^{n-k} - \mu(L)^{n-k} &= \prod_{i=1}^{n-k}\int_K g(x_i)\,dx - \prod_{i=1}^{n-k}\int_L g(x_i)\,dx \\ &= p_{n,n-k}\int_{G_{n,n-k}}\Bigg(\int_{K\cap F}\cdots\int_{K\cap F}\prod_{i=1}^{n-k} g(x_i)\,\big|\mathrm{conv}\{0,x_1,\dots,x_{n-k}\}\big|^k\,\overline{dx} \\ &\quad - \int_{L\cap F}\cdots\int_{L\cap F}\prod_{i=1}^{n-k} g(x_i)\,\big|\mathrm{conv}\{0,x_1,\dots,x_{n-k}\}\big|^k\,\overline{dx}\Bigg)\,dF \\ &= p_{n,n-k}\int_{G_{n,n-k}}\Bigg(\int_{P_{n-k}(K,L;F)}\prod_{i=1}^{n-k} g(x_i)\,\big|\mathrm{conv}\{0,x_1,\dots,x_{n-k}\}\big|^k\,\overline{dx}\Bigg)\,dF,\end{aligned} \tag{6.42}$$

where

$$P_{n-k}(K,L;F) = (K\cap F)^{n-k}\setminus(L\cap F)^{n-k}.$$

Note that

$$\big|\mathrm{conv}\{0,x_1,\dots,x_{n-k}\}\big|^k \leqslant |K\cap F|^k$$

for all $(x_1,\dots,x_{n-k}) \in P_{n-k}(K,L;F)$ by the convexity of $K\cap F$ and the assumption that $0\in K$. Therefore,

$$\begin{aligned}&\mu(K)^{n-k} - \mu(L)^{n-k} \\ &\leqslant p_{n,n-k}\int_{G_{n,n-k}} |K\cap F|^k\Bigg(\int_{P_{n-k}(K,L;F)} g(x_1)\cdots g(x_{n-k})\Bigg)\,dF\end{aligned}$$

$$= p_{n,n-k} \int_{G_{n,n-k}} |K \cap F|^k (\mu(K \cap F)^{n-k} - \mu(L \cap F)^{n-k})\, dF$$
$$\leqslant \max_{F \in G_{n,n-k}} (\mu(K \cap F)^{n-k} - \mu(L \cap F)^{n-k}) \cdot p_{n,n-k} \int_{G_{n,n-k}} |K \cap F|^k \, dF. \tag{6.43}$$

From Grinberg's inequality we have that

$$\int_{G_{n,n-k}} |K \cap F|^k \, d\nu_{n,n-k}(F) \leqslant \tau_{n,k}^{-kn} |K|^{\frac{k(n-k)}{n}}. \tag{6.44}$$

Using also (6.41) we see that

$$\mu(K)^{n-k} - \mu(L)^{n-k}$$
$$\leqslant (c_0 \sqrt{n-k})^{k(n-k)} |K|^{\frac{k(n-k)}{n}} \max_{F \in G_{n,n-k}} (\mu(K \cap F)^{n-k} - \mu(L \cap F)^{n-k}), \tag{6.45}$$

as claimed. □

Remark 6.3.2. Theorem 6.3.1 implies the slicing inequality

$$\mu(K) \leqslant (c_0 \sqrt{n-k})^k |K|^{\frac{k}{n}} \max_{F \in G_{n,n-k}} \mu(K \cap F) \tag{6.46}$$

of Theorem 5.4.1 for every convex body K with $0 \in K$ and any measure μ. Also, if we consider a measure μ with density supported in $K \setminus L$ in (6.46), we get the following measure difference inequality:

$$\mu(K) - \mu(L) \leqslant (c_0 \sqrt{n-k})^k |K|^{\frac{k}{n}} \max_{F \in G_{n,n-k}} (\mu(K \cap F) - \mu(L \cap F)), \tag{6.47}$$

under the assumptions of Theorem 6.3.1.

Using the same method we may also give a different generalization of (6.15).

Theorem 6.3.3. *Let $1 \leqslant k \leqslant n-1$, and let K and L be bounded Borel sets in $\mathbb{R}^n$ with $L \subset K$. Then,*

$$(|K| - |L|)^{\frac{n-k}{n}} \geqslant \tau_{n,k} \min_{F \in G_{n,n-k}} (|K \cap F| - |L \cap F|). \tag{6.48}$$

Note that Theorem 6.3.3 holds true for an arbitrary pair of bounded Borel sets $L \subseteq K$ and it no longer involves the distance d_k and the parameter $M(\overline{L})$. Actually, the constant $\tau_{n,k}$ is sharp, as one can check from the example of the ball $K = B_2^n$ and $L = \beta B_2^n$, where $\beta \to 0$. Nevertheless, it is formally not stronger than Theorem 6.1.7 because the quantity $|K|^{\frac{n-k}{n}} - |L|^{\frac{n-k}{n}}$ is smaller than $(|K| - |L|)^{\frac{n-k}{n}}$.

We deduce Theorem 6.3.3 from a more general statement for arbitrary measures.

Theorem 6.3.4. *Let $1 \leqslant k \leqslant n-1$, and let K and L be two bounded Borel sets in $\mathbb{R}^n$ such that $L \subset K$. Let μ a measure in $\mathbb{R}^n$ with bounded density g. Then,*

$$(\mu(K)-\mu(L))^{\frac{n-k}{n}} \geqslant \tau_{n,k} \frac{1}{\|g\|_\infty^{\frac{k}{n}}} \left(\int_{G_{n,n-k}} (\mu(K\cap F)-\mu(L\cap F))^{\frac{n}{n-k}} \, d\nu_{n,n-k}(F) \right)^{\frac{n-k}{n}}.$$

In particular,

$$(\mu(K)-\mu(L))^{\frac{n-k}{n}} \geqslant \tau_{n,k} \frac{1}{\|g\|_\infty^{\frac{k}{n}}} \min_{F\in G_{n,n-k}} (\mu(K\cap F)-\mu(L\cap F)).$$

For the proof of Theorem 6.3.4 we will use the following generalized Blaschke–Petkantschin formula.

Lemma 6.3.5. *Let $1 \leqslant q \leqslant s \leqslant n$. There exists a constant $p(n,s,q) > 0$ such that, for every non-negative bounded Borel measurable function $f : (\mathbb{R}^n)^q \to \mathbb{R}$*

$$\int_{\mathbb{R}^n} \cdots \int_{\mathbb{R}^n} f(x_1,\ldots,x_q)\, dx_1 \cdots dx_q = p(n,s,q) \int_{G_{n,s}} \int_F \cdots \int_F f(x_1,\ldots,x_q) \left|\operatorname{conv}\{0,x_1,\ldots,x_q\}\right|^{n-s} d\overline{x}\, dF. \tag{6.49}$$

The exact value of the constant $p(n,s,q)$ is

$$p(n,s,q) = (q!)^{n-s} \frac{\kappa_n \cdots \kappa_{n-q+1}}{\kappa_s \cdots \kappa_{s-q+1}}. \tag{6.50}$$

We will also use some basic facts about Sylvester-type functionals. Recall from Chapter 5 that if C is a bounded Borel set of positive measure in $\mathbb{R}^m$, then, for every $p > 0$, the normalized p-th moment of the expected volume of the random simplex $\operatorname{conv}\{0,x_1,\ldots,x_m\}$ is defined by

$$S_p(C) = \left(\frac{1}{|C|^{m+p}} \int_C \cdots \int_C |\operatorname{conv}\{0,x_1,\ldots,x_m\}|^p \, dx_m \cdots dx_1 \right)^{1/p}. \tag{6.51}$$

It was proved by Pfiefer [460] (see also [197]) that

$$S_p(C) \geqslant S_p(B_2^m).$$

More generally, for any Borel probability measure ν on $\mathbb{R}^m$, for any $1 \leqslant q \leqslant m$ and every $p > 0$, we define

$$S_{p,q}(\nu) = \left(\int_{\mathbb{R}^m} \cdots \int_{\mathbb{R}^m} |\mathrm{conv}\{0, x_1, \ldots, x_q\}|^p \, d\nu(x_q) \cdots d\nu(x_1) \right)^{1/p}. \tag{6.52}$$

A generalization of Pfiefer's result appears in [146]. Let ν be a measure on $\mathbb{R}^n$ with a bounded non-negative measurable density g. Then,

$$S_{p,q}^p(\nu) \geqslant \frac{\|g\|_1^{q+\frac{pq}{m}}}{\omega_m^{q+\frac{pq}{m}} \|g\|_\infty^{\frac{pq}{m}}} S_{p,q}^p(\mathbb{1}_{B_2^m}). \tag{6.53}$$

Proof of Theorem 6.3.4. Let $u(x) = g(x)\mathbb{1}_K(x)$ and $v(x) = g(x)\mathbb{1}_L(x)$. Using Lemma 6.3.5, we start by writing

$$\begin{aligned}
\mu(K) - \mu(L) &= \int_{\mathbb{R}^n} u(x)\,dx - \int_{\mathbb{R}^n} v(x)\,dx \\
&= p(n, n-k, 1) \int_{G_{n,n-k}} \left(\int_{K\cap F} g(x)\,|x|^k\,dx - \int_{L\cap F} g(x)\,|x|^k\,dx \right) d\nu_{n,n-k}(F) \\
&= p(n, n-k, 1) \int_{G_{n,n-k}} \int_{(K\cap F)\setminus(L\cap F)} g(x)\,|x|^k\,dx\,d\nu_{n,n-k}(F).
\end{aligned} \tag{6.54}$$

(Note that $|\mathrm{conv}\{0, x\}| = |x|$, the Euclidean norm of x). For every F set $C_F = (K\cap F)\setminus(L\cap F)$ and consider the measure ν_F with density g on C_F. Applying (6.53) with $p = k$, $q = 1$ and $m = n-k$ we have

$$\begin{aligned}
\mu(K) - \mu(L) &\geqslant p(n, n-k, 1) \int_{G_{n,n-k}} S_{k,1}^k(\nu_F)\,d\nu_{n,n-k}(F) \\
&\geqslant p(n, n-k, 1) \int_{G_{n,n-k}} \frac{\|g\,|_{C_F}\|_1^{1+\frac{k}{n-k}}}{\omega_{n-k}^{1+\frac{k}{n-k}} \|g\,|_{C_F}\|_\infty^{\frac{k}{n-k}}} S_k^k(\mathbb{1}_{B_2^{n-k}})\,d\nu_{n,n-k}(F) \\
&= \frac{p(n, n-k, 1)}{\omega_{n-k}^{\frac{n}{n-k}}} S_2^k(\mathbb{1}_{B_2^{n-k}}) \int_{G_{n,n-k}} \frac{\|g\,|_{C_F}\|_1^{\frac{n}{n-k}}}{\|g\,|_{C_F}\|_\infty^{\frac{k}{n-k}}}\,d\nu_{n,n-k}(F).
\end{aligned} \tag{6.55}$$

Note that

$$p(n, n-k, 1) = \frac{n\omega_n}{(n-k)\omega_{n-k}}$$

and

$$S_{k,1}^k(\mathbb{1}_{B_2^{n-k}}) = \int_{B_2^{n-k}} |x|^k\,dx = \frac{n-k}{n}\omega_{n-k}.$$

Therefore,

$$\frac{p(n, n-k, 1)}{\omega_{n-k}^{\frac{n}{n-k}}} S_2^k(\mathbb{1}_{B_2^{n-k}}) = \frac{\omega_n}{\omega_{n-k}^{\frac{n}{n-k}}} = \tau_{n,k}^{\frac{n}{n-k}}.$$

On the other hand, for any $F \in G_{n,n-k}$ we have

$$\|g|_{C_F}\|_1 = \mu(K \cap F) - \mu(L \cap F)$$

and

$$\|g|_{C_F}\|_\infty \leqslant \|g\|_\infty.$$

Combining the above we get

$$\mu(K) - \mu(L) \geqslant \tau_{n,k}^{\frac{n}{n-k}} \frac{1}{\|g\|_\infty^{\frac{k}{n-k}}} \int_{G_{n,n-k}} (\mu(K \cap F) - \mu(L \cap F))^{\frac{n}{n-k}} \, d\nu_{n,n-k}(F)$$

and the result follows. □

Remark 6.3.6. Theorem 6.3.3 is an immediate consequence of Theorem 6.3.4. It corresponds to the case $g \equiv \mathbf{1}$, for which we clearly have $\|g\|_\infty = 1$.

7 Surface area of sections and projections

In this chapter, we start by exploring the rich interplay between the volume of convex bodies in $\mathbb{R}^n$ and the volumes of their projections onto coordinate hyperplanes and sections by lower-dimensional subspaces. We begin with the classical Loomis–Whitney inequality, which provides an upper bound for the volume of a convex body in terms of the geometric mean of the volumes of its projections onto coordinate hyperplanes. The dual form of this inequality, known as Meyer's inequality, relates the volume of a body to the volumes of its hyperplane sections. These foundational results serve as a springboard for more general inequalities, including the Bollobás–Thomason uniform cover inequality, which accounts for projections and sections over s-uniform covers and weighted families of subspaces. Variants of these inequalities, both local and restricted, are also discussed, and dual inequalities for sections are obtained via methods based on L_p-centroid bodies. We emphasize the role of geometric Brascamp–Lieb inequalities, Berwald's inequality, and the properties of log-concave functions in establishing a unified framework that connects projections, sections, and the volume of convex bodies with precise quantitative estimates.

A significant part of the chapter is devoted to the study of the surface area of projections. The surface area of a convex body, denoted $S(K)$, is considered alongside its projections onto hyperplanes. The minimal surface area parameter ∂_K, defined via linear transformations that normalize the body, provides a way to measure how projections relate to the overall surface area. For general convex bodies, upper and lower bounds for the surface area of projections are derived, showing that in certain classical positions, such as minimal surface area, isotropic, or John's and Löwner's positions, these bounds are asymptotically sharp. Key tools in these results include Cauchy's surface area formula, Aleksandrov inequalities for quermassintegrals, and volume estimates for projection bodies and their polars.

We also investigate the possibility of slicing inequalities for surface area, in analogy with the classical slicing problem for volume. While one might hope that the surface area of a body could be universally bounded by the maximal surface area of its hyperplane sections, it is shown that no such universal bound is possible. This is illustrated using origin-symmetric ellipsoids with extreme axis ratios, which serve as counterexamples. More generally, inequalities for higher-order quermassintegrals also fail to hold universally. Despite these negative results, reverse inequalities can be established for bodies in certain classical positions, and ellipsoids provide a clear illustration of extremal behavior.

Finally, we examine the average volume of sections of convex bodies, denoted $\mathrm{as}(K)$ for central hyperplane sections and $\mathrm{as}_r(K)$ for r-codimensional sections. For intersection bodies, the average section volume can be controlled in terms of the maximal volume of lower-dimensional sections, with sharp constants achieved for Euclidean balls. More generally, for centered convex bodies and origin-symmetric star bodies, one can establish upper bounds for the average section volume in terms of maximal sections of codi-

https://doi.org/10.1515/9783111360508-007

mension k, up to constants depending on n and k. These results connect to the isotropic constant problem and demonstrate that the controlling constants remain bounded when k is proportional to the dimension. Extensions to higher-codimension sections are also discussed, showing that the structure of star bodies allows for precise control of the average section volume relative to sections of fixed codimension.

7.1 Coordinate projections and sections of convex bodies

A classical inequality of Loomis and Whitney [385] estimates the volume $|K|$ of a convex body K, or more generally a compact set, in $\mathbb{R}^n$ by the geometric mean of the volumes $|P_{e_i^\perp}(K)|$ of its orthogonal projections onto $e_i^\perp$, where $\{e_1,\dots,e_n\}$ is an orthonormal basis of $\mathbb{R}^n$. One has

$$|K|^{n-1} \leqslant \prod_{i=1}^{n} |P_{e_i^\perp}(K)|, \tag{7.1}$$

with equality if and only if K is an orthogonal parallelepiped such that $\pm e_i$ are the normal vectors of its facets. A dual inequality, in which the volume of K is estimated by the volumes of the sections $K\cap e_i^\perp$, was proved by Meyer in [410]. For every convex body K in $\mathbb{R}^n$ one has

$$|K|^{n-1} \geqslant \frac{n!}{n^n} \prod_{i=1}^{n} |K\cap e_i^\perp|, \tag{7.2}$$

with equality if and only if $K = \mathrm{conv}\{\pm\lambda_1 e_1,\dots,\pm\lambda_n e_n\}$ for some $\lambda_i > 0$.

An extension of the Loomis–Whitney inequality was suggested and proved by Bollobás and Thomason. To describe it, let us define what we mean by an s-uniform cover of $[n] = \{1,\dots,n\}$. For every non-empty $\tau \subset [n]$ we set $F_\tau = \mathrm{span}\{e_j : j\in\tau\}$ and $E_\tau = F_\tau^\perp$. Given an integer $s\geqslant 1$ we say that the (not necessarily distinct) sets $\sigma_1,\dots,\sigma_r \subseteq [n]$ form an s-uniform cover of $[n]$ if every $j\in[n]$ belongs to exactly s of the sets σ_i. The uniform cover inequality (see [88]) provides an upper bound for the volume of a compact set in terms of the volumes of its coordinate projections that correspond to an s-uniform cover of $[n]$.

Theorem 7.1.1. *Let $r\geqslant 1$ and let $(\sigma_1,\dots,\sigma_r)$ be an s-uniform cover of $[n]$. For every compact subset K of $\mathbb{R}^n$, which is the closure of its interior, we have*

$$|K|^s \leqslant \prod_{i=1}^{r} |P_{F_{\sigma_i}}(K)|. \tag{7.3}$$

Note that the Loomis–Whitney inequality (7.1) corresponds to the case of the $(n-1)$-uniform cover $(\sigma_1,\dots,\sigma_n)$ of $[n]$, where $\sigma_i = [n]\setminus\{i\}$. We shall show that Theorem 7.1.1 can be obtained by a straightforward application of Barthe's multidimensional geometric Brascamp–Lieb inequality [57].

Theorem 7.1.2. *Let $m, n \in \mathbb{N}$. For $j = 1, \dots m$, let F_j be a d_j-dimensional subspace of $\mathbb{R}^n$ and P_j be the orthogonal projection onto F_j. If*

$$I_n = \sum_{j=1}^{m} c_j P_j$$

for some $c_1, \dots, c_m > 0$, then for all non-negative integrable functions $f_j : F_j \to \mathbb{R}$ we have

$$\int_{\mathbb{R}^n} \prod_{j=1}^{m} f_j^{c_j}(P_j x)\, dx \leqslant \prod_{j=1}^{m} \Big(\int_{F_j} f_j \Big)^{c_j} \tag{7.4}$$

and

$$\int_{\mathbb{R}^n}^{*} \sup \Big\{ \prod_{j=1}^{m} f_j^{c_j}(x_j) : x = \sum_{j=1}^{m} c_j x_j,\ x_j \in F_j \Big\}\, dx \geqslant \prod_{j=1}^{m} \Big(\int_{F_j} f_j \Big)^{c_j}, \tag{7.5}$$

where $\int^*$ denotes the outer integral.

In fact, using Barthe's theorem we may prove a more general result, for which we need to introduce the notion of an s-uniform cover by general subspaces. Given $s > 0$, we say that the subspaces $F_1, \dots, F_r$ form an s-uniform cover of $\mathbb{R}^n$ with weights $c_1, \dots, c_r > 0$ if

$$sI_n = \sum_{i=1}^{r} c_i P_{F_i}. \tag{7.6}$$

Note that if $(\sigma_1, \dots, \sigma_r)$ is an s-uniform cover of $[n]$, then the projections $P_i := P_{F_{\sigma_i}}$ satisfy

$$sI_n = \sum_{i=1}^{r} P_i.$$

Therefore, Theorem 7.1.1 is indeed an immediate consequence of the next theorem, using all $c_i = 1$.

Theorem 7.1.3. *Let $F_1, \dots, F_r$ be subspaces with $\dim(F_i) = d_i$ that form an s-uniform cover of $\mathbb{R}^n$ with weights $c_1, \dots, c_r > 0$. For every compact subset K of $\mathbb{R}^n$ we have*

$$|K|^s \leqslant \prod_{i=1}^{r} |P_{F_i}(K)|^{c_i}. \tag{7.7}$$

Proof. Given a compact subset K of $\mathbb{R}^n$ we define $f_i : F_i \to [0,\infty)$ by $f_i = \mathbb{1}_{P_{F_i}(K)}$. It is easily checked that

$$\mathbb{1}_K(x) \leqslant \prod_{i=1}^{r} f_i^{\frac{c_i}{s}}(P_{F_i}(x))$$

for all $x \in \mathbb{R}^n$. Since $I_n = \sum_{i=1}^r \frac{c_i}{s} P_{F_i}$, by Theorem 7.1.2 we get

$$\begin{aligned}|K| = \int_{\mathbb{R}^n} \mathbb{1}_K(x)\,dx &\leqslant \int_{\mathbb{R}^n} \prod_{i=1}^{r} f_i^{\frac{c_i}{s}}(P_{F_i}(x))\,dx \leqslant \prod_{i=1}^{r}\Big(\int_{F_i} f_i\Big)^{\frac{c_i}{s}}\\ &= \prod_{i=1}^{r} |P_{F_i}(K)|^{\frac{c_i}{s}},\end{aligned}$$

which shows that $|K|^s \leqslant \prod_{i=1}^r |P_{F_i}(K)|^{c_i}$, as claimed. □

An inequality that bounds the volume of K from below, and, in particular, includes Meyer's inequality (7.2), was proved by Liakopoulos in [377]. It may be viewed as the dual version of Theorem 7.1.1 and its proof again employs the multidimensional geometric reverse Brascamp–Lieb inequality of Barthe. The starting point is a functional inequality. Below, we denote by $\mathcal{F}(\mathbb{R}^n)$ the class of log-concave integrable functions $f : \mathbb{R}^n \to [0,\infty)$.

Theorem 7.1.4. *Let $f \in \mathcal{F}(\mathbb{R}^n)$ with $f(0) = 1$ and $F_1, \ldots, F_r$ be subspaces of $\mathbb{R}^n$ that form an s-uniform cover of $\mathbb{R}^n$ with weights $c_1, \ldots, c_r > 0$. Then,*

$$n^n \int_{\mathbb{R}^n} f(y)^n\,dy \geqslant \prod_{i=1}^{r}\Big(\int_{F_i} f(x_i)\,dx_i\Big)^{c_i/s}.$$

Proof. Denote $d_i = \dim(F_i)$. Let $z \in \mathbb{R}^n$ and $x_i \in F_i$, $i \in [r]$ such that $z = \sum_{i=1}^r \frac{c_i}{s} x_i$. Then,

$$\frac{z}{n} = \sum_{i=1}^{r} \frac{c_i d_i}{sn} \cdot \frac{x_i}{d_i}$$

and since $\sum_{i=1}^r \frac{c_i d_i}{sn} = 1$ and f is log-concave we have that

$$f(z/n) \geqslant \prod_{l=1}^{r} f(x_i/d_i)^{\frac{c_i d_i}{ns}}.$$

Since $f(0) = 1$, for every $i \in [r]$ we see that $f(x_i/d_i) \geqslant f(x_i)^{1/d_i} f(0)^{1-1/d_i} = f(x_i)^{1/d_i}$. It follows that

$$f(z/n) \geqslant \prod_{i=1}^{r} f(x_i)^{\frac{1}{d_i}\cdot\frac{c_i d_i}{ns}} = \prod_{i=1}^{r} f(x_i)^{\frac{c_i}{ns}}$$

and hence

$$f(z/n)^n \geqslant \prod_{i=1}^{r} f(x_i)^{c_i/s}.$$

This shows that

$$f(z/n)^n \geqslant \sup\left\{\prod_{i=1}^{r} f(x_i)^{c_i/s} : z = \sum_{i=1}^{r} \frac{c_i}{s} x_i, x_i \in F_i\right\}.$$

Then, by Theorem 7.1.2 we have that

$$\begin{aligned}\int_{\mathbb{R}^n} f(z/n)^n \, dz &\geqslant \int_{\mathbb{R}^n}^{*} \sup\left\{\prod_{i=1}^{r} f(x_i)^{c_i/s} : z = \sum_{i=1}^{r} \frac{c_i}{s} x_i, x_i \in F_i\right\} dz \\ &\geqslant \prod_{i=1}^{r} \left(\int_{F_i} f(x_i)\, dx_i\right)^{c_i/s}.\end{aligned}$$

Making the change of variables $y = z/n$ we conclude the proof. □

A consequence of Theorem 7.1.4 is the following general uniform cover inequality for sections of a convex body.

Theorem 7.1.5. *Let K be a convex body in $\mathbb{R}^n$ with $0 \in \operatorname{int}(K)$ and $F_1, \dots, F_r$ be subspaces of $\mathbb{R}^n$, with $\dim(F_i) = d_i$, that form an s-uniform cover of $\mathbb{R}^n$ with weights $c_1, \dots, c_r > 0$. Then,*

$$|K|^s \geqslant \frac{1}{(n!)^s} \prod_{i=1}^{r} (d_i! \, |K \cap F_i|)^{c_i}.$$

Proof. The function $f(y) = e^{-\|y\|_K}$ is log-concave and $f(0) = 1$. Direct computation shows that

$$\begin{aligned}n^n \int_{\mathbb{R}^n} f(y)^n \, dy = n^n \int_{\mathbb{R}^n} e^{-n\|y\|_K} \, dy = n^n \int_{\mathbb{R}^n} e^{-\|y\|_{\frac{1}{n}K}} \, dy \\ = n^n \, n! \left|\frac{1}{n} K\right| = n! \, |K|\end{aligned}$$

and, for every $i \in [r]$

$$\int_{F_i} f(x_i) \, dx_i = \int_{F_i} e^{-\|x_i\|_K} \, dx_i = \int_{F_i} e^{-\|x_i\|_{K \cap F_i}} \, dx_i = d_i! \, |K \cap F_i|.$$

Applying Theorem 7.1.4 for f we get

$$n!|K| \geqslant \prod_{i=1}^{r} \left(d_i!\,|K \cap F_i|\right)^{c_i/s}$$

and the theorem follows. □

To get a counterpart of Theorem 7.1.1 we apply Theorem 7.1.5 in the following way: let $(\sigma_1, \dots, \sigma_r)$ be an s-uniform cover of $[n]$. Setting $F_i = F_{\sigma_i} = \mathrm{span}(\{e_j : j \in \sigma_i\})$, $i \in [r]$, we have $sI_n = \sum_{i=1}^{r} P_{F_i}$. Thus, we immediately see that if K is a convex body in $\mathbb{R}^n$ with $0 \in \mathrm{int}(K)$ and $(\sigma_1, \dots, \sigma_r)$ is an s-uniform cover of $[n]$, then

$$|K|^s \geqslant \frac{1}{(n!)^s} \prod_{i=1}^{r} |\sigma_i|!\,|K \cap F_i|. \tag{7.8}$$

To recover Meyer's inequality, we use the particular case $F_i = e_i^{\perp}$, $i \in [n]$, so that we have $(n-1)\,I_n = \sum_{i=1}^{n} P_{e_i^{\perp}}$, and applying (7.8) with $s = n-1$ and $|\sigma_i| = \dim(F_i) = n-1$ we get

$$|K|^{n-1} \geqslant \frac{n!}{n^n} \prod_{i=1}^{n} |K \cap e_i^{\perp}|$$

for any convex body K in $\mathbb{R}^n$ with $0 \in \mathrm{int}(K)$.

Restricted variants of the Loomis–Whitney inequality and of the uniform cover inequality of Theorem 7.1.1 were obtained in [111]. The starting point is the following inequality from [213]: If $i \neq j \in \{1, \dots, n\}$ and $P_{ij}(K) = P_{E_{ij}}(K)$, where $E_{ij} = \mathrm{span}\{e_i, e_j\}^{\perp}$, then

$$|P_i(K)|\,|P_j(K)| \geqslant \frac{n}{2(n-1)}\,|K|\,|P_{ij}(K)|. \tag{7.9}$$

This inequality may be viewed as a restricted (or "local") version of the Loomis–Whitney inequality, in the sense that it gives a lower estimate for the geometric mean of just two coordinate hyperplane projections of a convex body.

A generalization of (7.9) was proved in [111], using the uniform cover inequality of Theorem 7.1.1: If $r > s \geqslant 1$, σ is a subset of $[n]$ with cardinality $|\sigma| = d < n$ and $(\sigma_1, \dots, \sigma_r)$ is an s-uniform cover of σ, then for every convex body K in $\mathbb{R}^n$ we have

$$\prod_{i=1}^{r} |P_{E_{\sigma_i}}(K)| \geqslant \gamma(n,d,s,r)\,|P_{E_\sigma}(K)|^s\,|K|^{r-s}, \tag{7.10}$$

where

$$\gamma(n,d,s,r) = \binom{n}{d}^{r-s} \binom{n - \frac{sd}{r}}{n-d}^{-r}. \tag{7.11}$$

Note that (7.9) corresponds to the special case $d = r = 2$, $k = 1$, and $s = 1$. The case $k = 1$, $d = r$, and $s = 1$ has been also studied by Soprunov and Zvavitch in [513] who also present an example that shows that the constant

$$\gamma(n,r,1,r) = \binom{n}{r}^{r-1}\binom{n-1}{n-r}^{-r} = \left(\frac{n}{r}\right)^{r}\binom{n}{r}^{-1}$$

is optimal.

Below, we present more precise versions of (7.10) and (7.11), in which the multiplicative constant depends on the cover that was obtained in [399].

Theorem 7.1.6. *Let $r > s \geqslant 1$, let $\sigma \subseteq [n]$ with cardinality $|\sigma| = d < n$, and let $(\sigma_1,\ldots,\sigma_r)$ with $|\sigma_i| = d_i$ be an s-uniform cover of σ. For every convex body K in $\mathbb{R}^n$ we have*

$$\prod_{i=1}^{r} |P_{E_{\sigma_i}}(K)| \geqslant \gamma(n,d,d_i,s,r)|P_{E_\sigma}(K)|^{s}\,|K|^{r-s},$$

where

$$\gamma(n,d,d_i,s,r) = \binom{n}{d}^{r-s}\left(\prod_{i=1}^{r}\binom{n-d_i}{n-d}\right)^{-1}.$$

Note that in the case where $d_i = ds/r$ the constant in (7.10) is equal to the constant provided by (7.11).

For the proof of Theorem 7.1.6, besides the uniform cover inequality we need the next classical inequality of Berwald [72].

Lemma 7.1.7. *Let C be a convex body in $\mathbb{R}^m$ and let $\phi : C \to \mathbb{R}^+$ be a concave function. Then, for every $0 < p < q$*

$$\left(\binom{m+q}{m}\frac{1}{|C|}\int_C |\phi(x)|^q\,dx\right)^{1/q} \leqslant \left(\binom{m+p}{m}\frac{1}{|C|}\int_C |\phi(x)|^p\,dx\right)^{1/p}.$$

Proof of Theorem 7.1.6. Let $r > s \geqslant 1$, let $\sigma \subseteq [n]$ with cardinality $|\sigma| = d < n$ and let $(\sigma_1,\ldots,\sigma_r)$ be an s-uniform cover of σ. Note that if $|\sigma_i| = d_i$, then

$$ds = d_1 + \cdots + d_r.$$

For every $y \in P_{E_\sigma}(K)$ we define the sets

$$K(y) = \{t \in F_\sigma : y + t \in K\}$$

and

$$K_i(y) = \{t \in F_{\sigma\setminus\sigma_i} : y + t \in P_{E_{\sigma_i}}(K)\}.$$

Then, $K_i(y)$ is the orthogonal projection of $K(y)$ onto $F_{\sigma\setminus\sigma_i}$. Since $(\sigma_1,\ldots,\sigma_r)$ is an s-uniform cover of σ, we have that $(\sigma\setminus\sigma_1,\ldots,\sigma\setminus\sigma_r)$ is an $(r-s)$-uniform cover of σ. From Theorem 7.1.1 we get

$$|K| = \int_{P_{E_\sigma}(K)} |K(y)|\,dy \leqslant \int_{P_{E_\sigma}(K)} \prod_{i=1}^{r} |K_i(y)|^{\frac{1}{r-s}}\,dy.$$

For every $i = 1, \dots, r$ we define $f_i : P_{E_\sigma}(K) \to [0, \infty)$ with

$$f_i(y) = |K_i(y)|^{\frac{1}{d-d_i}}.$$

Hölder's inequality implies that

$$|K| = \int_{P_{E_\sigma}(K)} \prod_{i=1}^{r} (f_i(y))^{\frac{d-d_i}{r-s}}\,dy \leqslant \prod_{i=1}^{r} \Bigg(\int_{P_{E_\sigma}(K)} (f_i(y))^d\,dy \Bigg)^{\frac{d-d_i}{d(r-s)}}.$$

By the Brunn–Minkowski inequality, the functions f_i are concave, therefore we may apply Berwald's inequality (Lemma 7.1.7) to write

$$\begin{aligned}
&\left(\binom{n}{n-d} \frac{1}{|P_{E_\sigma}(K)|} \int_{P_{E_\sigma}(K)} (f_i(y))^d\,dy \right)^{\frac{1}{d}} \\
&\quad \leqslant \left(\binom{n-d_i}{n-d} \frac{1}{|P_{E_\sigma}(K)|} \int_{P_{E_\sigma}(K)} (f_i(y))^{d-d_i}\,dy \right)^{\frac{1}{d-d_i}}.
\end{aligned}$$

It follows that

$$\begin{aligned}
|K|^{r-s} &\leqslant \prod_{i=1}^{r} \Bigg(\int_{P_{E_\sigma}(K)} (f_i(y))^d\,dy \Bigg)^{\frac{d-d_i}{d}} \\
&\leqslant \frac{1}{\binom{n}{n-d}^{r-s}} \frac{1}{|P_{E_\sigma}(K)|^s} \prod_{i=1}^{r} \binom{n-d_i}{n-d} \int_{P_{E_\sigma}(K)} (f_i(y))^{d-d_i}\,dy \\
&= \frac{1}{\binom{n}{n-d}^{r-s}} \frac{1}{|P_{E_\sigma}(K)|^s} \prod_{i=1}^{r} \binom{n-d_i}{n-d} |P_{E_{\sigma_i}}(K)|,
\end{aligned}$$

which is the assertion of the theorem. □

We move on to the natural question of whether it is possible to have an inequality for sections, which is dual to (7.9). More precisely, the question is if, for every centered convex body K in $\mathbb{R}^n$ and every $i \neq j \in \{1, \dots, n\}$

$$|K \cap e_i^\perp|\,|K \cap e_j^\perp| \leqslant c_0 |K \cap E_{ij}|\,|K|, \tag{7.12}$$

where $c_0 > 0$ is an absolute constant. We shall exploit the main properties of the family of the L_p-centroid bodies $Z_p(K)$ of K to show that this question has an affirmative answer. In a few words, through a duality argument, the question about coordinate sections of K is translated into a question about coordinate projections of some projection of a suitable centroid body of K, and then one may use the Loomis–Whitney inequality (or some extension of it) to complete the proof. Generalizing the method and making full use of the uniform cover inequality of Bollobás and Thomason, one can prove more general inequalities of this form, in the spirit of Theorem 7.1.6.

Theorem 7.1.8. *Let $r > s \geqslant 1$, let $\sigma \subseteq [n]$ with cardinality $|\sigma| = d < n$, and let $(\sigma_1, \ldots, \sigma_r)$ be an s-uniform cover of σ. Let also $d_i = |\sigma_i|$. For every centered convex body K in $\mathbb{R}^n$ we have*

$$\prod_{i=1}^{r} |K \cap E_{\sigma_i}| \leqslant \frac{(c_0 d)^{ds}}{d_1^{d_1} \cdots d_r^{d_r}} |K \cap E_\sigma|^s |K|^{r-s}, \tag{7.13}$$

where $c_0 > 0$ is an absolute constant.

Note that under the assumptions of Theorem 7.1.8 we have $d_1 + \cdots + d_r = ds$, and hence

$$d_1^{d_1} \ldots d_r^{d_r} \geqslant \left(\frac{ds}{r}\right)^{ds}$$

by Jensen's inequality. Therefore, the result may be written in the simpler form

$$\prod_{i=1}^{r} |K \cap E_{\sigma_i}| \leqslant \left(\frac{c_0 r}{s}\right)^{ds} |K \cap E_\sigma|^s |K|^{r-s}.$$

This is equivalent to (7.13) if all the sets σ_i have the same cardinality $k = \frac{ds}{r}$. Our starting point (7.12) corresponds to the special case $d = r = 2$, $k = 1$, and $s = 1$.

In the next lemma, we collect some basic facts about the L_p-centroid bodies of a centered convex body K of volume 1 in $\mathbb{R}^n$. All of them were discussed in Section 2.3.

Lemma 7.1.9 (L_p-centroid bodies). *There exist absolute constants $c_i > 0$ such that, for every centered convex body K of volume 1 in $\mathbb{R}^n$, for every $1 \leqslant k \leqslant n-1$, $q > p \geqslant 1$, and $F \in G_{n,k}$, we have*

$$Z_q(K) \subseteq \frac{c_1 q}{p} Z_p(K) \tag{7.14}$$

and

$$c_2 \leqslant |K \cap F^\perp|^{\frac{1}{k}} |P_F(Z_k(K))|^{\frac{1}{k}} \leqslant c_3. \tag{7.15}$$

Moreover, if $p \geqslant n$, then we have that

$$Z_p(K) \supseteq c_4 Z_\infty(K),$$

where $Z_\infty(K) = \mathrm{conv}\{K, -K\}$.

Besides (7.14) and (7.15) we will need the following: For every centered convex body K of volume 1 in $\mathbb{R}^n$, for every $p \geqslant 1$ and every $u \in S^{n-1}$

$$c_5 h_{Z_p(K)}(u) \leqslant \frac{1}{|K \cap u^\perp|} \leqslant c_6 p h_{Z_p(K)}(u), \tag{7.16}$$

where $c_5, c_6 > 0$ are absolute constants. This inequality is a direct consequence of Theorem 2.1.5 and (7.14).

Proof of Theorem 7.1.8. Let $r > s \geqslant 1$, let $\sigma \subseteq [n]$ with cardinality $|\sigma| = d < n$ and let $(\sigma_1, \ldots, \sigma_r)$ be an s-uniform cover of σ. Note that if $|\sigma_i| = d_i$, then $ds = d_1 + \cdots + d_r$.

By homogeneity, we may assume that $|K| = 1$. Starting from (7.15) we may write

$$c_2 \leqslant |K \cap E_{\sigma_i}|^{\frac{1}{d_i}} |P_{F_{\sigma_i}}(Z_{d_i}(K))|^{\frac{1}{d_i}} \leqslant c_3$$

for all i, and hence

$$\prod_{i=1}^r |K \cap E_{\sigma_i}| \leqslant c_3^{d_1+\cdots+d_r} \prod_{i=1}^r |P_{F_{\sigma_i}}(Z_{d_i}(K))|^{-1} = c_3^{ds} \prod_{i=1}^r |P_{F_{\sigma_i}}(Z_{d_i}(K))|^{-1},$$

so we need a lower bound for the product

$$\prod_{i=1}^r |P_{F_{\sigma_i}}(Z_{d_i}(K))|.$$

From (7.14) we have

$$Z_d(K) \subseteq \frac{c_1 d}{d_i} Z_{d_i}(K)$$

for all $i = 1, \ldots, r$, which gives

$$\begin{aligned}\prod_{i=1}^r |P_{F_{\sigma_i}}(Z_d(K))| &\leqslant \prod_{i=1}^r \left(\frac{c_1 d}{d_i}\right)^{d_i} \prod_{i=1}^r |P_{F_{\sigma_i}}(Z_{d_i}(K))| \\ &= \frac{(c_1 d)^{ds}}{d_1^{d_1} \cdots d_r^{d_r}} \prod_{i=1}^r |P_{F_{\sigma_i}}(Z_{d_i}(K))|.\end{aligned}$$

Now, since $(\sigma_1, \ldots, \sigma_r)$ is an s-uniform cover of σ, applying the uniform cover inequality of Bollobás and Thomason to the convex body $P_{F_\sigma}(Z_d(K))$ we get

$$|P_{F_\sigma}(Z_d(K))|^s \leqslant \prod_{i=1}^r |P_{F_{\sigma_i}}(Z_d(K))|.$$

Next, using again (7.15), we see that

$$|P_{F_\sigma}(Z_d(K))|^s \geqslant c_2^{ds}|K \cap E_\sigma|^{-s}.$$

Combining the above, we have

$$\prod_{i=1}^{r} |K \cap E_{\sigma_i}| \leqslant \frac{(c_1c_3d)^{ds}}{d_1^{d_1}\cdots d_r^{d_r}}|P_{F_\sigma}(Z_d(K))|^{-s} \leqslant \frac{(c_0d)^{ds}}{d_1^{d_1}\cdots d_r^{d_r}}|K \cap E_\sigma|^s,$$

where $c_0 = c_1c_3/c_2$, and the result follows. □

7.2 Surface area of projections

The starting point of this section is a pair of inequalities about the surface area of hyperplane projections of projection bodies. In [314] it was proved that if Z is a projection body in $\mathbb{R}^n$, then

$$|Z|^{\frac{1}{n}} \min_{\xi \in S^{n-1}} S(P_{\xi^\perp}(Z)) \leqslant b_n S(Z), \tag{7.17}$$

where $S(A)$ denotes the surface area of A, $P_{\xi^\perp}(Z)$ stands for the orthogonal projection of Z to the hyperplane $\xi^\perp$ perpendicular to a vector $\xi \in S^{n-1}$, and

$$b_n = \frac{(n-1)\omega_{n-1}}{n\omega_n^{\frac{n-1}{n}}} = \frac{n-1}{n}\tau_{n,1}^{-1}, \tag{7.18}$$

where ω_m is the volume of the Euclidean unit ball B_2^m in $\mathbb{R}^m$. Note that $b_n \approx 1$ and that (7.17) is sharp; there is equality if $Z = B_2^n$. Conversely, in [317] it was proved that if Z is a projection body in $\mathbb{R}^n$ that is a dilate of a body in an isotropic position, then

$$|Z|^{\frac{1}{n}} \max_{\xi \in S^{n-1}} S(P_{\xi^\perp}(Z)) \geqslant c(\log n)^{-2} S(Z), \tag{7.19}$$

where $c > 0$ is an absolute constant.

We discuss similar inequalities for the surface area of hyperplane projections of an arbitrary convex body K in $\mathbb{R}^n$. In what follows, we denote by ∂_K the minimal surface area parameter of K, defined by

$$\partial_K := \min\{S(T(K))/|T(K)|^{\frac{n-1}{n}} : T \in GL_n\}. \tag{7.20}$$

We say that K has minimal surface area if $S(K) \leqslant S(T(K))$ for every $T \in SL_n$. Petty ([457], see also [212]) proved that K has minimal surface area if and only if the surface area measure $S(K,\cdot)$ of K satisfies the isotropic condition

$$S(K) = n \int_{S^{n-1}} \langle \xi, \theta\rangle^2 \, dS(K,\xi) \tag{7.21}$$

for every $\theta \in S^{n-1}$. From the isoperimetric inequality one gets $S(K) \geqslant n\omega_n^{\frac{1}{n}}|K|^{\frac{n-1}{n}}$. The reverse isoperimetric inequality of K. Ball [44] states that if K has minimal surface area and volume 1, then $S(K) \leqslant S(C_n) = 2n$ in the symmetric case, where C_n is a cube of volume 1 in $\mathbb{R}^n$, and $S(K) \leqslant S(\Delta_n) \leqslant c_0 n$ in the non-symmetric case, where Δ_n is a regular simplex of volume 1 in $\mathbb{R}^n$ and $c_0 > 0$ is an absolute constant. It follows that $c_1\sqrt{n} \leqslant \partial_K \leqslant c_2 n$ for every convex body K in $\mathbb{R}^n$, where $c_1, c_2 > 0$ are absolute constants.

A general estimate in the spirit of (7.17) is given by the following theorem from [214].

Theorem 7.2.1. *There exists an absolute constant $c_1 > 0$ such that, for every convex body K in $\mathbb{R}^n$*

$$|K|^{\frac{1}{n}} \min_{\xi\in S^{n-1}} S(P_{\xi^\perp}(K)) \leqslant \frac{2b_n\partial_K}{n\omega_n^{\frac{1}{n}}} S(K) \leqslant \frac{c_1\partial_K}{\sqrt{n}} S(K). \tag{7.22}$$

This inequality is sharp, e. g., for the Euclidean unit ball. Note that $c_1\partial_K/\sqrt{n} \leqslant c\sqrt{n}$ for every convex body K in $\mathbb{R}^n$, and hence we have the general upper bound

$$|K|^{\frac{1}{n}} \min_{\xi\in S^{n-1}} S(P_{\xi^\perp}(K)) \leqslant c\sqrt{n}\, S(K).$$

The proof of Theorem 7.2.1 involves the projection body ΠK of a convex body K; recall that this is the symmetric convex body whose support function is defined by

$$h_{\Pi K}(\xi) = |P_{\xi^\perp}(K)|, \quad \xi \in S^{n-1}. \tag{7.23}$$

From Cauchy's formula we know that

$$|P_{\xi^\perp}(K)| = \frac{1}{2}\int_{S^{n-1}} |\langle \xi,\theta\rangle| \, dS(K,\theta) \tag{7.24}$$

for every $\xi \in S^{n-1}$. Let K be a convex body of volume 1 in $\mathbb{R}^n$. Then, the volume of ΠK and of its polar body $\Pi^* K$ satisfy the bounds (see [212])

$$\left(\frac{\partial_K}{n}\right)^n \leqslant |\Pi K| \leqslant \omega_n \left(\frac{\omega_{n-1}\partial_K}{n\omega_n}\right)^n \tag{7.25}$$

and

$$\omega_n \left(\frac{n\omega_n}{\omega_{n-1}\partial_K}\right)^n \leqslant |\Pi^* K| \leqslant \frac{4^n n^n}{n!\,\partial_K^n}. \tag{7.26}$$

All these inequalities are sharp, as one can see from the examples of the ball and the cube.

A second main ingredient in the proof of Theorem 7.2.1 is the next lemma.

Lemma 7.2.2. *Let K be a convex body in $\mathbb{R}^n$. For every $u \in S^{n-1}$ we have*

$$\frac{S(K)}{|K|} \geqslant \frac{n}{2(n-1)} \frac{S(P_{u^\perp}(K))}{|P_{u^\perp}(K)|}.$$

Proof. From (7.9) we have that

$$|P_{\theta^\perp}(K)|\,|P_{u^\perp}(K)| \geqslant \frac{n}{2(n-1)}|K|\,|P_{\langle\theta,u\rangle^\perp}(K)|$$

for every θ on the unit sphere $S(u^\perp)$ of $u^\perp$. We integrate both sides with respect to the rotationally invariant probability measure σ_u on $S(u^\perp)$ and use Cauchy's surface area formula; the right-hand side gives

$$\begin{aligned}\frac{n}{2(n-1)}|K| \int_{S(u^\perp)} |P_{\langle\theta,u\rangle^\perp}(K)|\sigma_u(d\theta) &= \frac{n}{2(n-1)}|K|\, c_{n-1}S(P_{u^\perp}(K)) \\ &= \frac{n}{2(n-1)}c_{n-1}|K|\, S(P_{u^\perp}(K)),\end{aligned}$$

where $c_{n-1} = \omega_{n-2}/(n-1)\omega_{n-1}$, and from the left-hand side we get

$$\begin{aligned}&|P_{u^\perp}(K)| \int_{S(u^\perp)} |P_{\theta^\perp}(K)|\sigma_u(d\theta) \\ &\quad= |P_{u^\perp}(K)|\,\frac{1}{2}\int_{S(u^\perp)}\int_{S^{n-1}} |\langle\theta,\xi\rangle|\, dS(K,\xi)\sigma_u(d\theta) \\ &\quad= |P_{u^\perp}(K)|\,\frac{1}{2}\int_{S^{n-1}}\int_{S(u^\perp)} |\langle\theta,\xi\rangle|\sigma_u(d\theta)\, dS(K,\xi) \\ &\quad= |P_{u^\perp}(K)|\, c_{n-1}\int_{S^{n-1}} \sqrt{1-\langle\xi,u\rangle^2}\, dS(K,\xi) \\ &\quad\leqslant c_{n-1}|P_{u^\perp}(K)|\, S(K).\end{aligned}$$

It follows that

$$|P_{u^\perp}(K)|\, S(K) \geqslant \frac{n}{2(n-1)}|K|\, S(P_{u^\perp}(K)).$$ □

Proof of Theorem 7.2.1. From Lemma 7.2.2 we have that

$$|K|\, S(P_{\xi^\perp}(K)) \leqslant \frac{2(n-1)}{n} S(K)\,|P_{\xi^\perp}(K)| \tag{7.27}$$

for every $\xi \in S^{n-1}$. Therefore,

$$|K| \min_{\xi\in S^{n-1}} S(P_{\xi^\perp}(K)) \leqslant \frac{2(n-1)}{n} S(K) \min_{\xi\in S^{n-1}} |P_{\xi^\perp}(K)|. \tag{7.28}$$

Next, we observe that

$$\min_{\xi\in S^{n-1}} |P_{\xi^\perp}(K)| = \min_{\xi\in S^{n-1}} h_{\Pi K}(\xi) = r(\Pi K), \tag{7.29}$$

where $r(\Pi K)$ is the inradius of ΠK. From (7.25) it follows that

$$|\Pi K|^{\frac{1}{n}} \leqslant \frac{\omega_{n-1}\partial_K}{n\omega_n^{\frac{n-1}{n}}} |K|^{\frac{n-1}{n}} \tag{7.30}$$

and hence

$$r(\Pi K) \leqslant \mathrm{vrad}(\Pi K) \leqslant \frac{\omega_{n-1}\partial_K}{n\omega_n} |K|^{\frac{n-1}{n}}. \tag{7.31}$$

Going back to (7.28) we see that

$$|K| \min_{\xi\in S^{n-1}} S(P_{\xi^\perp}(K)) \leqslant \frac{2(n-1)\omega_{n-1}\partial_K}{n^2\omega_n} S(K)\,|K|^{\frac{n-1}{n}} \tag{7.32}$$

and this proves (7.22). □

Note. It would be interesting to decide whether there exist convex bodies K such that

$$|K|^{\frac{1}{n}} \min_{\xi\in S^{n-1}} S(P_{\xi^\perp}(K)) \geqslant c\sqrt{n}\, S(K).$$

This would show that Theorem 7.2.1 is asymptotically sharp. Note that in the case of the Euclidean ball one has

$$|B_2^n|^{\frac{1}{n}} \min_{\xi\in S^{n-1}} S(P_{\xi^\perp}(B_2^n)) \geqslant c\, S(B_2^n).$$

The proof of Theorem 7.2.1 made essential use of the fact that

$$\min_{\xi\in S^{n-1}} |P_{\xi^\perp}(K)| \leqslant c\sqrt{n}\,|K|^{\frac{n-1}{n}} \tag{7.33}$$

for an absolute constant $c > 0$. In the case where K is a projection body, one can see that (7.33) holds true with $c\sqrt{n}$ replaced by b_n.

Lemma 7.2.3. *Let Z be a zonoid in $\mathbb{R}^n$. Then,*

$$\min_{\xi\in S^{n-1}} |P_{\xi^\perp}(Z)| \leqslant \frac{nb_n}{n-1} |Z|^{\frac{n-1}{n}}. \tag{7.34}$$

Proof. We write $Z = \Pi K$ for some convex body K. Recall the volume formula for zonoids (see, e. g., [506])

$$|Z| = V_{n-1}(Z, \Pi K) = V_{n-1}(K, \Pi Z) = \frac{1}{n} \int_{S^{n-1}} h_{\Pi Z}(\xi)\, dS(K, \xi)$$
$$= \frac{1}{n} \int_{S^{n-1}} \big|P_{\xi^\perp}(Z)\big|\, dS(K, \xi). \tag{7.35}$$

Therefore,

$$|Z| \geqslant \frac{S(K)}{n} \min_{\xi \in S^{n-1}} \big|P_{\xi^\perp}(Z)\big|. \tag{7.36}$$

On the other hand, by (7.25) we have

$$|Z|^{\frac{1}{n}} = |\Pi K|^{\frac{1}{n}} \leqslant \frac{\omega_{n-1}\partial_K}{n\omega_n^{\frac{n-1}{n}}} |K|^{\frac{n-1}{n}} \leqslant \frac{b_n S(K)}{n-1},$$

where we used the definition (7.18) of the constant b_n and the fact that $S(K) \geqslant \partial_K |K|^{\frac{n-1}{n}}$ by the definition (7.20) of the minimal surface area parameter ∂_K. Then,

$$\frac{S(K)}{n} \min_{\xi \in S^{n-1}} \big|P_{\xi^\perp}(Z)\big| \leqslant |Z|^{\frac{1}{n}} |Z|^{\frac{n-1}{n}} \leqslant \frac{b_n S(K)}{n-1} |Z|^{\frac{n-1}{n}} \tag{7.37}$$

and the result follows. □

Remark 7.2.4. Since every projection of a zonoid is a zonoid, a simple induction argument leads to the following general result. If Z is a zonoid in $\mathbb{R}^n$, then for every $1 \leqslant k \leqslant n-1$ we have

$$\min_{F \in G_{n,k}} \big|P_F(Z)\big| \leqslant \tau_{n,n-k}^{-1} |Z|^{\frac{k}{n}}. \tag{7.38}$$

Recall that

$$\tau_{n,n-k} = \omega_n^{\frac{k}{n}} / \omega_k.$$

The example of the ball shows that Lemma 7.2.3 and (7.38) are sharp. In fact, they correspond to the known fact that the Shephard problem has a positive answer for zonoids (see Section 3.6).

Lemma 7.2.3 leads to an alternative proof of (7.17) with a weaker (by a factor of 2) constant.

Theorem 7.2.5. *Let Z be a projection body in $\mathbb{R}^n$. Then,*

$$|Z|^{\frac{1}{n}} \min_{\xi\in S^{n-1}} S(P_{\xi^\perp}(Z)) \leqslant 2b_n\, S(Z). \tag{7.39}$$

Proof. We have

$$\begin{aligned}|Z| \min_{\xi\in S^{n-1}} S(P_{\xi^\perp}(Z)) &\leqslant \frac{2(n-1)}{n} S(Z) \min_{\xi\in S^{n-1}} |P_{\xi^\perp}(Z)| \\ &\leqslant \frac{2(n-1)}{n}\frac{nb_n}{n-1} S(Z)\,|Z|^{\frac{n-1}{n}},\end{aligned} \tag{7.40}$$

by Lemma 7.2.3. Dividing by $|Z|^{\frac{n-1}{n}}$ we get the result. □

It should be noted that there are convex bodies that are not projection bodies but their minimal surface area parameter ∂_K is of the order of $\sqrt{n}$; an example is given by B_1^n, the unit ball of ℓ_1^n. On the other hand, there exist projection bodies whose minimal surface area parameter is of the order of n; an example is given by the cube. Thus, the estimates of Theorem 7.2.1 and Theorem 7.2.5 complement each other.

Question 7.2.6. It would be interesting to decide whether in the case of zonoids one has

$$|Z|\, S(P_{\xi^\perp}(Z)) \leqslant \frac{(n-1)}{n} S(Z)\,|P_{\xi^\perp}(Z)|$$

for every $\xi\in S^{n-1}$. This improvement of (7.9) (for the class of zonoids) would give a sharp version of Theorem 7.2.5.

Assuming that K is in the minimal surface area position we have a converse of Theorem 7.2.1.

Theorem 7.2.7. *Let K be a convex body in $\mathbb{R}^n$ that is in the minimal surface area position. Then,*

$$|K|^{\frac{1}{n}} \min_{\xi\in S^{n-1}} S(P_{\xi^\perp}(K)) \geqslant \frac{c}{\sqrt{n}} S(K), \tag{7.41}$$

where $c > 0$ is an absolute constant.

Proof. For every $1\leqslant k\leqslant n-1$ we set $V_{n-k}(K) := V((K, n-k), (B_2^n, k))$; this is just a different notation for the k-th quermassintegral $W_k(K)$ of K. Recall also that the sequence $Q_k(K) = (\omega_k^{-1}\int_{G_{n,k}} |P_F(K)|\, d\nu_{n,k}(F))^{1/k}$ is decreasing in k. We shall use the fact (see [213, Theorem 1.2]) that for any convex body K in $\mathbb{R}^n$ and any $\xi\in S^{n-1}$ one has

$$\frac{V_{n-2}(K)}{2V_{n-1}(K)} \leqslant \frac{V_{n-2}(P_{\xi^\perp}(K))}{|P_{\xi^\perp}(K)|}.$$

Note that

$$V_{n-2}(K) = \omega_n \left(Q_{n-2}(K)\right)^{n-2} \geqslant \omega_n \left(Q_{n-1}(K)\right)^{n-2} = \omega_n^{\frac{1}{n-1}} \left(V_{n-1}(K)\right)^{\frac{n-2}{n-1}},$$

while

$$nV_{n-1}(K) = S(K), \quad (n-1)V_{n-2}(P_{\xi^\perp}(K)) = S(P_{\xi^\perp}(K))$$

and if K is in the minimal surface area position by (7.21) we also have

$$\begin{aligned}|P_{\xi^\perp}(K)| &= \frac{1}{2}\int_{S^{n-1}} |\langle u,\xi\rangle|\, dS(K,u) \geqslant \frac{1}{2}\int_{S^{n-1}} \langle u,\xi\rangle^2\, dS(K,u) \\ &= \frac{S(K)}{2n} = \frac{V_{n-1}(K)}{2}.\end{aligned}$$

Combining the above, we get

$$S(P_{\xi^\perp}(K)) \geqslant \frac{(n-1)\omega_n^{\frac{1}{n-1}}}{4} \left(V_{n-1}(K)\right)^{\frac{n-2}{n-1}}.$$

Therefore, we have

$$\begin{aligned}|K|^{\frac{1}{n}} \min_{\xi\in S^{n-1}} S(P_{\xi^\perp}(K)) &\geqslant \frac{(n-1)\omega_n^{\frac{1}{n-1}}}{4} \left(V_{n-1}(K)\right)^{\frac{n-2}{n-1}} |K|^{\frac{1}{n}} \\ &= \frac{(n-1)\omega_n^{\frac{1}{n-1}}}{4n^{\frac{n-2}{n-1}}} \left(S(K)\right)^{\frac{n-2}{n-1}} |K|^{\frac{1}{n}}.\end{aligned}$$

Since $S(K) = \partial_K |K|^{\frac{n-1}{n}}$, we get

$$|K|^{\frac{1}{n}} \min_{\xi\in S^{n-1}} S(P_{\xi^\perp}(K)) \geqslant \frac{(n-1)\omega_n^{\frac{1}{n-1}}}{4n^{\frac{n-2}{n-1}}\partial_K^{\frac{1}{n-1}}} S(K).$$

This proves Theorem 7.2.7. □

Remark 7.2.8. Theorem 7.2.7 is sharp. In [400] it is proved that there exists an unconditional convex body K_0 of volume 1 in $\mathbb{R}^n$ that has minimal surface area and satisfies

$$\min_{\xi\in S^{n-1}} |P_{\xi^\perp}(K_0)| \leqslant \frac{c_1}{\sqrt{n}},$$

where $c_1 > 0$ is an absolute constant. From (7.28) we see that

$$|K_0| \min_{\xi\in S^{n-1}} S(P_{\xi^\perp}(K_0)) \leqslant \frac{2(n-1)}{n} S(K_0) \min_{\xi\in S^{n-1}} |P_{\xi^\perp}(K_0)| \leqslant \frac{2c_1}{\sqrt{n}} S(K_0)$$

and since $|K_0| = 1$ we get the following.

Proposition 7.2.9. *There exists an unconditional convex body K_0 in $\mathbb{R}^n$ that has minimal surface area and satisfies*

$$|K_0|^{\frac{1}{n}} \min_{\xi\in S^{n-1}} S(P_{\xi^\perp}(K_0)) \leqslant \frac{c}{\sqrt{n}} S(K_0),$$

where $c > 0$ is an absolute constant.

In the next result, we replace $\min_{\xi\in S^{n-1}} S(P_{\xi^\perp}(K))$ by the expectation of $S(P_{\xi^\perp}(K))$ on the sphere.

Theorem 7.2.10. *Let K be a convex body in $\mathbb{R}^n$. Then,*

$$|K| \int\limits_{S^{n-1}} S(P_{\xi^\perp}(K))\, d\sigma(\xi) \leqslant \frac{2(n-1)\omega_{n-1}}{n^2\omega_n} S(K)^2 \leqslant \frac{c_2}{\sqrt{n}} S(K)^2, \tag{7.42}$$

where $c_2 > 0$ is an absolute constant.

Proof. From Lemma 7.2.2, we have

$$|K|\, S(P_{\xi^\perp}(K)) \leqslant \frac{2(n-1)}{n} S(K)\, |P_{\xi^\perp}(K)| \tag{7.43}$$

for every $\xi \in S^{n-1}$. Integrating on S^{n-1} and using the identity

$$S(K) = \frac{n\omega_n}{\omega_{n-1}} \int\limits_{S^{n-1}} |P_{\xi^\perp}(K)|\, d\sigma(\xi), \tag{7.44}$$

we get

$$\begin{aligned} |K| \int\limits_{S^{n-1}} S(P_{\xi^\perp}(K))\, d\sigma(\xi) &\leqslant \frac{2(n-1)}{n} S(K) \int\limits_{S^{n-1}} |P_{\xi^\perp}(K)|\, d\sigma(\xi) \\ &= \frac{2(n-1)}{n}\frac{\omega_{n-1}}{n\omega_n} S(K)^2. \end{aligned} \tag{7.45}$$

Since

$$\frac{2(n-1)}{n}\frac{\omega_{n-1}}{n\omega_n} = \frac{2b_n}{n\omega_n^{\frac{1}{n}}} \approx \frac{1}{\sqrt{n}}, \tag{7.46}$$

we get the result. □

Now, let us assume that K is in the minimal surface area, isotropic or John's position, or it is symmetric and in Löwner's position. Then, we know that

$$S(K) \leqslant c_0 n\, |K|^{\frac{n-1}{n}}, \tag{7.47}$$

where $c_0 > 0$ is an absolute constant. From Theorem 7.2.10 we get the following.

Theorem 7.2.11. *Let K be a convex body in $\mathbb{R}^n$. If K is in the minimal surface area, isotropic or John's position, or it is symmetric and in Löwner's position, then*

$$|K|^{\frac{1}{n}} \int_{S^{n-1}} S(P_{\xi^\perp}(K))\, d\sigma(\xi) \leqslant c_2 \sqrt{n}\, S(K), \tag{7.48}$$

where $c_2 > 0$ is an absolute constant.

Note. If K is symmetric and in the minimal mean width position, then we get a weaker (by a $\log n$ term) result

$$|K|^{\frac{1}{n}} \int_{S^{n-1}} S(P_{\xi^\perp}(K))\, d\sigma(\xi) \leqslant c_2 \sqrt{n}(\log n)\, S(K), \tag{7.49}$$

where $c_2 > 0$ is an absolute constant.

Moving to lower bounds, our analogue of (7.19) is the following theorem.

Theorem 7.2.12. *Let K be a convex body in $\mathbb{R}^n$. Then,*

$$\int_{S^{n-1}} S(P_{\xi^\perp}(K))\, d\sigma(\xi) \geqslant c_3 \,(S(K))^{\frac{n-2}{n-1}}, \tag{7.50}$$

where $c_3 > 0$ is an absolute constant.

Proof. We write

$$\begin{aligned}
&\int_{S^{n-1}} S(P_{\xi^\perp}(K))\, d\sigma(\xi)\\
&\quad= \frac{(n-1)\omega_{n-1}}{\omega_{n-2}} \int_{S^{n-1}} \int_{S^{n-1}\cap\xi^\perp} |P_{\langle\xi,\theta\rangle^\perp}(K)|\, d\sigma_{\xi^\perp}(\theta)\, d\sigma(\xi)\\
&\quad= \frac{(n-1)\omega_{n-1}}{\omega_{n-2}} \int_{G_{n,n-2}} |P_F(K)|\, d\nu_{n,n-2}(F).
\end{aligned} \tag{7.51}$$

From the Aleksandrov inequalities it follows that

$$\begin{aligned}
&\left(\frac{1}{\omega_{n-2}} \int_{G_{n,n-2}} |P_F(K)|\, d\nu_{n,n-2}(F)\right)^{\frac{1}{n-2}}\\
&\quad\geqslant \left(\frac{1}{\omega_{n-1}} \int_{S^{n-1}} |P_{\xi^\perp}(K)|\, d\sigma(\xi)\right)^{\frac{1}{n-1}} = \left(\frac{S(K)}{n\omega_n}\right)^{\frac{1}{n-1}},
\end{aligned} \tag{7.52}$$

which gives

$$\int_{S^{n-1}} S(P_{\xi^\perp}(K))\,d\sigma(\xi) \geqslant \frac{(n-1)\omega_{n-1}}{(n\omega_n)^{\frac{n-2}{n-1}}}\,(S(K))^{\frac{n-2}{n-1}} \geqslant c_3\,(S(K))^{\frac{n-2}{n-1}}, \tag{7.53}$$

where $c_3 > 0$ is an absolute constant. □

Now, let us assume that K is in the minimal surface area, minimal mean width, isotropic, John or Löwner position. Then, we know that, e. g.

$$S(K) \leqslant c_0 n^2\,|K|^{\frac{n-1}{n}}, \tag{7.54}$$

where $c_0 > 0$ is an absolute constant. It follows that

$$(S(K))^{\frac{1}{n-1}} \leqslant (c_0 n^2)^{\frac{1}{n-2}}\,|K|^{\frac{1}{n}} \leqslant c_4|K|^{\frac{1}{n}}, \tag{7.55}$$

where $c_4 > 0$ is an absolute constant. Then,

$$|K|^{\frac{1}{n}}\,(S(K))^{\frac{n-2}{n-1}} \geqslant c_4^{-1}S(K). \tag{7.56}$$

Thus, we have proved the following.

Theorem 7.2.13. *Let K be a convex body in $\mathbb{R}^n$. If K is in the minimal surface area, minimal mean width, isotropic, John's or Löwner's position, then*

$$|K|^{\frac{1}{n}} \int_{S^{n-1}} S(P_{\xi^\perp}(K))\,d\sigma(\xi) \geqslant c_5\,S(K), \tag{7.57}$$

where $c_5 > 0$ is an absolute constant.

Note. In particular, under the assumptions of Theorem 7.2.13 we have that

$$|K|^{\frac{1}{n}}\,\max_{\xi\in S^{n-1}} S(P_{\xi^\perp}(K)) \geqslant c\,S(K). \tag{7.58}$$

Note that (7.58) is stronger than (7.19); moreover, for bounds of this type there is no need to assume that K is a projection body. In fact, our proof of Theorem 7.2.12 shows that (7.57) continues to hold as long as

$$(S(K))^{\frac{1}{n-1}} \leqslant c|K|^{\frac{1}{n}} \tag{7.59}$$

for an absolute constant $c > 0$. This is a mild condition that is satisfied not only by the classical positions but also by all *reasonable* positions of K.

7.3 Slicing inequalities for surface area

In this section, we discuss the question of whether it is possible to have a version of the slicing inequality for the surface area instead of volume. One possible version of the question can be stated as follows: Is it true that there exists a constant ζ_n depending (or not) on the dimension n so that

$$S(K) \leqslant \zeta_n |K|^{\frac{1}{n}} \max_{\xi \in S^{n-1}} S(K \cap \xi^{\perp}) \tag{7.60}$$

for every origin-symmetric convex body K in $\mathbb{R}^n$? A lower-dimensional slicing problem may be also formulated; for any $2 \leqslant k \leqslant n-1$ one may ask for a constant $\zeta_{n,k}$ such that

$$S(K) \leqslant \zeta_{n,k}^{k} |K|^{\frac{k}{n}} \max_{H \in G_{n,n-k}} S(K \cap H) \tag{7.61}$$

for every origin-symmetric convex body K in $\mathbb{R}^n$. Moreover, one may replace surface area by any other quermassintegral and pose the corresponding question.

It was proved in [110] that it is not possible to have an inequality such as (7.60).

Theorem 7.3.1. *For any $n \geqslant 2$ and any $\zeta > 0$ there exists an origin-symmetric convex body K in $\mathbb{R}^n$ such that*

$$S(K) > \zeta\, |K|^{\frac{1}{n}} \max_{\xi \in S^{n-1}} S(P_{\xi^{\perp}}(K)) \geqslant \zeta\, |K|^{\frac{1}{n}} \max_{\xi \in S^{n-1}} S(K \cap \xi^{\perp}).$$

In fact, the examples are given by ellipsoids. They are based on the next result that answers a natural question and might be useful in other situations too.

Theorem 7.3.2. *Let $\mathcal{E}$ be an origin-symmetric ellipsoid in $\mathbb{R}^n$ and write $a_1 \leqslant a_2 \leqslant \cdots \leqslant a_n$ for the lengths and $e_1, e_2, \ldots, e_n$ for the corresponding directions of its semi-axes. If $1 \leqslant k \leqslant n-1$, then for any $H \in G_{n,k}$ and any $0 \leqslant j < k$ we have that*

$$W_j(\mathcal{E} \cap F_k) \leqslant W_j(\mathcal{E} \cap H) \leqslant W_j(P_H(\mathcal{E})) \leqslant W_j(\mathcal{E} \cap E_k),$$

where $F_k = \mathrm{span}\{e_1, \ldots, e_k\}$ and $E_k = \mathrm{span}\{e_{n-k+1}, \ldots, e_n\}$. In particular, for every $\xi \in S^{n-1}$

$$S(\mathcal{E} \cap \xi^{\perp}) \leqslant S(P_{\xi^{\perp}}(\mathcal{E})) \leqslant S(\mathcal{E} \cap e_1^{\perp}).$$

The analogue of Theorem 7.3.2 for the volume of sections and projections of ellipsoids is known to be true (for a proof, see [299] and [143]). With the same notation, for all $1 \leqslant k \leqslant n-1$ one has

$$\min_{H \in G_{n,k}} |\mathcal{E} \cap H| = \min_{H \in G_{n,k}} |P_H(\mathcal{E})| = \omega_k \prod_{i=1}^{k} a_i$$

and

$$\max_{H\in G_{n,k}} |\mathcal{E}\cap H| = \max_{H\in G_{n,k}} |P_H(\mathcal{E})| = \omega_k \prod_{i=n-k+1}^{n} a_i.$$

For the proof of Theorem 7.3.2 we will use the following form of Cauchy's interlacing theorem (see [474, p. 64]).

Theorem 7.3.3. *Let A be a symmetric $n\times n$ matrix and consider the $k\times k$ matrix $B = PAP^*$, where $k \leqslant n$ and P is the orthogonal projection onto a subspace of dimension k. If the eigenvalues of A are $\lambda_1 \leqslant \lambda_2 \leqslant \cdots \leqslant \lambda_n$, and those of B are $\mu_1 \leqslant \mu_2 \leqslant \cdots \leqslant \mu_k$, then for all $i \leqslant k$ we have*

$$\lambda_i \leqslant \mu_i \leqslant \lambda_{n-k+i}.$$

Now, let $\mathcal{E}$ be an origin-symmetric ellipsoid in $\mathbb{R}^n$ and let $a_1 \leqslant \cdots \leqslant a_n$ be the lengths of its principal semi-axes. We can write $\mathcal{E} = \{x \in \mathbb{R}^n : \langle Ax, x\rangle \leqslant 1\}$, where A is an $n\times n$ symmetric positive definite matrix. The relations between the eigenvalues $\lambda_1(A) \leqslant \cdots \leqslant \lambda_n(A)$ of A and the lengths of the principal semi-axes of the ellipsoid $\mathcal{E}$ are given by

$$a_j = \frac{1}{\sqrt{\lambda_{n-j+1}(A)}}.$$

A k-dimensional section of $\mathcal{E}$ can be obtained by restriction onto a k-dimensional subspace H

$$\mathcal{E}\cap H = \{x \in H : \langle Ax, x\rangle \leqslant 1\}.$$

Let $b_1 \leqslant \cdots \leqslant b_k$ be the lengths of the principal semi-axes of $\mathcal{E}\cap H$. If $\{u_1,\ldots,u_k\}$ is an orthonormal basis of H, then we can write any $x \in H$ as

$$x = y_1u_1 + \cdots + y_ku_k$$

for some vector $y = (y_1,\ldots,y_k) \in \mathbb{R}^k$. Thus, we can write $x = Uy$, where U is an $n\times k$ matrix with columns u_i for $1 \leqslant i \leqslant k$. Using this language we can write

$$\mathcal{E}\cap H = \{y \in \mathbb{R}^k : \langle AUy, Uy\rangle \leqslant 1\} = \{y \in \mathbb{R}^k : \langle U^*AUy, y\rangle \leqslant 1\}.$$

By our previous observations we conclude that the j-th principal semi-axis of $\mathcal{E}\cap H$ is given by

$$b_j = (\lambda_{k-j+1}(U^*AU))^{-1/2}.$$

We can now use Theorem 7.3.3 for $i = k - j + 1$ to get

$$\lambda_{k-j+1}(A) \leqslant \lambda_{k-j+1}(U^*AU) \leqslant \lambda_{n-j+1}(A),$$

which implies that

$$a_j \leqslant b_j \leqslant a_{n-k+j}.$$

Therefore, we obtain the following geometric consequence of Theorem 7.3.3, which may be viewed as a generalization of Rayleigh's formula.

Lemma 7.3.4. *Let $\mathcal{E}$ be an origin-symmetric ellipsoid in $\mathbb{R}^n$ and write $a_1 \leqslant a_2 \leqslant \cdots \leqslant a_n$ for the lengths of its semi-axes. If H is a k-dimensional subspace of $\mathbb{R}^n$, then $\mathcal{E} \cap H$ is an origin-symmetric ellipsoid and its semi-axes $b_1 \leqslant b_2 \leqslant \cdots \leqslant b_k$ satisfy*

$$a_j \leqslant b_j \leqslant a_{n-k+j}$$

for all $1 \leqslant j \leqslant k$.

Proof of Theorem 7.3.2. We may assume that $\{e_1, \dots, e_n\}$ is the standard basis of $\mathbb{R}^n$ and write

$$\mathcal{E} = \left\{ x \in \mathbb{R}^n : \sum_{j=1}^{n} \frac{x_j^2}{a_j^2} \leqslant 1 \right\}.$$

Consider the ellipsoid

$$\mathcal{E}' := \left\{ x \in \mathbb{R}^n : \sum_{j=1}^{n-k+1} \frac{x_j^2}{a_{n-k+1}^2} + \sum_{j=n-k+2}^{n} \frac{x_j^2}{a_j^2} \leqslant 1 \right\}.$$

We clearly have $\mathcal{E} \subseteq \mathcal{E}'$. Then, for any k-dimensional subspace H we have that $\mathcal{E} \cap H \subseteq \mathcal{E}' \cap H$, and hence Kubota's formula implies that

$$W_j(\mathcal{E} \cap H) \leqslant W_j(\mathcal{E}' \cap H)$$

for all $0 \leqslant j < k$. On the other hand, if $b_1 \leqslant b_2 \leqslant \cdots \leqslant b_k$ are the lengths of the semi-axes of the ellipsoid $\mathcal{E}' \cap H$, then Lemma 7.3.4 shows that

$$a_{n-k+1} \leqslant b_1 \leqslant a_{n-k+1} \leqslant b_2 \leqslant a_{n-k+2} \leqslant \cdots \leqslant b_k \leqslant a_n,$$

therefore $a_{n-k+1} = b_1$ and $b_j \leqslant a_{n-k+j}$ for all $1 \leqslant j \leqslant k$. Thus, all the semi-axes of $\mathcal{E}' \cap H$ are smaller than or equal to the corresponding ones of $\mathcal{E}' \cap E_k$, which implies that

$$W_j(\mathcal{E}' \cap H) \leqslant W_j(\mathcal{E}' \cap E_k)$$

for all $1 \leqslant j \leqslant k$. Combining the above we get

$$W_j(\mathcal{E} \cap H) \leqslant W_j(\mathcal{E}' \cap H) \leqslant W_j(\mathcal{E}' \cap E_k) = W_j(\mathcal{E} \cap E_k),$$

where the last equality follows from the observation that $\mathcal{E}' \cap E_k = \mathcal{E} \cap E_k$. The proof of the inequality $W_j(\mathcal{E} \cap F_k) \leqslant W_j(\mathcal{E} \cap H)$ is similar.

For the proof of the corresponding result for projections we may use a duality argument. Given two ellipsoids $\mathcal{E}_1$ and $\mathcal{E}_2$, with semi-axes $a_1 \leqslant \cdots \leqslant a_n$ and $b_1 \leqslant \cdots \leqslant b_n$, respectively, we will write $\mathcal{E}_1 \preceq \mathcal{E}_2$ if $a_i \leqslant b_i$ for all i. Using this notation, what we have proved is that

$$\mathcal{E} \cap F_k \preceq \mathcal{E} \cap H \preceq \mathcal{E} \cap E_k$$

for every $H \in G_{n,k}$. Now, we start with the ellipsoid $\mathcal{E}^\circ$. Since the lengths of the semi-axes of $\mathcal{E}^\circ$ are the reciprocals of the ones of $\mathcal{E}$, we see that

$$\mathcal{E}^\circ \cap E_k \preceq \mathcal{E}^\circ \cap H \preceq \mathcal{E}^\circ \cap F_k$$

and hence their corresponding polars satisfy

$$P_{F_k}(\mathcal{E}) \preceq P_H(\mathcal{E}) \preceq P_{E_k}(\mathcal{E}).$$

The result follows from these observations. □

Remark 7.3.5. A formula that is related to this discussion is proved in [272]. If $\mathcal{F}$ is an ellipsoid in $\mathbb{R}^k$, then

$$W_{k-j}(\mathcal{F}) = \frac{|\mathcal{F}|}{\omega_k}\, W_j(\mathcal{F}^\circ) \tag{7.62}$$

for every $1 \leqslant j \leqslant k-1$. Let $\mathcal{E}$ be an ellipsoid in $\mathbb{R}^n$, and let $1 \leqslant k \leqslant n-1$ and $H \in G_{n,k}$. Keeping the notation E_k and F_k as above, and applying (7.62) for the ellipsoid $\mathcal{E} \cap H$, for every $1 \leqslant j \leqslant k-1$ we see that

$$\frac{W_{k-j}(\mathcal{E} \cap H)}{|\mathcal{E} \cap H|} = \frac{1}{\omega_k}\,|W_j(P_H(\mathcal{E}^\circ))| \leqslant \frac{1}{\omega_k}\,|W_j(P_{F_k}(\mathcal{E}^\circ))| = \frac{W_{k-j}(\mathcal{E} \cap F_k)}{|\mathcal{E} \cap F_k|}.$$

In other words, the ratio $W_{k-j}(\mathcal{E} \cap H)/|\mathcal{E} \cap H|$ is maximized when $H = F_k$, and similarly it is minimized when $H = E_k$. Analogously, applying (7.62) for the ellipsoid $P_H(\mathcal{E})$, for every $1 \leqslant j \leqslant k-1$ we see that

$$\frac{W_{k-j}(P_H(\mathcal{E}))}{|P_H(\mathcal{E})|} = \frac{1}{\omega_k}\,|W_j(\mathcal{E}^\circ \cap H)| \leqslant \frac{1}{\omega_k}\,|W_j(\mathcal{E}^\circ \cap F_k)| = \frac{W_{k-j}(P_{F_k}\mathcal{E})}{|P_{F_k}(\mathcal{E})|}.$$

In other words, the ratio $W_{k-j}(P_H(\mathcal{E}))/|P_H(\mathcal{E})|$ is also maximized when $H = F_k$, and similarly it is minimized when $H = E_k$.

We move on now to the proof of Theorem 7.3.1 and of the more general Theorem 7.3.7.

Proof of Theorem 7.3.1. We shall use the next formula of Rivin (see [479]): If $\mathcal{E}$ is an ellipsoid in $\mathbb{R}^n$ with semi-axes $a_1 \leqslant \cdots \leqslant a_n$ in the directions of $e_1, \ldots, e_n$, then

$$S(\mathcal{E}) = n\,|\mathcal{E}| \int_{S^{n-1}} \left(\sum_{i=1}^{n} \frac{\xi_i^2}{a_i^2} \right)^{1/2} d\sigma(\xi).$$

Recall also that for any norm $\|\cdot\|$ on $\mathbb{R}^n$ we have that

$$\mathbb{E}\|G\| = d_n \int_{S^{n-1}} \|\xi\|\, d\sigma(\xi),$$

where G is a standard gaussian random vector and $d_n \sim \sqrt{n}$.

Now, assume that there exists a constant $\zeta_n > 0$ such that we have the following inequality for ellipsoids:

$$S(\mathcal{E}) \leqslant \zeta_n |\mathcal{E}|^{1/n} \max_{\xi \in S^{n-1}} S(\mathcal{E} \cap \xi^\perp). \tag{7.63}$$

From Theorem 7.3.2, we know that the maximum is attained for the section $\mathcal{E} \cap e_1^\perp$. Then, we have

$$\max_{\xi \in S^{n-1}} S(\mathcal{E} \cap \xi^\perp) = S(\mathcal{E} \cap e_1^\perp) = (n-1)\,|\mathcal{E} \cap e_1^\perp| \int_{S^{n-2}} \left(\sum_{i=2}^{n} \frac{\xi_i^2}{a_i^2} \right)^{1/2} d\sigma(\xi).$$

We may assume that $\prod_{i=1}^n a_i = 1$. Then, we can rewrite (7.63) as

$$n\omega_n \frac{1}{d_n} \mathbb{E}\left(\left(\sum_{i=1}^{n} \frac{g_i^2}{a_i^2} \right)^{1/2} \right) \leqslant \zeta_n \omega_n^{1/n} (n-1)\omega_{n-1} \frac{1}{a_1} \frac{1}{d_{n-1}} \mathbb{E}\left(\left(\sum_{i=2}^{n} \frac{g_i^2}{a_i^2} \right)^{1/2} \right).$$

Since $x \mapsto (\sum_{i=1}^n \frac{x_i^2}{a_i^2})^{1/2}$ is a seminorm, using Hölder's and Khintchine's inequalities for this seminorm in Gauss space we get

$$\frac{\mathbb{E}((\sum_{i=1}^n \frac{g_i^2}{a_i^2})^{1/2})}{\mathbb{E}((\sum_{i=2}^n \frac{g_i^2}{a_i^2})^{1/2})} \geqslant c \left(\frac{\mathbb{E}(\sum_{i=1}^n \frac{g_i^2}{a_i^2})}{\mathbb{E}(\sum_{i=2}^n \frac{g_i^2}{a_i^2})} \right)^{1/2} = c \left(\frac{\sum_{i=1}^n \frac{1}{a_i^2}}{\sum_{i=2}^n \frac{1}{a_i^2}} \right)^{1/2}$$

and hence

$$\zeta_n \geqslant c \frac{n\omega_n^{\frac{n-1}{n}}}{(n-1)\omega_{n-1}} \frac{d_{n-1}}{d_n} a_1 \left(\frac{\sum_{i=1}^n \frac{1}{a_i^2}}{\sum_{i=2}^n \frac{1}{a_i^2}} \right)^{1/2}$$

$$= c\frac{n\omega_n^{\frac{n-1}{n}}}{(n-1)\omega_{n-1}}\frac{d_{n-1}}{d_n}\left(\frac{1+\sum_{i=2}^n \frac{a_1^2}{a_i^2}}{\sum_{i=2}^n \frac{1}{a_i^2}}\right)^{1/2}.$$

Now, choose $a_2 = \cdots = a_n = r$ and $a_1 = r^{-(n-1)}$. Then,

$$\left(\frac{1+\sum_{i=2}^n \frac{a_1^2}{a_i^2}}{\sum_{i=2}^n \frac{1}{a_i^2}}\right)^{1/2} = \left(\frac{1+\frac{n-1}{r^{2n}}}{\frac{n-1}{r^2}}\right)^{1/2} = \left(\frac{1}{r^{2n-2}}+\frac{r^2}{n-1}\right)^{1/2} \to \infty,$$

as $r \to \infty$. So, we arrive at a contradiction, i. e., there can be no upper bound for ζ_n. □

Remark 7.3.6. Let us note here that a reverse inequality can be obtained at least when K is in some of the classical positions. In the previous section, we saw that for any convex body K in $\mathbb{R}^n$ and any $\xi \in S^{n-1}$ we have

$$\frac{S(P_{\xi^\perp}(K))}{|P_{\xi^\perp}(K)|} \leqslant \frac{2(n-1)}{n}\frac{S(K)}{|K|},$$

therefore

$$|K| \max_{\xi\in S^{n-1}} S(P_{\xi^\perp}(K)) \leqslant \frac{2(n-1)}{n} S(K) \max_{\xi\in S^{n-1}} |P_{\xi^\perp}(K)|.$$

Since we trivially have

$$|P_{\xi^\perp}(K)| = \frac{1}{2}\int_{S^{n-1}} |\langle \xi, \theta\rangle| \, dS(K,\theta) \leqslant \frac{1}{2}S(K),$$

we see that

$$|K| \max_{\xi\in S^{n-1}} S(P_{\xi^\perp}(K)) \leqslant \frac{n-1}{n}(S(K))^2.$$

On the other hand, we have already mentioned that if K is in some classical position, then a reverse isoperimetric inequality of the form $S(K) \leqslant cn|K|^{\frac{n-1}{n}}$ holds true (with an extra $\log n$-term in the minimal mean width position). Combining the above we see that, in this case

$$|K|^{\frac{1}{n}} \max_{\xi\in S^{n-1}} S(P_{\xi^\perp}(K)) \leqslant cn\, S(K)$$

for some absolute constant $c > 0$.

For the more general question, where surface area is replaced by any quermassintegral, we may exploit a formula from [431] for the j-quermassintegrals of ellipsoids of revolution, i. e., ellipsoids of the form

$$\mathcal{E}_{r,s} = \left\{ x \in \mathbb{R}^m : \sum_{i=1}^{m-1} \frac{x_i^2}{r^2} + \frac{x_m^2}{s^2} \leqslant 1 \right\}.$$

For every $j = 0, 1, \ldots, m$ one has

$$W_j(\mathcal{E}_{r,s}) = \omega_m r^{m-j} \int_{S^{m-1}} \left(\frac{s^2}{r^2} \sum_{i=1}^{m-j} \theta_i^2 + \sum_{i=m-j+1}^{m} \theta_i^2 \right)^{1/2} d\sigma(\theta). \tag{7.64}$$

Theorem 7.3.7. *Let $n \geqslant 2$, $1 \leqslant k \leqslant n$ and $0 \leqslant j \leqslant n-k-1$. For every $\zeta > 0$ there exists a convex body K in $\mathbb{R}^n$ such that*

$$W_j(K) > \zeta\, |K|^{\frac{k}{n}} \max_{F \in G_{n,n-k}} W_j(P_F(K)) \geqslant \zeta\, |K|^{\frac{k}{n}} \max_{F \in G_{n,n-k}} W_j(K \cap F).$$

Proof. Assume that for some $n \geqslant 2$, $1 \leqslant k \leqslant n$, and $0 \leqslant j \leqslant n-k$ there exists a constant $C(n,k,j) > 0$ such that

$$W_j(K) \leqslant C(n,k,j) |K|^{\frac{k}{n}} \max_{F \in G_{n,n-k}} W_j(K \cap F). \tag{7.65}$$

Then, for any $r > 1 > s$ with $r^{n-1}s = 1$ consider the ellipsoid $\mathcal{E}_{r,s}$. Recall that

$$\max_{F \in G_{n,n-k}} W_j(\mathcal{E}_{r,s} \cap F) = W_j(\mathcal{E}_{r,s} \cap F_{n-k}),$$

where $F_{n-k} = \mathrm{span}\{e_1, \ldots, e_{n-k}\}$. Note that $|\mathcal{E}_{r,s}| = \omega_n$ and that $\mathcal{E}_{r,s} \cap F_{n-k}$ is a ball of radius r. Using (7.64) and assuming that (7.65) holds true, we see that

$$\omega_n r^{n-j} \int_{S^{n-1}} \left(\frac{s^2}{r^2} \sum_{i=1}^{n-j} \theta_i^2 + \sum_{i=m-j+1}^{m} \theta_i^2 \right)^{1/2} d\sigma(\theta) \leqslant C(n,k,j) \omega_n^{\frac{k}{n}} \omega_{n-k} r^{n-k-j}.$$

Since

$$\begin{aligned}
&\int_{S^{n-1}} \left(\frac{s^2}{r^2} \sum_{i=1}^{n-j} \theta_i^2 + \sum_{i=m-j+1}^{m} \theta_i^2 \right)^{1/2} d\sigma(\theta) \\
&\approx \left(\int_{S^{n-1}} \left(\frac{s^2}{r^2} \sum_{i=1}^{n-j} \theta_i^2 + \sum_{i=m-j+1}^{m} \theta_i^2 \right) d\sigma(\theta) \right)^{1/2} \\
&= \left(\frac{n-j}{n} \frac{s^2}{r^2} + \frac{j}{n} \right)^{1/2} = \left(\frac{n-j}{n} \frac{1}{r^{2n}} + \frac{j}{n} \right)^{1/2},
\end{aligned}$$

we must have

$$r^k\left(\frac{n-j}{n}\frac{1}{r^{2n}}+\frac{j}{n}\right)^{1/2} \leqslant c_1 C(n,k,j)\frac{\omega_{n-k}}{\omega_n^{\frac{n-k}{n}}}$$

for every $r > 1$, which leads to a contradiction if we let $r \to \infty$. □

7.4 Average volume of sections of convex bodies

Let K be a centered convex body in $\mathbb{R}^n$. We denote by $\mathrm{as}(K)$ the average volume of the central hyperplane sections of K

$$\mathrm{as}(K) = \int_{S^{n-1}} |K \cap \xi^\perp|\, d\sigma(\xi). \tag{7.66}$$

If E is an m-dimensional subspace of $\mathbb{R}^n$, then $\mathrm{as}(K \cap E)$ stands for the average $(m-1)$-dimensional volume of hyperplane sections of the body $K \cap E$ in $\mathbb{R}^m$. More generally, for any $1 \leqslant r \leqslant n-1$ we define

$$\mathrm{as}_r(K) = \int_{G_{n,n-r}} |K \cap E|\, d\nu_{n,n-r}(E). \tag{7.67}$$

Thus, $\mathrm{as}_r(K)$ is the average volume of r-codimensional central sections of K; note that $\mathrm{as}(K) = \mathrm{as}_1(K)$.

It was proved in [314] that if K is an intersection body in $\mathbb{R}^n$, then

$$\mathrm{as}(K) \leqslant b_{n,1}\, |K|^{\frac{1}{n}} \max_{\xi \in S^{n-1}} \mathrm{as}(K \cap \xi^\perp), \tag{7.68}$$

where

$$b_{n,1} := \omega_{n-1}/(\omega_{n-2}\omega_n^{\frac{1}{n}}) \approx 1.$$

Note that (7.68) is sharp: it becomes an equality for $K = B_2^n$. Inequality (7.68) appeared in [314] as an application of the following stability result in the Busemann–Petty problem: if K, L are symmetric star bodies in $\mathbb{R}^n$ with K an intersection body, $\varepsilon > 0$, and $|K \cap \xi^\perp| \leqslant |L \cap \xi^\perp| + \varepsilon$ for every $\xi \in S^{n-1}$, then $|K|^{\frac{n-1}{n}} \leqslant |L|^{\frac{n-1}{n}} + c\varepsilon$, where c is an absolute constant.

The article [112] discusses similar inequalities for the average volume of hyperplane sections of an arbitrary centered convex body K in $\mathbb{R}^n$.

Question 7.4.1. Let $1 \leqslant k \leqslant n-2$ and define $\phi_{n,k}$ as the smallest constant $\phi > 0$ such that for every centered, convex body K in $\mathbb{R}^n$ we have

$$\mathrm{as}(K) \leqslant \phi^k |K|^{\frac{k}{n}} \max_{E \in G_{n,n-k}} \mathrm{as}(K \cap E). \tag{7.69}$$

Is it true that $\sup_{n,k} \phi_{n,k} < \infty$?

Question 7.4.1 is closely related to the isotropic constant problem that we discussed in Chapter 2. We shall present two methods, leading to Theorem 7.4.2 and Theorem 7.4.5, which are independent. The first method allows us to work with (not necessarily symmetric) centered convex bodies and provides an affirmative answer to Question 7.4.1.

Theorem 7.4.2. *Let K be a centered, convex body in $\mathbb{R}^n$. Then, for every $1 \leqslant k \leqslant n-2$*

$$\mathrm{as}(K) \leqslant c_1^k\, |K|^{\frac{k}{n}} \max_{E\in G_{n,n-k}} \mathrm{as}(K\cap E), \tag{7.70}$$

where $c_1 > 0$ is an absolute constant.

The proof of Theorem 7.4.2 is based on the following fact.

Theorem 7.4.3. *Let K be a centered, convex body in $\mathbb{R}^n$. For every $1 \leqslant k \leqslant n-1$ and $E \in G_{n,n-k}$ we have*

$$|K\cap E|\, \mathrm{as}(K) \leqslant c_2^k\, \mathrm{as}(K\cap E)\, |K|, \tag{7.71}$$

where $c_2 > 0$ is an absolute constant.

Proof. We need the next well-known properties of the parameter M. If D is a symmetric convex body in $\mathbb{R}^m$, then for every $1 \leqslant s \leqslant m-1$ and $F \in G_{m,s}$ we have that

$$M(D\cap F) = \int_{S_F} \|\xi\|_D\, d\sigma_F(\xi) \leqslant c_1\sqrt{m/s} \int_{S^{m-1}} \|\xi\|_D\, d\sigma(\xi) = c_1\sqrt{m/s}\, M(D), \tag{7.72}$$

where $c_1 > 0$ is an absolute constant. It is also known that

$$\int_{S^{m-1}} \rho_D(\theta)\, d\sigma(\theta) = \int_{S^{m-1}} \|\theta\|_D^{-1}\, d\sigma(\theta) \approx \frac{1}{M(D)}. \tag{7.73}$$

For a proof of (7.72) and (7.73) see [27, Section 5.2.1] and [27, Theorem 5.8.7], respectively.

We consider an orthonormal basis $\{e_1, \dots, e_k\}$ of $E^\perp$ and any unit vector $\xi \in E$. From Theorem 7.1.8 we know that

$$|K\cap E|\, |K\cap \xi^\perp| \leqslant c_0^{k+1} |K\cap E\cap \xi^\perp|\, |K|. \tag{7.74}$$

Integrating (7.74) with respect to $\xi \in S_E$, we see that

$$\begin{aligned} |K\cap E| \int_{S_E} |K\cap \xi^\perp|\, d\sigma_E(\xi) &\leqslant c_0^{k+1} \int_{S_E} |(K\cap E)\cap \xi^\perp|\, d\sigma_E(\xi)\, |K| \\ &= c_0^{k+1}\, \mathrm{as}(K\cap E)\, |K|. \end{aligned} \tag{7.75}$$

Applying (7.73) for the symmetric convex body $IK \cap E$ yields

$$\int_{S_E} |K \cap \xi^\perp| d\sigma_E(\xi) = \int_{S_E} \rho_{IK}(\xi)\, d\sigma_E(\xi) \approx \frac{1}{M(IK \cap E)} \tag{7.76}$$

and hence, using (7.72) with $m = n$ and $s = n - k$ and applying (7.73) for the body IK this time, we obtain

$$\begin{aligned} \int_{S_E} |K \cap \xi^\perp| d\sigma_E(\xi) &\geqslant \frac{c\sqrt{n-k}}{\sqrt{n}} \frac{1}{M(IK)} \approx \frac{\sqrt{n-k}}{\sqrt{n}} \int_{S^{n-1}} \rho_{IK}(\xi)\, d\sigma(\xi) \\ &= \frac{\sqrt{n-k}}{\sqrt{n}} \int_{S^{n-1}} |K \cap \xi^\perp|\, d\sigma(\xi) = \frac{\sqrt{n-k}}{\sqrt{n}} \mathrm{as}(K). \end{aligned} \tag{7.77}$$

Therefore,

$$|K \cap E|\, \mathrm{as}(K) \leqslant \frac{c_1\sqrt{n}}{\sqrt{n-k}} c_0^{k+1}\, \mathrm{as}(K \cap E)\, |K| \leqslant c_2^k\, \mathrm{as}(K \cap E)\, |K| \tag{7.78}$$

for every $E \in G_{n,n-k}$. □

Proof of Theorem 7.4.2. Let K be a centered convex body in $\mathbb{R}^n$ and fix $1 \leqslant k \leqslant n-1$. From Theorem 7.4.3 we know that for every $E \in G_{n,n-k}$ we have

$$|K \cap E|\, \mathrm{as}(K) \leqslant c_2^k\, \mathrm{as}(K \cap E)\, |K|, \tag{7.79}$$

where $c_2 > 0$ is an absolute constant. Therefore,

$$\max_{E \in G_{n,n-k}} |K \cap E|\, \mathrm{as}(K) \leqslant c_2^k \max_{E \in G_{n,n-k}} \mathrm{as}(K \cap E)\, |K|. \tag{7.80}$$

On the other hand, using (3.20) (the lower bound for the dual quermassintegrals of K) we see that

$$\max_{E \in G_{n,n-k}} |K \cap E| \geqslant \left(\int_{G_{n,n-k}} |K \cap E|^n\, d\nu_{n,n-k}(E) \right)^{\frac{1}{n}} \geqslant c_4^k\, |K|^{\frac{n-k}{n}}. \tag{7.81}$$

Going back to (7.80), we see that

$$c_4^k |K|^{\frac{n-k}{n}}\, \mathrm{as}(K) \leqslant c_2^k |K| \max_{E \in G_{n,n-k}} \mathrm{as}(K \cap E) \tag{7.82}$$

and this proves the theorem. □

The second method allows us to work with origin-symmetric (not necessarily convex) star bodies and to consider even, continuous densities in place of volume. Writing the volume in (7.69) in polar coordinates puts this question in the form of a "maximal" Hölder-type inequality for the radial function ρ_K of the body K

$$\int_{S^{n-1}} \rho_K^{n-1}(\theta)\,d\theta \leqslant c^k \Big(\frac{1}{n}\int_{S^{n-1}} \rho_K^n(\theta)\,d\theta\Big)^{\frac{k}{n}} \max_{E\in G_{n,n-k}} \int_{S^{n-1}\cap E} \rho_K^{n-k-1}(\theta)\,d\theta, \tag{7.83}$$

where we use $d\theta$ for the non-normalized rotationally invariant measure on the sphere, and c is an absolute constant.

We shall generalize (7.68) using as a parameter the outer volume ratio distance $d_{\mathrm{ovr}}(K, \mathcal{BP}_k^n)$ of an origin-symmetric star body K to the class $\mathcal{BP}_k^n$ of generalized k-intersection bodies. Our estimates are based on the next more general theorem that is valid for the larger class of origin-symmetric star bodies in $\mathbb{R}^n$ and for any even, continuous density on S^{n-1}.

Theorem 7.4.4. *Let $1 \leqslant k < n-1$, let K be an origin-symmetric star body in $\mathbb{R}^n$, and let g be a non-negative even, continuous function on S^{n-1}. Then,*

$$\int_{S^{n-1}} \rho_K^{n-1}(\theta)g(\theta)\,d\theta \leqslant c_{n,k}^k d_{\mathrm{ovr}}^k(K, \mathcal{BP}_k^n)|K|^{\frac{k}{n}} \max_{E\in G_{n,n-k}} \int_{S^{n-1}\cap E} \rho_K^{n-k-1}(\theta)g(\theta)\,d\theta, \tag{7.84}$$

$$c_{n,k}^k = \frac{n\omega_n^{\frac{n-k}{n}}}{(n-k)\omega_{n-k}}$$

and one can check that $c_{n,k} \simeq 1$.

Proof. In what follows, the radial sum $K\tilde{+}D$ of two star bodies K and D is defined by

$$\rho_{K\tilde{+}D} = \rho_K + \rho_D.$$

Let $\varepsilon > 0$. Let f be a non-negative, even, continuous function on $\mathbb{R}^n$ such that $f(\rho_K(\theta)\theta) = g(\theta)$ for all $\theta \in S^{n-1}$. For every $E \in G_{n,n-k}$, we have

$$\begin{aligned} &\int_{(K\tilde{+}\varepsilon B_2^n)\cap E} f(x)\,dx - \int_{K\cap E} f(x)\,dx \\ &\quad\leqslant \max_{F\in G_{n,n-k}} \Big(\int_{(K\tilde{+}\varepsilon B_2^n)\cap F} f(x)\,dx - \int_{K\cap F} f(x)\,dx\Big). \end{aligned} \tag{7.85}$$

Note that $\rho_{K\tilde{+}\varepsilon B_2^n} = \rho_K + \varepsilon$. Expressing the integrals in polar coordinates we get

$$\begin{aligned} &R_{n-k}\Big(\int_{\rho_K(\cdot)}^{\rho_K(\cdot)+\varepsilon} r^{n-k-1} f(r\cdot)\,dr\Big)(E) \\ &\quad\leqslant \max_{F\in G_{n,n-k}} \Big(\int_{S^{n-1}\cap F} \int_{\rho_K(\theta)}^{\rho_K(\theta)+\varepsilon} r^{n-k-1} f(r\theta)\,dr\,d\theta\Big). \end{aligned} \tag{7.86}$$

Let $D \in \mathcal{BP}_k^n$ such that $K \subset D$. Integrating the latter inequality in E over $G_{n,n-k}$ equipped with the measure μ_D corresponding to D, we obtain

$$\int_{S^{n-1}} \rho_D^k(\theta) \int_{\rho_K(\theta)}^{\rho_K(\theta)+\varepsilon} r^{n-k-1} f(r\theta)\,dr\,d\theta$$
$$\leqslant \mu_D(G_{n,n-k}) \cdot \max_{F\in G_{n,n-k}} \left(\int_{S^{n-1}\cap F} \int_{\rho_K(\theta)}^{\rho_K(\theta)+\varepsilon} r^{n-k-1} f(r\theta)\,dr\,d\theta \right). \tag{7.87}$$

We divide both sides by ε and send ε to zero. Note that we can interchange the limit with the maximum, because the convergence is uniform with respect to F. Thus, we get

$$\int_{S^{n-1}} \rho_D^k(\theta)\rho_K^{n-k-1}(\theta) f(\rho_K(\theta)\theta)\,d\theta$$
$$\leqslant \mu_D(G_{n,n-k}) \cdot \max_{F\in G_{n,n-k}} \left(\int_{S^{n-1}\cap F} \rho_K^{n-k-1}(\theta) f(\rho_K(\theta)\theta)\,d\theta \right). \tag{7.88}$$

Since $K \subset D$, the integral on the left-hand side can be estimated from below by

$$\int_{S^{n-1}} \rho_K^{n-1}(\theta) f(\rho_K(\theta)\theta)\,d\theta.$$

To estimate $\mu_D(G_{n,n-k})$ from above, we observe that $1 = \frac{1}{\kappa_{n-k}} R_{n-k}\mathbf{1}(E)$ for every $E \in G_{n,n-k}$, and apply Hölder's inequality

$$\mu_D(G_{n,n-k}) = \frac{1}{\kappa_{n-k}} \int_{G_{n,n-k}} R_{n-k}\mathbf{1}(E)\,d\mu_D(E) = \frac{1}{\kappa_{n-k}} \int_{S^{n-1}} \|\theta\|_D^{-k}\,d\theta$$
$$\leqslant \frac{1}{\kappa_{n-k}} \kappa_n^{\frac{n-k}{n}} \left(\int_{S^{n-1}} \|\theta\|_D^{-n}\,d\theta \right)^{\frac{k}{n}} = \frac{1}{\kappa_{n-k}} \kappa_n^{\frac{n-k}{n}} n^{\frac{k}{n}} |D|^{\frac{k}{n}}. \tag{7.89}$$

These estimates show that

$$\int_{S^{n-1}} \rho_K^{n-1}(\theta) f(\rho_K(\theta)\theta)\,d\theta$$
$$\leqslant \frac{1}{\kappa_{n-k}} \kappa_n^{\frac{n-k}{n}} n^{\frac{k}{n}} |D|^{\frac{k}{n}} \max_{F\in G_{n,n-k}} \left(\int_{S^{n-1}\cap F} \rho_K^{n-k-1}(\theta) f(\rho_K(\theta)\theta)\,d\theta \right). \tag{7.90}$$

Finally, we choose D so that $|D|^{1/n} \leqslant (1+\delta) d_{\mathrm{ovr}}(K, \mathcal{BP}_k^n)|K|^{1/n}$, and then send δ to zero. □

Choosing $g \equiv 1$ we see that (7.84) implies the following answer to Question 7.4.1 in the case of origin-symmetric star bodies.

Theorem 7.4.5. *Let $1 \leqslant k \leqslant n-2$, and let K be an origin-symmetric star body in $\mathbb{R}^n$. Then,*

$$\mathrm{as}(K) \leqslant b_{n,k}^k d_{\mathrm{ovr}}^k(K, \mathcal{BP}_k^n)\,|K|^{\frac{k}{n}} \max_{E\in G_{n,n-k}} \mathrm{as}(K\cap E), \tag{7.91}$$

where

$$b_{n,k}^k = \frac{\omega_{n-1}}{\omega_{n-k-1}\omega_n^{\frac{k}{n}}}$$

and one can check that $b_{n,k} \simeq 1$. In other words, $\phi_{n,k} \leqslant c\, d_{\mathrm{ovr}}(K, \mathcal{BP}_k^n)$.

Using the available bounds for $d_{\mathrm{ovr}}(K, \mathcal{BP}_k^n)$ we see that $\phi_{n,k}$ remains bounded as long as k is proportional to n. More precisely, we have the following.

Theorem 7.4.6. *For every origin-symmetric convex body K, for every $1 \leqslant k \leqslant n-1$ and every even non-negative continuous function g on S^{n-1}*

$$\int_{S^{n-1}} \rho_K^{n-1}(\theta)g(\theta)\,d\theta \leqslant (c_1 h(n/k))^k\,|K|^{\frac{k}{n}} \max_{E\in G_{n,n-k}} \int_{S^{n-1}\cap E} \rho_K^{n-k-1}(\theta)g(\theta)\,d\theta, \tag{7.92}$$

where $c_1 > 0$ is an absolute constant and $h(t) = \sqrt{t}\log(et)$, $t \geqslant 1$. In particular,

$$\phi_{n,k} \leqslant c_1\sqrt{n/k}\,\log(en/k). \tag{7.93}$$

Theorem 7.4.4 also allows us to prove an analogue of Theorem 7.4.5 for the quantities $\mathrm{as}_r(K)$.

Theorem 7.4.7. *Let $1 \leqslant k < n-2$ and $1 \leqslant r < n-k$. For any origin-symmetric star body K in $\mathbb{R}^n$ we have*

$$\mathrm{as}_r(K) \leqslant \phi_{n,k,r}^k d_{\mathrm{ovr}}^k(K, \mathcal{BP}_k^n)\,|K|^{\frac{k}{n}} \max_{E\in G_{n,n-k}} \mathrm{as}_r(K\cap E), \tag{7.94}$$

where

$$\phi_{n,k,r}^k = \frac{\omega_{n-r}}{\omega_{n-k-r}\omega_n^{\frac{k}{n}}}$$

and one can check that $\phi_{n,k,r} \simeq \sqrt{\frac{n}{n-r}}$.

Proof. We choose $g(\theta) = \rho_K^{-r+1}(\theta)$ in Theorem 7.4.4 to get

$$\int_{S^{n-1}} \rho_K^{n-r}(\theta)\,d\theta \leqslant \tau_{n,k}\, d_{\mathrm{ovr}}^k(K, \mathcal{BP}_k^n)\,|K|^{\frac{k}{n}} \max_{E\in G_{n,n-k}} \int_{S^{n-1}\cap E} \rho_K^{n-k-r}(\theta)\,d\theta. \tag{7.95}$$

Then, we apply the formula

$$\mathrm{as}_r(K) = \omega_{n-r} \int_{S^{n-1}} \rho_K^{n-r}(\theta)d\sigma(\theta), \tag{7.96}$$

which generalizes (7.92) and is easily verified in the same way. □

7.5 Further reading

7.5.1 Inequalities for the quermassintegrals of sections of convex bodies

Liakopoulos [378] explores the possibility to obtain some positive results for variants of the question whether one can have an inequality of the form

$$W_j(K) \leqslant \alpha_{n,k,j}^k |K|^{\frac{k}{n}} \max_{F\in G_{n,n-k}} W_j(K\cap F) \tag{7.97}$$

for $j \geqslant 0$ and $k \geqslant 1$ with $k+j \leqslant n-1$, which we studied in Section 7.3. He shows that if K is a centered convex body in $\mathbb{R}^n$, then, for every $j \geqslant 0$, $k \geqslant 1$ with $k+j \leqslant n-1$

$$\begin{aligned} a_{n,k,j}\left(\frac{n-k-j+1}{n+1}\right)^{n-k-j} W_j(K) &\leqslant \int_{G_{n,n-k}} |P_{F^\perp}(K)| W_j(K\cap F)\, d\nu_{n,n-k}(F) \\ &\leqslant a_{n,k,j}\binom{n-j}{k} W_j(K), \end{aligned}$$

where $a_{n,k,j} = \frac{\omega_{n-k}\omega_{n-j}}{\omega_{n-k-j}\omega_n}$.

Using the left-hand side inequality, one immediately gets

$$a_{n,k,j}\left(\frac{n-k-j+1}{n+1}\right)^{n-k-j} W_j(K) \leqslant s_k(K)\cdot \max_{F\in G_{n,n-k}} W_j(K\cap F),$$

where

$$s_k(K) := \int_{G_{n,k}} |P_F(K)|\, d\nu_{n,k}(F).$$

By Kubota's formula, we have $s_k(K) = \frac{\omega_k}{\omega_n} W_{n-k}(K)$, which yields that if K is a centered convex body in $\mathbb{R}^n$, then, for every $j \geqslant 0$, $k \geqslant 1$ with $k+j \leqslant n-1$

$$\gamma_{n,k,j}\left(\frac{n-k-j+1}{n+1}\right)^{n-k-j} W_j(K) \leqslant W_{n-k}(K)\cdot \max_{F\in G_{n,n-k}} W_j(K\cap F), \tag{7.98}$$

where $\gamma_{n,k,j} = \frac{\omega_{n-k}\omega_{n-j}}{\omega_{n-k-j}\omega_k}$.

This estimate, which is in the spirit of the results from [110], shows that we can have some version of (7.97) if we make the additional assumption that the natural parameter $W_{n-k}^{1/k}(K)/|K|^{1/n}$ of the body is well bounded.

A different variant of (7.98) is obtained in [110], which involves the parameter

$$t(K) = \left(\frac{|K|}{|r(K)B_2^n|}\right)^{\frac{1}{n}},$$

where $r(K)$ is the inradius of K, i. e., the largest value of $r > 0$ for which there exists $x_0 \in K$ such that $x_0 + rB_2^n \subseteq K$. It is proved in [110] that if K is a convex body in $\mathbb{R}^n$ with $0 \in \operatorname{int}(K)$ then, for all $j, k \geqslant 1$ with $k + j \leqslant n - 1$ we have that

$$\frac{W_j(K)}{|K|} \leqslant t(K)^j \max_{F \in G_{n,n-k}} \frac{W_j(K \cap F)}{|K \cap F|}. \tag{7.99}$$

In [378], the inequality (7.99) is improved in two ways. First, $\frac{W_j(K)}{|K|}$ is compared to the average of $\frac{W_j(K\cap F)}{|K\cap F|}$ over all $F \in G_{n,n-k}$. Second, it is shown that the two quantities are equivalent up to constants depending only on the dimensions n, k, and j, with no dependence on any other parameter of the body: If K is a centrally symmetric convex body in $\mathbb{R}^n$, then, for any $j \geqslant 0$, $k \geqslant 1$ with $k + j \leqslant n - 1$

$$a_{n,k,j}\binom{n}{k}^{-1}\frac{W_j(K)}{|K|} \leqslant \int_{G_{n,n-k}} \frac{W_j(K \cap F)}{|K \cap F|}\, d\nu_{n,n-k}(F) \leqslant a_{n,k,j}\binom{n-j}{k}\frac{W_j(K)}{|K|}.$$

The same paper studies a variant of the slicing problem. It was proved in [110] that if K is a convex body in $\mathbb{R}^n$, then

$$S(K) \leqslant A_n S(K)^{\frac{1}{n-1}} \max_{\xi \in S^{n-1}} S(K \cap \xi^{\perp}), \tag{7.100}$$

where $A_n > 0$ is a constant depending only on n. Actually, the result is first proved for an arbitrary ellipsoid and then it is extended to any convex body, using John's theorem. Here, a direct proof is given for a more general result, showing that an inequality analogous to (7.100) holds for any k and j, where k is the codimension of the subspaces and j is the order of the quermassintegral that we consider. Note that (7.100) corresponds to the case $k = j = 1$ of the next theorem: If K is a centrally symmetric convex body in $\mathbb{R}^n$, then, for every $j \geqslant 0$, $k \geqslant 1$ with $k + j \leqslant n - 1$

$$W_j(K)^{n-k-j} \leqslant a_{n,k,j} \max_{F \in G_{n,n-k}} W_j(K \cap F)^{n-j},$$

where $a_{n,k,j} > 0$ is a constant depending only on n, k and j.

This theorem is a consequence of another double-sided inequality. Using the generalized Blaschke–Petkantschin formula and integral-geometric results of Dann, Paouris, and Pivovarov from [146] Liakopoulos shows that if K is a centrally symmetric convex body in $\mathbb{R}^n$, then for all $j \geqslant 0$, $k \geqslant 1$ with $k + j \leqslant n - 1$ we have that

$$c_{n,k,j}|K|^{n-k} W_{k+j}(K) \leqslant \int_K \cdots \int_K W_j^{(n-k)}(\mathrm{conv}\{0, x_1, \ldots, x_{n-k}\})\, dx_{n-k} \cdots dx_1$$
$$\leqslant \delta_{n,k,j}|K|^{n-k} W_{k+j}(K),$$

where $W_j^{(n-k)}$ denotes the j-th quermassintegral in dimension $(n-k)$ and $c_{n,k,j}$, $\delta_{n,k,j}$ are constants depending only on n, k, and j.

8 Discrete slicing inequalities

In this chapter, we explore the interface between classical convex geometry and discrete structures, focusing on how properties of convex bodies manifest when intersected with the integer lattice. While many phenomena in the continuous setting are well understood and intuitive, their discrete counterparts often exhibit subtle and sometimes counterintuitive behavior, revealing deep connections between convexity, lattice structure, and number theory. The foundational results come from the Geometry of Numbers, with Minkowski's theorems on lattice points in origin-symmetric convex bodies and the theory of successive minima providing the basic tools to quantify how convex bodies interact with the lattice at various scales.

We begin by introducing general lattices, their duals, and related concepts such as lattice planes, lattice hyperplanes, and the lattice width of a convex body, which measures the minimal number of lattice planes intersecting the body. These notions are closely tied to the covering radius, which describes how a scaled convex body combined with the lattice can fill space, leading naturally to the flatness constant. The flatness constant captures the inherent discreteness of convex bodies and has been a focal point of recent advances, especially in connection with estimating lattice point counts and understanding discrete analogues of classical geometric inequalities.

Central to the chapter is the study of discrete analogues of slicing problems. The discrete slicing problem asks whether the number of lattice points in a convex body can be bounded in terms of the maximum number of lattice points in a hyperplane section and the volume of the body. Various approaches are explored, ranging from two-dimensional cases, where Minkowski's first theorem and Pick's formula give explicit bounds, to higher-dimensional settings where the discrete John theorem and probabilistic Fourier-analytic methods provide estimates with explicit dependence on dimension and volume. These methods highlight the interplay between geometric shape, lattice structure, and probabilistic techniques.

We also present discrete versions of the Brunn–Minkowski theory. Unlike the continuous case, simple translations of classical inequalities fail in $\mathbb{Z}^n$, requiring new combinatorial or functional formulations. Strategies include the use of initial segments of integer sets, set extensions, and sup-convolutions to restore convex-like behavior in the discrete setting. These discrete Brunn–Minkowski inequalities link lattice point enumeration with classical volume growth, showing that appropriately augmented Minkowski sums or extensions of sets can reproduce inequalities analogous to those in the continuous setting.

A further theme is the study of discrete correlation inequalities. Starting from the four functions theorem and related FKG-type inequalities, we present a unifying framework that generalizes discrete Brunn–Minkowski results to functions on finitely generated abelian groups. This approach relies on translation-equivariant and exclusive operations that allow combinatorial induction, and it naturally recovers continuous analogues such as the Prékopa–Leindler inequality through limiting arguments. By connect-

https://doi.org/10.1515/9783111360508-008

ing discrete and continuous perspectives, these results demonstrate that many classical correlation inequalities have a common underlying combinatorial structure.

Throughout the chapter, we emphasize useful tools for estimating the number of lattice points in convex bodies. These include bounds in terms of lattice width, covering radius, and volume, as well as approximation formulas linking discrete and continuous measures. These provide a rich framework for understanding how convexity and arithmetic structure interact, laying the groundwork for further studies in discrete tomography, high-dimensional lattice geometry, and the broader interface between geometry and number theory.

8.1 Lattices and convex bodies

A main question in the Geometry of Numbers is to provide estimates on the volume or other geometric parameters of a set S in $\mathbb{R}^n$ that ensure that S contains points of the integral lattice $\mathbb{Z}^n$. Minkowski's first theorem provides an answer if the given set is an origin-symmetric convex body.

Theorem 8.1.1 (Minkowski's first theorem). *Let K be an origin-symmetric convex body in $\mathbb{R}^n$ such that $|K| \geqslant 2^n$. Then, K contains at least one non-zero element of $\mathbb{Z}^n$.*

Minkowski also obtained a stronger result, introducing the so-called successive minima of an origin-symmetric convex body K in $\mathbb{R}^n$. For any $1 \leqslant i \leqslant n$, the i-th successive minimum of K is defined as follows:

$$\lambda_i(K) := \min\{\lambda > 0 : \dim(K \cap \mathbb{Z}^n) \geqslant i\}.$$

Note that $\lambda_1(K) \leqslant \lambda_2(K) \leqslant \cdots \leqslant \lambda_n(K)$ and that $\lambda_1(K)$ is the smallest $\lambda > 0$ for which λK contains a non-zero point in $\mathbb{Z}^n$, which shows that

$$\lambda_1(K)^n |K| \leqslant 2^n,$$

an equivalent way to state Minkowski's first theorem. Minkowski (see, e. g., [240, Theorem 23.1]) proved a very strong extension of Theorem 8.1.1.

Theorem 8.1.2 (Minkowski's second theorem). *Let K be an origin-symmetric convex body in $\mathbb{R}^n$. Then,*

$$\frac{2^n}{n!} \leqslant \lambda_1(K)\lambda_2(K)\cdots\lambda_n(K)\,|K| \leqslant 2^n.$$

The lower bound in Theorem 8.1.2 is obtained if we use the fact that K contains the absolute convex hull of $\lambda_i v_i$ for some $v_i \in \mathbb{Z}^n \setminus \{0\}$, with equality if $K = \operatorname{conv}\{\pm e_1, \ldots, \pm e_n\}$ is the standard cross-polytope in $\mathbb{R}^n$.

In what follows, we shall need to consider more general lattices. Recall that a lattice $\Lambda \subset \mathbb{R}^n$ is a discrete subgroup of $\mathbb{R}^n$. The rank d of Λ is the dimension of the linear subspace of $\mathbb{R}^n$ spanned by the elements of Λ. One can show that Λ has rank d if and only if there exist linearly independent vectors $v_1, \ldots, v_d \in \mathbb{R}^n$ such that

$$\Lambda = \{k_1 v_1 + \cdots + k_d v_d : k_i \in \mathbb{Z} \text{ for all } 1 \leqslant i \leqslant d\}.$$

In particular, every lattice of rank d is finitely generated and isomorphic to $\mathbb{Z}^d$ via an invertible linear transformation from span(Λ) to $\mathbb{Z}^d$. The determinant det(Λ) of Λ is the d-dimensional volume of a fundamental domain of the action of Λ on span(Λ).

If rank(Λ) $= d$ and K is a d-dimensional origin-symmetric convex body contained in span(Λ), then, for $1 \leqslant i \leqslant d$, we may define the i-th successive minimum of K with respect to Λ setting

$$\lambda_i(K, \Lambda) = \min\{\lambda \geqslant 0 : \dim(\lambda K \cap \Lambda) \geqslant i\}.$$

The successive minima are related to volume by Minkowski's second theorem

$$\lambda_1(K, \Lambda) \cdots \lambda_d(K, \Lambda)|K| \leqslant 2^d \det(\Lambda). \tag{8.1}$$

For any convex body K in $\mathbb{R}^n$ we denote by

$$G_\Lambda(K) = |K \cap \Lambda|$$

the lattice point enumerator of K with respect to lattice points of Λ (in this chapter we write $|A|$ for the cardinality of a finite set of points that is not necessarily full-dimensional). In the case, where $\Lambda = \mathbb{Z}^n$ is the standard integral lattice, we simply write

$$G(K) = G_{\mathbb{Z}^n}(K) = |K \cap \mathbb{Z}^n|.$$

The polar lattice of Λ is the lattice

$$\Lambda^* = \{y \in \text{span}(\Lambda) : \langle x, y\rangle \in \mathbb{Z} \text{ for all } x \in \Lambda\}$$

in span(Λ). In particular, we have $(\mathbb{Z}^n)^* = \mathbb{Z}^n$. Note that $\det(\Lambda^*) = 1/\det(\Lambda)$.

A k-dimensional plane will be called a lattice plane if it contains $k+1$ affinely independent points of Λ. The orthogonal complement $L^\perp$ of a k-dimensional lattice plane L containing the origin is an $(n-k)$-dimensional lattice plane of Λ^*. The $(\dim(\Lambda)-1)$-dimensional lattice planes are called lattice hyperplanes and can be parameterized via the primitive vectors of Λ^*, i. e., such a lattice hyperplane H is given by

$$H(y, \beta) = \{x \in \text{span}(\Lambda) : \langle x, y\rangle = \beta\},$$

where $y \in \Lambda^* \setminus \{0\}$ is a generator of the 1-dimensional lattice $\operatorname{span}\{y\} \cap \Lambda^*$ and $\beta \in \mathbb{Z}$. The lattice Λ can be decomposed as

$$\Lambda = \bigcup_{\beta\in\mathbb{Z}} (\Lambda \cap H(y,\beta)),$$

where none of the sections is empty.

The lattice width of a convex body $K \subset \operatorname{span}(\Lambda)$ with respect to Λ is given by

$$w_\Lambda(K) = \min_{y\in\Lambda^*} \max_{x_1,x_2\in K} \langle x_1 - x_2, y\rangle. \tag{8.2}$$

The lattice width can be also expressed in terms of the first successive minimum with respect to the polar lattice: one can check that

$$w_\Lambda(K) = \lambda_1((K-K)^\circ, \Lambda^*),$$

where $K-K$ is the difference body of K and $(K-K)^\circ$ is its polar body. From (8.2) we see that $w_\Lambda(K)$ describes, up to a rounding, the minimum number of parallel lattice planes intersecting K. In particular, we have

$$G_\Lambda(K) \leqslant (w_\Lambda(K)+1)G_\Lambda(K \cap H(\overline{y}^*, \overline{\beta})), \tag{8.3}$$

where $\overline{y}^* \in \lambda_1((K-K)^\circ, \Lambda^*)(K-K)^\circ \cap \Lambda^*$ and $\overline{\beta}$ maximizes $G_\Lambda(K \cap H(\overline{y}^*, \beta))$ among all $\beta \in \mathbb{Z}$. If $\dim(K\cap\Lambda) = \dim(\Lambda)$, then we have $w_\Lambda(K) \geqslant 1$ and, thus, (8.3) yields

$$G_\Lambda(K) \leqslant 2\,w_\Lambda(K)G_\Lambda(K \cap H(\overline{y}^*, \overline{\beta})) \tag{8.4}$$

for such convex bodies.

One more useful lattice parameter is the covering radius $\mu_\Lambda(K)$ of K. It is commonly defined as

$$\mu_\Lambda(K) = \min\{\mu \geqslant 0 : \mu K + \Lambda = \operatorname{span}(\Lambda)\}.$$

Due to a result of Khintchine [281], there exists a constant depending only on the dimension n of Λ that bounds the product $w_\Lambda(K)\mu_\Lambda(K)$ from above for all convex bodies $K \subset \operatorname{lin}(\Lambda)$. The smallest number $\operatorname{flt}(n)$ with the property that

$$w_\Lambda(K)\mu_\Lambda(K) \leqslant \operatorname{flt}(n) \tag{8.5}$$

always holds is the so-called flatness constant. An upper bound $\operatorname{flt}(n) = O(n^{4/3}(\log n)^a)$ was given by Rudelson [490], building upon [50]. To this day, the best known estimate for $\operatorname{flt}(n)$ is

$$\operatorname{flt}(n) \leqslant O(n(\log n)^3). \tag{8.6}$$

This was proved by Reis and Rothvoss in [477]. On the other hand, it is easy to see that $\mathrm{flt}(n) \geqslant n$ and the current best lower bound is due to a recent result of Mayrhofer, Schade, and Weltge [404], showing that $\mathrm{flt}(n) \geqslant 2n - o(n)$. Moreover, $\mathrm{flt}(n)$ is monotonous in n, as can be seen by extending an n-dimensional lattice Λ to an $(n+1)$-dimensional lattice $\overline{\Lambda}$ via $\overline{\Lambda} = \Lambda \oplus \mathbb{Z}e_{n+1}$ and replacing $K \subseteq \mathrm{lin}(\Lambda)$ by $\overline{K} = K \times [0, \mathrm{flt}(n)] \cdot e_{n+1}$: then, $w_{\overline{\Lambda}}(\overline{K}) = w_\Lambda(K)$ and $\mu_{\overline{\Lambda}}(\overline{K}) = \mu_\Lambda(K)$.

Since the lattice width and covering radius are translation invariant, their definition and properties extend naturally to affine lattices $\Lambda + t$, where $\Lambda \subset \mathbb{R}^n$ is a lattice and $t \in \mathbb{R}^n$, together with convex bodies $K \subset t + \mathrm{span}(\Lambda)$.

Another key ingredient of the proofs that we present in the next section is the following result that has been obtained recently in [189, Proposition 1.6] (where the lower bound was proven independently by Dadush [141]):

$$(1 - \mu_\Lambda(K))^n \frac{|K|}{\det(\Lambda)} \leqslant G_\Lambda(K) \leqslant \frac{|K|}{\det(\Lambda)} (1 + \mu_\Lambda(K))^n. \tag{8.7}$$

For the lower bound, it is necessary to assume that $\mu_\Lambda(K) \leqslant 1$. Although (8.7) is stated in [189] only for the case $\Lambda = \mathbb{Z}^n$, the above generalization follows easily by applying a linear isomorphism that maps $\mathbb{Z}^n$ to Λ.

For the sake of brevity, if $\Lambda = \mathbb{Z}^n$, we write $\mu_\Lambda(K) = \mu(K)$ and, likewise, $\lambda_i(K, \Lambda) = \lambda_i(K)$, $w_\Lambda(K) = w(K)$, and $G_\Lambda(K) = G(K)$. Affine planes containing the origin will be called central planes.

8.2 The discrete slicing problem

In this section, we discuss the following question, asked by Koldobsky, aiming at a discrete analogue of the slicing inequality.

Problem 8.2.1. Does there exist a constant d_n such that

$$G(K) \leqslant d_n \max_{\xi \in S^{n-1}} G(K \cap \xi^\perp)\, |K|^{\frac{1}{n}}$$

for all origin-symmetric convex bodies K in $\mathbb{R}^n$ that contain n linearly independent lattice points?

We note here that we require that K contains n linearly independent lattice points, i. e., $\dim(K \cap \mathbb{Z}^n) = n$, in order to eliminate the degenerate case of a body (for example, take a box $[-\delta, \delta]^{n-1} \times [-20, 20]$) whose maximal section contains all lattice points in the body, but whose volume may be taken almost equal to 0 if we let $\delta \to 0$.

We start with the answer for the 2-dimensional case. The solution is based on Minkowski's first theorem and on Pick's theorem (see [523, Theorem 3.28] and [68, Theorem 2.8]). Recall that a polygon is called integral if it can be described as the convex hull of lattice points.

Theorem 8.2.2 (Pick's theorem). *Let P be an integral 2-dimensional convex polygon. Then, $A = I + \frac{1}{2}B - 1$, where $A = |P|$ is the area of the polygon, I is the number of lattice points in the interior of P, and B is the number of lattice points on the boundary of P.*

Using Theorem 8.1.1 and Theorem 8.2.2 we show that the constant d_2 in Problem 8.2.1 can be chosen independently of the origin-symmetric convex body K in $\mathbb{R}^2$.

Theorem 8.2.3. *Let K be an origin-symmetric convex body in $\mathbb{R}^2$ with $\dim(K \cap \mathbb{Z}^2) = 2$. Then,*

$$G(K) \leqslant 5 \max_{\xi \in S^1} G(K \cap \xi^\perp)\, |K|^{\frac{1}{2}}.$$

Proof. Let $s = \sqrt{|K|/4}$. Since $|\frac{1}{s}K| = 4$, from Theorem 8.1.1 we see that there exists a non-zero vector $u \in \mathbb{Z}^2 \cap \frac{1}{s}K$. Then, $su \in K$ and

$$G(L_u \cap K) \geqslant 2\lfloor s \rfloor + 1,$$

where $\lfloor s \rfloor$ is the integer part of s, and L_u is the line containing u and the origin.

Next, consider $P = \operatorname{conv}(K \cap \mathbb{Z}^2)$, i. e., the convex hull of the integral points inside K. Clearly, P is an integral 2-dimensional convex polygon, and Theorem 8.2.2 shows that

$$|P| = I + \frac{1}{2}B - 1 \geqslant \frac{I+B}{2} - \frac{1}{2},$$

because $I \geqslant 1$. It follows that

$$G(P) = I + B \leqslant 2|P| + 1 \leqslant \frac{5}{2}|P|,$$

where we also use the fact that every origin-symmetric integral convex polygon has area at least 2. Then, we check that

$$\begin{aligned} G(K) = G(P) &\leqslant \frac{5}{2}|P| \leqslant \frac{5}{2}|K| \leqslant \frac{5}{2}(2s)\,|K|^{\frac{1}{2}} \leqslant 5\,(2\lfloor s \rfloor + 1)\,|K|^{\frac{1}{2}} \\ &\leqslant 5 \max_{\xi \in S^1} G(K \cap \xi^\perp)\,|K|^{\frac{1}{2}} \end{aligned}$$

and the proof is complete. □

A first answer to Problem 8.2.1 was given by Alexander, Henk, and Zvavitch in [9]. Their approach is based on a discrete version of John's theorem, due to Tao and Vu [522], and shows that the constant d_n can be chosen independent of the body K and as small as $O(n)^{7n/2}$. We first recall the definition of a generalized arithmetic progression (see [522, 523] for more details).

Definition 8.2.4. Let H be an additive group, let $N = (N_1, \dots, N_n)$ be an n-tuple of non-negative integers and $v = (v_1, \dots, v_n) \in H^n$. A generalized symmetric arithmetic progression P is a triplet (N, v, n). We also define

$$\operatorname{im}(P) := [-N, N] \cdot v,$$

where

$$[-N, N] \cdot v := \{\langle k, v\rangle = k_1 v_1 + \cdots + k_n v_n : k = (k_1, \dots, k_n) \in [-N_j, N_j]^n \cap \mathbb{Z}^n\}.$$

The progression is called proper if the map $k \mapsto \langle k, v\rangle$ is injective. Then, $v = (v_1, \dots, v_n)$ is called its basis and n its rank.

With this definition, the discrete John theorem [522, Theorem 1.6] can be formulated as follows.

Theorem 8.2.5. *Let K be an origin-symmetric convex body in $\mathbb{R}^n$. There exists a symmetric proper generalized arithmetic progression $P \subset \mathbb{Z}^n$, such that* $\operatorname{rank}(P) \leqslant n$ *and*

$$(O(n)^{-3n/2} K) \cap \mathbb{Z}^n \subset \operatorname{im}(P) \subset K \cap \mathbb{Z}^n. \tag{8.8}$$

Moreover,

$$O(n)^{-7n/2} G(K) \leqslant G(P). \tag{8.9}$$

Using Theorem 8.2.5, we can give a first answer to Koldobsky's question.

Theorem 8.2.6. *For any origin-symmetric convex body K in $\mathbb{R}^n$ with* $\dim(K \cap \mathbb{Z}^n) = n$, *we have that*

$$G(K) \leqslant O(n)^{7n/2} \max_{\xi \in S^{n-1}} G(K \cap \xi^{\perp}) |K|^{\frac{1}{n}}. \tag{8.10}$$

Proof. We apply the discrete John's theorem to get a symmetric proper generalized arithmetic progression $P = (N, v, n)$, as in Theorem 8.2.5. Note that if $\operatorname{rank}(P) < n$, then there exists a hyperplane $\xi^{\perp}$ such that $P \subset \xi^{\perp}$ and hence, using (8.9), we get

$$O(n)^{-7n/2} G(K) \leqslant G(P) \leqslant G(K \cap \xi^{\perp}).$$

Since $\dim(K \cap \mathbb{Z}^n) = n$, we easily check that $|K| \geqslant 2^n/n!$, therefore $|K|^{\frac{1}{n}} > 2/n$. This implies that

$$G(K) \leqslant O(n)^{7n/2}\, G(K \cap \xi^{\perp})\, |K|^{\frac{1}{n}}.$$

Next, we consider the case $\operatorname{rank}(P) = n$. Without loss of generality we may assume that $N_1 \geqslant N_2 \geqslant \cdots \geqslant N_n \geqslant 1$. We define $\xi^{\perp} = \operatorname{span}\{v_1, \dots, v_{n-1}\}$ and from (8.9) we get

$$
\begin{aligned}
G(K) &\leqslant O(n)^{7n/2}G(P) \leqslant O(n)^{7n/2}\prod_{i=1}^{n}(2N_i+1)\\
&= O(n)^{7n/2}(2N_n+1)\prod_{i=1}^{n-1}(2N_i+1)\\
&\leqslant O(n)^{7n/2}\Bigl(\prod_{i=1}^{n}(2N_i+1)\Bigr)^{\frac{1}{n}}G(K\cap\xi^{\perp}),
\end{aligned}
\tag{8.11}
$$

where we use the fact that $N_n = \min N_j$ as well as (8.8) to claim that

$$G(K\cap\xi^{\perp}) \geqslant \prod_{i=1}^{n-1}(2N_i+1).$$

Now, we estimate the volume covered by the progression P. Consider a fundamental parallelepiped

$$\Pi = \{a_1v_1 + \cdots + a_nv_n : a_i \in [0,1),\ \text{for all } i = 1,\dots,n\}.$$

We set $X = [-N, N-1]\cdot v$ and observe that

$$K \supseteq \bigcup_{x\in X}(x+\Pi).$$

Indeed, from the inclusion $\mathrm{im}(P) \subset K\cap\mathbb{Z}^n$ we see that the vertices of $x+\Pi$ belong to $K\cap\mathbb{Z}^n$ for all $x\in X$ and then convexity implies that $x+\Pi\subset K$ for all $x\in X$. Since $v_1,\dots,v_n$ are independent vectors in $\mathbb{Z}^n$ we must have $\det(v_1,\dots,v_n)\geqslant\det(\mathbb{Z}^n)=1$, therefore

$$|K| \geqslant \Bigl(\prod_{i=1}^{n}2N_i\Bigr)\det(v_1,\dots,v_n) \geqslant \prod_{i=1}^{n}2N_i.$$

Going back to (8.11) we see that

$$
\begin{aligned}
G(K) &\leqslant O(n)^{7n/2}\Bigl(\prod_{i=1}^{n}(2N_i+1)\Bigr)^{\frac{1}{n}}G(K\cap\xi^{\perp})\\
&\leqslant O(n)^{7n/2}\Bigl(\prod_{i-1}^{n}(2N_i)\Bigr)^{\frac{1}{n}}G(K\cap\xi^{\perp})\\
&\leqslant O(n)^{7n/2}G(K\cap\xi^{\perp})|K|^{\frac{1}{n}}
\end{aligned}
$$

and the proof is complete. □

Now, our goal is to improve the previous estimate. A probabilistic-harmonic analysis approach to the discrete slicing problem was proposed by Regev [475]. We will show that there is a distribution of directions ξ for which $G(K \cap \xi^{\perp})$ is large enough, with a positive probability. For any $s > 0$ consider the function $\rho_s(x) : \mathbb{R}^n \to (0,1]$, defined by $\rho_s(x) = e^{-\pi|x|^2/s^2}$, and for a countable set $A \subset \mathbb{R}^n$ let

$$\rho_s(A) = \sum_{x \in A} e^{-\pi|x|^2/s^2}.$$

We will use the Poisson summation formula (see, for example, [66, Lemma 17.2]). Consider a function $f : \mathbb{R}^n \to \mathbb{C}$ and let

$$\widehat{f}(y) = \int_{\mathbb{R}^n} e^{-2\pi i\langle x,y\rangle} f(x)dx \quad (y \in \mathbb{R}^n)$$

be the Fourier transform of f. Assume that f and $\widehat{f}$ decay sufficiently fast, i. e., there exist positive constants C, δ such that $|f(x)|, |\widehat{f}(x)| \leqslant \frac{C}{1+|x|^{n+\delta}}$ for all $x \in \mathbb{R}^n$ (note that this condition is trivially satisfied by ρ_s). Then, for any lattice Λ in $\mathbb{R}^n$

$$\sum_{x \in \Lambda} f(x) = \frac{1}{\det(\Lambda)} \sum_{y \in \Lambda^*} \widehat{f}(y). \tag{8.12}$$

Using the fact that $\widehat{\rho}_s(y) = s^n \rho_{1/s}(y)$ together with (8.12) we get

$$\rho_s(\Lambda) = \sum_{x \in \Lambda} \rho_s(x) = \frac{1}{\det(\Lambda)} \sum_{y \in \Lambda^*} \widehat{\rho}_s(y) = \frac{s^n}{\det(\Lambda)} \rho_{1/s}(\Lambda^*).$$

In particular, we see that

$$\rho_s(\Lambda) \geqslant \det(\Lambda)^{-1} s^n. \tag{8.13}$$

We may use again (8.12) to show that $\rho_s(\Lambda + a) \leqslant \rho_s(\Lambda)$ for all $a \in \mathbb{R}^n$. Indeed, using $\widehat{\rho}_s(y) = s^n \rho_{1/s}(y) > 0$, we get

$$\rho_s(\Lambda + a) = \frac{1}{\det(\Lambda)} \sum_{y \in \Lambda^*} \widehat{\rho}_s(y) e^{-2\pi i\langle a,y\rangle} \leqslant \frac{1}{\det(\Lambda)} \sum_{y \in \Lambda^*} \widehat{\rho}_s(y) = \rho_s(\Lambda). \tag{8.14}$$

Let $Z_{A,s}$ be a random vector taking values in a countable set A, such that

$$\mathbb{P}(Z_{A,s} = x) = \frac{\rho_s(x)}{\rho_s(A)}.$$

Using (8.13) we get

$$\mathbb{P}(Z_{\Lambda,s} = 0) = \frac{1}{\rho_s(\Lambda)} \leqslant \frac{\det(\Lambda)}{s^n}. \tag{8.15}$$

Finally, if we pick a lattice $\Lambda \subset \mathbb{R}^n$, $x \in \Lambda$ and $s > 0$, we claim that

$$\mathbb{P}(\langle Z_{\Lambda^*,s}, x\rangle = 0) \geqslant \mathbb{P}(Z_{\mathbb{Z}\setminus|x|,s} = 0) = \rho_{s|x|}(\mathbb{Z})^{-1} \geqslant c\min\{1, (s|x|)^{-1}\}. \tag{8.16}$$

We may assume that $x \neq 0$ and note that by definition $\langle Z_{\Lambda^*,s}, x\rangle$ can take only integer values. Fix some $k \in \mathbb{Z}$ and consider the set of points $y \in \Lambda^*$ such that $\langle x, y\rangle = k$. Note that if this set is empty, then its ρ_s mass is clearly zero. Now, consider the case where this set is not empty. The affine subspace $\{y \in \mathbb{R}^n : \langle x, y\rangle = k\}$ is a shift of $x^\perp$ at distance $k/|x|$ in the direction of $x/|x|$. Note that such a shift will not necessarily move points of Λ^* into itself, and thus an additional shift inside $\{y \in \mathbb{R}^n : \langle x, y\rangle = k\} \cap \Lambda^*$ may be required. Thus, using the product structure of ρ_s and (8.14) we get

$$\rho_s(\{y \in \mathbb{R}^n : \langle x, y\rangle = k\} \cap \Lambda^*) \leqslant \rho_s(k/|x|)\rho_s(\Lambda^* \cap x^\perp).$$

Dividing the above inequality by $\rho_s(\Lambda^*)$ and summing over $k \in \mathbb{Z}$ we get (8.16).

Regev [475] proved the next theorem.

Theorem 8.2.7. *Let K be an origin-symmetric convex body in $\mathbb{R}^n$. Then,*

$$G(K) \leqslant \max_{\xi \in S^{n-1}} G(K \cap \xi^\perp) \max\{1, cn|K|^{\frac{1}{n-1}}\}.$$

Note that this theorem improves the bound on d_n from [9], when the volume of $|K|$ is smaller, then C^{n^3} and provides a polynomial bound on d_n for bodies of volume smaller than n^{cn^2}.

Proof. Using John's theorem we see that there is a linear transformation such that $\det(T) = 1$ and $T(K) \subset n|T(K)|^{1/n}B_2^n$. Without loss of generality, we consider the body $T(K)$ instead of K and the lattice $\Lambda = T(\mathbb{Z}^n)$ instead of $\mathbb{Z}^n$. We may also assume that $|K| \geqslant n^{-n}$, otherwise $K \subset \delta B_2^n$, where $0 < \delta < 1$ and using $\det(\Lambda) = 1$ we get that $K \cap \Lambda$ is not full-dimensional and the statement is trivial.

Now, we will select the direction $\xi \in \Lambda^* \setminus \{0\}$, using a probabilistic approach. Let $\xi = Z_{\Lambda^*,s}$, where $s \geqslant 1$ will be chosen later. Then, we may apply (8.16) to claim that for any fixed $x \in K \cap \Lambda$ we have

$$\mathbb{P}(x \in \xi^\perp) \geqslant c\min\{1, (s|x|)^{-1}\} \geqslant c\min\{1, (sn|K|^{1/n})^{-1}\} = \frac{c}{sn|K|^{1/n}}.$$

Then,

$$\mathbb{E}\left(\frac{G_\Lambda(K \cap \xi^\perp)}{G_\Lambda(K)}\right) = \mathbb{E}\left(\frac{|\{x \in K \cap \Lambda : \langle x, \xi\rangle = 0\}|}{G_\Lambda(K)}\right) \geqslant \frac{c}{sn|K|^{1/n}}.$$

The above inequality allows us to select ξ for which $|\{x \in K \cap \Lambda : \langle x, \xi\rangle = 0\}|$ is large. To finish the proof, we need to choose $s \geqslant 1$ to guarantee that ξ can be selected not to be equal to 0. Using (8.15) we see that $\mathbb{P}(\xi = 0) \leqslant s^{-n}$. Thus, we need to pick $s \geqslant 1$ such that $s^{-n} \leqslant c(sn|K|^{1/n})^{-1}$, i. e., $s = C|K|^{\frac{1}{n(n-1)}} \geqslant 1$, where $C > 0$ is a large enough absolute constant. This completes the proof. □

It was proved in [9] that for every origin-symmetric convex body K in $\mathbb{R}^n$ one may find an affine hyperplane A so that

$$G(K)^{\frac{n-1}{n}} \leqslant O(n)\, G(K \cap A).$$

In fact, in the same work, it was shown that for every convex body K in $\mathbb{R}^n$ one may find an affine hyperplane A so that

$$G(K)^{\frac{n-1}{n}} \leqslant O(n^2)\, G(K \cap A).$$

The next result from [188] refines this inequality by replacing n^2 by the flatness constant $\mathrm{flt}(n)$.

Theorem 8.2.8. *Let K be a convex body in $\mathbb{R}^n$, $n \geqslant 2$. For any $1 \leqslant k \leqslant n-1$ there exists a k-dimensional affine plane A in $\mathbb{R}^n$ such that*

$$G(K)^{\frac{k}{n}} \leqslant O(\mathrm{flt}(n))^{n-k} G(K \cap A). \tag{8.17}$$

For the proof, we need a number of preparatory results.

Lemma 8.2.9. *Let K be a convex body in $\mathbb{R}^n$, $n \geqslant 2$, and let $1 \leqslant m \leqslant n-1$. Then,*

$$(\lambda_1((K-K)^\circ)\cdots\lambda_m((K-K)^\circ))^n \leqslant \left(n!\left(\frac{8}{\pi}\right)^n |K|\right)^m.$$

Proof. Let $\lambda_i^* = \lambda_i((K-K)^\circ)$, $1 \leqslant i \leqslant n$. Note that

$$(\lambda_1^* \cdots \lambda_m^*)^n \leqslant (\lambda_1^* \cdots \lambda_n^*)^m,$$

therefore Minkowski's second theorem implies that

$$(\lambda_1^* \cdots \lambda_m^*)^n \leqslant \frac{2^n}{|(K-K)^\circ|}.$$

Applying the reverse Blaschke–Santaló inequality for $K-K$, and taking into account the Rogers–Shephard inequality $|K-K| \leqslant 4^n|K|$, we conclude the proof. □

Proposition 8.2.10. *Let K be a convex body and Λ a lattice in $\mathbb{R}^n$. Then,*

$$G_\Lambda(K) \leqslant \prod_{i=1}^n \lfloor \lambda_i((K-K)^\circ, \Lambda^*) + 1 \rfloor$$

and the inequality is the best possible.

Proof. We set $\lambda_i^* := \lambda_i((K-K)^\circ, \Lambda^*)$. Let $v_1, \ldots, v_n \in \Lambda^*$ be linearly independent vectors with $v_i \in \lambda_i^* \cdot (K-K)^\circ$, $1 \leqslant i \leqslant n$. Then, by the definition of the polar body and the symmetry of $K-K$, we have that

$$K - K \subseteq \{x \in \mathbb{R}^n : |\langle v_i, x\rangle| \leqslant \lambda_i^*,\ 1 \leqslant i \leqslant n\}.$$

Therefore,

$$h_K(v_i) + h_K(-v_i) = h_{K-K}(v_i) \leqslant \lambda_i^*, \text{ for } 1 \leqslant i \leqslant n. \tag{8.18}$$

On the other hand, the parallelotope

$$P = \{x \in \mathbb{R}^n : h_K(-v_i) \leqslant \langle v_i, x\rangle \leqslant h_K(v_i),\ 1 \leqslant i \leqslant n\},$$

contains K, and since the map $\Lambda \to \mathbb{Z}^n$, which is defined by $a \mapsto (\langle v_1, a\rangle, \dots, \langle v_n, a\rangle)^t$, is injective by the linear independence of $v_1, \dots, v_n \in \Lambda^*$, we see that

$$G_\Lambda(P) \leqslant \prod_{i=1}^{n} \lfloor h_K(v_i) + h_K(-v_i) + 1\rfloor.$$

Then, the result follows from (8.18). □

The case $k = 1$ of Theorem 8.2.8 is covered by the next result of Rabinowitz [472].

Proposition 8.2.11. *For any convex body K in $\mathbb{R}^n$ there exists a line ℓ such that*

$$G(K)^{\frac{1}{n}} \leqslant G(K \cap \ell).$$

Proof. Let S be the set of integer points in K. We may assume that $|S| \geqslant 2$ and then $m^n < |S| \leqslant (m+1)^n$ for some integer $m \geqslant 1$. We consider the coordinates of points modulo m. Since there are m^n distinct ordered n-tuples of integers modulo m, we can find $x = (x_1, \dots, x_n)$ and $x' = (x_1', \dots, x_n')$ in S that are congruent modulo m. That is, $x_i \equiv x_i' \pmod m$ for all $i = 1, \dots, n$. We consider the set T of points

$$x_k = \left(x_1 + \frac{x_1' - x_1}{m}k, \dots, x_n + \frac{x_n' - x_n}{m}k\right), \quad k = 0, 1, \dots, m.$$

Note that T is a set of $m+1$ collinear points that belong to $\mathbb{Z}^n$ because $m \mid x_i' - x_i$ for all $1 \leqslant i \leqslant n$. Since the points $x_0 = x$ and $x_m = x'$ belong to S, we see that $T \subset [x_0, x_m] \subseteq S$. Finally,

$$G(K)^{\frac{1}{n}} \leqslant m + 1 \leqslant G(K \cap \ell),$$

where ℓ is the line defined by $x_0 = x$ and $x_n = x'$. □

With these tools, we are able to prove Theorem 8.2.8.

Proof of Theorem 8.2.8. The proof goes by induction on the dimension, so we may assume that $\dim(K \cap \mathbb{Z}^n) = n$. The case $k = 1$ follows from Proposition 8.2.11, so we assume that $k \geqslant 2$ and distinguish two cases depending on the lattice width of K.

Case 1. Assume that $w(K) \leqslant \mathrm{flt}(n) + n$. Let $H = H(\overline{y}^*, \overline{\beta})$ as in (8.4). Then,

$$G(K) \leqslant O(w(K))\, G(K \cap H) = O(\mathrm{flt}(n))\, G(K \cap H). \tag{8.19}$$

If $k = n - 1$ this proves the theorem, so we may assume that $k < n - 1$. By induction, we may find a k-dimensional affine plane A in H so that

$$G(K \cap H)^{\frac{k}{n-1}} \leqslant O(\mathrm{flt}(n-1))^{n-1-k} G(K \cap A) \tag{8.20}$$

and hence (8.19) and the monotonicity of the flatness constant give

$$G(K) \leqslant O(\mathrm{flt}(n))^{1+\frac{(n-1)(n-k-1)}{k}} G(K \cap A)^{\frac{n-1}{k}} \leqslant O(\mathrm{flt}(n))^{\frac{n(n-k)}{k}} G(K \cap A)^{\frac{n}{k}}.$$

It follows that

$$G(K)^{\frac{k}{n}} \leqslant O(\mathrm{flt}(n))^{n-k} G(K \cap A).$$

Case 2. Assume that $w(K) \geqslant \mathrm{flt}(n) + n$. Then, from (8.5) we see that

$$\mu(K) \leqslant \frac{\mathrm{flt}(n)}{\mathrm{flt}(n) + n} < 1. \tag{8.21}$$

Let $y_1, \ldots, y_{n-k} \in \mathbb{Z}^n \setminus \{0\}$ be linearly independent vectors with

$$y_i \in \lambda_i((K-K)^\circ)(K-K)^\circ \cap \mathbb{Z}^n, \quad 1 \leqslant i \leqslant n-k$$

and let $L = \mathrm{span}\{y_1, \ldots, y_{n-k}\}$. Consider the orthogonal projection $\widetilde{K} = P_L(K)$ of K onto L and the lattice $\widetilde{\Lambda} = P_L(\mathbb{Z}^n)$.

Taking polars in the subspace L, we see that

$$(\widetilde{K} - \widetilde{K})^\circ = (K-K)^\circ \cap L \quad \text{and} \quad \widetilde{\Lambda}^* = \mathbb{Z}^n \cap L.$$

Hence, by the choice of L we have

$$\lambda_i((\widetilde{K} - \widetilde{K})^\circ, \widetilde{\Lambda}^*) = \lambda_i((K-K)^\circ) \geqslant \lambda_1((K-K)^\circ) = w(K) \geqslant 1, \quad 1 \leqslant i \leqslant n-k.$$

From Proposition 8.2.10 it follows that

$$G_{\widetilde{\Lambda}}(\widetilde{K}) \leqslant \prod_{i=1}^{n-k}(\lambda_i((\widetilde{K} - \widetilde{K})^\circ, \widetilde{\Lambda}^*) + 1) \leqslant 2^{n-k} \prod_{i=1}^{n-k} \lambda_i((\widetilde{K} - \widetilde{K})^\circ, \widetilde{\Lambda}^*). \tag{8.22}$$

Then, Lemma 8.2.9 shows that

$$G_{\widetilde{\Lambda}}(\widetilde{K}) \leqslant O(n)^{n-k} |K|^{\frac{n-k}{n}}. \tag{8.23}$$

Since $\mu(K) \leqslant \frac{\mathrm{flt}(n)}{\mathrm{flt}(n)+n} < 1$ by (8.21), we can use (8.7) to write

$$|K| \leqslant (1-\mu(K))^{-n} G(K) \leqslant \left(\frac{\mathrm{flt}(n)+n}{n}\right)^{n} G(K). \tag{8.24}$$

Inserting this estimate into (8.23) we get

$$G_{\widetilde{\Lambda}}(\widetilde{K}) \leqslant O(\mathrm{flt}(n))^{n-k} G(K)^{\frac{n-k}{n}}. \tag{8.25}$$

It follows that

$$\begin{aligned} G(K) &= \sum_{x \in \widetilde{K} \cap \widetilde{\Lambda}} G(K \cap (x + L^{\perp})) \\ &\leqslant G_{\widetilde{\Lambda}}(\widetilde{K}) \max_{x \in \widetilde{\Lambda}} G(K \cap (x + L^{\perp})) \\ &\leqslant O(\mathrm{flt}(n))^{n-k} G(K)^{\frac{n-k}{n}} \max_{x \in \widetilde{\Lambda}} G(K \cap (x + L^{\perp})) \end{aligned}$$

and hence

$$G(K)^{\frac{k}{n}} \leqslant O(\mathrm{flt}(n))^{n-k} G(K \cap A),$$

where $A = \widetilde{x} + L^{\perp}$ for some $\widetilde{x} \in \widetilde{\Lambda}$ is a k-dimensional lattice plane such that $G(K \cap A) = \max_{x \in \widetilde{\Lambda}} G(K \cap (x + L^{\perp}))$. □

Theorem 8.2.8 shows that we can always find a k-dimensional affine plane such that $G(K \cap A)$ is large. Our aim is to move to a parallel central plane L with the property that $G(K \cap L)$ does not deviate too much from $G(K \cap A)$. We start with some preliminary observations about the volume of the sections of a convex body with parallel planes. If K is an origin-symmetric convex body in $\mathbb{R}^n$, then the classical Brunn concavity principle states that for any k-dimensional plane A

$$|K \cap A| \leqslant |K \cap (A - A)|, \tag{8.26}$$

where $A - A$ is the central plane parallel to A passing through the origin (see, e. g., [27, Theorem 1.2.1]). For centered convex bodies we have the following analogue of (8.26), due to Grünbaum [241] for $k = n - 1$ and to Fradelizi [177] for general k:

$$|K \cap A| \leqslant \left(\frac{n+1}{k+1}\right)^{k} |K \cap (A - A)|. \tag{8.27}$$

In the case $k = n-1$, the constant in the above inequality is bounded from above by e. In the discrete setting one has to replace this factor by 2^{n-1}, even in the symmetric case, as one can check from the example $K = \mathrm{conv}(\pm([0,1]^{n-1} \times \{1\}))$ with $A = \{x \in \mathbb{R}^n : x_n = 1\}$ (see [9] or [187]). The next proposition shows that we can always find a central plane that contains “many” lattice points.

Proposition 8.2.12. *Let $n \geqslant 2$, $k \in \{1, \ldots, n-1\}$, and let A be a k-dimensional plane in $\mathbb{R}^n$. For every centered convex body K in $\mathbb{R}^n$ there exists a k-dimensional central plane $L \subset \mathbb{R}^n$ such that*

$$G(K \cap A) \leqslant O\left(\max\left\{\left(\frac{n+1}{k+1}\right)^k, kn\,\mathrm{flt}(k)\right\}\right) G(K \cap L). \tag{8.28}$$

Also, for every origin-symmetric convex body K in $\mathbb{R}^n$ there exists a k-dimensional central plane $L \subset \mathbb{R}^n$ such that

$$G(K \cap A) \leqslant O\left(\frac{\mathrm{flt}(k)k}{\log(k+1)}\right) G(K \cap L). \tag{8.29}$$

In both bounds, if $k \in O(1)$, the asymptotic order of the constant is the same as in the corresponding continuous inequalities (8.26) and (8.27), respectively. Moreover, for $k = n-1$ the maximum in (8.28) is of order $O(\mathrm{flt}(n)n^2)$.

Proof. Let $\Lambda = \mathbb{Z}^n \cap A$. We may assume that Λ is a k-dimensional affine lattice.

We start with the centered case. We first assume that

$$\mu_\Lambda(K \cap A) \leqslant \frac{1}{(k+1)\,(n+2)}. \tag{8.30}$$

Let $L = A - A$ and let $\Lambda_0 = \mathbb{Z}^n \cap L$. Then, $\Lambda = t + \Lambda_0$ for some $t \in \mathbb{R}^n$. Since K is centered, we have $(K - K) \subseteq (n+1)K$ (see [505, Lemma 2.3.3]), and hence

$$(K \cap A) - (K \cap A) \subseteq (n+1)(K \cap L). \tag{8.31}$$

The covering radius is translation invariant and homogeneous of degree (-1), therefore

$$\mu_{\Lambda_0}(K \cap L) \leqslant (n+1)\mu_\Lambda(K \cap A) < 1. \tag{8.32}$$

Combining (8.7) with (8.27), and using (8.32) and the fact that $\det\Lambda = \det\Lambda_0$, we obtain

$$\begin{aligned} G(K \cap A) &\leqslant \frac{|K \cap A|}{\det(\Lambda)}(1 + \mu(K \cap A, \Lambda))^k \\ &\leqslant \left(\frac{n+1}{k+1}\right)^k \frac{|K \cap L|}{\det(\Lambda)}(1 + \mu(K \cap A, \Lambda))^k \\ &\leqslant \left(\frac{n+1}{k+1}\right)^k \left(\frac{1 + \mu_\Lambda(K \cap A)}{1 - \mu_{\Lambda_0}(K \cap L)}\right)^k G(K \cap L) \\ &\leqslant \left(\frac{n+1}{k+1}\right)^k \left(\frac{1 + \mu_\Lambda(K \cap A)}{1 - (n+1)\mu_\Lambda(K \cap A)}\right)^k G(K \cap L). \end{aligned} \tag{8.33}$$

We have assumed that (8.30) holds, and this implies that

$$\left(\frac{1+\mu_\Lambda(K\cap A)}{1-(n+1)\mu_\Lambda(K\cap A)}\right)^k \leqslant \left(\frac{(k+1)(n+2)+1}{(k+1)(n+2)-(n+1)}\right)^k$$
$$= \left(\frac{(k+1)(n+2)+1}{k(n+2)+1}\right)^k \leqslant \left(\frac{k+1}{k}\right)^k \leqslant e.$$

Combining this inequality with (8.33) we get

$$G(K\cap A) \leqslant O\left(\left(\frac{n+1}{k+1}\right)^k\right) G(K\cap L). \tag{8.34}$$

This is the first term in the estimate of (8.28). Now, assume that (8.30) does not hold. Then, we use (8.5) to write

$$w(K\cap A) \leqslant \mathrm{flt}(k)\,(k+1)\,(n+2). \tag{8.35}$$

In the special case where $\dim(K\cap A\cap\mathbb{Z}^n) < k$, we just consider the central plane $L = \mathrm{lin}(K\cap A\cap\mathbb{Z}^n)$ whose dimension is at most k. Then, we clearly have $G(K\cap A) \leqslant G(K\cap L)$, and we can finish the proof by extending L to a k-dimensional central plane, if needed. So, we may assume that $K\cap A$ contains k affinely independent lattice points. Then, (8.4) implies that there exists an affine $(k-1)$-dimensional lattice plane $\widetilde{A}\subset A$ such that

$$G(K\cap A) \leqslant O(w(K\cap A))G(K\cap\widetilde{A}) \leqslant O(\mathrm{flt}(k)\,k\,n)G(K\cap \mathrm{span}\widetilde{A}). \tag{8.36}$$

If $0\in\widetilde{A}\subset A$, then A is a central plane itself and the statement of the proposition is obvious. Otherwise, we have that $\mathrm{span}\widetilde{A}$ is a k-dimensional central plane. Taking the maximum of the upper bounds in (8.34) and (8.36) completes the proof in the centered case.

We move on now to the symmetric case. For the proof of (8.29) we replace the threshold value in (8.30). Assume first that

$$\mu_\Lambda(K\cap A) \leqslant \frac{\log(k+1)}{3(k+1)}. \tag{8.37}$$

Since K is symmetric, we have $\frac{1}{2}((K\cap A)-(K\cap A)) \subseteq K\cap L$, where $L = A-A$ is the central plane parallel to A. Therefore,

$$\mu_{\Lambda_0}(K\cap L) \leqslant 2\mu_\Lambda(K\cap A).$$

As in the proof of (8.33), using now Brunn's concavity principle (8.26) instead of Fradelizi's inequality (8.27), we see that

$$G(K\cap A) \leqslant \left(\frac{1+\mu_\Lambda(K\cap A)}{1-2\mu_\Lambda(K\cap A)}\right)^k G(K\cap L)$$
$$= \left(1+\frac{3\mu_\Lambda(K\cap A)}{1-2\mu_\Lambda(K\cap A)}\right)^k G(K\cap L)$$

$$= \left(1 + \frac{1}{k} \cdot \frac{3k}{1/\mu_\Lambda(K \cap A) - 2}\right)^k G(K \cap L)$$
$$\leqslant \exp(3k\mu_\Lambda(K \cap A))G(K \cap L), \tag{8.38}$$

where in the last step we have also used the fact that the function $x \mapsto \log(1 + x)$ is concave. Then, the assumption (8.37) implies that

$$G(K \cap A) \leqslant (k + 1)G(K \cap L). \tag{8.39}$$

If $\mu_\Lambda(K \cap A) \geqslant \frac{\log(k+1)}{3(k+1)}$, then the argument that we used for the proof of (8.28) gives

$$G(K \cap A) \leqslant O(w_\Lambda(K \cap A))G(K \cap L') \leqslant O\left(\frac{\mathrm{flt}(k)}{\mu_\Lambda(K \cap A)}\right)G(K \cap L')$$
$$\leqslant O\left(\frac{\mathrm{flt}(k)k}{\log(k + 1)}\right)G(K \cap L') \tag{8.40}$$

for some central k-dimensional plane $L' \subset \mathbb{R}^n$. Since the bound in (8.39) is smaller than the one in (8.40), the result follows. □

Combining Theorem 8.2.8 with Proposition 8.2.12, we immediately get the main estimate of Freyer and Henk in [188].

Theorem 8.2.13. *Let K be a convex body in $\mathbb{R}^n$ with barycenter at the origin and let $1 \leqslant k \leqslant n - 1$. There exists a k-dimensional subspace L of $\mathbb{R}^n$ such that*

$$G(K)^{\frac{k}{n}} \leqslant O(\mathrm{flt}(n))^{n-k} O\left(\max\left\{\left(\frac{n+1}{k+1}\right)^k, \mathrm{flt}(k)kn\right\}\right) G(K \cap L).$$

Proof. From Theorem 8.2.8, there exists a k-dimensional affine plane A in $\mathbb{R}^n$ such that

$$G(K)^{\frac{k}{n}} \leqslant O(\mathrm{flt}(n))^{n-k} G(K \cap A).$$

Applying Proposition 8.2.12, we find a k-dimensional subspace L of $\mathbb{R}^n$ with

$$G(K \cap A) \leqslant O\left(\max\left\{\left(\frac{n+1}{k+1}\right)^k, \mathrm{flt}(k)kn\right\}\right) G(K \cap L).$$

Combining these estimates we obtain the theorem. □

From Theorem 8.2.13 we can deduce a polynomial upper bound for d_n.

Theorem 8.2.14. *Let K be an origin-symmetric convex body in $\mathbb{R}^n$ such that $\dim(K \cap \mathbb{Z}^n) = n$. There exists a central hyperplane H in $\mathbb{R}^n$ such that*

$$G(K) \leqslant O\left(\frac{n^2\mathrm{flt}(n)}{\log(n + 1)}\right) G(K \cap H)|K|^{\frac{1}{n}}.$$

In particular,

$$G(K) \leqslant O(n^3(\log n)^2)G(K \cap H)|K|^{\frac{1}{n}}.$$

Proof. Lemma 8.2.9 with $m = 1$ shows that

$$\mathrm{w}(K) \leqslant O(n)|K|^{\frac{1}{n}}.$$

From (8.4) we see that there exists an affine hyperplane A such that

$$G(K) \leqslant O(\mathrm{w}(K))G(K \cap A) \leqslant O(n)G(K \cap A)|K|^{\frac{1}{n}}. \tag{8.41}$$

Applying Proposition 8.2.12 to $G(K \cap A)$, we find a central hyperplane H such that

$$G(K \cap A) \leqslant O\left(\frac{n\mathrm{flt}(n)}{\log n}\right)G(K \cap H) \leqslant O(n^2(\log n)^2)G(K \cap H),$$

taking into account (8.6). Combining the above we get the theorem. □

One might conjecture that the right order in Theorem 8.2.14 is linear in the dimension. This would strengthen Regev's result (Theorem 8.2.7) that shows that $d_n = O(n)$ if the volume of K is bounded by C^{n^2}.

8.3 Discrete Brunn–Minkowski inequality

In this section, we discuss the discrete Brunn–Minkowski inequality. It is not hard to see that in the discrete setting we cannot expect such an inequality in the classical form; if $A = \{0\}$ and B is any finite subset of $\mathbb{Z}^n$, then we clearly have $|A + B|^{1/n} < |A|^{1/n} + |B|^{1/n}$. Therefore, a discrete Brunn–Minkowski-type inequality should have a different form. An example is given by the inequality

$$|A + B| \geqslant |A| + |B| - 1, \tag{8.42}$$

which holds true for any pair of finite subsets A, B of $\mathbb{Z}^n$ (see [523, Chapter 2]).

Another idea is to introduce suitable modifications of the sets involved in the inequality. For example, Gardner and Gronchi proved in [198] that if A, B are finite subsets of the integer lattice $\mathbb{Z}^n$, and the affine dimension of B is equal to n, then

$$|A + B| \geqslant |D^B_{|A|} + D^B_{|B|}|, \tag{8.43}$$

where $D^B_{|A|}$, $D^B_{|B|}$ are the so-called B-initial segments: for any $m \in \mathbb{N}$, D^B_m is the set of the first m points of $\mathbb{Z}^n_+$ in the so-called B-order, which is a particular order defined on $\mathbb{Z}^n_+$ depending only on the cardinality of B (see [198] for precise definitions and details). Using (8.43) one can prove the discrete Brunn–Minkowski-type inequalities

$$|A+B|^{1/n} \geqslant |A|^{1/n} + \frac{1}{(n!)^{1/n}}\left(|B|-n\right)^{1/n}$$

and, assuming that $|B| \leqslant |A|$

$$|A+B| \geqslant |A| + (n-1)|B| + \left(|A|-n\right)^{(n-1)/n}\left(|B|-n\right)^{1/n} - \frac{n(n-1)}{2}.$$

These inequalities improve previous results of Ruzsa from [492] and [493].

An alternative way to obtain a discrete Brunn–Minkowski-type inequality that is closer to its classical form is to augment the sets that are involved in the problem. Hernández Cifre, Iglesias, and Yepes Nicolás used in [258] an extension $\bar{A}$ defined for any non-empty finite set $A \subset \mathbb{Z}^n$ as follows: for $n = 1$ the set $\bar{A}$ is $A \cup \{\max(A) + 1\}$, and for $n > 1$ the set $\bar{A}$ is obtained by first adding the maximal cardinality section of A, and then applying the corresponding extension to every section of the latter new set. Using this technique they proved the following discrete Brunn–Minkowski inequality: If A, B are two non-empty finite subsets of $\mathbb{Z}^n$, then

$$|\bar{A}+B|^{1/n} \geqslant |A|^{1/n} + |B|^{1/n} \tag{8.44}$$

and this inequality is sharp.

Note that for any pair of non-empty finite sets $A, B \subset \mathbb{R}^n$, by (8.42) we have that

$$\begin{aligned} \left|(1-\lambda)A+\lambda B\right| &\geqslant \left|(1-\lambda)A\right| + |\lambda B| - 1 = |A| + |B| - 1 \\ &\geqslant (1-\lambda)|A| + \lambda|B| \geqslant \left((1-\lambda)|A|^{1/n} + \lambda|B|^{1/n}\right)^n. \end{aligned} \tag{8.45}$$

However, the quantities $|A|$, $|B|$ on the right-hand side are reduced by the factors $(1-\lambda)$ and λ, while the sets $(1-\lambda)A$ and λB on the left-hand side have the same cardinality as A and B, respectively. This shows that (8.45) may be true but does not have a strong meaning from a geometric point of view. Therefore, it is more natural to ask for a discrete version of the Brunn–Minkowski inequality in terms of the lattice point enumerator G for compact sets in $\mathbb{R}^n$ and not in terms of the cardinality of finite subsets in $\mathbb{R}^n$. Again, the choice $\lambda = 1/2$, $K = [0, m-\varepsilon]^n$, and $L = [0, m+\varepsilon/2]^n$, with $m \in \mathbb{N}$ and $0 < \varepsilon < 1$, gives

$$G\left(\frac{K+L}{2}\right)^{1/n} = m < m + \frac{1}{2} = \frac{G(K)^{1/n} + G(L)^{1/n}}{2}, \tag{8.46}$$

which means that we cannot expect a Brunn–Minkowski-type inequality for G in the classical form.

In view of the above, the idea is to consider a suitable extension of $(1-\lambda)K + \lambda L$, as in (8.44). More precisely, given two compact sets $K, L \subset \mathbb{R}^n$ that intersect $\mathbb{Z}^n$ and some $\lambda \in (0,1)$, we would like to define a "minimal" set M_λ such that $(1-\lambda)K + \lambda L \subset M_\lambda$ and

$$G(M_\lambda)^{1/n} \geqslant (1-\lambda)G(K)^{1/n} + \lambda G(L)^{1/n}.$$

We shall sketch the proof of a theorem of this type, which is due to Iglesias, Yepes Nicolás, and Zvavitch [264].

Theorem 8.3.1. *Let $\lambda \in (0,1)$ and let $K, L \subset \mathbb{R}^n$ be non-empty bounded sets. Then,*

$$G((1-\lambda)K + \lambda L + (-1,1)^n)^{1/n} \geqslant (1-\lambda)G(K)^{1/n} + \lambda G(L)^{1/n}. \tag{8.47}$$

Moreover, this inequality is sharp.

The proof of Theorem 8.3.1 relies on its functional analogue. Recall that the classical Brunn–Minkowski inequality can be also obtained in this way; one can use the Borell–Brascamp–Lieb inequality, originally proved in [93] and [109]. In order to state the latter, we need to introduce the p-mean $\mathcal{M}_p^\lambda(\cdot,\cdot)$ of two non-negative numbers, where p is a parameter varying in $\mathbb{R} \cup \{\pm\infty\}$. If $p \in \mathbb{R}$ and $p \neq 0$, then for any $a, b > 0$ we set

$$\mathcal{M}_p^\lambda(a,b) = \left((1-\lambda)a^p + \lambda b^p\right)^{1/p}.$$

For $p = 0$ we define $\mathcal{M}_0^\lambda(a,b) = a^{1-\lambda}b^\lambda$. Finally, we set $\mathcal{M}_\infty^\lambda(a,b) = \max\{a,b\}$ and $\mathcal{M}_{-\infty}^\lambda(x,y) = \min\{a,b\}$, and if $ab = 0$, we define $\mathcal{M}_p^\lambda(a,b) = 0$ for all $p \in \mathbb{R} \cup \{\pm\infty\}$.

Theorem 8.3.2 (Borell–Brascamp–Lieb inequality). *Let $\lambda \in (0,1)$. Let $-1/n \leqslant p \leqslant \infty$ and let $f, g, h : \mathbb{R}^n \longrightarrow \mathbb{R}^+$ be measurable functions such that*

$$h((1-\lambda)x + \lambda y) \geqslant \mathcal{M}_p^\lambda(f(x), g(y))$$

for all $x, y \in \mathbb{R}^n$. Then,

$$\int_{\mathbb{R}^n} h(x)\,dx \geqslant \mathcal{M}_{\frac{p}{np+1}}^\lambda\left(\int_{\mathbb{R}^n} f(x)\,dx, \int_{\mathbb{R}^n} g(x)\,dx\right). \tag{8.48}$$

The case $p = 0$ of Theorem 8.3.2 is the Prékopa–Leindler inequality, which gives immediately the Brunn–Minkowski inequality for compact sets $K, L \subset \mathbb{R}^n$ of positive volume in the form $|(1-\lambda)K + \lambda L| \geqslant |K|^{1-\lambda}|L|^\lambda$. We can get the Brunn–Minkowski inequality in its classical form if we apply (8.48) with $p = \infty$ to the indicator functions $f = \mathbb{1}_K$, $g = \mathbb{1}_L$, and $h = \mathbb{1}_{(1-\lambda)K+\lambda L}$.

A discrete analogue of Theorem 8.3.2 is proved in [264]. For any function $\phi : \mathbb{R}^n \to \mathbb{R}^+$, we denote by $\phi^\diamond : \mathbb{R}^n \longrightarrow \mathbb{R}^+$ the function defined by

$$\phi^\diamond(z) = \sup_{u \in (-1,1)^n} \phi(z+u) \quad \text{for all } z \in \mathbb{R}^n.$$

The function ϕ is the Asplund product $\star$ of the functions ϕ and $\chi_{(-1,1)^n}$ (known also as the sup-convolution), and can be viewed as the functional analogue of the Minkowski sum in the setting of log-concave functions (see [505, Section 9.5] and the references therein). Indeed,

$$\phi^{\diamond}(z) = \sup_{u\in(-1,1)^n} \phi(z+u) = \sup_{u\in\mathbb{R}^n} \phi(z+u)\chi_{(-1,1)^n}(-u)$$
$$= \sup_{u_1+u_2=z} \phi(u_1)\chi_{(-1,1)^n}(u_2) = (\phi \star \chi_{(-1,1)^n})(z).$$

The discrete Borell–Brascamp–Lieb inequality is the next theorem.

Theorem 8.3.3. *Let $\lambda \in (0,1)$ and let $K, L \subset \mathbb{R}^n$ be non-empty, bounded sets. Let $-1/n \leqslant p \leqslant \infty$ and let $f, g, h : \mathbb{R}^n \to \mathbb{R}^+$ be functions such that*

$$h((1-\lambda)x + \lambda y) \geqslant \mathcal{M}_p^{\lambda}(f(x), g(y))$$

for all $x \in K, y \in L$. Then,

$$\sum_{z\in M\cap\mathbb{Z}^n} h^{\diamond}(z) \geqslant \mathcal{M}^{\lambda}_{\frac{p}{np+1}}\Bigg(\sum_{x\in K\cap\mathbb{Z}^n} f(x), \sum_{y\in L\cap\mathbb{Z}^n} g(y)\Bigg), \tag{8.49}$$

where $M = (1-\lambda)K + \lambda L + (-1,1)^n$.

Theorem 8.3.3 immediately implies the next discrete Brunn–Minkowski inequality for the lattice enumerator.

Theorem 8.3.4. *Let $\lambda \in (0,1)$ and let $K, L \subset \mathbb{R}^n$ be non-empty, bounded sets. Let $-1/n \leqslant p \leqslant \infty$. Then,*

$$G((1-\lambda)K + \lambda L + (-1,1)^n) \geqslant \mathcal{M}^{\lambda}_{\frac{p}{np+1}}(G(K), G(L)). \tag{8.50}$$

For the proof we just consider the functions $f = \mathbb{1}_K, g = \mathbb{1}_L$, and $h = \mathbb{1}_{(1-\lambda)K+\lambda L}$, and note that $h^{\diamond} = \mathbb{1}_{(1-\lambda)K+\lambda L+(-1,1)^n}$. Then, we apply Theorem 8.3.3. Note that the inequality of Theorem 8.3.4 is sharp. Equality may be attained if we consider $K = L = [0,m]^n$ with $m \in \mathbb{N}$. Then, $G((1-\lambda)K + \lambda L + (-1,1)^n) = G(K) = G(L) = (m+1)^n$.

Applying (8.50) for $p = \infty$ we get (8.47) for bounded sets $K, L \subset \mathbb{R}^n$ with $G(K)G(L) > 0$. To complete the proof of Theorem 8.3.1, it is then enough to deal with the case in which one of the sets, say L, has no integer points. This is carefully done in [264] with the help of an analogue of Theorem 8.3.3.

8.4 Discrete correlation inequalities

The starting point of this section is the four functions theorem of Ahlswede and Daykin [7] that is closely related to correlation inequalities such as the (FKG) inequality of Fortuin, Kasteleyin, and Ginibre, an important result that is usually formulated under the additional assumption that f, g, h, k are all supported in the discrete cube $\{0,1\}^n$ and is used in the analysis of various models in probability theory and statistical physics (see, e. g., Grimmett [238, 239]).

Theorem 8.4.1. *Let $f,g,h,k : \mathbb{Z}^n \to [0,\infty)$ be four functions such that*

$$f(x)g(y) \leqslant h(x \wedge y)k(x \vee y)$$

for all $x = (x_1,\dots,x_n)$ and $y = (y_1,\dots,y_n) \in \mathbb{Z}^n$, where $(x \wedge y)_i = \min\{x_i,y_i\}$ and $(x \vee y)_i = \max\{x_i,y_i\}$. Then,

$$\Big(\sum_{x\in\mathbb{Z}^n} f(x)\Big)\Big(\sum_{x\in\mathbb{Z}^n} g(x)\Big) \leqslant \Big(\sum_{x\in\mathbb{Z}^n} h(x)\Big)\Big(\sum_{x\in\mathbb{Z}^n} k(x)\Big).$$

We shall discuss the next theorem of Halikias, Klartag, and Slomka from [251].

Theorem 8.4.2. *Let $\lambda \in [0,1]$ and let $f,g,h,k : \mathbb{Z}^n \to [0,\infty)$ be four functions such that*

$$f(x)g(y) \leqslant h(\lfloor \lambda x + (1-\lambda)y\rfloor)k(\lceil (1-\lambda)x + \lambda y\rceil)$$

for all $x,y \in \mathbb{Z}^n$, where $\lfloor x\rfloor = (\lfloor x_1\rfloor,\dots \lfloor x_n\rfloor)$ and $\lceil x\rceil = (\lceil x_1\rceil,\dots,\lceil x_n\rceil)$. Then,

$$\Big(\sum_{x\in\mathbb{Z}^n} f(x)\Big)\Big(\sum_{x\in\mathbb{Z}^n} g(x)\Big) \leqslant \Big(\sum_{x\in\mathbb{Z}^n} h(x)\Big)\Big(\sum_{x\in\mathbb{Z}^n} k(x)\Big).$$

Starting from the case $\lambda = 1/2$, $h = k$ of Theorem 8.4.2 and using a standard limiting argument (see [232, Section 2.3]) we obtain the case $\lambda = 1/2$ of the Brunn–Minkowski inequality in its multiplicative form

$$\left|\frac{A+B}{2}\right| \geqslant \sqrt{|A|\,|B|}.$$

A proof of Theorem 8.4.2 for the case $n = 1$, $\lambda = 1/2$, which involves stochastic analysis and admits a straightforward generalization to the more general case described above, was given in [292]. An alternative proof employing optimal transport theory was given by Gozlan, Roberto, Samson, and Tetali in [232].

Here, following [251], we shall describe a unified proof of Theorem 8.4.1 and Theorem 8.4.2. The starting observation is that the operations $T : \mathbb{Z}^n \times \mathbb{Z}^n \to \mathbb{Z}^n$ defined by

$$x \wedge y, x \vee y, \left\lfloor \frac{x+y}{2}\right\rfloor, \left\lceil \frac{x+y}{2}\right\rceil, \lfloor \lambda x + (1-\lambda)y\rfloor, \lceil \lambda x + (1-\lambda)y\rceil, \dots, \tag{8.51}$$

where $x,y \in \mathbb{Z}^n$ and $0 < \lambda < 1$, have the following two properties:

(C1) they are translation equivariant, namely they satisfy

$$T(x+z,y+z) = T(x,y) + z \tag{8.52}$$

for all $z \in \mathbb{Z}^n$, and

(C2) they are monotone in the sense of Knothe with respect to some decomposition of $\mathbb{Z}^n$ into a direct sum of groups $\mathbb{Z}^n = G_1 \times \cdots \times G_k$. This means that $T = (T_1, \ldots, T_k)$, where for each $i \in \{1, \ldots, k\}$ we have that $T_i : (G_1 \times \cdots \times G_i) \times (G_1 \times \cdots \times G_i) \to G_i$ (in other words, $T_i(x, y)$ depends only on the first i coordinates of its arguments $x, y \in G_1 \times \cdots \times G_k$) and there exists a total additive ordering $\ll_i$ on G_i such that $T_i^{(a,b)} : G_i \times G_i \to G_i$ defined by $T_i^{(a,b)}(x, y) = T_i((a, x), (b, y))$ for $a, b \in G_1 \times \cdots \times G_{i-1}$ satisfies

$$x_1 \ll_i x_2,\ y_1 \ll_i y_2 \quad \Longrightarrow \quad T_i^{(a,b)}(x_1, y_1) \ll_i T_i^{(a,b)}(x_2, y_2) \tag{8.53}$$

for all $a, b \in G_1 \times \cdots \times G_{i-1}$ and $x_1, x_2, y_1, y_2 \in G_i$.

Maps that satisfy a condition similar to (C2) were used by Knothe [302] in his proof of the Brunn–Minkowski inequality. Recall that a total ordering $\ll$ on an abelian group G is a binary relation that is reflexive, anti-symmetric, and transitive, such that for any distinct x, y, either $x \ll y$ or else $y \ll x$. An ordering $\ll$ is additive if for all x, y, z

$$x \ll y \quad \Longrightarrow \quad x + z \ll y + z.$$

The existence of a total additive ordering on a finitely generated abelian group G implies that G is isomorphic to $\mathbb{Z}^\ell$ for some ℓ.

Using the standard cartesian decomposition of $\mathbb{Z}^n = \mathbb{Z} \times \cdots \times \mathbb{Z}$ into one-dimensional groups, each of which is equipped with the standard order on $\mathbb{Z}$, we verify that all the examples in (8.51) satisfy (C1) and (C2). In these examples, each T_i is a function from $G_i \times G_i$ to G_i.

The next theorem implies both Theorem 8.4.1 and Theorem 8.4.2 (see also Borell [93, Theorem 2.1] for Brunn–Minkowski-type inequalities for operations other than the Minkowski sum with monotonicity properties).

Theorem 8.4.3. *Let $T : \mathbb{Z}^n \times \mathbb{Z}^n \to \mathbb{Z}^n$ be a map that satisfies* (C1) *and* (C2). *If $f, g, h, k : \mathbb{Z}^n \to [0, \infty)$ are four functions such that*

$$f(x)g(y) \leqslant h(T(x, y))k(x + y - T(x, y))$$

for all $x, y \in \mathbb{Z}^n$, then

$$\Big(\sum_{x \in \mathbb{Z}^n} f(x)\Big)\Big(\sum_{x \in \mathbb{Z}^n} g(x)\Big) \leqslant \Big(\sum_{x \in \mathbb{Z}^n} h(x)\Big)\Big(\sum_{x \in \mathbb{Z}^n} k(x)\Big).$$

In fact, we shall prove a more general result, in which we replace the monotonicity assumption (C2) by an “exclusion” property, which requires no ordering at all. We shall formulate this property, as well as the next theorem, in greater generality, with $\mathbb{Z}^n$ replaced by a finitely generated abelian group G, and $T : \mathbb{Z}^n \times \mathbb{Z}^n \to \mathbb{Z}^n$ replaced by

$T : G \times G \to G$. Any such G is isomorphic to $\mathbb{Z}^n \times (\mathbb{Z}/p_1\mathbb{Z}) \times \cdots \times (\mathbb{Z}/p_k\mathbb{Z})$ for some powers of primes $p_1, \ldots, p_k$.

We say that an operation $T : G \times G \to G$ is exclusive if for every finite set $A \subseteq G$ with at least two elements, and all $z \in G$, there exist distinct $x, y \in A$ such that for $A_1 = A \setminus \{x\}$, $A_2 = A \setminus \{y\}$, and $A_3 = A \setminus \{x, y\}$, the following conditions are satisfied:

(E1) $\{T(x, z-y), T(y, z-x)\} \not\subseteq T(A_i, z - A_i)$ for $i = 1, 2$; and

(E2) $\{T(x, z-y), T(y, z-x)\} \cap T(A_3, z-A_3) = \varnothing$, where $T(A_i, z-A_i) = \{T(u, z-v) : u, v \in A_i\}$.

In the next theorem, we replace (C2) by the following property: There exists a decomposition of abelian groups $G = G_1 \times \cdots \times G_k$ such that

(C3) $T = (T_1, \ldots, T_k)$ with $T_i : (G_1 \times \cdots \times G_i) \times (G_1 \times \cdots \times G_i) \to G_i$ for each i.

(C4) For all $i \in \{1, \ldots, k\}$ and $a, b \in G_1 \times \cdots \times G_{i-1}$ the operation $T_i^{(a,b)} : G_i \times G_i \to G_i$ defined by $T_i^{(a,b)}(x, y) = T_i((a, x), (b, y))$ is exclusive.

With these definitions we prove the following.

Theorem 8.4.4. *Let* $(G, +)$ *be a finitely generated abelian group, and* $T : G \times G \to G$ *satisfy* (C1), (C3), *and* (C4). *If* $f, g, h, k : G \to [0, \infty)$ *are four functions such that*

$$f(x)g(y) \leqslant h(T(x,y))k(x + y - T(x,y))$$

for all $x, y \in G$, *then*

$$\Big(\sum_{x\in G} f(x)\Big)\Big(\sum_{x\in G} g(x)\Big) \leqslant \Big(\sum_{x\in G} h(x)\Big)\Big(\sum_{x\in G} k(x)\Big).$$

The next lemma shows that Theorem 8.4.3 follows immediately from Theorem 8.4.4.

Lemma 8.4.5. *Let* $T : \mathbb{Z}^n \times \mathbb{Z}^n \to \mathbb{Z}^n$ *be a map that satisfies* (C1) *and* (C2). *Then, T satisfies* (C3) *and* (C4).

Proof. We first show that T_1 is exclusive. To this end, let $m \geqslant 2$ and $z \in G_1$. Let $A = \{x_1, \ldots, x_m\} \subseteq G_1$, where $x_1 \prec_1 \cdots \prec_1 x_m$. Here, $a \prec_1 b$ means that $a \ll_1 b$ and $a \neq b$. Set $x = x_1, y = x_m$, and recall that $A_1 = A \setminus \{x\}, A_2 = A \setminus \{y\}$, and $A_3 = A \setminus \{x, y\}$. For a finite subset $S \subseteq G_1$ we write $\max S$ and $\min S$ for the maximal and minimal elements with respect to the total order $\ll_1$. By properties (C1) and (C2) we have

$$\begin{aligned}\max_{v,w\in A_1} T_1(w, z-v) - \min_{v,w\in A_1} T_1(v, z-w) &= T_1(x_m, z-x_2) - T_1(x_2, z-x_m)\\ &= x_m - x_2 \prec_1 x_m - x_1 = T_1(x_m, z-x_1) - T_1(x_1, z-x_m).\end{aligned}$$

Therefore, $\{T_1(x_1, z-x_n), T_1(x_m, z-x_1)\} \not\subseteq T_1(A_1, z-A_1)$. A similar argument shows that the same holds for A_2. This verifies (E1) in the definition of an exclusive map.

To verify (E2) note that if $m = 2$, then $A_3 = \varnothing$, and hence (E2) is trivially true. Otherwise, we let

$$v = \min\{x_{i+1} - x_i\,;\, i \in \{1, \ldots, m-1\}\} > 0$$

and observe that

$$\begin{aligned}\max_{v,w\in A_3} T_1(v, z-w) &= T_1(x_{m-1}, z-x_2) \ll T_1(x_m - v, z - x_1 - v)\\ &= T_1(x_m, z-x_1) - v \prec T_1(x_m, z-x_1),\end{aligned}$$

which shows that $T_1(x_m, z-x_1) \notin T_1(A_3, z-A_3)$. Similarly, $T_1(x_1, z-x_m) \notin T_1(A_3, z-A_3)$, and this verifies (E2). Hence, T_1 is an exclusive map.

We proceed by induction on the number of groups k in the decomposition of G given in conditions (C3) and (C4). For $k = 1$, we have already verified that the statement holds for $T = T_1$.

Let $k \geqslant 2$ and assume that the statement holds for a decomposition into $k-1$ groups. Fix $a_1, b_1 \in G_1$ and let $G' = G_2 \times \cdots \times G_k$. For $i \in \{2, \ldots, k\}$, define $T_i' : (G_2 \times \cdots \times G_i) \times (G_2 \times \cdots \times G_i) \to G_i$ by $T_i'(x,y) = T_i((a_1, x), (b_1, y))$ and $T' : G' \times G' \to G'$ by $T' = (T_2', \ldots, T_k')$. Note that T' satisfies (C1) and (C2) with respect to this decomposition. By the induction hypothesis, T' satisfies (C3) and (C4).

This means that for all $i \in \{2, \ldots, k\}$ and $a = (a_1, a'), b = (b_1, b') \in G_1 \times \cdots \times G_{i-1}$, the map $T_i^{(a,b)} = T_i'^{\,(a',b')}$ is exclusive. Since a_1 and b_1 are arbitrary, it follows that $T_i^{(a,b)}$ is exclusive for all $i \in \{1, \ldots, k\}$ and $a, b \in G_1 \times \cdots \times G_{i-1}$, and thus T satisfies (C3) and (C4). □

We move on now to the proof of Theorem 8.4.4, starting with the particular case where T itself is exclusive. We shall use the following elementary fact: If a, b, c, d are non-negative reals that satisfy $ab \leqslant cd$ and $\max\{a,b\} \leqslant \max\{c,d\}$, then

$$a + b \leqslant c + d. \tag{8.54}$$

Let $f, g, h, k : G \to [0, \infty)$ be four functions such that

$$f(x)g(y) \leqslant h(T(x,y))k(x+y-T(x,y)) \tag{8.55}$$

for all $x, y \in G$. For any $j, z \in G$ we define $F_z(j) = f(j)g(z-j)$ and $H_z(j) = h(j)k(z-j)$. Note that (8.55) implies that

$$F_z(j) \leqslant H_z(T(j, z-j)). \tag{8.56}$$

We claim that for all $i, j, z \in G$ we have

$$F_z(i)F_z(j) \leqslant H_z(T(i, z-j))H_z(T(j, z-i)). \tag{8.57}$$

Indeed, by (8.55) and (C1) we have

$$F_z(i)F_z(j) = f(i)g(z-i)f(j)g(z-j) = f(i)g(z-j)f(j)g(z-i)$$

$$\begin{aligned}&\leqslant h(T(i,z-j))k(z+(i-j)-T(i,z-j))\\&\quad\times h(T(j,z-i))k(z+(j-i)-T(j,z-i))\\&=h(T(i,z-j))k(z-T(i,z-j))h(T(j,z-i))k(z-T(j,z-i))\\&=H_z(T(i,z-j))H_z(T(j,z-i)).\end{aligned}$$

Proposition 8.4.6. *Let $(G,+)$ be a finitely generated abelian group, and let $T:G\times G\to G$ be an exclusive operation that satisfies (C1). If $f,g,h,k:G\to[0,\infty)$ are four functions such that*

$$f(x)g(y)\leqslant h(T(x,y))k(x+y-T(x,y))$$

for all $x,y\in G$, then

$$\Big(\sum_{x\in G}f(x)\Big)\Big(\sum_{x\in G}g(x)\Big)\leqslant\Big(\sum_{x\in G}h(x)\Big)\Big(\sum_{x\in G}k(x)\Big).$$

Proof. The conclusion of the proposition can be equivalently written as

$$\sum_{j,z\in G}F_z(j)\leqslant\sum_{j,z\in G}H_z(j).$$

It is sufficient to prove that for any fixed $z\in G$ and every finite set $A\subseteq G$

$$\sum_{j\in A}F_z(j)\leqslant\sum_{j\in T(A,z-A)}H_z(j). \tag{8.58}$$

We shall verify this inequality by induction on $n=|A|$. For $n=0$ the statement is trivially true, because the empty sum equals zero. For $n=1$ the statement holds by (8.56).

So, we assume that $n\geqslant 2$ and that the statement holds for all $m\leqslant n-1$. Let $A\subseteq G$ with $|A|=n$. By assumption, there exist distinct $x,y\in A$ such that the conditions (E1) and (E2) in the definition of an exclusive map are satisfied. By switching x with y if necessary, we may assume that $F_z(x)\leqslant F_z(y)$. Then, (8.57) shows that

$$F_z(x)F_z(y)\leqslant H_z(T(x,z-y))H_z(T(y,z-x)). \tag{8.59}$$

We distinguish two cases. Assume first that

$$F_z(y)\geqslant\max\{H_z(T(x,z-y)),H_z(T(y,z-x))\}.$$

Then, from (8.59) we see that

$$F_z(x)\leqslant\min\{H_z(T(x,z-y)),H_z(T(y,z-x))\}. \tag{8.60}$$

By the induction hypothesis for $A_1=A\setminus\{x\}$ we have

$$\sum_{j\in A_1} F_z(j) \leqslant \sum_{j\in T(A_1,z-A_1)} H_z(j). \tag{8.61}$$

Adding (8.60) and (8.61), and using (E1) we get (8.58).

Now, consider the case where

$$F_z(y) \leqslant \max\{H_z(T(x,z-y)), H_z(T(y,z-x))\}.$$

Since $F_z(x) \leqslant F_z(y)$, we may apply (8.59) and (8.54) to write

$$F_z(x) + F_z(y) \leqslant H_z(T(x,z-y)) + H_z(T(y,z-x)). \tag{8.62}$$

Note that $T(x,z-y) \neq T(y,z-x)$ because $T(y,z-x) - T(x,z-y) = y - x \neq 0$. Therefore, using (8.62), the induction hypothesis for $A_3 = A \setminus \{x,y\}$ and condition (E2) in the definition of an exclusive map, we obtain (8.58). This completes the proof. □

Proof of Theorem 8.4.4. The proof goes by induction on the number k of groups in the decomposition of G given in (C3) and (C4). For $k = 1$, the result follows from Proposition 8.4.6. So, we assume that $k \geqslant 2$ and that the statement holds true for $k-1$.

We define $G' = G_2 \times \cdots \times G_k$. For any $a, b \in G_1$ and $x', y' \in G'$ we set

$$f^a(x') = f(a,x'), \quad g^a(x') = g(a,x'), \quad h^a(x') = h(a,x'), \quad k^a(x') = k(a,x').$$

Fix $a, b \in G_1$. For $i \in \{2,\ldots,k\}$ we define $T_i' : (G_2 \times \cdots \times G_i) \times (G_2 \times \cdots \times G_i) \to G_i$ by $T_i'(x,y) = T_i((a,x),(b,y))$, and $T' : G' \times G' \to G'$ by $T' = (T_2',\ldots,T_k')$. Note that T' satisfies (C1), (C3), and (C4) with respect to this decomposition. The assumption of the theorem implies that

$$f^a(x')g^b(y') \leqslant h^{T_1(a,b)}(T'(x',y'))k^{a+b-T_1(a,b)}(x'+y'-T'(x',y'))$$

for all $x', y' \in G'$. By the induction hypothesis, we have that

$$\Big(\sum_{x'\in G'} f^a(x')\Big)\Big(\sum_{x'\in G'} g^b(x')\Big) \leqslant \Big(\sum_{x'\in G'} h^{T_1(a,b)}(x')\Big)\Big(\sum_{x'\in G'} k^{a+b-T_1(a,b)}(x')\Big). \tag{8.63}$$

For every $a \in G_1$ we set

$$F(a) = \sum_{x'\in G'} f^a(x'), \quad G(a) = \sum_{x'\in G'} g^a(x')$$

and

$$H(a) = \sum_{x'\in G'} h^a(x') \quad \text{and} \quad K(a) = \sum_{x'\in G'} k^a(x').$$

Then, (8.63) shows that for all $a, b \in G_1$

$$F(a)G(b) \leqslant H(T_1(a,b))K(a+b-T_1(a,b)).$$

Since $T_1 : G_1 \times G_1 \to G_1$ is an exclusive map satisfying (C1), we may apply Proposition 8.4.6 to get

$$\Big(\sum_{a\in G_1} F(a)\Big)\Big(\sum_{a\in G_1} G(a)\Big) \leqslant \Big(\sum_{a\in G_1} H(a)\Big)\Big(\sum_{a\in G_1} K(a)\Big).$$

This completes the proof. □

Next, we add some comments on the relation of the inequalities that we presented in this section with the results of the previous section and known volume inequalities. The first observation is that a standard limiting argument, which is presented in [232, Section 2.3], leads from the case $\lambda = \frac{1}{2}$ and $h = k$ of Theorem 8.4.2 to the classical Prékopa–Leindler inequality. For $\lambda \neq 1/2$, a similar limiting argument leads to a weighted variant of the Prékopa–Leindler inequality, which is due to Cordero-Erausquin and Maurey [137].

Theorem 8.4.7. *Let $\lambda \in [0,1]$. If $f,g,h,k : \mathbb{R}^n \to [0,\infty)$ are measurable functions such that*

$$f(x)g(y) \leqslant h(\lambda x + (1-\lambda)y)k((1-\lambda)x + \lambda y)$$

for all $x,y \in \mathbb{R}^n$, then

$$\Big(\int_{\mathbb{R}^n} f(x)\,dx\Big)\Big(\int_{\mathbb{R}^n} g(x)\,dx\Big) \leqslant \Big(\int_{\mathbb{R}^n} h(x)\,dx\Big)\Big(\int_{\mathbb{R}^n} k(x)\,dx\Big).$$

Note that for $\lambda = 1/2$ and $h = k$, Theorem 8.4.7 coincides with the Prékopa–Leindler inequality. Similar arguments lead to the proof of the next theorem, which is in the spirit of Theorem 8.4.7.

Theorem 8.4.8. *Let $f,g,h,k : \mathbb{R}^n \to [0,\infty)$ be Borel measurable functions such that*

$$f(x)g(y) \leqslant h(x\wedge y)k(x\vee y) \tag{8.64}$$

for all $x,y \in \mathbb{R}^n$. Then,

$$\Big(\int_{\mathbb{R}^n} f(x)\,dx\Big)\Big(\int_{\mathbb{R}^n} g(x)\,dx\Big) \leqslant \Big(\int_{\mathbb{R}^n} h(x)\,dx\Big)\Big(\int_{\mathbb{R}^n} k(x)\,dx\Big). \tag{8.65}$$

One can also replace the operations $x\wedge y$ and $x\vee y$ in (8.64) by the operations $\min(x,y)$ and $\max(x,y)$ with respect to the standard lexicographic order on $\mathbb{R}^n$. The conclusion of (8.65) in this case is a consequence of Theorem 8.4.3.

The results of this section are also related to the work of Ollivier and Villani [449] who proved a Brunn–Minkowski-type inequality with curvature terms on the discrete cube. A simplified version of their inequality, without curvature, states that if A, B are subsets of $\{0,1\}^n$, then

$$|M| \geqslant \sqrt{|A|\,|B|}, \tag{8.66}$$

where M is the set of all midpoints of pairs (a,b) with $a \in A$ and $b \in B$. A point $m = (m_1,\dots,m_n) \in \{0,1\}^n$ is called a midpoint of two points $a,b \in \{0,1\}^n$ if $m_i = a_i$ whenever $a_i = b_i$, and

$$|\{1 \leqslant j \leqslant n : m_j = a_j\}| = |\{1 \leqslant j \leqslant n : m_j = b_j\}| + \epsilon,$$

with $\epsilon \in \{-1,0,1\}$, i. e., essentially half of the remaining coordinates of m coincide with those of a and the other half with those of b.

It is proved in [251] that (8.66) holds for a much smaller subset of midpoints. Consider the operation $T(a,b) = (T_1(a,b),\dots,T_n(a,b))$ where

$$T_i(a,b) = \begin{cases} a_i, & |\{j \leqslant i : a_j \neq b_j\}| \text{ is odd} \\ b_i, & \text{otherwise.} \end{cases}$$

It is clear that $T(a,b)$ is one of the midpoints of a and b in the sense of [449], and the same is true for the point $a+b-T(a,b)$. We denote

$$M_1^- = \bigcup_{a\in A, b\in B} T(a,b), \quad M_1^+ = \bigcup_{a\in A, b\in B} (a+b-T(a,b)), \quad M_1 = M_1^- \cup M_1^+$$

and consider the four functions $f = \mathbb{1}_A, g = \mathbb{1}_B$, and $h = k = \mathbb{1}_{M_1}$. Applying Theorem 8.4.3 with this operation T we get

$$\sqrt{|A|\,|B|} \leqslant \sqrt{|M_1^-|\,|M_1^+|} \leqslant |M_1|.$$

This inequality is stronger than (8.66) because $M_1 \subset M$. Moreover, the inequality holds true for other choices of the map T, and for subsets of $\mathbb{Z}^n$ instead of subsets of $\{0,1\}^n$.

Finally, it is interesting to make the comparison with the result of the previous section: we saw that if $\lambda \in [0,1]$ and K, L are bounded non-empty sets in $\mathbb{R}^n$, then

$$G(\lambda K + (1-\lambda)L + (-1,1)^n)^{1/n} \geqslant \lambda G(K)^{1/n} + (1-\lambda)G(L)^{1/n}. \tag{8.67}$$

Theorem 8.4.2 implies a multiplicative version of this inequality for $\lambda = 1/2$: Let $f = \mathbb{1}_K, g = \mathbb{1}_L$ be the indicator functions of K and L, and let $h = \mathbb{1}_{\frac{1}{2}(K+L)+(-1,0]^n}$ and $k = \mathbb{1}_{\frac{1}{2}(K+L)+[0,1)^n}$. Note that for every $x \in K$ and $y \in L$, we have

$$\left\lfloor \frac{x+y}{2} \right\rfloor \in \frac{K+L}{2} - [0,1)^n \text{ and } \left\lceil \frac{x+y}{2} \right\rceil \in \frac{K+L}{2} + [0,1)^n,$$

therefore $f(x)g(y) \leqslant h(\lfloor \frac{x+y}{2} \rfloor)k(\lceil \frac{x+y}{2} \rceil)$ for all $x, y \in \mathbb{Z}^n$. Then, Theorem 8.4.2 shows that

$$\begin{aligned}\sqrt{G(K)G(L)} &\leqslant \sqrt{G\left(\frac{K+L}{2} + \left[0,1\right)^n\right)G\left(\frac{K+L}{2} + (-1,0\right]^n)} \\ &\leqslant G\left(\frac{K+L}{2} + (-1,1)^n\right).\end{aligned}$$

We can obtain the same from (8.67) and the arithmetic-geometric means inequality.

8.5 Further reading

8.5.1 Discrete tomography

Discrete tomography is the area that focuses on the reconstruction of finite subsets of discrete spaces (typically subsets of $\mathbb{Z}^n$) from a small number of projections, called X-rays. An X-ray in a given direction records the number of points of the set lying on each line parallel to that direction. Formally, given a finite subset A of $\mathbb{Z}^n$ and a direction u, the X-ray $X_u A$ assigns to each line ℓ parallel to u the value

$$X_u A(\ell) = |A \cap \ell|.$$

The central problem is to determine whether A can be uniquely reconstructed from its X-rays in a given set of directions, and if so, how to compute it efficiently.

Discrete tomography has important applications in areas where objects have an underlying discrete or atomic structure and only limited directional data is available: materials science and crystallography, electron tomography, image processing, medical imaging, coding theory, and data transmission, but also combinatorics, and discrete geometry.

The discrete theory is fundamentally different from continuous tomography due to combinatorial constraints. It employs techniques from combinatorics, geometry, and computer science, and is characterized by challenges related to non-uniqueness and computational complexity.

Gardner and Gritzmann [200] provide a deep and systematic investigation of when finite subsets of $\mathbb{Z}^n$ can be uniquely reconstructed from X-rays. They develop criteria for when a class of finite sets is uniquely determined by X-rays in a given set of directions and study how many and which directions are sufficient for uniqueness, showing that two directions are generally insufficient, while more directions can guarantee uniqueness under additional constraints. A key concept introduced is that of *switching components*, which describes local modifications that preserve X-rays, explaining non-uniqueness. The paper also discusses reconstruction algorithms and highlights the gap between theoretical uniqueness and computational feasibility. In [201], the authors continue this study and show that the reconstruction problem is computationally difficult, and in particular, deciding consistency of X-ray data is NP-hard for certain sets of directions. They also show that small perturbations in X-ray data can lead to large differences in reconstructions, indicating ill-posedness.

In a different direction, the work of Gardner, Gronchi, and Zong [199] investigates structural properties of finite lattice sets in $\mathbb{Z}^n$ through three fundamental operations: sums, projections, and sections, with a particular focus on the discrete covariogram. The covariogram encodes overlap information between a set and its translates and serves as a key tool in discrete tomography and geometric reconstruction. Given a finite set $A \subset \mathbb{Z}^n$, its covariogram is defined as

$$g_A(x) = |A \cap (A + x)|, \quad x \in \mathbb{Z}^n.$$

The support of the covariogram is exactly the difference set $A - A$. This connects the covariogram to additive combinatorics and highlights its role as a second-order structural invariant. Distinct lattice sets may share the same covariogram. The paper provides constructions and characterizations of such ambiguities, extending earlier non-uniqueness results in discrete tomography. Under certain geometric or combinatorial restrictions (e. g., convexity or special configurations), the covariogram determines the set uniquely up to translation and reflection. The authors analyze how projections (X-rays) and sections of lattice sets relate to the covariogram, establishing connections between different types of geometric data used in reconstruction problems. The paper also draws parallels with the continuous covariogram problem for convex bodies, emphasizing similarities and key differences between discrete and continuous settings.

8.5.2 Flatness constant

Finding the relation between the geometry of a convex set and the number of integer points contained in the set is always a non-trivial task. One can see this, for example, from the history of Khintchine's flatness theorem. Recall that the covering radius $\mu_\Lambda(K)$ of a convex body $K \subset \operatorname{lin}(\Lambda)$ is defined as

$$\mu_\Lambda(K) = \min\{\mu \geqslant 0 : \mu K + \Lambda = \operatorname{span}(\Lambda)\}.$$

A result of Khintchine [281] states that the product $w_\Lambda(K)\mu_\Lambda(K)$, where $w_\Lambda(K)$ is the lattice width of K, is bounded from above by a constant that depends only on the dimension n. The smallest number $\operatorname{flt}(n)$ for which

$$w_\Lambda(K)\mu_\Lambda(K) \leqslant \operatorname{flt}(n)$$

always holds is the flatness constant. Rudelson [490] established the upper bound $\operatorname{flt}(n) = O(n^{4/3}(\log n)^a)$ after work of Banaszczyk, Litvak, Pajor, and Szarek [50] who had obtained the bound $\operatorname{flt}(n) = O(n^{3/2}\sqrt{\log n})$. On the other hand, it is known that $\operatorname{flt}(n) \geqslant n$ and the current best lower bound of Mayrhofer, Schade, and Weltge [404] shows that $\operatorname{flt}(n) \geqslant 2n - o(n)$.

Kannan and Lovász [274] considered a quantity $\mu_{\mathrm{KL}}(\Lambda, K)$ that denotes the best volume-based lower bound on the covering radius $\mu_\Lambda(K)$ of a convex body K with respect to a lattice Λ, and proved that $\mu_\Lambda(K) \leqslant n \cdot \mu_{\mathrm{KL}}(\Lambda, K)$. The subspace flatness conjecture of Dadush [141] claims that an $O(\log(2n))$ factor suffices, which would match the lower bound from the work of Kannan and Lovász. Reis and Rothvoss [477] settled this conjecture up to a constant in the exponent by proving that $\mu_\Lambda(K) \leqslant O(\log^3(2n))\,\mu_{\mathrm{KL}}(\Lambda, K)$. The proof is based on the reverse Minkowski theorem of Regev and Stephens-Davidowitz [476]. Following the work of Dadush [141, 142], they also obtained a $(\log(2n))^{O(n)}$-time randomized algorithm to solve integer programs in n variables. Another implication of this result is a near-optimal bound $O(n\log^3(2n))$ for the flatness constant $\operatorname{flt}(n)$.

8.5.3 Discrete version of Minkowski's second theorem

Minkowski's second theorem states that for any lattice Λ and any convex body K in $\mathbb{R}^n$, the following inequality holds:

$$|K| \leqslant 2^n \frac{\det\Lambda}{\lambda_1 \cdots \lambda_n},$$

where $\lambda_1 \leqslant \cdots \leqslant \lambda_n$ are the successive minima of K with respect to Λ. Betke, Henk, and Wills [74] conjectured the following discrete version of Minkowski's second theorem: if K is an origin-symmetric convex body in $\mathbb{R}^n$, then

$$G(K \cap \Lambda) \leqslant \prod_{i=1}^{n} \left\lfloor \frac{2}{\lambda_i} + 1 \right\rfloor.$$

They confirmed it in dimension $n = 2$ and in general dimension n they showed that it holds up to a factor of roughly $n!$. Henk [256] later improved this to a factor of 2^{n-1}. Malikiosis [398] then proved the conjecture for arbitrary convex bodies in dimension at most 3, and in general dimension n he showed that

$$G(K \cap \Lambda) \leqslant \frac{4}{e} 3^{\frac{n-1}{2}} \prod_{i=1}^{n} \left\lfloor \frac{2}{\lambda_i} + 1 \right\rfloor \tag{8.68}$$

and even

$$G(K \cap \Lambda) \leqslant \frac{4}{e}(40/9)^{\frac{n-1}{3}} \prod_{i=1}^{n} \left\lfloor \frac{2}{\lambda_i} + 1 \right\rfloor, \tag{8.69}$$

if K is assumed origin-symmetric. Freyer and Lucas [189] proved that

$$G(K \cap \Lambda) \leqslant \prod_{i=1}^{n} \left(\frac{2}{\lambda_i} + n \right), \tag{8.70}$$

which improves on (8.68) in fixed dimension for small enough successive minima, and is the first bound that is strong enough to imply Minkowski's second theorem. On the other hand, (8.68) remains stronger than (8.70), for example, if the dimension goes to infinity with the successive minima bounded below.

More recently, Tointon [524] adapted a proof of Minkowski's second theorem due to Tao and Vu [523, Theorem 3.30] to obtain improvements to both (8.68) and (8.69) in all cases, and to (8.70) in some cases, such as when those successive minima that are less than 2 are all within a factor of n of one another. Moreover, like Freyer and Lucas's bound (8.70), these improvements are strong enough to imply Minkowski's second theorem.

Tointon proved that if $K \subseteq \mathbb{R}^n$ is an origin-symmetric convex body, then, if $\lambda_1 > 1$ one has $G(K \cap \Lambda) = 1$, otherwise, setting $k = \max\{j : \lambda_j \leqslant 1\}$, for any real numbers $\mu_1 \leqslant \cdots \leqslant \mu_k$ satisfying $0 < \mu_i \leqslant \lambda_i$ we have

$$G(K \cap \Lambda) \leqslant \frac{2^k (1 + \frac{\mu_k}{2})^k}{\mu_1 \cdots \mu_k}.$$

Setting each $\mu_i = \lambda_i$, this already improves on (8.69), but depending on the values of the successive minima, the flexibility to choose the μ_i's can improve the bound still further, particularly in high dimension if there is one successive minimum that is larger than a large number of the other ones. The assertion that $G(K \cap \Lambda) = 1$ if $\lambda_1 > 1$ is of course trivial, but separating that case makes the definition of k less cumbersome.

In the general setting, Tointon proved that if $K \subseteq \mathbb{R}^n$ is a convex body, then, if $\lambda_1 > 2$ one has $G(K \cap \Lambda) \leqslant 1$, otherwise, setting $k = \max\{j : \lambda_j \leqslant 2\}$, for any real numbers $\mu_1 \leqslant \cdots \leqslant \mu_k$ satisfying $0 < \mu_i \leqslant \lambda_i$ we have

$$G(K \cap \Lambda) \leqslant \frac{2^k (1 + \frac{\mu_k}{2})^k}{\mu_1 \cdots \mu_k}.$$

8.5.4 Lattice point enumerator

Apart from the discrete analogues of the Brunn–Minkowski inequality that we discussed in Section 8.3, there are various discrete counterparts of results from convex geometry for the lattice point enumerator. A discrete analogue of the Rogers–Shephard inequality was obtained by Alonso-Gutiérrez, Lucas, and Yepes Nicolás in [15]. For any non-empty convex bounded set K in $\mathbb{R}^n$

$$G(K-K) \leqslant \binom{2n}{n} G\left(K + \left(-\frac{3}{4}, \frac{3}{4}\right)^n\right). \tag{8.71}$$

Also, if we consider the Minkowski sum of two non-empty convex bounded sets $K, L \subset \mathbb{R}^n$, instead of $K - K$, we have

$$G(K+L)G(K\cap(-L)) \leqslant \binom{2n}{n} c_{K,L}\, G(K+(-1,1)^n)G(L+(-1,1)^n), \tag{8.72}$$

where

$$c_{K,L} = \frac{|K+L+(-\frac{1}{2},\frac{1}{2})^n|}{|K+L+(-1,1)^n|} \in (0,1).$$

Both (8.71) and (8.72) are asymptotically sharp, in the sense that, for any of these inequalities, there exist convex bodies containing the origin in such a way that the ratio between the right-hand and the left-hand sides, when applied to dilations of these sets scaled by a factor $r > 0$, tends to 1 as $r \to \infty$.

In the same work, a suitable discrete version of the Rogers–Shephard inequality (1.14) is proved for the lattice point enumerator. If $k \in \{1, \dots, n-1\}$ and $H = \operatorname{span}\{\varepsilon_1, \dots, \varepsilon_k\} \in G_{n,k}$ then, for any non-empty convex bounded set K in $\mathbb{R}^n$

$$G_{n-k}(P_{H^\perp}K)G_k(K\cap H) \leqslant \binom{n}{k} G_n(K+(-1,1)^n). \tag{8.73}$$

The same work provides a discrete analogue of Berwald's inequality. If $K \subset \mathbb{R}^n$ is a convex bounded set containing the origin and $f : K \to \mathbb{R}^+$ is a concave function with $f(0) = \|f\|_\infty$, then, for any $0 < p < q$

$$\left(\frac{\binom{n+q}{n}}{G(K)} \sum_{x\in K\cap\mathbb{Z}^n} f^q(x)\right)^{1/q} \leqslant \left(\frac{\binom{n+p}{n}}{G(K)} \sum_{x\in(K+(-1,1)^n)\cap\mathbb{Z}^n} (f^\diamond)^p(x)\right)^{1/p}, \tag{8.74}$$

where $f^\diamond$ is the Asplund product $\star$ of the functions f and $\chi_{(-1,1)^n}$ discussed in Section 8.3. The origin does not play any role in the latter theorem and the assumption on the origin could be substituted by $\|f\|_\infty = f(x_0)$ for some $x_0 \in \mathbb{Z}^n$. It is, however, needed in the proof, that the maximum of f is attained at some point in $\mathbb{Z}^n$.

8.5.5 Discrete correlation inequalities

Slomka [512] offers an alternative proof for the discrete Brunn-–Minkowski-type inequalities that we discussed in Section 8.4. His proof also implies somewhat stronger weighted versions of these inequalities. For all functions $f, g, h, k : \mathbb{Z}^n \to [0, \infty)$ and positive parameters $\alpha, \beta, \gamma, \delta$ such that for all $x, y \in \mathbb{Z}^n$

$$f(x)^\alpha g(y)^\beta \leqslant h(T(x,y))^\gamma k(x + y - T(x,y))^\delta,$$

we have

$$\Big(\sum_{x\in\mathbb{Z}^n} f(x)\Big)^\alpha \Big(\sum_{x\in\mathbb{Z}^n} g(x)\Big)^\beta \leqslant \Big(\sum_{x\in\mathbb{Z}^n} h(x)\Big)^\gamma \Big(\sum_{x\in\mathbb{Z}^n} k(x)\Big)^\delta,$$

assuming that $T : \mathbb{Z}^n \times \mathbb{Z}^n \to \mathbb{Z}^n$ is a translation invariant map that satisfies a certain monotonicity property (e. g., one can take $T(x,y) = \lfloor (x+y)/2 \rfloor$).

The proof generalizes ideas of Gozlan, Roberto, Samson, and Tetali [232] from the theory of measure transportation and provides new displacement convexity of entropy-type inequalities on the n-dimensional integer lattice.

9 Extremal problems

Given a convex body K in $\mathbb{R}^n$ and $1 \leqslant k \leqslant n-1$, one may pose the natural questions to find the minimal and maximal volume sections $K \cap F$ among all k-dimensional subspaces F in $\mathbb{R}^n$, as well as the minimal and maximal volume projections $P_F(K)$ among all k-dimensional subspaces F in $\mathbb{R}^n$. In this chapter, we discuss these classical questions for the sections and projections of classical convex bodies, such as the cube and the cross-polytope, as well as the family of ℓ_p^n-balls, which interpolate between the cross-polytope ($p = 1$) and the cube ($p = \infty$), with volumes depending sensitively on p and dimension n. This subject has attracted the interest of mathematicians for decades and a wide range of tools from probability and Fourier analysis have been employed in its study. We briefly discuss earlier approaches, which are also covered in the books [310, 113], and [28], and focus on more recent developments. A beautiful source of information is the survey article of Nayar and Tkocz [436] that provides a detailed account of classical results, recent advances, and further references.

For the cube, sections perpendicular to a coordinate axis have minimal volume, while diagonal directions maximize the section. In ℓ_p^n-balls, a similar phenomenon occurs with a subtle dependence on p: for $p < 2$, maximal-volume hyperplane sections align with coordinate axes, whereas for $p > 2$, the largest sections tend toward diagonal directions. Probabilistic representations, using independent random variables with densities related to $e^{-|t|^p}$, provide a powerful tool for understanding these sections and their dependence on p.

Projections of these bodies onto hyperplanes or lower-dimensional subspaces reveal dual behavior. For the cube, the projection onto a hyperplane is determined by summing the absolute values of the components of the normal vector, while for the cross-polytope, coordinate directions maximize the projection. More generally, in ℓ_p^n-balls, maximal-volume projections occur along coordinate directions for $1 < p < 2$, with minimal directions depending on majorization relations among the squared coordinates.

Another aspect of cubic sections is the perimeter of a hyperplane section, defined as the $(n-2)$-dimensional measure of the intersection of the hyperplane with the cube's boundary. The directions that maximize the perimeter coincide with those giving maximal section volume, typically along diagonals $(1, 1, 0, \ldots, 0)/\sqrt{2}$ in $\mathbb{R}^n$. The perimeter can be expressed through integral formulas, reflecting the contributions from each facet of the cube.

Overall, these investigations show a coherent picture: the geometry of high-dimensional convex bodies is governed by symmetry and the interplay between coordinate axes and diagonals. Extremal sections and projections, as well as perimeters of sections, exhibit patterns determined by the shape parameter p and the dimension. Probabilistic and analytic methods, including stochastic representations, majorization, and integral inequalities, allow precise characterization of these extremal phenomena, bridging combinatorial, geometric, and functional perspectives.

https://doi.org/10.1515/9783111360508-009

9.1 Extremal sections of the unit balls of p-norms

9.1.1 Sections of the cube

We start with the unit cube $Q_n = [-\frac{1}{2}, \frac{1}{2}]^n$. A sharp lower bound for the volume of hyperplane sections of Q_n was proved independently by Hadwiger [250] and Hensley [257].

Theorem 9.1.1. *For every $\xi \in S^{n-1}$ we have*

$$|Q_n \cap \xi^{\perp}| \geqslant 1,$$

with equality if and only if $\xi = \pm e_j$ for some $1 \leqslant j \leqslant n$.

Proof. Consider the function $f(t) = |Q_n \cap (\xi^{\perp} + t\xi)|$, $t \in \mathbb{R}$. Since Q_n is isotropic, we have

$$\int t^2 f(t)\, dt = \int_{Q_n} \langle x, \xi\rangle^2\, dx = \frac{1}{12}$$

for every $\xi \in S^{n-1}$. Moreover, $\|f\|_{\infty} = f(0) = |Q_n \cap \xi^{\perp}|$, because f is even and log-concave. So, it is enough to show that for every even probability density f we have

$$\|f\|_{\infty}^2 \int t^2 f(t)\, dt \geqslant \frac{1}{12}. \tag{9.1}$$

We have already discussed (9.1) in the proof of Theorem 2.1.5. If we define $h = \|f\|_{\infty} \mathbb{1}_{[-c,c]}$, where $c = (2\|f\|_{\infty})^{-1}$, then $\|h\|_{\infty} = \|f\|_{\infty}$ and

$$\int t^2(f(t) - h(t))\, dt = \int [(t^2 - c^2)(f(t) - h(t))]\, dt \geqslant 0,$$

because the integrant is non-negative. □

It is not hard to see that (9.1) continues to hold without the assumption that f is even. To see this, we define $g(t) = \frac{1}{2}(f(t) + f(-t))$ and note that $\|g\|_{\infty} \leqslant \|f\|_{\infty}$ while the second moments of f and g are the same. A proof of (9.1) can be already found in [430], while a more general sharp inequality for the p-norm of $x \mapsto \langle x, \xi\rangle$ was obtained by Ball in [38].

Sections with coordinate subspaces of lower dimension have again minimal volume. This is a result of Vaaler [530], which confirms a conjecture of Good (see [46]).

Theorem 9.1.2. *For every $1 \leqslant k \leqslant n-1$ and every $F \in G_{n,k}$ we have that*

$$|Q_n \cap F| \geqslant 1,$$

with equality if and only if F is a coordinate subspace.

Vaaler's proof of Theorem 9.1.2 uses log-concavity and the notion of peakedness, introduced by Kanter in [276]. It is described in detail in [28, Section 4.6], Given two probability measures μ and ν on $\mathbb{R}^n$ we say that ν is more peaked than μ, and then we write $\nu \succ \mu$, if $\nu(K) \geqslant \mu(K)$ for every symmetric convex set K in $\mathbb{R}^n$. One can check that this property is preserved by taking products and convolutions of even log-concave measures (see Corollaries 3.2 and 3.3 in [276]). We consider the uniform measure μ_{Q_n} on $Q_n = [-1/2, 1/2]^n$ in $\mathbb{R}^n$, which is even and log-concave and satisfies

$$\mu_{Q_n} = \mu_{Q_1} \otimes \cdots \otimes \mu_{Q_1}.$$

Similarly, the gaussian measure

$$\gamma_{n,\pi}(C) = \int_C \exp(-\pi|x|^2)\, dx$$

is an even and log-concave measure in $\mathbb{R}^n$. Clearly,

$$\gamma_{n,\pi} = \gamma_{1,\pi} \otimes \cdots \otimes \gamma_{1,\pi}.$$

For every symmetric interval $J_a = [-a, a]$, $a > 0$, we check that $\mu_{Q_1}(J_a) \geqslant \gamma_{1,\pi}(J_a)$. This implies that $\mu_{Q_n} \succ \gamma_{n,\pi}$ for every $n \in \mathbb{N}$. Now, given $F \in G_{n,k}$, consider an orthonormal basis $\{w_1, \ldots, w_k\}$ of F and the $n \times k$ matrix A with columns w_j, and note that

$$|K \cap F| = \int_{\mathbb{R}^k} \mathbb{1}_{Q_n}(Ax)\, dx.$$

It follows that

$$|K \cap F| \geqslant \int_{\mathbb{R}^k} e^{-\pi|Ax|^2/2}\, dx = \int_{\mathbb{R}^k} e^{-\pi\langle A^*Ax, x\rangle/2}\, dx = \frac{1}{\sqrt{\det(A^*A)}}.$$

Now, observe that

$$\mathrm{tr}(A^*A) = \sum_{j=1}^{k} |w_j|^2 = k.$$

By the arithmetic-geometric means inequality

$$\det(A^*A) \leqslant \left(\frac{\mathrm{tr}(A^*A)}{k}\right)^n = 1,$$

which shows that $|K \cap F| \geqslant 1$. A different proof of Vaaler's theorem, which is based on topological methods, appears in [8].

A general upper bound for the volume of hyperplane sections of the cube was given by Hensley in [257], who showed that $|Q_n \cap \xi^\perp| \leqslant 5$ for every $\xi \in S^{n-1}$. Hensley also conjectured that the sharp upper bound should be $\sqrt{2}$, attained when $\xi = (\frac{1}{\sqrt{2}}, \frac{1}{\sqrt{2}}, 0, \ldots, 0)$. This was proved by Ball in [38].

Theorem 9.1.3. *For every $\xi \in S^{n-1}$ we have*

$$|Q_n \cap \xi^{\perp}| \leqslant \sqrt{2},$$

with equality if and only if $\xi = (\pm e_i \pm e_j)/\sqrt{2}$ for some $1 \leqslant i < j \leqslant n$.

Sketch of the proof. Let $\xi \in S^{n-1}$ and let f be the density of $\langle X, \xi\rangle$, where X is a random vector uniformly distributed in Q_n. Then, the coordinates X_j of X are independent and uniformly distributed in $[-\frac{1}{2}, \frac{1}{2}]$. By the Fourier inversion formula

$$\begin{aligned}|Q_n \cap \xi^{\perp}| = f(0) &= \frac{1}{2\pi}\int_{-\infty}^{\infty} \widehat{f}(t)\,dt = \frac{1}{2\pi}\int_{-\infty}^{\infty}\prod_{j=1}^{n}\frac{\sin(\frac{1}{2}\xi_j t)}{\frac{1}{2}\xi_j t}\,dt \\ &= \frac{1}{\pi}\int_{-\infty}^{\infty}\prod_{j=1}^{n}\frac{\sin(\xi_j t)}{\xi_j t}\,dt.\end{aligned}$$

This formula appears already in Pólya's work [465], and was also used by Hensley. Applying Hölder's inequality with weights $p_j = \xi_j^{-2}$ we get

$$\begin{aligned}|Q_n \cap \xi^{\perp}| = \frac{1}{\pi}\int_{-\infty}^{\infty}\prod_{j=1}^{n}\frac{\sin(\xi_j t)}{\xi_j t}\,dt &\leqslant \prod_{j=1}^{n}\left(\frac{1}{\pi}\int_{-\infty}^{\infty}\left|\frac{\sin(\xi_j t)}{\xi_j t}\right|^{\xi_j^{-2}} dt\right)^{\xi_j^2} \\ &= \prod_{j=1}^{n}\left(\frac{1}{\pi}\frac{1}{\xi_j}\int_{-\infty}^{\infty}\left|\frac{\sin t}{t}\right|^{\xi_j^{-2}} dt\right)^{\xi_j^2}.\end{aligned} \tag{9.2}$$

This computation appears in the work of Haagerup [245], where he computes the best constants in Khintchine's inequalities. Then, Ball proves the integral inequality

$$\frac{1}{\pi}\int_{-\infty}^{\infty}\left|\frac{\sin t}{t}\right|^{p} dt \leqslant \sqrt{\frac{2}{p}} \tag{9.3}$$

for all $p \geqslant 2$, with equality if and only if $p = 2$. If we assume that $|\xi_j| \leqslant \frac{1}{\sqrt{2}}$ for all $1 \leqslant j \leqslant n$, then

$$\left(\frac{1}{\pi}\frac{1}{\xi_j}\int_{-\infty}^{\infty}\left|\frac{\sin t}{t}\right|^{\xi_j^{-2}} dt\right)^{\xi_j^2} \leqslant (\sqrt{2})^{\xi_j^2}$$

for every $1 \leqslant j \leqslant n$ and the result follows from (9.2), since $\sum_{j=1}^{n}\xi_j^2 = 1$.

On the other hand, if $|\xi_j| \geqslant \frac{1}{\sqrt{2}}$ for some $1 \leqslant j \leqslant n$, then we can check that $P_{e_j^{\perp}}(Q_n \cap \xi^{\perp}) = Q_n \cap e_j^{\perp}$, which implies that

$$|Q_n \cap \xi^\perp| = \frac{1}{|\xi_j|}|P_{e_j^\perp}(Q_n \cap \xi^\perp)| = \frac{1}{|\xi_j|}|Q_n \cap e_j^\perp| = \frac{1}{|\xi_j|} \leqslant \sqrt{2}.$$

This completes the proof. □

An alternative proof of (9.3) was given by Nazarov and Podkorytov in [437]. We shall not present the details of the proof of (9.3), as they can be also found in [310, Section 7.1] and [113, Section 4.5.3]. Various extensions, generalizations, discrete versions, and stability properties were obtained with this approach (see [33, 155, 278, 383, 408, 409]).

Chasapis, Nayar, and Tkocz have obtained in [128] a dimension-free stability result for both the lower and the upper bound.

Theorem 9.1.4. *Let $\xi \in S^{n-1}$ with $\xi_1 \geqslant \cdots \geqslant \xi_n \geqslant 0$. Then,*

$$1 + c_1|\xi - e_1|^2 \leqslant |Q_n \cap \xi^\perp| \leqslant \sqrt{2} - c_2\left|\xi - \frac{e_1 + e_2}{\sqrt{2}}\right|,$$

where $c_1, c_2 > 0$ are absolute constants.

The exponents 2 and 1 on the left- and the right-hand sides, respectively, are the best possible, as can be explicitly verified for $n = 2$. Melbourne and Roberto [408] have obtained, independently, a similar result.

The proof of Theorem 9.1.4 makes use of a formula that expresses the volume of sections in terms of negative moments of linear functionals, which was originally observed in [273]. If f is a density on $\mathbb{R}$, which is continuous at 0, then

$$f(0) = \lim_{q \to -1+} \frac{1+q}{2} \int |x|^q f(x)\, dx. \tag{9.4}$$

This allows us to reformulate Theorem 9.1.1 and Theorem 9.1.3 as follows:

$$1 \leqslant \lim_{q \to -1+} \frac{1+q}{2} \mathbb{E}\left|\sum_{j=1}^n \xi_j U_j\right|^q \leqslant \sqrt{2}$$

for every $\xi \in S^{n-1}$, where U_i are independent random variables uniformly distributed in $[-\frac{1}{2}, \frac{1}{2}]$. The next lemma holds for any star body and its higher codimensional sections.

Lemma 9.1.5. *Let K be a star-shaped body in $\mathbb{R}^n$ of volume 1. Let $1 \leqslant k \leqslant n-1$ and $F \in G_{n,n-k}$, and let X be a random vector uniformly distributed in K. Let $\|\cdot\|$ be a norm in $F^\perp$ and write B for its unit ball. Then,*

$$|K \cap F| = \lim_{q \to -k+} \frac{k+q}{k|B|} \mathbb{E}\|P_{F^\perp} X\|^q.$$

Proof. Let $f : F^\perp \to [0, +\infty)$ be the density of $P_{F^\perp} X$. Then,

$$|K \cap F| = f(0).$$

We identify $F^{\perp}$ with $\mathbb{R}^k$. Given $\varepsilon > 0$, the set $A_\varepsilon := \{x : f(x) < f(0) + \varepsilon\}$ contains a neighbourhood δB of 0. Then,

$$\frac{k+q}{k|B|}\int_{\mathbb{R}^k} \|x\|^q f(x)\,dx \leqslant (f(0)+\varepsilon)\frac{k+q}{k|B|}\int_{\delta K}\|x\|^q\,dx + \frac{k+q}{k|B|}\delta^q\int_{\mathbb{R}^k} f$$
$$= (f(0)+\varepsilon)\delta^{k+q} + \frac{k+q}{k|B|}\delta^q\int_{\mathbb{R}^k} f.$$

Letting $q \downarrow -k$ and then $\varepsilon \to 0^+$ we see that

$$\limsup_{q\to -k+}\frac{k+q}{k|B|}\mathbb{E}\|P_{F^{\perp}}X\|^q \leqslant |K\cap F|.$$

Choosing $0 < \varepsilon < f(0)$ and working in a similar way with the set $B_\varepsilon := \{x : f(x) > f(0)-\varepsilon\}$ that also contains a neighbourhood δB of 0, we see that

$$\frac{k+q}{k|B|}\int_{H^{\perp}} \|x\|^{-q} f(x)\,dx \geqslant (f(0)-\varepsilon)\frac{k+q}{k|B|}\int_{\delta K}\|x\|^q\,dx = (f(0)-\varepsilon)\delta^{k+q}$$

and the lemma follows. □

In [345] this formula was exploited to study hyperplane sections of the cube and obtained a useful expression for the above limit in this case.

Lemma 9.1.6. *Let $Y_1, Y_2, \ldots$ be independent random vectors uniformly distributed on the sphere S^2 in $\mathbb{R}^3$. For any $\xi \in S^{n-1}$ we have*

$$|Q_n \cap \xi^{\perp}| = \mathbb{E}\left|\sum_{j=1}^n \xi_j Y_j\right|^{-1}.$$

Proof. From Lemma 9.1.5 we see that

$$|Q_n \cap \xi^{\perp}| = \lim_{q\to -1+}\frac{1+q}{2}\mathbb{E}\left|\sum_{j=1}^n \xi_j X_j\right|^q,$$

where $X = (X_1, \ldots, X_n)$ is uniform on Q_n. Using the fact that $\langle Y_j, e_1\rangle$ has the same distribution as $2X_j$, one can check that

$$\frac{1+q}{2}\mathbb{E}\left|\sum_{j=1}^n \xi_j Y_j\right|^q = 2^{-1-q}\mathbb{E}\left|\sum_{j=1}^n \xi_j Y_j\right|^q$$

for every $q > -1$ (a proof of this claim is given in [127]). Letting $q \to -1^+$ we obtain the lemma. □

Lemma 9.1.6 allows us to prove the lower bound in Theorem 9.1.4.

Proof. We write

$$\left|\sum_{j=1}^{n}\xi_j Y_j\right|^2 = \sum_{i,j}\xi_i\xi_j\langle Y_i, Y_j\rangle = 1 + 2\sum_{i<j}\xi_i\xi_j\langle Y_i, Y_j\rangle.$$

The random variable $R = 2\sum_{i<j}\xi_i\xi_j\langle Y_i, Y_j\rangle$ has mean 0. Therefore,

$$\mathbb{E}(1+R)^{-1/2} \geqslant \mathbb{E}(1-R/2) = 1.$$

To obtain the result, we use the more precise inequality

$$(1+r)^{-1/2} \geqslant 1 - \frac{1}{2}r + \frac{1}{3}r^2 - \frac{5}{24}r^3, \quad r > -1$$

and estimate the expectations $\mathbb{E}(R^2)$ and $\mathbb{E}(R^3)$ that can be expressed explicitly in terms of the coordinates ξ_j of ξ. □

We move on now to upper bounds for the volume of lower-dimensional sections of the cube. The next theorem of Ball [41] provides two general bounds.

Theorem 9.1.7. *Let $1 \leqslant k \leqslant n$. For every $F \in G_{n,k}$ we have that*

$$|Q_n \cap F| \leqslant \min\left\{\left(\frac{n}{k}\right)^{k/2}, 2^{\frac{n-k}{2}}\right\}.$$

The first bound $(n/k)^{k/2}$ is optimal when k divides n. The second bound $2^{\frac{n-k}{2}}$ is better than the first one for $k \geqslant \frac{n}{2}$ and turns out to be sharp in this case. The proof of both bounds is based on Ball's geometric version of the Brascamp–Lieb inequality [109]. For the first bound, we apply the Brascamp–Lieb inequality to indicator functions of intervals, while for the second bound we apply it to the Fourier analytic formula. The bound $2^{\frac{n-k}{2}}$, which is exponential in the codimension $n-k$, is consistent with the general fact that if K is an isotropic convex body in $\mathbb{R}^n$, then

$$|K \cap F| \leqslant C^{n-k}$$

for every $F \in G_{n,n-k}$, which we discussed in Chapter 2.

The optimal upper bound remains an open question for many values of k and n. It is conjectured in [266, 466] that the maximal volume k-dimensional central section of the cube is attained at an affine cube. More precisely, given $1 \leqslant k \leqslant n$, write $n = k\ell + r$, where r is the remainder of the division of n by k and consider the following k orthogonal vectors in $\mathbb{R}^n$: $u_{j+1} = \sum_{s=1}^{\ell} e_{j\ell+s}$ $(0 \leqslant j < k-r)$ and $u_{k-r+j} = \sum_{s=1}^{\ell} e_{(d-r)\ell+(\ell+1)j+s}$ $(0 \leqslant j < r)$. Denote by F^* the k-dimensional subspace spanned by these vectors. Then,

$$Q_n \cap F^* = \left\{ \sum_{j=1}^{k} t_k u_k,\ |t_1|, \ldots, |t_k| \leqslant \frac{1}{2} \right\},$$

which is an affine cube of volume

$$|Q_n \cap F^*| = \prod_{j=1}^{k} |u_j| = \sqrt{\ell}^{\,k-r} \sqrt{\ell+1}^{\,r}.$$

Note that this quantity is equal to $(n/k)^{k/2}$ when k divides n and equal to $2^{\frac{n-k}{2}}$ when $k \geqslant n/2$.

Conjecture 9.1.8. *Let $1 \leqslant k \leqslant n-1$. For every $F \in G_{n,k}$ we have*

$$|Q_n \cap F| \leqslant |Q_n \cap F^*|.$$

Apart from the cases that are covered by Theorem 9.1.7, this conjecture has been confirmed for planar sections ($k = 2$) by Ivanov and Tsiutsiurupa [269], who developed local conditions for extremal subspaces.

9.1.2 Sections of the *p*-balls

Next, we present a number of results about the extremal sections of the unit ball B_p^n of ℓ_p^n, $0 \leqslant p < +\infty$, starting with a monotonicity result of Meyer and Pajor from [411].

Theorem 9.1.9. *Let $1 \leqslant k \leqslant n-1$. For every $F \in G_{n,k}$, the function*

$$p \mapsto \frac{|B_p^n \cap F|}{|B_p^k|}$$

is increasing on $[0, +\infty)$.

Since the value of this function at 2 is equal to 1 for every $F \in G_{n,k}$, Theorem 9.1.9 implies that

$$|B_p^n \cap F| \leqslant |B_p^k|, \quad 0 < p < 2,$$
$$|B_p^n \cap F| \geqslant |B_p^k|, \quad p > 2,$$

with equality if and only if F is a k-dimensional coordinate subspace.

Letting $p \to \infty$ we obtain Vaaler's Theorem 9.1.2 on the extremal sections of the cube. Vaaler's argument uses the notion of peakedness in order to compare the uniform and gaussian distributions. The key point in [411] is that one can extend this comparison to the whole family of p-measures with densities $\{e^{-c_p|x|^p}\}_{p>0}$. The article of Meyer and

Pajor treats the case $1 \leqslant p < +\infty$; the result was later extended to $p < 1$ independently by Barthe in [56] and Caetano in [122].

We shall outline a proof of Theorem 9.1.9 that is based on the normalized cone volume measure μ_K of a convex body K in $\mathbb{R}^n$, defined by

$$\mu_K(A) = \frac{|\mathrm{conv}(\{0\} \cup A)|}{|K|}$$

for every Borel set $A \subseteq \mathrm{bd}(K)$. A very useful general formula of Naor and Romik from [434] states that

$$\int h = n|K| \int_0^{\infty} r^{n-1} \int_{\mathrm{bd}(K)} h(rz)\, d\mu_K(z)\, dr \tag{9.5}$$

for any integrable function $h : \mathbb{R}^n \to \mathbb{R}$. For the proof of this identity we may assume that $h = \mathbb{1}_A$, where $A = [a,b] \cdot A_0$ and $A_0 \subset \mathrm{bd}(K)$, because these sets generate the sigma algebra of Borel sets in $\mathbb{R}^n$. For any $z \in \mathrm{bd}(K)$ and $r > 0$ we have $h(rz) = \mathbb{1}_{[a,b]}(r)\mathbb{1}_{A_0}(z)$. Thus, (9.5) reduces to

$$|A| = |K|\Big(\int_a^b nr^{n-1}\, dr\Big)\mu_K(A_0) = |K|(b^n - a^n)\mu_K(A_0) = \big|[a,b]A_0\big|, \tag{9.6}$$

which holds true.

The cone volume measure $\mu_{B_p^n}$ admits a probabilistic representation in terms of i. i. d. random variables, which was discovered by Rachev and Rüschendorf in [473] and independently by Schechtman and Zinn in [500]. We shall use a modification of the representation of the uniform measure on B_p^n obtained in [65] by Barthe, Guédon, Mendelson, and Naor, more precisely, the next generalization from [470].

Lemma 9.1.10. *Let $f : [0,\infty) \to [0,\infty)$ be a continuous function and let K be a symmetric convex body in $\mathbb{R}^n$. If Z is a random vector in $\mathbb{R}^n$ with density $f(\|x\|_K)$ and U is a random variable that is uniform in $[0,1]$ and independent of Z, then*
(i) $\frac{Z}{\|Z\|_K}$ *has distribution μ_K and $U^{1/n}\frac{Z}{\|Z\|_K}$ is uniformly distributed on K;*
(ii) $\frac{Z}{\|Z\|_K}$ *and $\|Z\|_K$ are independent.*

In the particular case $K = B_p^n$ we can take $Z = (Y_1, \dots, Y_n)$, where Y_i are independent random variables with density $(2\Gamma(1+\frac{1}{p}))^{-1}e^{-|t|^p}$.

Proof. From (9.5) we see that for any $\phi : \mathbb{R}^n \to \mathbb{R}$ and $\psi : \mathbb{R} \to \mathbb{R}$ we have

$$\mathbb{E}\Big(\phi\Big(\frac{Z}{\|Z\|_K}\Big)\psi(\|Z\|_K)\Big) = \int_{\mathbb{R}^n} \phi\Big(\frac{x}{\|x\|_K}\Big)\psi(\|x\|_K)f(\|x\|_K)\, dx$$

$$= n|K| \int_0^\infty \psi(r) f(r) r^{n-1}\, dr \int_{\mathrm{bd}(K)} \phi(z)\, d\mu_K(z).$$

Choosing $\phi, \psi \equiv 1$ we check that $n|K| \int_0^\infty f(r) r^{n-1}\, dr = 1$. Then, the choice $\psi \equiv 1$ gives

$$\mathbb{E}\left(\phi\left(\frac{Z}{\|Z\|_K}\right)\right) = \int_{\mathrm{bd}(K)} \phi(z)\, d\mu_K(z)$$

and the choice $\phi \equiv 1$ gives

$$\mathbb{E}(\psi(\|Z\|_K)) = n|K| \int_0^\infty \psi(r) f(r) r^{n-1}\, dr.$$

The first equation shows that $\frac{Z}{\|Z\|_K}$ has distribution μ_K. Moreover,

$$\mathbb{E}\left(\phi\left(\frac{Z}{\|Z\|_K}\right)\psi(\|Z\|_K)\right) = \mathbb{E}\left(\phi\left(\frac{Z}{\|Z\|_K}\right)\right)\mathbb{E}(\psi(\|Z\|_K)),$$

which shows (ii). Since $U^{1/n}$ has density nr^{n-1} on $[0,1]$, from (9.6) we see that

$$\frac{|A|}{|K|} = \mathbb{P}(U^{1/n} \in [a,b])\mathbb{P}\left(\frac{Z}{\|Z\|_K} \in A_0\right) = \mathbb{P}\left(U^{1/n}\frac{Z}{\|Z\|_K} \in A\right),$$

which shows that $U^{1/n}\frac{Z}{\|Z\|_K}$ is uniformly distributed on K. □

The next lemma comes from [128].

Lemma 9.1.11. *Let $p > 0$ and $Y_1, Y_2, \ldots$ be independent random variables with density $e^{-\beta_p^p |x|^p}$, where $\beta_p = 2\Gamma(1 + 1/p)$. Let $F \in G_{n,n-k}$ and let U be a $k \times n$ matrix U whose rows form an orthonormal basis of $F^\perp$. Let $v_1, \ldots, v_n \in \mathbb{R}^k$ denote the columns of U. Then,*

$$|B_p^n \cap F| = |B_p^{n-k}| \lim_{q \to -k+} \frac{k+q}{k|B|} \mathbb{E}\left\|\sum_{j=1}^n Y_j v_j\right\|^q,$$

where $\|\cdot\|$ is a norm on $\mathbb{R}^k$ with unit ball B.

Proof. Let $X = (X_1, \ldots, X_n)$ be a random vector uniformly distributed in B_p^n. Lemma 9.1.5 gives a formula similar to the one we want with the X_j in place of Y_j and without the factor $|B_p^{n-k}|$. Lemma 9.1.10 shows that if $Y = (Y_1, \ldots, Y_n)$ and $S = (\sum_{i=1}^n |Y_i|^p)^{1/p}$, then the random vector $\frac{Y}{S}$ is independent of S and moreover $U^{1/n}\frac{Y}{S}$ is uniformly distributed in B_p^n, where U is independent of Y_i and uniform on $[0,1]$. Therefore,

$$\mathbb{E}\left\|\sum_{j=1}^n X_j v_j\right\|^q = \mathbb{E}\left\|\sum_{j=1}^n U^{1/n}\frac{Y_j}{S}v_j\right\|^q = \mathbb{E}(U^{q/n})\,\frac{\mathbb{E}(S^q)}{\mathbb{E}(S^q)}\cdot\mathbb{E}\left\|\sum_{j=1}^n \frac{Y_j}{S}v_j\right\|^q$$
$$= \frac{\mathbb{E}(U^{q/n})}{\mathbb{E}(S^q)}\,\mathbb{E}\left\|\sum_{j=1}^n Y_j v_j\right\|^q. \qquad \square$$

It is now clear that combining Lemma 9.1.11 with the next theorem we obtain Theorem 9.1.9.

Theorem 9.1.12. *For $p > 0$, let $Y_1^{(p)}, Y_2^{(p)}, \ldots$ be independent random variables with density $e^{-\beta_p^p|x|^p}$, where $\beta_p = 2\Gamma(1+1/p)$. For any choice of vectors $v_1, \ldots, v_n$ in $\mathbb{R}^k$ and $-k < q < 0$, the function*

$$(p_1, \ldots, p_n) \mapsto \mathbb{E}\left\|\sum_{j=1}^n Y_j^{(p_j)} v_j\right\|^q$$

is increasing in each variable.

Proof. If $0 < p < p'$, then the density of $Y_1^{(p')}$ intersects the density of $Y_1^{(p)}$ exactly once and dominates it pointwise near the origin. Thus, $Y_1^{(p')}$ is more peaked than $Y_1^{(p)}$ and this implies that $\sum Y_j^{(p_j')} v_j$ is more peaked than $\sum Y_j^{(p_j)} v_j$, if $p_j \leqslant p_j'$. In particular, for every $t > 0$

$$\mathbb{P}\left(\left\|\sum_{j=1}^n Y_j^{(p_j)} v_j\right\| \leqslant t\right) \leqslant \mathbb{P}\left(\left\|\sum_{j=1}^n Y_j^{(p_j')} v_j\right\| \leqslant t\right)$$

and the result follows by integrating with respect to t. $\square$

Some additional results are known for the hyperplane sections of B_p^n when $0 < p < 2$. Meyer and Pajor showed in [411] that the minimal volume hyperplane sections of the cross-polytope B_1^n are the ones in the diagonal directions and conjectured that the same is true for every $0 < p < 2$. This was confirmed by Koldobsky in [304], who established a strong Schur-convexity-type result (see [75] for background information on majorization and Schur-convexity).

Theorem 9.1.13. *Let $0 < p < 2$. If aξ and η are unit vectors in $\mathbb{R}^n$ such that $(\eta_1^2, \ldots, \eta_n^2)$ majorizes $(\xi_1^2, \ldots, \xi_n^2)$, then*

$$|B_p^n \cap \xi^\perp| \leqslant |B_p^n \cap \eta^\perp|.$$

Theorem 9.1.13 helps us to identify the minimal and maximal volume sections of B_p^n using the fact that

$$\left(\frac{1}{n}, \ldots, \frac{1}{n}\right) \prec (\xi_1^2, \ldots, \xi_n^2) \prec (1, 0, \ldots, 0)$$

for every $\xi \in S^{n-1}$. The additional feature of the range $0 < p < 2$ is the fact that the Fourier transform of $e^{-|x|^p}$ is a non-negative function of the form $t \mapsto \int_0^\infty e^{-ut^2}\,d\mu(u)$.

More generally, we say that a random variable X is a symmetric gaussian mixture if X has the same distribution as RG for some non-negative random variable R and a standard gaussian random variable G, independent of R. Gaussian mixtures are continuous, i. e., have densities, and X is a gaussian mixture if and only if its density f is of the form

$$f(x) = \int_0^\infty e^{-tx^2}\,d\nu(t)$$

for some Borel measure ν on $[0,+\infty)$. By Bernstein's theorem, this is equivalent to the fact that $g(x) = f(\sqrt{x})$ is completely monotone; we have that $(-1)^n g^{(n)}(x) \geqslant 0$ for all $n \geqslant 0$ and $x > 0$. Gaussian mixtures are studied in [164]. A useful observation is that if $X_1 = R_1G_1, \ldots, X_n = R_nG_n$ are independent gaussian mixtures and $v_1, \ldots, v_n \in \mathbb{R}^k$, then, conditioned on the values of the R_j, the sum $\sum X_jv_j$ is a centered gaussian random vector in $\mathbb{R}^k$ with covariance matrix $\sum R_j^2 v_j v_j^t$.

The measure with density $e^{-\beta_p^p|x|^p}$ from Lemma 9.1.5 has a gaussian mixture form when $0 < p < 2$. As a consequence, we can evaluate the limit from Lemma 9.1.5.

Lemma 9.1.14. *Let $0 < p < 2$. There exist nonnegative i. i. d. random variables $R_1, R_2, \ldots$ such that for every $F \in G_{n,n-k}$ we have*

$$|B_p^n \cap F| = |B_p^{n-k}|\,\mathbb{E}\left(\det\left(\sum_{j=1}^n R_jv_jv_j^t\right)\right)^{-1/2},$$

where $v_1, \ldots, v_n$ are vectors in $\mathbb{R}^k$ such that the rows of the $k \times n$ matrix with columns $v_1, \ldots, v_n$ form an orthonormal basis of $F^\perp$.

Note. To describe the distribution of the R_j, for $0 < \alpha < 1$, we let g_α be the density of a standard positive α-stable random variable W_α, i. e., with Laplace transform $\mathbb{E}(e^{-uW_\alpha}) = e^{-u^\alpha}$ for $t > 0$, and let $V_1, V_2, \ldots$ be i. i. d. random variables with density $\frac{\sqrt{\pi}}{2\Gamma(1+1/p)} t^{-3/2} g_{p/2}(t^{-1})$. Then, $R_j = (\mathbb{E}V_j^{-1/2})^2 V_j$, see [164].

Proof. The random variables Y_j from Lemma 9.1.5 are gaussian mixtures, say $Y_j = T_jG_j$, for some non-negative random variables T_j and standard gaussians G_j, and all of them are independent. Then, conditioned on the T_j, the limit in Lemma 9.1.5 gives the density at 0 of $\sum Y_jv_j$, which is a centered gaussian random vector in $\mathbb{R}^k$ with covariance $\sum T_j^2 v_jv_j^t$, and hence its density at 0 is equal to $(2\pi)^{-k/2}(\det[\sum_{j=1}^n T_j^2 v_jv_j^t])^{-1/2}$. □

In the case of hyperplane sections, using this formula we can give a direct proof of the Schur-convexity theorem.

Proof of Theorem 9.1.13. We observe that if $H : \mathbb{R}^n \to \mathbb{R}$ is convex and permutation symmetric, then H is Schur-convex, namely $x \prec y$ implies $H(x) \leqslant H(y)$. To see this, we use the standard fact (see [75]) that there exist $(\lambda_\sigma)_{\sigma\in S_n}$, where S_n is the set of permutations of $\{1,\dots,n\}$ such that $\lambda_\sigma \geqslant 0$, $\sum_{\sigma\in S_n}\lambda_\sigma = 1$, and $x = \sum_{\sigma\in S_n}\lambda_\sigma y_\sigma$, where $y_\sigma = (y_{\sigma(1)},\dots,y_{\sigma(n)})$. Then,

$$H(x) = H\Big(\sum_{\sigma\in S_n}\lambda_\sigma y_\sigma\Big) \leqslant \sum_{\sigma\in S_n}\lambda_\sigma H(y_\sigma) = \sum_{\sigma\in S_n}\lambda_\sigma H(y) = H(y).$$

Now, given $\xi \in S^{n-1}$, from Lemma 9.1.14 we have that

$$|B_p^n \cap \xi^\perp| = |B_p^{n-1}|\,\mathbb{E}\Big(\sum_{j=1}^n a_j^2 R_j\Big)^{-1/2}.$$

Since $m(x) = x^{-1/2}$ is convex and the random variables R_j are i. i. d., the right-hand side is a convex and permutation symmetric function of $(\xi_1^2,\dots,\xi_n^2)$, which implies that it is also Schur-convex. □

The question to identify the maximal-volume hyperplane sections of B_p^n-balls for $2 < p < \infty$ remains open. Oleszkiewicz has proved in [447] that Ball's upper bound for the cube, Theorem 9.1.3, does not extend to all $p > 2$; it fails for all $p < 26.265\dots$ and n large enough. This follows from comparing the hyperplane section $(1,1,\dots,0)^\perp$ to the diagonal one $(1,1\dots,1)^\perp$ as $n \to \infty$.

Eskenazis, Nayar, and Tkocz [165] prove that if p is sufficiently large, then the extremal hyperplanes are Ball-type for any dimension $n \geqslant 3$: There exists $p_0 \in (26, 10^{15})$ such that for all $n \geqslant 2$ and all $p \geqslant p_0$ one has

$$\max_{\xi\in S^{n-1}} |B_p^n \cap \xi^\perp| = \Big|B_p^n \cap \Big(\frac{e_1+e_2}{\sqrt{2}}\Big)^\perp\Big|.$$

For the proof, they strengthen several of the existing tools and introduce a number of completely new ideas. Consider the normalized section function

$$A_{n,p}(\xi) = \frac{|B_p^n \cap \xi^\perp|}{|B_p^{n-1}|}, \quad 2 \leqslant p \leqslant \infty,\ \xi \in S^{n-1}$$

and given a unit vector $\xi = (\xi_1,\dots,\xi_n)$ with $\xi_1 \geqslant \xi_2 \geqslant \dots \geqslant \xi_n \geqslant 0$, set $\delta(\xi) = |\xi - \frac{e_1+e_2}{\sqrt{2}}|^2$. The authors distinguish two cases: $\sqrt{\delta(\xi)} \geqslant \delta_p$ and $\sqrt{\delta(\xi)} < \delta_p$, where $\delta_p \approx 1/p$. In the first case, when ξ is far from the extremizer $\frac{e_1+e_2}{\sqrt{2}}$, they show that there is a deficit in Ball's inequality for the case $p = \infty$

$$A_{n,\infty}(\xi) \leqslant \sqrt{2} - \kappa_\infty\sqrt{\delta(\xi)},$$

where $\kappa_\infty = 6 \cdot 10^{-5}$. The existence of such a constant κ_∞ is proved by Chasapis, Nayar, and Tkocz in [128] (see Theorem 9.1.4), but the authors need and prove a strengthened version, with this value for κ_∞. Moreover, they show that for all $p > 5$ one has

$$\max_{\xi \in S^{n-1}} |A_{n,p}(\xi) - A_{n,\infty}(\xi)| \leqslant \frac{5}{p}.$$

Combining the above, one gets

$$A_{n,p}(\xi) \leqslant \frac{5}{p} + \sqrt{2} - \kappa_\infty \sqrt{\delta(\xi)} \leqslant 2^{\frac{1}{2}-\frac{1}{p}} = A_{n,p}\left(\frac{e_1 + e_2}{\sqrt{2}}\right),$$

if $\sqrt{\delta(\xi)} \geqslant \frac{C}{p}$ for some absolute constant $c > 0$ (the value $C = 10^5$ works). In the second case, when ξ is close to the extremizer $\frac{e_1+e_2}{\sqrt{2}}$ in the sense that $\sqrt{\delta(\xi)} < \frac{C}{p}$, the authors express $|B_p^n \cap \xi^\perp|$ as a negative moment of an independent sum and use an inductive argument to reduce the problem to a two-dimensional one. Delicate estimates lead to the required estimate under the assumption that p is large enough.

It has been conjectured that in each dimension there is a unique phase-transition point (see [270, 342, 343])..

Conjecture 9.1.15. *For every $n \geqslant 3$, there is a unique $p_0(n)$ such that*

$$\max_{\xi \in S^{n-1}} |B_p^n \cap \xi^\perp| = |B_p^n \cap (1, \ldots, 1)^\perp|,$$

if $2 < p \leqslant p_0(n)$, and

$$\max_{\xi \in S^{n-1}} |B_p^n \cap \xi^\perp| = |B_p^n \cap (1, 1, 0, \ldots, 0)^\perp|,$$

if $p \geqslant p_0(n)$.

For lower-dimensional sections, there is a general bound of Barthe [56] that extends Theorem 9.1.7. As in the case of the cube, the proof exploits the Brascamp–Lieb inequalities. The result is again sharp when k divides n with the same extremizing subspaces.

Theorem 9.1.16 (Barthe). *Let $p \geqslant 2$ and $1 \leqslant k \leqslant n-1$. For every $F \in G_{n,k}$ we have*

$$|B_p^n \cap F| \leqslant \left(\frac{n}{k}\right)^{k(1/2-1/p)} |B_p^k|.$$

Note. We close this section with a number of related results.

(i) All known results on extremal-volume hyperplane sections of the ℓ_p-balls admit robust versions in the spirit of Theorem 9.1.4 (see [128]).
(ii) Nazarov (see [128] for the proof) has shown that minimal-area planar sections of the cross-polytope are attained at regular polygons: For every $n \geqslant 3$ and every 2-dimensional subspace F of $\mathbb{R}^n$, we have

$$|B_1^n \cap F| \geqslant \frac{n^2 \sin^3(\frac{\pi}{2n})}{\cos(\frac{\pi}{2n})},$$

with equality when $B_1^n \cap F$ is a regular $2n$-gon.

(iii) Vaaler conjectured in [530] that for every origin-symmetric isotropic convex body K in $\mathbb{R}^n$ and for every $1 \leqslant k \leqslant n-1$ and $F \in G_{n,k}$ we have

$$|K \cap F| \geqslant 1. \tag{9.7}$$

We have seen that this inequality holds true for the cube and that the constant 1 is optimal. The lower bound of Meyer and Pajor confirms (9.7) for $K = B_p^n$, where $2 < p < \infty$ (see [423]) and for $1 < p < 2$ in the hyperplane case $k = n-1$ (see [502]).

(iv) For more results on sections of the cube, including non-central sections, slabs and the complex case, see [10, 19, 55, 63, 108, 221, 346, 347, 349, 429, 448, 467, 468]. For more results related to ℓ_p-balls, see [161, 326, 380].

(v) For results on sections of the simplex, see [538, 539, 115, 150, 151, 172, 341, 521, 433, 344].

9.2 Extremal projections of the unit balls of p-norms

9.2.1 Projections of the cube

As in the previous section, we start with the case of the cube. Using Cauchy's formula

$$|P_{\xi^\perp}(K)| = \frac{1}{2} \int_{S^{n-1}} |\langle \theta, \xi \rangle| \, dS(K, \theta), \tag{9.8}$$

we can determine the extremal volume hyperplane projections of Q_n. Note that the $2n$ basis vectors $\{\pm e_j, j \leqslant n\}$ are the outer normals to the facets of Q_n, and hence the surface area measure of Q_n is the counting measure $\sum_{j=1}^n \delta_{\pm e_j}$. Then, for every $\xi \in S^{n-1}$

$$|P_{\xi^\perp}(Q_n)| = \sum_{j=1}^n |\xi_j|$$

and this shows that

$$1 \leqslant |P_{\xi^\perp}(Q_n)| \leqslant \sqrt{n},$$

with equality if and only if ξ is a basis vector or a main diagonal vector, respectively.

For lower-dimensional projections one can exploit the fact that zonotopes coincide with projections of a possibly rescaled and translated cube in a sufficiently high dimension, and a formula of Shephard [509] for the volume of zonotopes: If $v_1, \ldots, v_n$ are vectors in $\mathbb{R}^k$, then the volume of the zonotope $Z = \sum_{j=1}^n [0, v_j]$ is given by

$$|Z| = \sum_{1\leqslant j_1<\cdots<j_k\leqslant n} |\det[v_{j_1} \ \dots \ v_{j_k}]|,$$

where $[v_{j_1} \ \dots \ v_{j_k}]$ is the $k\times k$ matrix with columns $v_{j_1},\dots,v_{j_k}$. When the zonotope is the orthogonal projection of Q_n onto a subspace $F\in G_{n,k}$, we see that the vectors v_j can be taken as columns of the $k\times n$ matrix whose rows form an orthonormal basis of F. Then, the Cauchy–Binet formula implies the constraint

$$\sum_{1\leqslant j_1<\cdots<j_k\leqslant n} |\det[v_{j_1} \ \dots \ v_{j_k}]|^2 = 1$$

and, as in the case $k = n-1$ case, we can deduce the next upper and lower bounds for the volume of $P_F(Q_n)$. This argument is due to Chakerian and Filliman [125].

Theorem 9.2.1. *Fix $1\leqslant k\leqslant n-1$. For every $F\in G_{n,k}$ we have*

$$1\leqslant |P_F(Q_n)| \leqslant \min\left\{\binom{n}{k}^{1/2}, \frac{\omega_{k-1}^k}{\omega_k^{k-1}}\left(\frac{n}{k}\right)^{k/2}\right\}.$$

Note that the lower bound in Theorem 9.2.1 is sharp: we have equality if F is a coordinate subspace. The first term $\sqrt{\binom{n}{k}}$ of the upper bound on the right-hand side is sharp only when $k=1$ or $n-1$. The second term comes from the observation that the diameter of $P_F(Q_n)$ is bounded by the diameter of the cube $\sqrt{n}$, and hence, by the isodiametric inequality, it has volume at most $\omega_k(\sqrt{n}/2)^k$. Both bounds are of the order $n^{k/2}$ for a fixed k as $n\to\infty$, which is the right order, and the second term of the upper bound in Theorem 9.2.1 is asymptotically better than the first one. Ivanov in [265] has developed local conditions for maximizers of k-dimensional projections.

An interesting observation of McMullen [407] and, independently, Chakerian and Filliman [125], is that projections of Q_n onto orthogonal complementary subspaces have the same volume. This follows again from Shephard's formula and readily implies that bounds on the volumes of k-dimensional and $(n-k)$-dimensional projections of the cube are equivalent.

Theorem 9.2.2. *Let $1\leqslant k\leqslant n-1$. For every $F\in G_{n,k}$ we have*

$$|P_F(Q_n)| = |P_{F^\perp}(Q_n)|.$$

We move on now to projections of the cross-polytope. Since B_1^n has 2^n congruent simplicial facets of $(n-1)$-dimensional volume $\frac{\sqrt{n}}{(n-1)!}$ with outer-normals $\frac{1}{\sqrt{n}}(\pm1,\dots,\pm1)$, using Cauchy's formula (9.8), for every $\xi\in S^{n-1}$ we get

$$|P_{\xi^\perp}(B_1^n)| = \frac{1}{2(n-1)!}\sum_{\varepsilon\in\{-1,1\}^n} |\langle\xi,\varepsilon\rangle| = \frac{2^{n-1}}{(n-1)!}\mathbb{E}\left|\sum_{j=1}^n \xi_j\varepsilon_j\right|,$$

where ε_j are independent symmetric Bernoulli random variables. Therefore, the question is to find the best constants c, C for which

$$c\Big(\mathbb{E}\Big|\sum \xi_j\varepsilon_j\Big|^2\Big)^{1/2} \leqslant \mathbb{E}\Big|\sum \xi_j\varepsilon_j\Big| \leqslant C\Big(\mathbb{E}\Big|\sum \xi_j\varepsilon_j\Big|^2\Big)^{1/2} \tag{9.9}$$

holds true for all $\xi_1, \dots, \xi_n \in \mathbb{R}$. In other words, we need to determine the best constants in the classical Khintchine inequality [280]. From the Cauchy–Schwarz inequality

$$\mathbb{E}\Big|\sum \xi_j\varepsilon_j\Big| \leqslant \Big(\mathbb{E}\Big|\sum \xi_j\varepsilon_j\Big|^2\Big)^{1/2},$$

we see that $C = 1$. We have equality if and only if $a = \pm e_i$ for some $i \leqslant n$; this shows that the maximum-volume projections are on the coordinate subspaces. The sharp lower bound, which had been also conjectured by Littlewood [376] in a different context, is attained when $\xi = \frac{\pm e_i \pm e_j}{\sqrt{2}}$, $i \neq j$. This was proved by Szarek in [519], and several simplified and different proofs were given later in [245, 358, 525]. We summarize in the next theorem.

Theorem 9.2.3. *Let $n \geqslant 2$. For every $\xi \in S^{n-1}$ we have*

$$\frac{1}{\sqrt{2}}|B_1^{n-1}| \leqslant |P_{\xi^\perp}(B_1^n)| \leqslant |B_1^{n-1}|.$$

Below, we present two completely different Fourier-analytic proofs. The first proof is due to Haagerup [245] and is in the same spirit as Ball's subsequent argument [38] for hyperplane sections of the cube.

First proof of Theorem 9.2.3. We want to minimize $\mathbb{E}|\sum \xi_j\varepsilon_j|$ under the constraint $\sum \xi_j^2 = 1$. We may assume that all the coordinates of ξ are positive. If at least one of them is larger than $\frac{1}{\sqrt{2}}$, say $\xi_1 > \frac{1}{\sqrt{2}}$, then averaging over the other coefficients we get

$$\mathbb{E}\Big|\sum \xi_j\varepsilon_j\Big| \geqslant \mathbb{E}_{\varepsilon_1}\Big|\xi_1\varepsilon_1 + \mathbb{E}\sum_{j>1} \xi_j\varepsilon_j\Big| = \xi_1 > \frac{1}{\sqrt{2}}.$$

So, we may assume that $\xi_j \leqslant \frac{1}{\sqrt{2}}$ for all $1 \leqslant j \leqslant n$. We start from the formula

$$|x| = \frac{1}{\pi}\int_{\mathbb{R}} (1 - \cos(tx))t^{-2}\,dt, \quad x \in \mathbb{R}.$$

This gives that, for any integrable random variable X

$$\mathbb{E}|X| = \frac{1}{\pi}\int_{\mathbb{R}} (1 - \mathrm{Re}(\mathbb{E}e^{itX}))t^{-2}\,dt$$

(see also Lemmas 2.3 and 4.2 in [245] as well as [229, Lemma 3]). In particular, using independence and the identity $\mathbb{E}(e^{it\varepsilon_j}) = \cos t$, we get

$$\mathbb{E}\left|\sum \xi_j \varepsilon_j\right| = \frac{1}{\pi}\int_{\mathbb{R}}\left(1 - \prod_{j=1}^{n}\cos(t\xi_j)\right)t^{-2}\,dt.$$

By the arithmetic-geometric means inequality, we obtain the lower bound

$$\mathbb{E}\left|\sum \xi_j \varepsilon_j\right| \geqslant \sum \xi_j^2 F(\xi_j^{-2}),$$

where

$$F(s) = \frac{1}{\pi}\int_{\mathbb{R}}\left(1 - \left|\cos\left(\frac{t}{\sqrt{s}}\right)\right|^s\right)t^{-2}\,dt, \quad s > 0.$$

This reminds us of (9.2). Here, the function F can be expressed explicitly. Using the fact that $\sum_{n=-\infty}^{\infty}\frac{1}{(t+n\pi)^2} = \frac{1}{\sin^2 t}$, we see that

$$\begin{aligned}
F(s) &= \frac{1}{\pi\sqrt{s}}\int_{\mathbb{R}}(1 - |\cos t|^s)t^{-2}\,dt \\
&= \frac{1}{\pi\sqrt{s}}\sum_{n=-\infty}^{\infty}\int_{-\pi/2}^{\pi/2}(1 - (\cos t)^s)(t + n\pi)^{-2}\,dt \\
&= \frac{1}{\pi\sqrt{s}}\int_{-\pi/2}^{\pi/2}(1 - (\cos t)^s)\sin^{-2} t\,dt \\
&= \frac{2}{\sqrt{\pi s}}\frac{\Gamma(\frac{s+1}{2})}{\Gamma(\frac{s}{2})}.
\end{aligned}$$

From this expression, we can check that F is increasing on $(0, +\infty)$. To see this, note that $\lim_{s\to\infty} F(s) = \sqrt{\frac{2}{\pi}}$ by Stirling's formula, and

$$F(s+2) = \sqrt{\frac{s}{s+2}}\frac{s+1}{s}F(s) = (1 - 1/(s+1)^2)^{-1/2}F(s),$$

which can be iterated to give $F(s+2n) = F(s)\prod_{k=0}^{n-1}(1-1/(s+2k+1)^2)^{-1/2}$. Letting $n \to \infty$ we get

$$F(s) = \sqrt{\frac{2}{\pi}}\prod_{k=0}^{\infty}(1 - 1/(s+2k+1)^2)^{1/2},$$

which is clearly an increasing function of s. Using also the fact that $\xi_j \leqslant \frac{1}{\sqrt{2}}$ for all $1 \leqslant j \leqslant n$, we can complete the proof, writing

$$\mathbb{E}\left|\sum \xi_j \varepsilon_j\right| \geqslant \sum \xi_j^2 F(\xi_j^{-2}) \geqslant \sum \xi_j^2 F(2) = F(2) = \frac{1}{\sqrt{2}}.$$

This inequality becomes an equality when $n = 2$ and $\xi_1 = \xi_2 = \frac{1}{\sqrt{2}}$. □

The second proof (see [352, 358, 446]) uses Fourier analysis on the discrete cube $\{-1,1\}^n$. We refer to [444, Chapter 1] for basic background.

Second proof of Theorem 9.2.3. We work on $L_2(\{-1,1\}^n,\mathbb{R})$, equipped with the product probability measure on the discrete cube $\{-1,1\}^n$ and the inner product $\langle f,g\rangle = \mathbb{E}[f(\varepsilon)g(\varepsilon)]$, where $f,g:\{-1,1\}^n\to\mathbb{R}$. Let

$$f(x)=\Big|\sum_{j=1}^{n}\xi_j x_j\Big|,\quad x\in\{-1,1\}^n.$$

We consider the Fourier expansion of f with respect to the Walsh orthonormal system. Recall that the Walsh function w_S is defined by $w_S(x)=\prod_{j\in S}x_j$ for any $S\subset\{1,\dots,n\}$, and $w_\varnothing(x)\equiv 1$. We can write

$$f(x)=\sum_{S}b_S w_S(x),\quad b_S=\langle f,w_S\rangle.$$

Since f is even, we have that $b_S=0$ if $|S|$ is odd. We consider the Laplace operator $\mathcal{L}$ acting on $L_2(\{-1,1\}^n,\mathbb{R})$, defined by

$$(\mathcal{L}g)(x)=\frac{1}{2}\sum_{y\sim x}(g(y)-g(x)),$$

where the sum is over all neighbors y of x, i. e., the points in $\{-1,1\}^n$ differing from x in one coordinate. One can check that the Walsh functions are the eigenfunctions of $\mathcal{L}$. More precisely, $\mathcal{L}w_S=-|S|w_S$, and for even functions g we have the Poincaré-type inequality

$$\langle g,-\mathcal{L}g\rangle\geqslant 2\mathrm{Var}(g).$$

Now, we claim that $(-\mathcal{L}f)(x)\leqslant f(x)$ for every $x\in\{-1,1\}^n$. This claim follows from the triangle inequality: we observe that

$$\begin{aligned}&|-\xi_1x_1+\xi_2x_2+\cdots+\xi_nx_n|+|\xi_1x_1-\xi_2x_2+\cdots+\xi_nx_n|\\&\quad+\cdots+|\xi_1x_1+\xi_2x_2+\cdots-\xi_nx_n|\geqslant(n-2)|\xi_1x_1+\cdots+\xi_nx_n|.\end{aligned}$$

Going back to the Poincaré inequality, we get

$$2(\mathbb{E}f^2-(\mathbb{E}f)^2)\leqslant\langle f,-\mathcal{L}f\rangle\leqslant\langle f,f\rangle=\mathbb{E}f^2,$$

which gives $\mathbb{E}f\geqslant\frac{1}{\sqrt{2}}(\mathbb{E}f)^2$. This is exactly the assertion of the theorem. □

This second proof uses only the triangle inequality and hence extends to the case where the coefficients ξ_j are vectors in an arbitrary normed space. The details appear in [352] and [446].

A stability version of Theorem 9.2.3 has been derived by De, Diakonikolas, and Servedio in [148] (see also [408] for a local statement with explicit constants).

Regarding lower-dimensional projections, it is conjectured that in analogy to Vaaler's Theorem 9.1.2 for the cube, maximal-volume projections of the cross-polytope B_1^n occur at coordinate subspaces.

Conjecture 9.2.4. *Let* $1 \leqslant k \leqslant n-1$. *For every* $F \in G_{n,k}$ *we have*

$$|P_F(B_1^n)| \leqslant |B_1^k|.$$

This conjecture has been stated by Ivanov in [266], who has confirmed it for $k = 2, 3$ and arbitrary n, using perturbation methods for frames (see also [265]). Earlier, Filliman in [170], established the same for $k = 2$, using different, more algebraic methods from his work [171], also reducing the case of $k = 3$ with arbitrary n to $n \leqslant 42$. For minimal-volume projections, we have the following dual analogue of Conjecture 9.1.8 for the cube.

Conjecture 9.2.5 (Ivanov). *Let* $1 \leqslant k \leqslant n-1$. *If* F^* *is the k-dimensional subspace from Conjecture* 9.1.8, *then for every* $F \in G_{n,k}$ *we have*

$$|P_F(B_1^n)| \geqslant |P_{F^*}(B_1^n)|.$$

This has been confirmed for $k = 2$ by Ivanov in [266]. As in the case of the cube, the conjectured maximizer is an affine cross-polytope.

9.2.2 Projections of the p-balls

Sharp results for the volume of hyperplane projections of the ℓ_p-balls, when $p \geqslant 2$, were obtained by Barthe and Naor in [64]. They follow from the next monotonicity result.

Theorem 9.2.6. *For every* $\xi \in S^{n-1}$, *the function*

$$p \mapsto \frac{|P_{\xi^\perp}(B_p^n)|}{|B_p^{n-1}|}$$

is increasing on $[1, +\infty)$.

This theorem may be viewed as a counterpart of Theorem 9.1.9. It is not known if such a monotonicity result holds true for lower-dimensional subspaces.

The proof of Theorem 9.2.6 makes use of the next lemma from [434], which relates the surface area measure of an origin-symmetric convex body to its cone volume measure.

Lemma 9.2.7. *Let K be an origin-symmetric convex body in $\mathbb{R}^n$. Then, the normalized surface area measure σ_K on $\mathrm{bd}(K)$ is absolutely continuous with respect to μ_K and for almost all $x \in \mathrm{bd}(K)$ one has*

$$\frac{d\sigma_K}{d\mu_K}(x) = \frac{n\,|K|}{|\mathrm{bd}(K)|}|\nabla(\|\cdot\|_K)(x)|.$$

Proof. Recall that the normalized cone volume measure μ_K of K is defined by $\mu_K(A) = |\mathrm{conv}(\{0\} \cup A)|/|K|$ for every Borel set $A \subseteq \mathrm{bd}(K)$. We also consider the unnormalized cone measure $C(A) = |\mathrm{conv}(\{0\} \cup A)|$ on $\mathrm{bd}(K)$.

Note that if x is perpendicular to the surface of K, then $|x|\,dS(K,x) = n\,dC(x)$. If the angle between the surface and x is α, then $|\cos\alpha|\,|x|\,dS(K,x) = n\,dC(x)$. We have $|\cos\alpha| = |\langle n(x), x/|x|\rangle|$. Let $z = \nabla\|\cdot\|_K(x)$. If $x \in \mathrm{bd}(K)$, then

$$1 + \varepsilon = \|x + \varepsilon x\|_K \approx \|x\|_K + \varepsilon\langle z, x\rangle = 1 + \varepsilon\langle z, x\rangle,$$

which gives $\langle z, x\rangle = 1$. Also, z is a vector perpendicular to $\mathrm{bd}(K)$. Thus, $n(x) = z/|z|$. It follows that

$$|\cos\alpha| = \frac{1}{|x|}\left|\langle n(x), x\rangle\right| = \frac{|\langle z, x\rangle|}{|x|\,|z|}.$$

This gives

$$\begin{aligned}\frac{|\mathrm{bd}(K)|d\sigma_K(x)}{|\nabla(\|\cdot\|_K)(x)|} &= \frac{dS(K,x)}{|\nabla(\|\cdot\|_K)(x)|} = \frac{|\langle z, x\rangle|}{|z|}\,dS(K,x)\\ &= n\,dC(x) = n\,|K|\,d\mu_K(x)\end{aligned}$$

and the lemma follows. □

A consequence of Lemma 9.2.7 and Cauchy's formula is that

$$|P_{\xi^\perp}K| = \frac{n}{2}|K| \int\limits_{\mathrm{bd}(K)} |\langle(\nabla\|\cdot\|_K)(x), \xi\rangle|\,d\mu_K(x), \tag{9.10}$$

because $(\nabla\|\cdot\|_K)(x) = n(x)|(\nabla\|\cdot\|_K)(x)|$.

We shall also use Lemma 9.1.10 (for the volume measure $\mu_{B_p^n}$, this assertion is due to Rachev and Rüschendorf [473] and, independently, Schechtman and Zinn [500]). Combining (9.10) and Lemma 9.1.10 we get the next result.

Lemma 9.2.8. *For every $p > 1$ and every $\xi \in S^{n-1}$ we have that*

$$|P_{\xi^\perp}(B_p^n)| = \frac{|B_p^{n-1}|}{\mathbb{E}|X_1|}\mathbb{E}\left|\sum_{j=1}^n \xi_j X_j\right|, \tag{9.11}$$

where $X_1, \dots, X_n$ are i. i. d. random variables with density

$$f_p(x) = \frac{p}{2(p-1)\Gamma(1/p)}|x|^{\frac{2-p}{p-1}}e^{-|x|^{\frac{p}{p-1}}}.$$

Proof. By (9.10) and Lemma 9.1.10, for some constant $c_{p,n}$ we have

$$\begin{aligned}|P_{\xi^\perp}B_p^n| &= c_{p,n}\mathbb{E}\left|\sum_{i=1}^n \xi_i\left|\frac{Y_i}{S}\right|^{p-1}\operatorname{sgn}\left(\frac{Y_i}{S}\right)\right| \\ &= c_{p,n}\cdot\frac{\mathbb{E}S^{p-1}}{\mathbb{E}S^{p-1}}\cdot\mathbb{E}\left|\sum_{i=1}^n \xi_i\left|\frac{Y_i}{S}\right|^{p-1}\operatorname{sgn}(Y_i)\right| \\ &= \frac{c_{p,n}}{\mathbb{E}S^{p-1}}\cdot\mathbb{E}\left|\sum_{i=1}^n \xi_i|Y_i|^{p-1}\operatorname{sgn}(Y_i)\right|.\end{aligned}$$

It now suffices to observe that the density of $X_i = |Y_i|^{p-1}\operatorname{sgn}(Y_i)$ for $p > 1$ is f_p. We compute $c_{p,n}$ by taking $\xi = e_1$. □

The last ingredient in the proof of the monotonicity result of Theorem 9.2.6 is the so-called stochastic convex ordering. Given two symmetric random variables X and Y, we say that Y dominates X in the convex (or often called Choquet) stochastic ordering, if

$$\mathbb{E}(\phi(X)) \leqslant \mathbb{E}(\phi(Y))$$

for every even convex function $\phi : \mathbb{R} \to [0, +\infty]$. One can check that this property tensorizes and is also preserved under convolution: if Y dominates X and Z is a symmetric random variable, independent of X and Y, then $Y + Z$ dominates $X + Z$. What we really need is the following sufficient condition.

Lemma 9.2.9. *Let X and Y be two random variables such that $\mathbb{E}|X| = \mathbb{E}|Y|$. Assume that X and Y have even densities f and g, respectively, and there exist $0 < x_1 < x_2$ such that $\{t \geqslant 0 : g(t) < f(t)\} = (x_1, x_2)$. Then, Y dominates X in the convex stochastic ordering.*

Proof. Let $\phi : \mathbb{R} \to [0, +\infty]$ be an even, convex function. By the symmetry of X, Y and the assumption that $\mathbb{E}|X| = \mathbb{E}|Y|$, the inequality $\int \phi f \leqslant \int \phi g$ is equivalent to

$$\int_0^\infty (\phi(x) - \alpha x - \beta)(g(x) - f(x))\,dx \geqslant 0 \tag{9.12}$$

for some $\alpha, \beta \in \mathbb{R}$. We choose α, β so that the convex function $\psi(x) = \phi(x) - \alpha x - \beta$ vanishes at x_1 and x_2. Then, by convexity, $\psi \leqslant 0$ on (x_1, x_2), and $\psi \geqslant 0$ outside that interval. Therefore, the integrand in (9.12) is non-negative. □

Proof of Theorem 9.2.6. In view of (9.11), it is enough to show that the function

$$p \mapsto \frac{1}{\mathbb{E}|X_1^{(p)}|} \mathbb{E}\Big|\sum \xi_j X_j^{(p)}\Big|$$

is increasing on $[1, +\infty)$, where $X_j^{(p)}$ are i. i. d. random variables with density proportional to $|x|^{\frac{2-p}{p-1}} \exp(-|x|^{\frac{p}{p-1}})$. By the tensorization property of the convex stochastic ordering, it suffices to prove that for $1 \leqslant p < q$, $X_1^{(q)}/\mathbb{E}|X_1^{(q)}|$ dominates $X_1^{(p)}/\mathbb{E}|X_1^{(p)}|$. This is a consequence of Lemma 9.2.9. □

In the case where $p > 2$ we have the additional information that the X_j are gaussian mixtures. This enabled Barthe and Naor [64] to prove an analogue of Theorem 9.1.13. Their argument was later simplified in [164], where Fourier analytic arguments are avoided.

Theorem 9.2.10. *Let $p > 2$. For any $\xi, \eta \in S^{n-1}$ such that $(\eta_1^2, \dots, \eta_n^2)$ majorizes $(\xi_1^2, \dots, \xi_n^2)$, we have that*

$$|P_{\xi^\perp}(B_p^n)| \geqslant |P_{\eta^\perp}(B_p^n)|.$$

Proof. We shall use the formula (9.11) for hyperplane projections. If $p > 2$, then the density $f_p(t)$ of X_i is completely monotone, and hence the X_j are gaussian mixtures that can be expressed as $X_j = R_j G_j$ for some i. i. d. non-negative random variables R_j and standard gaussian random variables G_j, all of them independent. Conditioning on the R_j and adding the gaussians we get

$$\mathbb{E}\Big|\sum_{j=1}^n \xi_j X_j\Big| = \mathbb{E}\Big(\sum_{j=1}^n \xi_j^2 R_j^2\Big)^{1/2} \mathbb{E}|G_1|. \tag{9.13}$$

Then, the Schur-concavity result follows from the concavity of $h(t) = \sqrt{t}$, as in the corresponding proof for sections. □

The proof in [64] exploits the formula for the volume of hyperplane projections of a convex body as the Fourier transform of its curvature function (see [333, 332]). The proof that we presented above is from [164].

When $1 \leqslant p < 2$, the density of the random variables X_j in (9.11) is bimodal and it is not easy to understand the L_1-norm of their weighted sums. For $p = 1$ the X_j become discrete (symmetric random signs). In the range $0 < p < 1$, Cauchy's formula cannot be applied due to the lack of convexity and no non-trivial bounds are known. When $1 < p < 2$, Theorem 9.2.6 shows that the maximal-volume hyperplane projection is onto a coordinate subspace, but the minimal one is not known. Barthe and Naor in [64] have shown that the cross-polytope minimizer $(\frac{1}{\sqrt{2}}, \frac{1}{\sqrt{2}}, 0, \dots, 0)^\perp$ is "worse" than the diagonal one for every $p > p_0 = \frac{4}{3}$ in large enough dimensions (in particular, Oleszkiewicz has pointed out in [447] that there is no “formal duality” with sections, for there is not such a phase transition at $\frac{p_0}{p_0-1} = 4$).

Eskenazis, Nayar, and Tkocz [165] prove that there exists $p_0 \in (1+10^{-12}, \frac{4}{3})$ such that for all $n \geqslant 2$ and all $p \in [1, p_0]$ one has

$$\min_{\xi \in S^{n-1}} |P_{\xi^\perp}(B_p^n)| = |P_{(\frac{e_1+e_2}{\sqrt{2}})^\perp}(B_p^n)|.$$

Their proof is dual to the one for sections (see Section 9.1.2). In particular, since the argument requires a strong stability version of Theorem 9.2.3, they re-prove the theorem of De, Diakonikolas, and Servedio from [148] with an explicit constant.

For codimensions greater than 1, Theorem 9.1.9 gives a sharp lower bound: for every $p \geqslant 2$, $1 \leqslant k \leqslant n-1$, and $F \in G_{n,k}$ we have

$$|P_F(B_p^n)| \geqslant |B_p^n \cap F| \geqslant |B_p^k|,$$

with equality at coordinate subspaces. For $0 < p < 2$, using his reverse form of the Brascamp–Lieb inequality from [57], Barthe [58] has established the following lower bound.

Theorem 9.2.11. *Let $0 < p < 2$ and $1 \leqslant k \leqslant n-1$. For every $F \in G_{n,k}$ we have*

$$|P_F(B_p^n)| \geqslant \left(\frac{k}{n}\right)^{k(1/p-1/2)} |B_p^k|.$$

This is optimal when k divides n and $p \geqslant 1$; equality holds for the same subspaces as in Conjecture 9.1.8.

9.3 Perimeter of sections of the cube

Let $Q_n = [-\frac{1}{2}, \frac{1}{2}]^n$ be the centered cube of volume 1 in $\mathbb{R}^n$. For any $\xi \in S^{n-1}$ we define the perimeter of the cubic section by $\xi^\perp$ as

$$P_{n-2}(\xi) := |\mathrm{bd}(Q_n) \cap \xi^\perp|.$$

Answering a question of Pelczyński about the cubical hyperplane sections of maximal perimeter, König and Koldobsky in [348] showed that the extremal direction is the same as for the volume.

Theorem 9.3.1. *Let $n \geqslant 3$ and $\xi_{\max} := \frac{1}{\sqrt{2}}(1,1,0,\dots,0) \in \mathbb{R}^n$. For every unit vector $\xi \in S^{n-1}$ we have*

$$P_{n-2}(\xi) \leqslant P_{n-2}(\xi_{\max}) = 2((n-2)\sqrt{2}+1).$$

This bound is attained if $\xi = (\pm e_i \pm e_j)/\sqrt{2}$ for some $1 \leqslant i < j \leqslant n$. This theorem also leads to counterexamples to the perimeter version of the Busemann–Petty problem in dimensions $n \geqslant 14$.

For any $\xi \in S^{n-1}$ and $t \in \mathbb{R}$ let

$$A_{n-1}(\xi, t) = \left|Q_n \cap (\xi^\perp + (t/2)\xi)\right|$$

denote the volume of the hyperplane section of Q_n perpendicular to ξ and at distance $t/2$ from the origin. We also set $A_{n-1}(\xi) = A_{n-1}(\xi, 0)$. If ξ^* denotes the decreasing rearrangement of the sequence $(|\xi_k|)_{k=1}^n$, then, since volume is invariant under coordinate permutations and sign changes, we have that

$$A_{n-1}(\xi, t) = A_{n-1}(\xi^*, |t|).$$

So, we may assume that $\xi = (\xi_k)_{k=1}^n$ satisfies $\xi_1 \geqslant \cdots \geqslant \xi_n \geqslant 0$ as well as $t \geqslant 0$. We can express the parallel section function in the form

$$A_{n-1}(\xi, t) = \frac{2}{\pi} \int_0^\infty \prod_{k=1}^n \frac{\sin(\xi_k s)}{\xi_k s} \cos(ts)\, ds, \tag{9.14}$$

where we agree that if $\xi_k = 0$, then $\frac{\sin(\xi_k s)}{\xi_k s} = 1$. The proof of Theorem 9.3.1 is based on the next Fourier analytic formula for the perimeter.

Proposition 9.3.2. *For any $\xi = (\xi_k)_{k=1}^n \in S^{n-1}$ we have that*

$$P_{n-2}(\xi) = 2\sum_{k=1}^n \sqrt{1 - a_k^2}\, \frac{2}{\pi} \int_0^\infty \prod_{1 \leqslant j \neq k \leqslant n} \frac{\sin(\xi_j s)}{\xi_j s} \cos(\xi_k s)\, ds. \tag{9.15}$$

Proof. Let $\xi = (\xi_k)_{k=1}^n \in S^{n-1}$ with $\xi_1 \geqslant \cdots \geqslant \xi_n \geqslant 0$ and let $x \in \mathbb{R}^n$. We write $\xi = (\xi_1, \tilde{\xi})$ and $x = (x_1, \tilde{x})$, where $\tilde{\xi} = (\xi_k)_{k=2}^n$ and $\tilde{x} = (x_k)_{k=2}^n \in \mathbb{R}^{n-1}$. The $(n-1)$-dimensional hyperplane $\xi^\perp$ intersects the boundary of the cube in $2n$ typically non-central $(n-2)$-dimensional sections of an $(n-1)$-cube, corresponding to $x_j = \pm\frac{1}{2}, j = 1, \ldots, n$. For the case $x_1 = -\frac{1}{2}$, we need to calculate the volume

$$\left|\{\tilde{x} \in \mathbb{R}^{n-1} : \langle \tilde{x}, \tilde{\xi}\rangle = \xi_1/2\}\right|.$$

Let $\xi_j' := \frac{\xi_j}{\sqrt{1-\xi_1^2}}, j = 1, \ldots, n$, and $\tilde{\xi}' := (\xi_j')_{j=2}^n$. Then, $|\tilde{\xi}'|^2 = \sum_{j=2}^n {\xi_j'}^2 = 1$. Using (9.14), we see that

$$\begin{aligned}\left|\{\tilde{x} \in \mathbb{R}^{n-1} : \langle \tilde{x}, \tilde{\xi}\rangle = \xi_1/2\}\right| &= A_{n-2}(\tilde{\xi}', \xi_1') \\ &= \frac{2}{\pi} \int_0^\infty \prod_{j=2}^n \frac{\sin(\xi_j' r)}{\xi_j' r} \cos(\xi_1' r)\, dr\end{aligned}$$

$$= \sqrt{1-\xi_1^2}\,\frac{2}{\pi}\int_0^\infty \prod_{j=2}^n \frac{\sin(\xi_j s)}{\xi_j s}\cos(\xi_1 s)\,ds.$$

The same holds for $x_1 = +\frac{1}{2}$ and similarly for all $x_j = \pm\frac{1}{2}$. Therefore,

$$P_{n-2}(\xi) = 2\sum_{k=1}^n \sqrt{1-\xi_k^2}\,\frac{2}{\pi}\int_0^\infty \prod_{1\leqslant j\neq k\leqslant n} \frac{\sin(\xi_j s)}{\xi_j s}\cos(\xi_k s)\,ds,$$

which proves (9.15). □

The analysis of this formula involves new ingredients, most notably local conditions for constrained extrema, as well as subtle technical estimates around Ball's integral inequality. In Ball's result, the integral in (9.14) for $t = 0$ is estimated by using Hölder's inequality if $\xi_1 \leqslant \frac{1}{\sqrt{2}}$, which is natural since in the extremal case the integrand is nonnegative. In (9.15) we have a weighted sum of integrals where the integrands are nonpositive in the extremal case. Therefore, estimating $P_{n-2}(\xi)$ requires additional methods. One idea is to consider the perimeter estimate as a constrained optimization problem, in view of the following two results.

Proposition 9.3.3. *For any $\xi = (\xi_k)_{k=1}^n \in S^{n-1}$ and any $k \in \{1,\dots,n\}$ we define*

$$D_k(a) := \frac{2}{\pi}\int_0^\infty \prod_{1\leqslant j\neq k\leqslant n} \frac{\sin(\xi_j s)}{\xi_j s}\cos(\xi_k s)\,ds. \tag{9.16}$$

Then,

$$\sum_{k=1}^n D_k(\xi) = (n-1)A_{n-1}(\xi). \tag{9.17}$$

Proof. We use integration by parts and the observation that

$$\frac{d}{ds}\left(\frac{\sin(\xi_j s)}{\xi_j s}\right) = \frac{1}{s}\left(\cos(\xi_j s) - \frac{\sin(\xi_j s)}{\xi_j s}\right),$$

if all ξ_j are strictly positive. We have

$$\begin{aligned} D_1(\xi) &= \left[\frac{2}{\pi}\prod_{j=2}^n \frac{\sin(\xi_j s)}{\xi_j s}\frac{\sin(\xi_1 s)}{\xi_1}\right]_{s=0}^\infty \\ &\quad - \frac{2}{\pi}\int_0^\infty \sum_{j=2}^n \prod_{2\leqslant k\neq j\leqslant n} \frac{\sin(\xi_k s)}{\xi_k s}\left(\frac{\cos(\xi_j s)}{s} - \frac{\sin(\xi_j s)}{\xi_j s^2}\right)\frac{\sin(\xi_1 s)}{\xi_1}\,ds \\ &= (n-1)A_{n-1}(\xi) - \sum_{j=2}^n D_j(\xi) \end{aligned}$$

and hence $\sum_{j=1}^n D_j(\xi) = (n-1)A_{n-1}(\xi)$. If some ξ_k are zero, then the corresponding term $D_k(\xi)$ equals $A_{n-1}(\xi)$, and (9.17) follows by integration by parts only for those k with $\xi_k \neq 0$. □

Proposition 9.3.4. *For any $\xi = (\xi_k)_{k=1}^n \in S^{n-1}$ and any $k \in \{1,\dots,n\}$*

$$D_k(a) \leqslant A_{n-1}(\xi). \tag{9.18}$$

Proof. Without loss of generality we may assume that $k = 1$ and $\xi_1 > 0$. We do not need to consider the decreasing rearrangement ξ^* of ξ, but assume that $\xi_k \geqslant 0$ for all k. Again, we define $\xi_j' = \frac{\xi_j}{\sqrt{1-\xi_1^2}}$ for $j = 1,\dots,n$ and $\tilde{\xi}' = (\xi_j')_{j=2}^n \in \mathbb{R}^{n-1}$. Then, $\sum_{j=2}^n \xi_j'^2 = 1$, and a change of variables together with (9.16) in dimension $n-1$ show that

$$\begin{aligned} D_1(\xi) &= \frac{1}{\sqrt{1-\xi_1^2}} \frac{2}{\pi} \int_0^\infty \prod_{j=2}^n \frac{\sin(\xi_j' r)}{\xi_j' r} \cos(\xi_1' r)\, dr \\ &= \frac{1}{\sqrt{1-\xi_1^2}} \big|\{\tilde{x} \in Q_{n-1} : \langle \tilde{x}, \tilde{\xi}'\rangle = \xi_1'/2\}\big|. \end{aligned}$$

By the Brunn–Minkowski inequality, for any $t \in \mathbb{R}$ with $|t| \leqslant \xi_1'$ we have that

$$\big|\{\tilde{x} \in Q_{n-1} : \langle \tilde{x}, \tilde{\xi}'\rangle = \xi_1'/2\}\big| \leqslant \big|\{\tilde{x} \in Q_{n-1} : \langle \tilde{x}, \tilde{\xi}'\rangle = t/2\}\big|.$$

Therefore,

$$D_1(\xi) \leqslant \frac{1}{\xi_1'} \frac{1}{\sqrt{1-\xi_1^2}} \big|\{\tilde{x} \in Q_{n-1} : |\langle \tilde{x}, \tilde{\xi}'\rangle| \leqslant \xi_1'/2\}\big|.$$

Define $T : \mathbb{R}^{n-1} \to \mathbb{R}^n$ by $T(x) = (-\frac{\langle \tilde{x}, \tilde{\xi}'\rangle}{\xi_1'}, \tilde{x})$. Then, T maps the slab $\{\tilde{x} \in Q_{n-1} : |\langle \tilde{x}, \tilde{\xi}'\rangle| \leqslant \xi_1'/2\}$ in dimension $n-1$ into a central section of Q_n

$$\begin{aligned} T(\{\tilde{x} \in Q_{n-1} : |\langle \tilde{x}, \tilde{\xi}'\rangle| \leqslant \xi_1'/2\}) &= \{y = (x_1, \tilde{x}) \in Q_n : x_1\xi_1 + \langle \tilde{x}, \tilde{\xi}'\rangle = 0\} \\ &= \{y \in Q_n : \langle y, \xi\rangle = 0\}. \end{aligned}$$

Since $T^*T = I_n + \xi_1'^{-2}\langle \cdot, \tilde{\xi}'\rangle\, \tilde{\xi}'$, we see that T^*T has 1 as an eigenvalue of multiplicity $n-2$ and the other eigenvalue is $1 + \frac{1}{\xi_1'^2}$, with eigenvector $\tilde{\xi}'$ of norm 1, so that

$$\sqrt{\det(T^*T)} = \frac{1}{\xi_1'} \sqrt{\xi_1'^2 + 1} = \frac{1}{\xi_1'} \frac{1}{\sqrt{1-\xi_1^2}}.$$

Therefore,

$$D_1(\xi) \leqslant \sqrt{\det(T^*T)}\, \big|\{\tilde{x} \in Q_{n-1} : |\langle \tilde{x}, \tilde{\xi}'\rangle| \leqslant \xi_1'/2\}\big|$$

$$= \left|\{y \in Q_n : \langle y, \xi\rangle = 0\}\right| = A_{n-1}(\xi). \qquad \square$$

The proof of Proposition 9.3.4 also yields the following estimate for the parallel section function.

Corollary 9.3.5. *For any $\xi \in S^{n-1}$ and any $t > 0$ we have*

$$A_{n-1}(\xi, t) \leqslant \sqrt{2/(1+t^2)}.$$

Proof. Let $\xi \in S^{n-1}$ and $t > 0$. Put $\xi_j' = \frac{\xi_j}{\sqrt{1+t^2}}$ and $t' = \frac{t}{\sqrt{1+t^2}}$. Then, $(\xi', t') \in \mathbb{R}^{n+1}$ and $|(\xi', t')| = 1$. With a change of variables, from (9.14) we get $A_{n-1}(\xi, t) = \frac{1}{\sqrt{1+t^2}} D((\xi', t'))$, where

$$D((\xi', t')) = \frac{2}{\pi} \int_0^\infty \prod_{j=1}^n \frac{\sin(\xi_j' s)}{\xi_j' s} \cos(t's)\, ds.$$

Applying Proposition 9.3.4 to $(\xi', t') \in S^n$ we get $D((\xi', t')) \leqslant A_n((\xi', t')) \leqslant \sqrt{2}$. Therefore, $A_{n-1}(\xi, t) \leqslant \sqrt{2/(1+t^2)}$. $\square$

We can now prove Theorem 9.3.1, if $\xi_1 \notin (\sqrt{\sqrt{2}-1}, \frac{1}{\sqrt{\sqrt{2}+\frac{1}{2}}}) \simeq (0.643, 0.723)$, i. e., when ξ_1 is not close to $\frac{1}{\sqrt{2}}$.

Proof of Theorem 9.3.1. A direct computation shows that

$$\begin{aligned} D_1(\xi_{\max}) = D_2(\xi_{\max}) &= \frac{2}{\pi} \int_0^\infty \frac{\sin(s/\sqrt{2})}{s/\sqrt{2}} \cos(s/\sqrt{2})\, ds \\ &= \frac{1}{\sqrt{2}} \frac{2}{\pi} \int_0^\infty \frac{\sin(t)}{t}\, dt = \frac{1}{\sqrt{2}} \end{aligned}$$

and

$$D_j(\xi_{\max}) = \frac{2}{\pi} \int_0^\infty \left(\frac{\sin(s/\sqrt{2})}{s/\sqrt{2}}\right)^2 ds = \sqrt{2}\, \frac{2}{\pi} \int_0^\infty \left(\frac{\sin(t)}{t}\right)^2 dt = \sqrt{2}$$

for all $j > 2$. Hence, by (9.15)

$$P_{n-2}(\xi_{\max}) = 2((n-2)\sqrt{2} + 1). \tag{9.19}$$

Now, let $\xi = (\xi_k)_{k=1}^n \in S^{n-1}$ be arbitrary with $\xi_1 \geqslant \cdots \geqslant \xi_n \geqslant 0$. By Proposition 9.3.2 we have

$$\frac{1}{2}P_{n-2}(\xi) = \sum_{k=1}^{n}(1-\xi_k^2)\,D_k(\xi)$$

and using Propositions 9.3.3 and 9.3.4 we see that

$$\frac{1}{2}P_{n-2}(\xi) \leqslant \sup\left\{\sum_{k=1}^{n}\sqrt{1-\xi_k^2}\,C_k : 0 \leqslant C_k \leqslant A_{n-1}(\xi), \sum_{k=1}^{n} C_k = (n-1)A_{n-1}(\xi)\right\}.$$

Since also $(\sqrt{1-\xi_k^2})_{k=1}^n$ is increasing in k, the supremum is attained for increasing C_k as well and, in fact, for $C_1 = 0, C_2 = \cdots = C_k = A_{n-1}(\xi)$, which shows that

$$\frac{1}{2}P_{n-2}(\xi) \leqslant \sum_{k=2}^{n}\sqrt{1-\xi_k^2}\,A_{n-1}(\xi). \tag{9.20}$$

Since the function $\phi(x) = \sqrt{1-x}$ is concave, we have

$$\frac{1}{n-1}\sum_{k=2}^{n}\phi(\xi_k^2) \leqslant \phi\left(\frac{1}{n-1}\sum_{k=2}^{n}\xi_k^2\right) = \phi\left(\frac{1}{n-1}(1-\xi_1^2)\right).$$

Therefore,

$$\frac{1}{2}P_{n-2}(\xi) \leqslant (n-1)\sqrt{1-\frac{1-\xi_1^2}{n-1}}\,A_{n-1}(\xi) \leqslant \left(n-1-\frac{1-\xi_1^2}{2}\right)A_{n-1}(\xi). \tag{9.21}$$

If $\xi_1 \leqslant \frac{1}{\sqrt{2}}$, we use the fact that $A_{n-1}(\xi) \leqslant \sqrt{2}$ to get

$$\frac{1}{2}P_{n-2}(\xi) \leqslant (n-2)\sqrt{2} + \frac{3\sqrt{2}}{4}.$$

If $\xi_1 > \frac{1}{\sqrt{2}}$, we use the fact that $A_{n-1}(\xi) \leqslant \frac{1}{\xi_1}$ and we see again that

$$\frac{1}{2}P_{n-2}(\xi) \leqslant \left(n-\frac{3}{2}+\xi_1^2\right)\frac{1}{\xi_1} \leqslant (n-2)\sqrt{2} + \frac{3\sqrt{2}}{4}.$$

Since $\frac{3\sqrt{2}}{4} \simeq 1.0607 > 1$, this does not prove that $P_{n-2}(\xi) \leqslant P_{n-2}(\xi_{\max})$ for all $\xi \in S^{n-1}$. However, if ξ_1 satisfies $\xi_1 \leqslant \sqrt{\sqrt{2}-1} \simeq 0.643$, then (9.21) yields

$$\frac{1}{2}P_{n-2}(\xi) \leqslant \left(n-\frac{3}{2}+\frac{\sqrt{2}-1}{2}\right)\sqrt{2} = (n-2)\sqrt{2}+1 = \frac{1}{2}P_{n-2}(\xi_{\max}).$$

For $\xi_1 > \frac{1}{\sqrt{2}}$, the requirement that $(n-1)\sqrt{1-\frac{1-\xi_1^2}{n-1}\frac{1}{\xi_1}} \leqslant (n-2)\sqrt{2}+1$ is strongest for $n = 3$, in which case it means $\xi_1 \geqslant \frac{1}{\sqrt{\sqrt{2}+\frac{1}{2}}} \simeq 0.723$. Therefore, for any $\xi \in S^{n-1}$ with $\xi_1 \notin (\sqrt{\sqrt{2}-1}, \frac{1}{\sqrt{\sqrt{2}+\frac{1}{2}}})$ we have shown that $P_{n-2}(\xi) \leqslant P_{n-2}(\xi_{\max})$. □

To complete the proof of Theorem 9.3.1 we need to consider separately the case

$$\xi_1 \in \left(\sqrt{\sqrt{2}-1}, \frac{1}{\sqrt{\sqrt{2}+\frac{1}{2}}}\right), \tag{9.22}$$

where the estimate is off by at most $2(\frac{3\sqrt{2}}{4}-1) \simeq 0.121$. This discrepancy occurs because in (9.20) the extremals for the sum of weights and for the section function A occur for different sequences ξ. This could possibly be avoided, if one could show how the monotonicity of the sequence $(\xi_k)_{k=1}^n$ affects the size of the integrals $D_k(\xi)$, but a result of this type is not available. Instead, we will address the case of (9.22) by a different method. Ball's proof relies on the non-trivial estimate $f(p) \leqslant f(2) = 1$ for the function

$$f(p) := \sqrt{p/2}\,\frac{2}{\pi}\int_0^\infty \left|\frac{\sin(t)}{t}\right|^p dt.$$

König and Koldobsky establish additional information on the function f:

(i) $\lim_{p\to\infty} f(p) = \sqrt{\frac{3}{\pi}}$ and $f(\frac{9}{4}) < \sqrt{\frac{3}{\pi}}$;

(ii) $f(\sqrt{2}+\frac{1}{2}) < \frac{51}{50}$;

(iii) $f|_{[\sqrt{2}+\frac{1}{2},\frac{9}{4}]}$ is decreasing and convex.

Also, for all $p \geqslant \frac{9}{4}$, $f(p) \leqslant \sqrt{\frac{3}{\pi}}$.

Using the convexity of f and these estimates for $f(p)$ with $p = \sqrt{2}+1/2$ and $p = \frac{9}{4}$, one may improve the general estimate $A_{n-1}(\xi) \leqslant \sqrt{2}$ for certain sequences ξ with ξ_1 close to $\frac{1}{\sqrt{2}}$, and this essentially suffices to prove Theorem 9.3.1 in these cases. The property of f that makes this argument work is that it is strictly smaller near ∞ than in $p = 2$, where $f(2) = 1$.

10 Measure-theoretic geometric inequalities

In this chapter, we touch upon "functional convexity," the study of functional extensions of classical results in convex geometry.

We begin with the dimensional Brunn–Minkowski inequality and discuss its gaussian analogue, which has been established for origin-symmetric convex sets. This inequality neither trivially follows from nor implies Ehrhard's inequality; however, it does imply the log-concavity of gaussian measure for symmetric convex sets. Several related results also rely on convexity and symmetry, including the B-inequality of Cordero-Erausquin, Fradelizi, and Maurey, the S-inequality of Latala and Oleszkiewicz, and Royen's gaussian correlation inequality. These developments motivate further conjectures, such as the log-Brunn–Minkowski inequality for symmetric log-concave measures.

We then turn to volume product inequalities, highlighting the Blaschke–Santaló inequality and its functional analogue. These extend classical bounds on the product of the volumes of a convex body and its polar. We also examine Mahler's conjecture and its functional formulation, which propose conjectural lower bounds, along with partial results in symmetric, unconditional, and low-dimensional settings. Connections to transport and entropy inequalities are explored through property (τ) and Talagrand-type inequalities, illustrating how functional inequalities capture concentration phenomena.

Next, we derive functional Rogers–Shephard inequalities, generalizing classical bounds on difference bodies to the setting of log-concave functions. We also introduce mixed integrals, which extend the theory of mixed volumes and give rise to functional Alexandrov–Fenchel- and Brunn–Minkowski-type inequalities. Additional directions and results in "functional convexity" are briefly outlined in the further reading section.

Throughout the chapter, symmetry, convexity, and log-concavity play central roles. Recent advances on log-Brunn–Minkowski inequalities, stability results, and information-theoretic perspectives further highlight the rich and evolving interplay between geometry, analysis, and probability.

10.1 Gaussian Brunn–Minkowski inequality

Recall that the classical Brunn–Minkowski inequality asserts that for every pair of compact sets A, B in $\mathbb{R}^n$ and every $\lambda \in (0,1)$

$$|\lambda A + (1-\lambda)B|^{\frac{1}{n}} \geqslant \lambda|A|^{\frac{1}{n}} + (1-\lambda)|B|^{\frac{1}{n}}. \tag{10.1}$$

In this section, we discuss versions of the Brunn–Minkowski inequality in which the Lebesgue measure is replaced by some other measure. We start with the standard gaussian measure γ_n with density $(2\pi)^{-n/2}\exp(-|x|^2/2)$. A deep geometric inequality for γ_n is

https://doi.org/10.1515/9783111360508-010

Ehrhard's inequality [156], which states that for every pair of Borel measurable sets A, B in $\mathbb{R}^n$ and every $\lambda \in (0,1)$

$$\Phi^{-1}(\gamma_n(\lambda A + (1-\lambda)B)) \geqslant \lambda\Phi^{-1}(\gamma_n(A)) + (1-\lambda)\Phi^{-1}(\gamma_n(B)), \tag{10.2}$$

where Φ^{-1} is the inverse of the gaussian distribution function $\Phi(x) = \gamma_1((-\infty, x])$. Ehrhard's inequality (10.2) implies the gaussian isoperimetric inequality, which asserts that half-spaces have minimal gaussian surface area among all sets of fixed measure. Ehrhard's original proof of (10.2) proceeded via gaussian symmetrization and required both sets A and B to be convex, an assumption that was later removed by Borell in [94] (see also [356] for a partial result). A presentation of all the above is given in [28, Section 3.2].

While Ehrhard's inequality (10.2) captures the optimal dimension-free convexity of the gaussian measure, the fact that it holds for general Borel subsets A, B of $\mathbb{R}^n$ indicates that stronger forms of measure-theoretic inequalities might hold under additional geometric properties of the underlying sets, such as convexity. In fact, Gardner and Zvavitch asked in [204, Question 7.1] whether the dimensional Brunn–Minkowski inequality (10.1) holds for γ_n if we assume that the underlying sets are convex and contain the origin. A counterexample to this statement was given by Nayar and Tkocz in [435], but this inequality is true for origin-symmetric convex sets. This was proved by Eskenazis and Moschidis in [162].

Theorem 10.1.1. *Let $n \in \mathbb{N}$. For every pair of origin-symmetric convex sets K, L in $\mathbb{R}^n$ and every $\lambda \in (0,1)$*

$$\gamma_n(\lambda K + (1-\lambda)L)^{\frac{1}{n}} \geqslant \lambda\gamma_n(K)^{\frac{1}{n}} + (1-\lambda)\gamma_n(L)^{\frac{1}{n}}. \tag{10.3}$$

Choosing K and L to be small coordinate boxes centered at the origin, we see that the exponent $\frac{1}{n}$ is optimal (i. e., maximal) in (10.3). It is interesting to note (see [204]) that (10.3) neither trivially follows nor implies Ehrhard's inequality (10.2) for origin-symmetric convex sets. If we restrict our attention to origin-symmetric convex sets, the dimensional Brunn–Minkowski inequality (10.3) implies the log-concavity of the gaussian measure. This is the fact that, more generally, for every pair of Borel sets A, B in $\mathbb{R}^n$ and every $\lambda \in (0,1)$

$$\gamma_n(\lambda A + (1-\lambda)B) \geqslant \gamma_n(A)^{\lambda}\gamma_n(B)^{1-\lambda}. \tag{10.4}$$

Several other important inequalities under convexity and symmetry assumptions have been established for the gaussian measure. The B-inequality of Cordero-Erausquin, Fradelizi, and Maurey [140], states that for every symmetric convex set K in $\mathbb{R}^n$, the map $\alpha(t) = \gamma_n(e^tK)$ is a log-concave function on $\mathbb{R}$. Note that by (10.4) the function $\beta(s) = \gamma_n(sK)$ on $\mathbb{R}^+$ is log-concave for every convex set K in $\mathbb{R}^n$, but the symmetry assumption is necessary in order to conclude the log-concavity of α. Other well-known inequalities in Gauss space, which crucially rely on the symmetry and convexity of the underlying sets, are the S-inequality of Latala and Oleszkiewicz [359] and Royen's gaussian correlation inequality [486]. An exposition of all these results is provided in [28, Chapter 3] (see also [357]).

In [336], Kolesnikov and Livshyts presented a strategy to address the Gardner–Zvavitch problem, by studying how inequalities of the form

$$\mu(\lambda K + (1-\lambda)L)^{\frac{\delta}{n}} \geqslant \lambda\mu(K)^{\frac{\delta}{n}} + (1-\lambda)\mu(L)^{\frac{\delta}{n}}, \quad \lambda \in (0,1) \tag{10.5}$$

behave infinitesimally when the convex bodies K and L are small perturbations of each other and then proving a local-to-global principle. To illustrate this technique, we briefly return to the B-inequality. It is observed in [140, p. 413]) that for every symmetric convex set K in $\mathbb{R}^n$, the function $\alpha(t) = \gamma_n(e^t K)$ is log-concave on $\mathbb{R}$ if and only if it is infinitesimally log-concave at $t = 0$, that is, $\alpha''(0)\alpha(0) \leqslant \alpha'(0)^2$. The latter inequality is equivalent to

$$\mathrm{Var}_{\gamma_K}(|x|^2) \leqslant \frac{1}{2}\int \big|\nabla|x|^2\big|^2 \, d\gamma_K(x), \tag{10.6}$$

where γ_K is the normalized restriction of γ_n on a symmetric convex set K with nonempty interior, that is, the measure given by $\gamma_K(A) = \gamma_n(A \cap K)/\gamma_n(K)$ for every Borel subset A of $\mathbb{R}^n$. The constant $\frac{1}{2}$ on the right-hand side of (10.6) is very important. The same Poincaré inequality with constant 1 is valid in much greater generality, following from a theorem of Brascamp and Lieb [109], which implies that if μ is a measure of the form $d\mu(x) = e^{-V(x)}\,dx$ with a potential whose Hessian satisfies $\nabla^2 V \geqslant \beta I_n$, then every twice continuously differentiable function $f : \mathbb{R}^n \to \mathbb{R}$ satisfies

$$\mathrm{Var}_\mu(f) \leqslant \frac{1}{\beta}\int |\nabla f|^2 \, d\mu. \tag{10.7}$$

Equivalently, (10.7) says that the spectral gap of such a measure is at least $\sqrt{\beta}$. In order to obtain the stronger inequality (10.6), Cordero-Erausquin, Fradelizi, and Maurey used, in a crucial way, the symmetry of both the measure γ_K and the function $x \mapsto |x|^2$. Similarly, the symmetry of the problem plays a crucial role in the proof of Theorem 10.1.1.

Recall that the generator of the Ornstein–Uhlenbeck semigroup is the elliptic differential operator $\mathcal{L}$ whose action on a smooth function $u : \mathbb{R}^n \to \mathbb{R}$ is given by

$$\mathcal{L}u(x) - \Delta u(x) - \sum_{i=1}^{n} x_i \partial_i u(x), \quad x \in \mathbb{R}^n. \tag{10.8}$$

We also denote by $\|A\|_{\mathrm{HS}}$ the Hilbert–Schmidt norm of a matrix A, i. e., $\|A\|_{\mathrm{HS}}^2 = \sum_{i,j} a_{ij}^2$. In [336], Kolesnikov and Livshyts proved the following local-to-global principle (see also [132, 134, 339] for closely related infinitesimal versions of other Brunn–Minkowski-type inequalities).

Proposition 10.1.2. *Let $n \in \mathbb{N}$ and let $\delta \in [0,1]$ be such that for every symmetric convex set K in $\mathbb{R}^n$, every twice continuously differentiable, even function $u : K \to \mathbb{R}$ with $\mathcal{L}u = 1$ on K satisfies*

$$\int (\|\nabla^2 u\|_{\mathrm{HS}}^2 + |\nabla u|^2)\, d\gamma_K \geqslant \frac{\delta}{n}. \tag{10.9}$$

Then, for every pair of symmetric convex sets K, L in $\mathbb{R}^n$ and every $\lambda \in (0,1)$

$$\gamma_n(\lambda K + (1-\lambda)L)^{\frac{\delta}{n}} \geqslant \lambda \gamma_n(K)^{\frac{\delta}{n}} + (1-\lambda)\gamma_n(L)^{\frac{\delta}{n}}. \tag{10.10}$$

The main result of [336] was a proof of (10.9) with $\delta = \frac{1}{2}$ for all convex sets K containing the origin. Then, a local-to-global principle for such convex sets (similar to Proposition 10.1.2) implies the corresponding gaussian Brunn–Minkowski inequality (10.10) with $\delta = \frac{1}{2}$. Eskenazis and Moschidis obtained the following refinement of the inequality of Kolesnikov and Livshyts for origin-symmetric convex sets and symmetric (i. e., even) solutions of the equation $\mathcal{L}u = 1$ on K. Note that Proposition 10.1.2 and the next theorem immediately imply Theorem 10.1.1.

Theorem 10.1.3. *For every $n \in \mathbb{N}$ and every symmetric convex set K in $\mathbb{R}^n$, every twice continuously differentiable, even function $u : \mathbb{R}^n \to \mathbb{R}$ with $\mathcal{L}u = 1$ on K, satisfies*

$$\int (\|\nabla^2 u\|_{\mathrm{HS}}^2 + |\nabla u|^2)\, d\gamma_K \geqslant \frac{1}{n}. \tag{10.11}$$

Proof. For every $n \times n$ matrix A we set $\widehat{A} = A - \frac{\mathrm{tr}(A)}{n} I_n$, where I_n is the identity matrix. In particular, for any twice continuously differentiable function $u : K \to \mathbb{R}$, where K is a symmetric convex set, we write

$$\widehat{\nabla}^2 u := \nabla^2 u - \frac{\Delta u}{n} I_n \tag{10.12}$$

for the traceless part of its Hessian. By orthogonality, we have the pointwise identity

$$\|\nabla^2 u\|_{\mathrm{HS}}^2 = \|\widehat{\nabla}^2 u\|_{\mathrm{HS}}^2 + \frac{(\Delta u)^2}{n}. \tag{10.13}$$

Let $r : \mathbb{R}^n \to \mathbb{R}$ be the function defined by $r(x) = \frac{|x|^2}{2n}$. Note that $\widehat{\nabla}^2 r \equiv 0$ and $\Delta r \equiv 1$. From (10.13) we get

$$\begin{aligned} \|\widehat{\nabla}^2 u\|_{\mathrm{HS}}^2 &= \|\widehat{\nabla}^2 (u-r)\|_{\mathrm{HS}}^2 \\ &= \|\nabla^2 (u-r)\|_{\mathrm{HS}}^2 - \frac{(\Delta(u-r))^2}{n} \\ &= \|\nabla^2 (u-r)\|_{\mathrm{HS}}^2 - \frac{(\Delta u - 1)^2}{n}. \end{aligned} \tag{10.14}$$

Combining (10.13) and (10.14) we see that

$$\|\nabla^2 u\|_{\mathrm{HS}}^2 = \|\nabla^2(u-r)\|_{\mathrm{HS}}^2 + \frac{2}{n}\Delta u - \frac{1}{n}. \tag{10.15}$$

Taking into account that $\mathcal{L}u(x) = \Delta u(x) - \sum_{i=1}^n x_i\partial_i u(x) = 1$, we can then write

$$\|\nabla^2 u(x)\|_{\mathrm{HS}}^2 = \|\nabla^2(u-r)(x)\|_{\mathrm{HS}}^2 + \frac{2}{n}\sum_{i=1}^n x_i\partial_i u(x) + \frac{1}{n} \tag{10.16}$$

for all $x \in K$. For a fixed $i \in \{1,\dots,n\}$ the partial derivative $\partial_i(u-r)$ is an odd function on K and thus has expectation 0 with respect to γ_K. Therefore, the Brascamp–Lieb inequality (10.7) gives

$$\begin{aligned}\sum_{j=1}^n \int (\partial_j\partial_i(u-r))^2\, d\gamma_K &\geqslant \mathrm{Var}_{\gamma_K}(\partial_i(u-r))\\ &= \int \Big(\partial_i u(x) - \frac{x_i}{n}\Big)^2\, d\gamma_K(x).\end{aligned} \tag{10.17}$$

Summing (10.17) over $i \in \{1,\dots,n\}$, we get

$$\begin{aligned}\int \|\nabla^2(u-r)\|_{\mathrm{HS}}^2\, d\gamma_K &\geqslant \sum_{i=1}^n \int \Big(\partial_i u(x) - \frac{x_i}{n}\Big)^2\, d\gamma_K(x)\\ &= \int |\nabla u(x)|^2 - \frac{2}{n}\sum_{i=1}^n x_i\partial_i u(x) + \frac{|x|^2}{n^2}\, d\gamma_K(x).\end{aligned} \tag{10.18}$$

Combining (10.16) and (10.18), we finally deduce that

$$\int \|\nabla^2 u\|_{\mathrm{HS}}^2 + |\nabla u|^2\, d\gamma_K \geqslant \int 2|\nabla u(x)|^2 + \frac{|x|^2}{n^2} + \frac{1}{n}\, d\gamma_K(x) \geqslant \frac{1}{n} \tag{10.19}$$

and the proof is complete. □

Remark 10.1.4. The only property of K and u that was used in the proof of Theorem 10.1.3 is the fact that $\int \nabla(u-r)\, d\gamma_K = 0$, which allows us to use the Brascamp–Lieb inequality.

The exponent $1/n$ in the gaussian Brunn–Minkowski inequality cannot be improved. However, Ehrhard's inequality suggests that one might strengthen it by replacing the function $t \mapsto t^{\frac{1}{n}}$ by a more general function of the gaussian measure of the sets. Note that if $\zeta_n : [0,1] \to \mathbb{R}$ satisfies

$$\zeta_n(\gamma_n(\lambda A + (1-\lambda)B)) \geqslant \lambda\zeta_n(\gamma_n(A)) + (1-\lambda)\zeta_n(\gamma_n(B)) \tag{10.20}$$

for every pair of Borel sets A, B in $\mathbb{R}^n$, then choosing A and B to be half-spaces of the form $A = \{x \in \mathbb{R}^n : x_1 \leqslant a\}$ and $B = \{x \in \mathbb{R}^n : x_1 \leqslant b\}$, we get that $\zeta_n \circ \Phi$ is concave, where Φ is the gaussian distribution function $\Phi(x) = \gamma_1((-\infty, x])$. An interesting question is to find an "optimal" increasing function $\xi_n : [0,1] \to \mathbb{R}$ such that for every origin-symmetric convex sets K, L in $\mathbb{R}^n$ and every $\lambda \in (0,1)$, the inequality

$$\xi_n(\gamma_n(\lambda K + (1-\lambda)L)) \geqslant \lambda \xi_n(\gamma_n(K)) + (1-\lambda)\xi_n(\gamma_n(L)) \tag{10.21}$$

is satisfied. Such an inequality should be a joint refinement of Ehrhard's inequality (10.2) and of the dimensional Brunn–Minkowski inequality (10.3) that becomes an equality for some non-trivial pairs of symmetric convex sets K, L in $\mathbb{R}^n$. In [204, p. 5350], Gardner and Zvavitch presented an argument of Barthe that implies that (10.21) is not satisfied for $\xi_n = \Psi_n^{-1}$, where $\Psi_n(r) = \gamma_n(rB_2^n)$ and $rB_2^n = \{x \in \mathbb{R}^n : |x| \leqslant r\}$ is the Euclidean ball of radius r. Eskenazis and Moschidis [162] prove the following theorem.

Theorem 10.1.5. *Fix $n \in \mathbb{N}$ and let $\sigma_n : [0,1] \to \mathbb{R}$ be a strictly increasing function such that, for all $r > 0$*

$$1 + \frac{\sigma_n''(\Psi_n(r))\Psi_n(r)}{\sigma_n'(\Psi_n(r))} = \frac{2}{n} - \frac{c_n}{n^2\Psi_n(r)} r^n e^{-r^2/2}, \tag{10.22}$$

where $c_n^{-1} = 2^{\frac{n}{2}-1}\Gamma(n/2)$. Then, for every pair of origin-symmetric convex sets K, L in $\mathbb{R}^n$ and every $\lambda \in (0,1)$

$$\sigma_n(\gamma_n(\lambda K + (1-\lambda)L)) \geqslant \lambda \sigma_n(\gamma_n(K)) + (1-\lambda)\sigma_n(\gamma_n(L)). \tag{10.23}$$

One can check that the function $y \mapsto \sigma_n(y^n)$ is convex, and hence (10.23) is indeed an improvement of the dimensional Brunn–Minkowski inequality (10.3). Moreover, the strict convexity of $y \mapsto \sigma_n(y^n)$ settles the equality cases of the dimensional Brunn–Minkowski inequality (10.3) for the gaussian measure. For any $n \in \mathbb{N}$ and $\lambda \in (0,1)$, if K, L are origin-symmetric convex sets in $\mathbb{R}^n$ such that

$$\gamma_n(\lambda K + (1-\lambda)L)^{\frac{1}{n}} = \lambda \gamma_n(K)^{\frac{1}{n}} + (1-\lambda)\gamma_n(L)^{\frac{1}{n}}, \tag{10.24}$$

then $K = L$.

10.2 Volume product

In this section, we discuss measure-theoretic versions of the Blaschke–Santaló inequality and its reverse (the Bourgain–Milman inequality). A beautiful source of information is the survey article of Fradelizi, Meyer, and Zvavitch [185] that provides a detailed account of classical results, recent advances, and further references.

10.2.1 Functional versions of the Blaschke–Santaló inequality

The volume product of a convex body K in $\mathbb{R}^n$ with $0 \in \operatorname{int}(K)$ is the quantity

$$\mathcal{P}(K) = |K|\,|K^{\circ}|.$$

Recall that the classical Blaschke–Santaló inequality states that if K is a convex body in $\mathbb{R}^n$ with barycenter at the origin, then

$$\mathcal{P}(K) \leqslant \mathcal{P}(B_2^n) = \omega_n^2.$$

The following general functional version of the Blaschke–Santaló inequality was proved by Ball [37] for f even, by Fradelizi and Meyer [179] for f log-concave, and by Lehec [366] in the general case.

Theorem 10.2.1. *Let $f : \mathbb{R}^n \to \mathbb{R}^+$ be an integrable function. There exists $z \in \mathbb{R}^n$ such that for any $\rho : \mathbb{R}^+ \to \mathbb{R}^+$ and any measurable function $g : \mathbb{R}^n \to \mathbb{R}^+$ with*

$$f(x+z)g(y) \leqslant \rho(\langle x,y\rangle)^2 \quad \text{for all } x,y \in \mathbb{R}^n \text{ satisfying } \langle x,y\rangle > 0,$$

one has

$$\int f(x)\,dx \int g(y)\,dy \leqslant \Big(\int \rho(|x|^2)\,dx\Big)^2.$$

If f is even, one can take $z = 0$.

We present two arguments that prove Theorem 10.2.1 under additional assumptions. We need the next variant of the Prékopa–Leindler inequality.

Lemma 10.2.2. *Let $\alpha, \beta, \gamma : \mathbb{R}^+ \to \mathbb{R}^+$ be measurable functions such that $\alpha(s)\beta(t) \leqslant \gamma(\sqrt{st})^2$ for every $s,t > 0$. Then, $\int_{\mathbb{R}^+} \alpha(t)\,dt \int_{\mathbb{R}^+} \beta(t)\,dt \leqslant (\int_{\mathbb{R}^+} \gamma(t)\,dt)^2$.*

Proof. Define $f,g,h : \mathbb{R} \to \mathbb{R}$ by $f(x) = \alpha(e^x)e^x$, $g(x) = \beta(e^x)e^x$ and $h(x) = \gamma(e^x)e^x$. Then, $f(x)g(y) \leqslant h(\frac{x+y}{2})$ for all $x,y \in \mathbb{R}$. By the Prékopa–Leindler inequality we get $\int_{\mathbb{R}} f(x)\,dx \int_{\mathbb{R}} g(x)\,dx \leqslant (\int_{\mathbb{R}} h(x)\,dx)^2$, and the lemma follows with a change of variables. □

For the first proof of Theorem 10.2.1 we need to further assume that f is even and ρ is decreasing. Then, we may use the argument of Ball from [37].

First proof of Theorem 10.2.1. For any $s,t > 0$, let $K_s = \{f \geqslant s\}$ and $L_t = \{g \geqslant t\}$. The hypothesis on f and g implies that $L_t \subset \rho^{-1}(\sqrt{st})K_s^{\circ}$. Since f is even, K_s is symmetric. Consider the functions $\alpha(s) = |K_s|$ and $\beta(t) = |L_t|$. From the Blaschke–Santaló inequality we have that, for every $s,t > 0$

$$\alpha(s)\beta(t) = |K_s|\,|L_t| \leqslant (\rho^{-1}(\sqrt{st}))^n |K_s|\,|K_s^{\circ}| \leqslant (\rho^{-1}(\sqrt{st}))^n |B_2^n|^2.$$

We define $\gamma(t) = (\rho^{-1}(t))^{n/2}|B_2^n|$ and apply Lemma 10.2.2. Note that

$$\int_{\mathbb{R}^n} f(x)\,dx = \int_0^{+\infty} \alpha(s)\,ds$$

and similarly for β and γ, hence the theorem follows. □

In the next proof, we do not require that f is even, but we make the assumption that f is log-concave. This argument is due to Fradelizi and Meyer [179].

Second proof of Theorem 10.2.1. We consider Ball's body $K_f(z)$ associated to f, defined by

$$K_f(z) = \left\{x \in \mathbb{R}^n;\ \int_0^{+\infty} r^{n-1} f(z+rx)\,dr \geqslant 1\right\},$$

with radial function

$$\varrho_{K_f(z)}(x) = \left(\int_0^{+\infty} r^{n-1} f(z+rx)\,dr\right)^{\frac{1}{n}}$$

for $x \in \mathbb{R}^n \setminus \{0\}$. Ball has proved in [40] that $K_f(z)$ is convex.

Given $x, y \in \mathbb{R}^n$ with $\langle x,y\rangle > 0$, for every $r \geqslant 0$ we define $\alpha(r) = r^{n-1}f(z+rx)$, $\beta(r) = r^{n-1}g(rx)$ and $\gamma(r) = r^{n-1}\rho(r^2\langle x,y\rangle)$. From Lemma 10.2.2 we see that

$$\int_0^{+\infty} r^{n-1} f(z+rx)\,dr \int_0^{+\infty} r^{n-1} g(rx)\,dr \leqslant \left(\int_0^{+\infty} r^{n-1}\rho(r^2\langle x,y\rangle)\,dr\right)^2.$$

This shows that

$$\langle x,y\rangle \leqslant \frac{c_n(\rho)}{r_{K_f(z)}(x)\, r_{K_g(0)}(y)},$$

where $c_n(\rho) := (\int_0^{+\infty} r^{n-1}\rho(r^2)\,dr)^{2/n}$. Equivalently, $K_g(0) \subset c_n(\rho)K_f(z)^\circ$. Moreover,

$$\int_{\mathbb{R}^n} f(x)\,dx = n\,|K_f(z)|$$

for every $z \in \mathrm{supp}(f)$. Using Brouwer's fixed point theorem, Fradelizi and Meyer show that for some $z \in \mathbb{R}^n$, the center of mass of $K_f(z)$ is at the origin. Then, the result follows from the Blaschke–Santaló inequality. □

Barthe, Böröczky, and Fradelizi used the same method in [59] to prove stability versions of the functional form of the Blaschke–Santaló inequality.

Applying Theorem 10.2.1 with $\rho = \mathbb{1}_{[0,1]}$ and $f = \mathbb{1}_K$, we recover the Blaschke–Santaló inequality for convex sets. Applying the same theorem with $\rho(t) = e^{-t/2}$, we obtain a proof of the following functional Blaschke–Santaló inequality for the Legendre transform, which is due to Artstein-Avidan, Klartag, and V. Milman [29] (see also [367] for another proof). Recall that the Legendre transform $\mathcal{L}\psi$ of a function $\psi : \mathbb{R}^n \to \mathbb{R} \cup \{+\infty\}$ is defined by

$$\mathcal{L}\psi(y) = \sup_{x\in\mathbb{R}^n}(\langle x,y\rangle - \psi(x)), \quad \text{for } y \in \mathbb{R}^n. \tag{10.25}$$

Theorem 10.2.3. *Let $\varphi : \mathbb{R}^n \to \mathbb{R} \cup \{+\infty\}$ such that $0 < \int e^{-\varphi} < +\infty$. For $x, y \in \mathbb{R}^n$ we set $\varphi_y(x) := \varphi(x + y)$. There exists $z \in \mathbb{R}^n$ such that*

$$\int_{\mathbb{R}^n} e^{-\varphi(x)}\,dx \int_{\mathbb{R}^n} e^{-\mathcal{L}(\varphi_z)(y)}\,dy \leqslant \left(\int_{\mathbb{R}^n} e^{-\frac{|x|^2}{2}}\,dx\right)^2 = (2\pi)^n,$$

with equality if and only $\varphi_z(x) = |Ax|^2$ for some invertible linear map A and some $z \in \mathbb{R}^n$.

We present below two proofs of this theorem. The first one, due to Artstein-Avidan, Klartag, and V. Milman [29] assumes that φ is even.

First proof. For every $m \geqslant 1$ we consider the function $f_m(x) = (1 - \frac{\varphi(x)}{m})_+^m$ and the convex body

$$K_m(f_m) := \{(x,y) \in \mathbb{R}^{n+m} : |y| \leqslant f_m(\sqrt{m}x)^{1/m}\}.$$

When $m \to +\infty$, we have that $f_m \to e^{-\varphi}$ and

$$m^{\frac{n}{2}}\frac{|K_m(f_m)|}{|B_2^m|} = \int_{\mathbb{R}^n} f_m(x)\,dx \to \int_{\mathbb{R}^n} e^{-\varphi(x)}\,dx.$$

Moreover, $(K_m(f_m))^\circ = K_m(\mathcal{L}_m f_m)$, where

$$(\mathcal{L}_m f_m)(y) = \inf_x \frac{(1 - \frac{\langle x,y\rangle}{m})_+^m}{f_m(x)}.$$

Note that, when $m \to +\infty$

$$\mathcal{L}_m(f_m)(y) \to e^{-\mathcal{L}\varphi(y)}$$

and

$$m^{\frac{n}{2}}\frac{|K_m(\mathcal{L}_m\varphi)|}{|B_2^m|} = \int_{\mathbb{R}^n} \mathcal{L}_m f_m(x)\,dx \to \int_{\mathbb{R}^n} e^{-\mathcal{L}\varphi(x)}\,dx.$$

Applying the Blaschke–Santaló inequality to the bodies $K_m(f_m)$ we obtain the result. □

The second proof, due to Lehec [367], uses induction on the dimension.

Second proof. For $n = 1$, choose $z \in \mathbb{R}$ so that

$$\int_z^{+\infty} e^{-\varphi(t)}\,dt = \int_{\mathbb{R}} e^{-\varphi(t)}\,dt/2.$$

For all $s, t \geqslant 0$ we have $\varphi_z(s) + \mathcal{L}(\varphi_z)(t) \geqslant st$. Then, the functions $\alpha(s) = e^{-\varphi_z(s)}$, $\beta(t) = e^{-\mathcal{L}(\varphi_z)(t)}$, and $\gamma(u) = e^{-u^2/2}$ satisfy the condition $\alpha(s)\beta(t) \leqslant \gamma(\sqrt{st})^2$ for every $s, t \geqslant 0$. It follows from Lemma 10.2.2 that

$$\int_0^{+\infty} e^{-\varphi_z(t)}\,dt \int_0^{+\infty} e^{-\mathcal{L}(\varphi_z(t))}\,dt = \int_{\mathbb{R}^+} \alpha(t)\,dt \int_{\mathbb{R}^+} \beta(t)\,dt \leqslant \Big(\int_{\mathbb{R}^+} \gamma(u)\,du\Big)^2 = \frac{\pi}{2}. \tag{10.26}$$

This inequality also holds on $\mathbb{R}_-$; adding the two inequalities, we get the result.

For the inductive step, suppose that the result holds for n. Let $\varphi : \mathbb{R}^{n+1} \to \mathbb{R} \cup \{+\infty\}$. If $\bar{x} \in \mathbb{R}^{n+1}$, we denote $\bar{x} = (x, s) \in \mathbb{R}^n \times \mathbb{R}$. Let

$$\mathcal{P}(\varphi) := \min_z \int_{\mathbb{R}^{n+1}} e^{-\varphi(\bar{x})}\,d\bar{x} \int_{\mathbb{R}^{n+1}} e^{-\mathcal{L}(\varphi_z)(\bar{x})}\,d\bar{x}.$$

Note that $\mathcal{P}(\varphi \circ A) = \mathcal{P}(\varphi)$ for every invertible affine transformation of $\mathbb{R}^n$. Translating φ in the e_{n+1} direction, we may assume that

$$\int_{\{s:s>0\}} \int e^{-\varphi(x,s)}\,dx\,ds = \int_{\{s:s<0\}} \int e^{-\varphi(x,s)}\,dx\,ds.$$

Define $b_+(\varphi)$ and $b_-(\varphi)$ in $\mathbb{R}^{n+1}$ by

$$b_+(\varphi) = \frac{\int_{\{s:s>0\}} \int (x,s) e^{-\varphi(x,s)}\,dx\,ds}{\int_{\{s:s>0\}} \int e^{-\varphi}\,dx\,ds} \quad\text{and}\quad b_-(\varphi) = \frac{\int_{\{s:s<0\}} \int (x,s) e^{-\varphi(x,s)}\,dx\,ds}{\int_{\{s:s<0\}} \int e^{-\varphi}\,dx\,ds}.$$

Since $\langle b_+(\varphi), e_{n+1}\rangle > 0$ and $\langle b_-(\varphi), e_{n+1}\rangle < 0$, there exists a unique point z in $[b_-(\varphi), b_+(\varphi)] \cap e_{n+1}^{\perp}$. By translating φ in the remaining directions, we may assume that $z = 0$. Let A be the linear invertible map defined by $Ax = x$ for $x \in \mathbb{R}^n$ and $Ae_{n+1} = b_+(\varphi)$. Then, $b_+(\varphi \circ A) = A^{-1} b_+(\varphi) = e_{n+1}$. Changing φ into $\varphi \circ A$, we may assume that $b_+(\varphi) = e_{n+1}$. We define $\Phi, \Psi : \mathbb{R}^n \to \mathbb{R} \cup \{+\infty\}$ by

$$e^{-\Phi(x)} = \int_0^{+\infty} e^{-\varphi(x,t)}\,dt \quad\text{and}\quad e^{-\Psi(x)} = \int_0^{+\infty} e^{-\mathcal{L}\varphi(x,t)}\,dt.$$

Since $b_+(\varphi) = e_{n+1}$, we get $\int_{\mathbb{R}^n} xe^{-\Phi(x)}\,dx = \int_{t>0}\int_{\mathbb{R}^n} xe^{-\varphi(x,t)}\,dx\,dt = 0$. Hence, $b(\Phi) = 0$. From the induction hypothesis and Remark 10.2.4 below, it follows that

$$\int_{\mathbb{R}^n} e^{-\Phi(x)}\,dx \int_{\mathbb{R}^n} e^{-\mathcal{L}\Phi(y)}\,dy \leqslant (2\pi)^n. \tag{10.27}$$

For every $x, y \in \mathbb{R}^n$ and $s, t \in \mathbb{R}$, let $\varphi^x(s) = \varphi(x,s)$ and $(\mathcal{L}\varphi)^y(t) = \mathcal{L}\varphi(y,t)$. Applying again Lemma 10.2.2 as in (10.26), we get

$$\int_0^{+\infty} e^{-\varphi^x(s)}\,ds \int_0^{+\infty} e^{-\mathcal{L}(\varphi^x)(t)}\,dt \leqslant \frac{\pi}{2}.$$

Since $\varphi^x(s) + (\mathcal{L}\varphi)^y(t) \geqslant \langle x,y\rangle + st$, one has $(\mathcal{L}\varphi)^y(t) - \langle x,y\rangle \geqslant \mathcal{L}(\varphi^x)(t)$. Thus, for $x, y \in \mathbb{R}^n$

$$e^{-\Phi(x)-\Psi(y)} = \int_0^{+\infty} e^{-\varphi^x(s)}\,ds \int_0^{+\infty} e^{-(\mathcal{L}\varphi)^y(t)}\,dt \leqslant \frac{\pi}{2}e^{-\langle x,y\rangle}.$$

This implies that $e^{-\Psi(y)} \leqslant \frac{\pi}{2}e^{-\mathcal{L}\Phi(y)}$. Using (10.27), we get

$$\int_{\mathbb{R}^n} e^{-\Phi(x)}\,dx \int_{\mathbb{R}^n} e^{-\Psi(y)}\,dy \leqslant \frac{\pi}{2}(2\pi)^n,$$

that is,

$$\int_0^{+\infty}\int_{\mathbb{R}^n} e^{-\varphi(x,s)}\,dx\,ds \int_0^{+\infty}\int_{\mathbb{R}^n} e^{-\mathcal{L}\varphi(y,t)}\,dy\,dt \leqslant \frac{\pi}{2}(2\pi)^n.$$

Adding this to the analogous bound for $s < 0$ and using $\int_{\{s:s>0\}}\int e^{-\varphi(x,s)}\,dx\,ds = \int_{\{s:s<0\}}\int e^{-\varphi(x,s)}\,dx\,ds$, we conclude the proof. □

Remark 10.2.4. Lehec observes that if the "barycenter"

$$b(\varphi) := \int xe^{-\varphi(x)}\,dx \Big/ \int e^{-\varphi} = 0,$$

then

$$\int e^{-\varphi}\int e^{-\mathcal{L}\varphi} \leqslant (2\pi)^n.$$

Indeed, for any z, one has $\mathcal{L}(\varphi_z)(y) = \mathcal{L}\varphi(y) - \langle y,z\rangle$. It follows that $\mathcal{L}((\mathcal{L}\varphi)_z)(y) = \mathcal{L}\mathcal{L}\varphi(y) - \langle y,z\rangle \leqslant \varphi(y) - \langle y,z\rangle$. Using Jensen's inequality and $b(\varphi) = 0$, we get

$$\int e^{-\mathcal{L}((\mathcal{L}\varphi)_z)} \geqslant \int e^{-\varphi(y)+\langle y,z\rangle}\,dy \geqslant e^{\langle b(\varphi),z\rangle}\int e^{-\varphi} = \int e^{-\varphi}.$$

Applying Theorem 10.2.3 to $\mathcal{L}\varphi$, we see that there exists z such that

$$\int e^{-\varphi} \int e^{-\mathcal{L}\varphi} \leqslant \int e^{-\mathcal{L}((\mathcal{L}\varphi)_z)} \int e^{-\mathcal{L}\varphi} \leqslant (2\pi)^n.$$

Lehec also observed that this gives a new proof of the following result of Lutwak [393].

Proposition 10.2.5. *For any star body $K \subset \mathbb{R}^n$ with barycenter at 0 we have that*

$$|K|\,|K^\circ| \leqslant |B_2^n|^2.$$

Proof. Let $\varphi(x) = \frac{\|x\|_K^2}{2}$. Since

$$\int_{\mathbb{R}^n} xe^{-\frac{\|x\|_K^2}{2}}\,dx = \int_{\mathbb{R}^n} x \int_{\|x\|_K}^{+\infty} te^{-\frac{t^2}{2}}\,dt\,dx = \int_0^{+\infty} t^{n+1}e^{-\frac{t^2}{2}}\,dt \int_K x\,dx = 0,$$

one has $b(\varphi) = 0$. Moreover, for any $y \in \mathbb{R}^n$ we have $\mathcal{L}\varphi(y) = \sup_x \langle x,y\rangle - \frac{\|x\|_K^2}{2} = \frac{\|y\|_{K^\circ}^2}{2}$ and $\int_{\mathbb{R}^n} e^{-\frac{\|x\|_K^2}{2}}\,dx = 2^{\frac{n}{2}}\Gamma(\frac{n}{2}+1)|K|$. □

Remark 10.2.6. Various L_p-versions of the functional Blaschke–Santaló inequalities have been given (see, for instance, [248]). Also, Blaschke–Santaló-type inequalities were established in the study of extremal general affine surface area [219, 550, 259]. A consequence of the Blaschke–Santaló inequality appears in [535].

10.2.2 Functional version of Mahler's conjecture

We move now to lower bounds for the volume product $\mathcal{P}(K) = |K|\,|K^\circ|$ of a convex body K in $\mathbb{R}^n$ with $0 \in \mathrm{int}(K)$. It is conjectured that

$$\mathcal{P}(K) \geqslant \mathcal{P}(\Delta_n) = \frac{(n+1)^{n+1}}{(n!)^2}, \tag{10.28}$$

where Δ_n is a simplex in $\mathbb{R}^n$, with equality if and only if $K = \Delta_n$. In the symmetric case, it has been conjectured that for every origin-symmetric convex body K in $\mathbb{R}^n$ one has

$$\mathcal{P}(K) \geqslant \mathcal{P}(B_\infty^n) = \frac{4^n}{n!}, \tag{10.29}$$

where $B_\infty^n = [-1,1]^n$, with equality if and only if K is a Hanner polytope. Recall that in dimension 1, Hanner polytopes are the symmetric segments, and then inductively, if Hanner polytopes have been defined in all dimensions up to $n-1$, a Hanner polytope in dimension n is the ℓ_∞-sum or the ℓ_1-sum of two lower-dimensional Hanner polytopes.

The first conjecture was formulated explicitly by Mahler in [397], but not the second one. We shall refer to both (10.28) and (10.29) as Mahler conjectures.

Functional analogues of the Mahler conjectures were proposed in [181]. Let $\varphi : \mathbb{R}^n \to \mathbb{R} \cup \{+\infty\}$ be a convex function. The domain of φ is the convex set $\mathrm{dom}(\varphi) := \{x \in \mathbb{R}^n; \varphi(x) < +\infty\}$. If $0 < \int e^{-\varphi} < +\infty$, we define the functional volume product of φ as follows:

$$\mathcal{P}(\varphi) = \min_z \int_{\mathbb{R}^n} e^{-\varphi(x)}\,dx \int_{\mathbb{R}^n} e^{-\mathcal{L}(\varphi_z)(y)}\,dy.$$

If φ is even, then it is easily checked that this minimum is attained when $z = 0$.

Conjecture 10.2.7. *Let $n \geqslant 1$ and let $\varphi : \mathbb{R}^n \to \mathbb{R} \cup \{+\infty\}$ be a convex function such that $0 < \int e^{-\varphi} < +\infty$. Then,*

$$\int_{\mathbb{R}^n} e^{-\varphi(x)}\,dx \int_{\mathbb{R}^n} e^{-\mathcal{L}\varphi(y)}\,dy \geqslant e^n,$$

with equality if and only if there exist a constant $c > 0$ and an invertible linear map T such that

$$e^{-\varphi(Tx)} = c\prod_{i=1}^{n} e^{-x_i}\mathbb{1}_{[-1,+\infty)}(x_i).$$

If we make the additional assumption that φ is even, then

$$\int_{\mathbb{R}^n} e^{-\varphi(x)}\,dx \int_{\mathbb{R}^n} e^{-\mathcal{L}\varphi(y)}\,dy \geqslant 4^n,$$

with equality if and only if there exist a constant $c > 0$, two complementary subspaces F_1 and F_2 and two Hanner polytopes $K_1 \subset F_1$ and $K_2 \subset F_2$ such that for all $(x_1, x_2) \in F_1 \times F_2$

$$e^{-\varphi(x_1+x_2)} = ce^{-\|x_1\|_{K_1}}\mathbb{1}_{K_2}(x_2).$$

Remark 10.2.8. With a different duality for a convex function φ, different forms of the Blaschke–Santaló and inverse Santaló inequality were obtained in [25, 176]. Another extension of the Blaschke–Santaló inequality and of its functional form was considered in [260], where duality is combined with the study of inequalities related to monotone non-trivial Minkowski endomorphisms.

In the next theorem, we gather partial results regarding Conjecture 10.2.7.

Theorem 10.2.9. *Let $n \geqslant 1$ and $\varphi : \mathbb{R}^n \to \mathbb{R} \cup \{+\infty\}$ be a convex function such that $0 < \int e^{-\varphi} < +\infty$. Then, Conjecture 10.2.7 holds for $n = 1$ and also for all $n \geqslant 1$, if there exists an invertible affine map T such that $\mathrm{dom}(\varphi \circ T) = \mathbb{R}^n_+$ and $\varphi \circ T$ is non-decreasing on*

P, in the sense that if $x_i \leqslant y_i$ *for all* $1 \leqslant i \leqslant n$*, then* $(\varphi \circ T)(x_1, \ldots, x_n) \leqslant (\varphi \circ T)(y_1, \ldots, y_n)$*. The "even part" of Conjecture* 10.2.7 *holds if* $n = 1$ *or* $n = 2$*. It holds also for all* $n \geqslant 1$ *if* φ *is unconditional, in the sense that there exists an invertible linear map T such that* $(\varphi \circ T)(x_1, \ldots, x_n) = (\varphi \circ T)(|x_1|, \ldots, |x_n|)$ *for all* $(x_1, \ldots, x_n) \in \mathbb{R}^n$*.*

For $n = 1$, Conjecture 10.2.7 was proved in two different ways in [182, 181]. The case of non-decreasing convex functions on the positive octant was also settled in [181]. For unconditional convex functions on $\mathbb{R}^n$, the even part of Conjecture 10.2.7 was established in two different ways in [180, 181], with the case of equality in [186]. In particular, this gives a complete answer for the general case $n = 1$. For $n = 2$, it was proved in [183].

Remark 10.2.10. There is a strong link between the conjectured inequalities (10.28) and (10.29) for convex bodies and their functional counterparts. It was observed in [181] that given a symmetric convex body K in $\mathbb{R}^n$, if $\varphi_K(x) = \|x\|_K$, then $e^{-\mathcal{L}\varphi_K} = \mathbb{1}_{K^\circ}$, and integrating on level sets we get $\mathcal{P}(\varphi_K) = n!\mathcal{P}(K)$. Therefore, if Conjecture 10.2.7 holds for φ_K, then (10.29) holds for K. Conversely, if (10.29) holds in $\mathbb{R}^n$ for every dimension n, then, given an even, convex function $\varphi : \mathbb{R}^n \to \mathbb{R} \cup \{+\infty\}$, we can apply it in dimension $n + m$ to the convex sets

$$K_m(\varphi) = \left\{(x, y) \in \mathbb{R}^n \times \mathbb{R}^m; \|y\|_\infty \leqslant \left(1 - \frac{\varphi(mx)}{m}\right)_+\right\}.$$

Using the identity

$$|K_m(\varphi)| = \frac{2^m}{m^n} \int\limits_{\mathbb{R}^n} \left(1 - \frac{\varphi(x)}{m}\right)_+^m dx$$

and the fact that

$$K_m(\varphi)^\circ = \left\{(x, y) \in \mathbb{R}^n \times \mathbb{R}^m; \|y\|_1 \leqslant \inf_{\varphi(x') \leqslant m} \frac{(1 - \langle x, x' \rangle)_+}{1 - \frac{\varphi(x')}{m}}\right\},$$

one can prove (see [181]) that when $m \to +\infty$, the inequality $\mathcal{P}(K_m(\varphi)) \geqslant \frac{4^{n+m}}{(n+m)!}$ gives $\mathcal{P}(\varphi) \geqslant 4^n$. In a similar way, if Conjecture 10.2.7 holds in dimension $n + 1$, then, given a convex body K in $\mathbb{R}^n$ with Santaló point at the origin, we apply it to the function $\varphi : \mathbb{R}^n \times \mathbb{R} \to \mathbb{R} \cup \{+\infty\}$ defined by

$$e^{-\varphi(x,t)} = \mathbb{1}_{[-n-1,+\infty)}(t)\mathbb{1}_{(t+n+1)K}(x)e^{-t}.$$

The Legendre transform of φ is

$$e^{-\mathcal{L}\varphi(y,s)} = \mathbb{1}_{(-\infty,1]}(s)\mathbb{1}_{(1-s)K^\circ}(y)e^{(n+1)(s-1)}$$

and

$$\mathcal{P}(\varphi) = \frac{(n!)^2 e^{n+1}}{(n+1)^{n+1}} |K|\,|K^\circ|.$$

This proves that if Conjecture 10.2.7 holds for φ, then

$$\mathcal{P}(K) \geqslant \mathcal{P}(\Delta_n) = \frac{(n+1)^{n+1}}{(n!)^2},$$

which is exactly (10.28) for K. Finally, it is shown in [181] that one can adapt the arguments for even functions to prove that, given a convex function $\varphi : \mathbb{R}^n \to \mathbb{R} \cup \{+\infty\}$ one can apply (10.29) to a well-chosen sequence of bodies $\Delta_m(\varphi)$ in dimension $n+m$ and prove Conjecture 10.2.7 for φ when $m \to +\infty$.

It was also proved in [227] that if $\mathcal{P}(\varphi)$ is minimal, then φ does not have a positive Hessian at any point. Asymptotic estimates hold too: in the even case, it was proved in [299] that for some constant $c > 0$, one has for all even convex functions φ and all $n \geqslant 1$, $\mathcal{P}(\varphi) \geqslant c^n$. This was generalized to all convex functions in [181].

10.2.3 Volume product and transport inequalities

The property (τ) was introduced by Maurey in [403]: Let μ be a measure on $\mathbb{R}^n$ and $c : \mathbb{R}^n \times \mathbb{R}^n \to \mathbb{R}^+$ be a lower semi-continuous function (called a cost function); we say that the couple (μ, c) satisfies property (τ) if for any continuous and bounded function $f : \mathbb{R}^n \to \mathbb{R}$ we have

$$\int_{\mathbb{R}^n} e^{-f(x)}\, d\mu(x) \int_{\mathbb{R}^n} e^{Q_c f(y)}\, d\mu(y) \leqslant 1,$$

where

$$Q_c f(y) = \inf_x (f(x) + c(x,y)) \text{ for } y \in \mathbb{R}^n.$$

Maurey [403] showed that if γ_n is the standard gaussian probability measure on $\mathbb{R}^n$, with density $(2\pi)^{-n/2} e^{-|x|^2/2}$, and $c_2(x,y) = \frac{1}{2}|x-y|^2$, then as a consequence of the Prékopa–Leindler inequality, $(\gamma_n, \frac{c_2}{2})$ satisfies property (τ).

Artstein-Avidan, Klartag, and V. Milman [29] made the observation that the functional form of the Blaschke–Santaló inequality for the Legendre transform (Theorem 10.2.3) is equivalent to an improved property (τ) for even functions: we say that the pair (γ_n, c_2) satisfies the even property (τ) if for any even function f we have that

$$\int_{\mathbb{R}^n} e^{-f(x)}\, d\gamma_n(x) \int_{\mathbb{R}^n} e^{Q_{c_2} f(y)}\, d\gamma_n(y) \leqslant 1. \tag{10.30}$$

The equivalence of the even property (τ) with Theorem 10.2.3 follows from the fact that if $\varphi(x) = f(x) + \frac{|x|^2}{2}$, then

$$-\mathcal{L}\varphi(y) = \inf_x\left(f(x) + \frac{|x|^2}{2} - \langle x, y\rangle\right) = Q_{c_2}f(y) + \frac{|y|^2}{2}.$$

A direct proof of (10.30) was then given by Lehec in [365]. Remark 10.2.4 implies that (10.30) holds as soon as $\int_{\mathbb{R}^n} xe^{-f(x)}\,d\gamma_n(x) = 0$.

Moreover, it is shown in [231, Proposition 8.2] that there is a general equivalence between property (τ) and symmetrized forms of the so-called transport–entropy inequality. This type of inequalities was introduced by Talagrand [520], who showed that, for every probability measure ν on $\mathbb{R}^n$, one has

$$W_2^2(\nu, \gamma_n) \leqslant 2H(\nu|\gamma_n), \tag{10.31}$$

where W_2 is the Kantorovich–Wasserstein distance defined by

$$W_2^2(\nu, \gamma_n) = \inf\left\{\int_{\mathbb{R}^n\times\mathbb{R}^n} |x - y|^2\,d\pi(x, y); \pi \in \Pi(\nu, \gamma_n)\right\}$$

and $\Pi(\nu, \gamma_n)$ is the set of probability measures on $\mathbb{R}^n \times \mathbb{R}^n$ whose first marginal is ν and second marginal is γ_n, while H is the relative entropy defined for $d\nu = fd\gamma_n$ by

$$H(\nu|\gamma_n) = -\int_{\mathbb{R}^n} f\log f d\gamma_n.$$

Using this type of equivalence between property (τ) and transport-entropy inequalities, Fathi [166] proved the following symmetrized form of Talagrand's transport-entropy inequality: if ν_1 (or ν_2) is centered, in the sense that $\int x\,d\nu_1(x) = 0$, then

$$W_2^2(\nu_1, \nu_2) \leqslant 2(H(\nu_1|\gamma_n) + H(\nu_2|\gamma_n)). \tag{10.32}$$

In fact, Talagrand showed that (10.32) is equivalent to the functional form of the Blaschke–Santaló inequality (Theorem 10.2.3). Applying (10.32) to $\nu_1 = \gamma_n$, one recovers Talagrand's inequality (10.31). In his proof, Fathi used a reverse logarithmic Sobolev inequality for log-concave functions established in [32] under some regularity assumptions, removed later with a simplified proof in [123].

In a similar way, Gozlan [230] gave equivalent transport–entropy forms of Conjecture 10.2.7 and of the Bourgain–Milman inequality. In [184] new proofs of the one-dimensional case of Conjecture 10.2.7 are also provided.

10.3 Functional Rogers–Shephard-type inequalities

The difference body inequality of Rogers and Shephard states that if K is a convex body in $\mathbb{R}^n$, then

$$|K - K| \leqslant \binom{2n}{n}|K|, \tag{10.33}$$

with equality attained if and only if K is a simplex. This inequality is a special case of the following fact, which is valid for any pair of convex bodies L, T in $\mathbb{R}^n$:

$$|L - T|\,|L \cap T| \leqslant \binom{2n}{n}|L|\,|T|. \tag{10.34}$$

Note that (10.34), in turn, can be deduced from the next inequality about sections and projections: For any convex body K in $\mathbb{R}^n$ and any $F \in G_{n,k}$ we have

$$\left|P_{F^\perp}(K)\right|\,|K \cap F| \leqslant \binom{n}{k}|K|. \tag{10.35}$$

Indeed, to see that (10.35) implies (10.34) we consider the convex body $K = L \times T$ in $\mathbb{R}^{2n}$ and the subspace $F = \{(x, x) : x \in \mathbb{R}^n\}$, and we observe that $K \cap F$ is a linear image of $L \cap T$ and $P_{F^\perp}(K)$ is a linear image of $L - T$. Applying (10.35) to the pair (K, F) and computing the respective volumes we conclude the proof.

Similarly, considering now the body $K = \mathrm{conv}(L \times \{0\}, \{0\} \times T)$ in $\mathbb{R}^{2n}$ and taking its section with F and projection onto $F^\perp$ we get another Rogers–Shephard-type inequality proved by Artstein, Einhorn, Florentin, and Ostrover in [31]

$$\left|\mathrm{conv}(L, -T)\right|\,\left|(L^\circ + T^\circ)^\circ\right| \leqslant |L|\,|T|, \tag{10.36}$$

which is a strengthening of the classical inequality of Rogers and Shephard

$$\left|\mathrm{conv}(L, -T)\right|\,|L \cap T| \leqslant 2^n|L|\,|T|. \tag{10.37}$$

There are many variants of functional analogues for these theorems. We present two functional analogues of (10.35) that can then translate, by a similar argument to the one presented above for bodies, to functional generalizations of (10.34) and of (10.37). Both of them appear in [18]. Through this example, we would like to stress the fact that a single geometric inequality can have more than one "reasonable" functional versions.

In what follows, if $f : \mathbb{R}^n \to \mathbb{R}^+$ is a measurable function, then, for any $E \in G_{n,k}$, the "section" of f with E is the restriction $f|_E$ of f onto E and the "geometric projection" or "shadow" of f onto E is the function

$$P_E f(x) := \max\{f(y) : y \in x + E^\perp\}, \quad x \in E,$$

where $E^\perp$ is the orthogonal subspace of E.

Theorem 10.3.1. *Let $f : \mathbb{R}^n \to \mathbb{R}^+$ be a log-concave function and let $E \in G_{n,k}$. Then,*

$$\int_{E^\perp} P_{E^\perp} f \int_E f \leqslant \binom{n}{k} \|f\|_\infty \int_{\mathbb{R}^n} f. \tag{10.38}$$

Choosing $f = \mathbb{1}_K$ we recover (10.35). An alternative functional generalization of the same inequality is the next theorem.

Theorem 10.3.2. *Let $f : \mathbb{R}^n \to \mathbb{R}^+$ be a log-concave function, let $E \in G_{n,k}$ and $\lambda \in [0,1]$. Then,*

$$(1-\lambda)^k \lambda^{n-k} \int_{E^\perp} (P_{E^\perp} f)^{1-\lambda} \int_E f^\lambda \leqslant \int_{\mathbb{R}^n} f.$$

Choosing $f(x) = \exp(-\|\cdot\|_K)$, and using the identity $\int_{\mathbb{R}^n} \exp(-\|x\|_K)\,dx = n!|K|$, we recover again (10.35).

For the proof of both theorems we use, as a first step, a symmetrization procedure for functions, which allows us to translate facts about sections to similar facts regarding projections. The symmetrization is a Schwarz symmetrization for the epigraph, with respect to a subspace $E \subset \mathbb{R}^n \subset \mathbb{R}^n \times \mathbb{R}$. The function $f : \mathbb{R}^n \to \mathbb{R}^+$ is replaced by a function $S_E f : \mathbb{R}^n \to \mathbb{R}^+$ such that for every $x \in E^\perp$ and every t for which the intersection of the t-level set of $S_E f$ with $x + E$ is non-empty, this level set is a Euclidean ball, centered at x, of the same volume as the intersection of the t-level set of f with $x + E$. Thus, in particular, the function $S_E f$ has the same integral as f on each fiber $x + E^\perp$, and also the same volume of level sets. Considering φ, where $f = \exp(-\varphi)$, we have that $S_E \varphi|_{E^\perp} = P_{E^\perp} \varphi$ because the minimal level set of this fiber has been symmetrized to be centered at $E^\perp$. Note that by [27, Chapter 1] this symmetrization procedure preserves convexity, and in particular, the function $S_E f$ is again log-concave.

Proof of Theorem 10.3.2. We define $f_1 = S_E f$, so that $f_1|_{E^\perp} = P_{E^\perp} f$ and $\int_E f_1 = \int_E f$. Without loss of generality, let E be the subspace of the first k coordinates, and write $\mathbb{R}^n = \mathbb{R}^k \times \mathbb{R}^{n-k}$. Given $z = (x,y) \in \mathbb{R}^k \times \mathbb{R}^{n-k}$, we write $z = (1-\lambda)(x/(1-\lambda), 0) + \lambda(0, y/\lambda)$ and use the log-concavity of f_1 to write

$$\begin{aligned}
\int f_1(z)\,dz &\geqslant \int_E \int_{E^\perp} f_1(x/(1-\lambda))^{(1-\lambda)} f_1(y/\lambda)^\lambda \,dx\,dy \\
&= \int_E f_1(x/(1-\lambda))^{(1-\lambda)}\,dx \int_{E^\perp} f_1(y/\lambda)^\lambda\,dy \\
&= (1-\lambda)^k \lambda^{n-k} \int_E f_1^{1-\lambda} \int_{E^\perp} f_1^\lambda.
\end{aligned}$$

In view of the connection between f and f_1, this completes the proof. □

Proof of Theorem 10.3.1. Without loss of generality, we may assume that $\|f\|_\infty = f(0) = 1$. Again, we define $f_1 = S_E f$. We set $g = f_1|_E = e^{-\varphi}$ and $h = f_1|_{E^\perp} = e^{-\psi}$. Then, φ, ψ are convex functions with $\varphi(0) = \psi(0) = 0$. We define a new convex function w on $\mathbb{R}^n$ by taking the convex hull of the epigraphs of φ and ψ in the orthogonal subspaces. In other words, we define

$$w(x,y) = \inf\{(1-t)\varphi(x/(1-t)) + t\psi(y/t) : t \in (0,1)\}.$$

Then,

$$\int \exp(-w) \geqslant \frac{1}{\binom{n}{k}} \int \exp(-\varphi) \int \exp(-\psi). \tag{10.39}$$

To prove (10.39), instead of the convex hull w we look at an even bigger, not convex, function $\tilde{w}$ that is obtained by taking the convex hulls of the level sets only. Note that if we had taken direct product instead of the convex hull, then we would get the same volume up to the factor $\binom{n}{k}$. The function corresponding to the direct product is simply $\max\{\varphi(x), \psi(y)\}$, and, in particular, it is convex. Therefore,

$$\int \exp(-w) \geqslant \int \exp(-\tilde{w}) = \frac{1}{\binom{n}{k}} \int \exp(-\max\{\varphi(x), \psi(y)\})\, dx\, dy.$$

Since φ, ψ are non-negative, the maximum is less than the sum and we get

$$\int \exp(-w) \geqslant \frac{1}{\binom{n}{k}} \int \exp(-\varphi(x))\, dx \int \exp(-\psi(y))\, dy.$$

Having proved (10.39) we are able to complete the proof of the theorem. Since f_1 is log-concave and greater than g and h, it is also greater than e^{-w}, and hence we have that

$$\int f \geqslant \int e^{-w} \geqslant \frac{1}{\binom{n}{k}} \int_E f_1 \int_{E^\perp} f_1,$$

which, by symmetrization and normalization, implies (10.38). □

The next application from [18] is a stronger geometric theorem of Rogers–Shephard type, which is a version of (10.35) when the two subspaces have an intersection. It supports the point of view that functional inequalities are, in some sense, more "powerful" than their geometric counterparts.

Theorem 10.3.3. *Let K be a convex body, $E \in G_{n,n-i}$ and $F \in G_{n,n-j}$, where $n-1 \geqslant i+j$, and $E^\perp \subset F$. Denote $k = n-i-j$, so that $1 \leqslant k \leqslant n-2$. Then,*

$$|P_E(K)|\,|K \cap F| \leqslant \binom{j+i}{i} |K| \max_{x \in \mathbb{R}^n} |P_E(K \cap (x+F))|.$$

Proof. We write $\mathbb{R}^n = E^\perp \oplus F^\perp \oplus H$, where $H^\perp = E^\perp \oplus F^\perp$. Note that $\dim(H) = k$. Next, we apply a Schwarz symmetrization of the body with respect to the subspace $E^\perp$ and let $K_1 = S_{E^\perp}K$. Consider the function $f(x) = |K_1 \cap (x + H)|$ on $H^\perp = E^\perp \oplus F^\perp$. We have that f is log-concave; in fact, it is $1/k$-concave by Brunn's principle. Then,

$$\begin{aligned}\|f\|_\infty \int_{H^\perp} f &= |K_1| \max_{x\in\mathbb{R}^n} |K_1 \cap (x + H)| \\ &= |K| \max_{x\in\mathbb{R}^n} |P_E K \cap (x + H)|.\end{aligned}$$

On the other hand

$$\int_{E^\perp} f \int_{F^\perp} P_{F^\perp} f = |K_1 \cap F|\,|K_1 \cap E| = |K \cap F|\,|P_E K|,$$

where we have used the fact that $E^\perp \subset F$, which implies that the volume of the section remains the same under symmetrization. Applying Theorem 10.3.1, which gives the binomial coefficient $\binom{j+i}{i}$, we obtain the result. □

10.4 Mixed integrals of functions

In this section, we introduce a notion of mixed volume (or mixed integral) for functions. A linear structure on log-concave functions under which the volume is polynomial was discovered by V. Milman and Rotem in [424]. In fact, this linear structure can be defined on a much larger class, but we restrict our attention here to the log-concave class.

Definition 10.4.1. Let f and g be two log-concave functions on $\mathbb{R}^n$. For any $\lambda > 0$ we define the addition operation $\oplus$ and the homothety operation $\odot$ by

$$(f \oplus g)(x) = \sup_{y+z=x} \min\{f(y), g(z)\}$$

and $(\lambda \odot f)(x) = f(\frac{x}{\lambda})$.

The geometric meaning of this definition is very simple. For any log-concave function h consider its t-level sets $\{h \geqslant t\}$, where $t > 0$. Then, the t-level set of $\lambda \odot f$ is the t-homothet of the t-level set of f, and the t-level set of $f \oplus g$ is the Minkowski sum of the t-level sets of f and g. One may similarly define these operations on the corresponding convex functions.

The next lemma shows that the operation $\oplus$ preserves log-concavity.

Lemma 10.4.2. *Let f and g be two log-concave functions on $\mathbb{R}^n$. Then, the function*

$$(f \oplus g)(x) = \sup_{y+z=x} \min\{f(y), g(z)\}$$

is log-concave.

Proof. Let $\varphi = -\log f$ and $\psi = -\log g$, and define

$$(\varphi \oplus \psi)(x) = -\log(f \oplus g)(x) = \inf_{y+z=x} \max\{\varphi(y), \psi(z)\}.$$

Fix $x_0, x_1 \in \mathbb{R}^n$ and $0 < \lambda < 1$. For every $\varepsilon > 0$ we can find $y_0 + z_0 = x_0, y_1 + z_1 = x_1$ such that

$$\begin{aligned}(\varphi \oplus \psi)(x_0) &\geqslant \max\{\varphi(y_0), \psi(z_0)\} - \varepsilon,\\ (\varphi \oplus \psi)(x_1) &\geqslant \max\{\varphi(y_1), \psi(z_1)\} - \varepsilon.\end{aligned}$$

If $x_\lambda = (1-\lambda)x_0 + \lambda x_1$, then we define $y_\lambda = (1-\lambda)y_0 + \lambda y_1$ and $z_\lambda = (1-\lambda)z_0 + \lambda z_1$. Note that $y_\lambda + z_\lambda = x_\lambda$ and hence, by convexity

$$\begin{aligned}\varphi(y_\lambda) &\leqslant (1-\lambda)\varphi(y_0) + \lambda\varphi(y_1)\\ &\leqslant (1-\lambda)\max\{\varphi(y_0), \psi(z_0)\} + \lambda\max\{\varphi(y_1), \psi(z_1)\}\\ &\leqslant (1-\lambda)(\varphi \oplus \psi)(x_0) + \lambda(\varphi \oplus \psi)(x_1) + \varepsilon.\end{aligned}$$

Similarly,

$$\psi(z_\lambda) \leqslant (1-\lambda)(\varphi \oplus \psi)(x_0) + \lambda(\varphi \oplus \psi)(x_1) + \varepsilon.$$

Therefore,

$$\begin{aligned}(\varphi \oplus \psi)(x_\lambda) &\leqslant \max\{\varphi(y_\lambda), \psi(z_\lambda)\}\\ &\leqslant (1-\lambda)(\varphi \oplus \psi)(x_0) + \lambda(\varphi \oplus \psi)(x_1) + \varepsilon.\end{aligned}$$

Since this is true for any ε, we see that $\varphi \oplus \psi$ is convex, which shows that $f \oplus g$ is log-concave. □

Using Lemma 10.4.2 and integration on level sets we immediately get the next theorem.

Theorem 10.4.3. *Let $f_1, f_2, \ldots, f_m$ be log-concave functions on $\mathbb{R}^n$. Then, the function $F : (\mathbb{R}^+)^m \to [0, \infty]$ defined by*

$$F(\lambda_1, \lambda_2, \ldots, \lambda_m) = \int [(\lambda_1 \circ f_1) \oplus (\lambda_2 \circ f_2) \oplus \cdots \oplus (\lambda_m \circ f_m)]$$

is a homogeneous polynomial of degree n, with non-negative coefficients, where each coefficient depends only on the functions whose index appears in the corresponding monomial.

In complete analogy with the case of convex bodies, this theorem is equivalent to the existence of a function V defined on n-tuples of log-concave functions and attaining values in $[0, \infty]$, which is invariant under permutations, multi-linear with respect to $\oplus$, and its diagonal is $\int_{\mathbb{R}^n} f(x)\,dx$. In other words, the integration operator on log-concave functions can be polarized. We will call the coefficient of such a monomial, denoted $V(f_1, f_2, \dots f_n)$, the mixed integral of $f_1, f_2, \dots, f_n$. It can be written as

$$V(f_1, f_2, \dots, f_n) = \int_0^\infty V(K_1^t, \dots, K_n^t)\, dt,$$

where K_j^t is the t-level set of f_j.

With this definition for the "mixed volume/integral" of log-concave functions, one can prove a series of inequalities in the spirit of Alexandrov and Fenchel. For example, defining $W_k(f) = V(f[n-k], \mathbb{1}_{B_2^n}[k])$, V. Milman and Rotem [424] prove the next theorem.

Theorem 10.4.4. *Let $e(x) = e^{-|x|}$. For every $f : \mathbb{R}^n \to [0,1]$ that is log-concave and any two integers $0 \leqslant k < m < n$, we have*

$$\left(\frac{W_k(f)}{W_k(e)}\right)^{\frac{1}{n-k}} \leqslant \left(\frac{W_m(f)}{W_m(e)}\right)^{\frac{1}{n-m}}.$$

The proof exploits decreasing rearrangement. Given a log-concave function $f : \mathbb{R}^n \to [0,1]$ with $f(0) = 1$ we consider its Schwarz symmetral $Sf : \mathbb{R}^n \to [0,1]$ that is defined so that for every t the set $\{x : Sf(x) \geqslant t\}$ is a Euclidean ball with the same volume as the set $\{x : f(x) \geqslant t\}$. The main step is to show the following.

Theorem 10.4.5. *Let $f_1, \dots, f_n : \mathbb{R}^n \to [0,1]$ be log-concave functions with compact level sets. Then,*

$$V(f_1, f_2, \dots, f_n) \geqslant V(Sf_1, Sf_2, \dots, Sf_n).$$

One can also obtain a Brunn–Minkowski inequality for this type of addition

$$S(f \oplus g) \supseteq S(f) + S(g),$$

which cannot be translated to a concavity result for the integral under this addition, as one may construct examples for which the "volume" of an "average" is less than the minimum of the "volumes".

Bobkov, Colesanti, and Fragalà [84] developed, independently and simultaneously, a theorem for quermassintegrals of log-concave or quasi-concave functions, with very different resulting inequalities for the corresponding notions. In particular, they proved the next theorem.

Theorem 10.4.6. *Let f, g be log-concave functions, $0 < \lambda < 1$ and $i = 0, 1, \dots, n-1$. Then,*

$$W_i(\lambda \circ f \star (1-\lambda) \circ g) \geqslant W_i(f)^\lambda W_i(g)^{1-\lambda}.$$

They also proved an inequality that implies the Alexandrov inequalities if we choose f to be the indicator function of a convex set.

Theorem 10.4.7. *For any quasi-concave function f and $0 \leqslant i \leqslant k \leqslant n-1$, we have that*

$$W_k(f) \geqslant c \cdot W_i(f^p)^{\frac{1}{p}},$$

where $p = \frac{n-i}{n-k}$ and $c = |B_2^n|^{1-\frac{1}{p}}$.

10.5 Further reading

10.5.1 Dimensional Brunn–Minkowski inequality

The dimensional Brunn–Minkowski conjecture asks whether, for any even log-concave measure on $\mathbb{R}^n$ and for every pair of symmetric convex bodies K, L and $\lambda \in [0,1]$, one has

$$\mu(\lambda K + (1-\lambda)L)^{\frac{1}{n}} \geqslant \lambda\mu(K)^{\frac{1}{n}} + (1-\lambda)\mu(L)^{\frac{1}{n}}. \tag{10.40}$$

The conjecture was first formulated by Gardner and Zvavitch [204] in the gaussian setting. They proved the gaussian version (10.3) of (10.40) in the special cases where the sets K, L are either coordinate boxes containing the origin or dilates of a fixed origin-symmetric convex set. These results were later generalized to broader classes of measures by Marsiglietti [401]. In [134], Colesanti, Livshyts, and Marsiglietti showed that (10.3) holds when both symmetric bodies K, L are small perturbations of the Euclidean ball. Moreover, Livshyts, Marsiglietti, Nayar, and Zvavitch [384] used a variant of the Prékopa–Leindler inequality [505, Theorem 7.1.2] to prove that (10.40) holds for all $\lambda \in (0,1)$, when μ is an unconditional product measure on $\mathbb{R}^n$ and K, L are ideals in $\mathbb{R}^n$. This result was later extended to weakly unconditional sets by Ritoré and Yepes Nicolás [478]. Böröczky and Kalantzopoulos [95] settled the Gardner–Zvavitch problem for a class of symmetric convex sets with many hyperplane symmetries. As discussed in Section 10.1, the conjecture of Gardner and Zvavitch was finally confirmed by Eskenazis and Moschidis [162].

Regarding the general dimensional Brunn–Minkowski conjecture, further progress was made by Cordero-Erausquin and Rotem [138], who verified (10.40) for a broad class of rotationally invariant log-concave measures, and by Kolesnikov and Livshyts [336], who established the conjecture for hereditarily convex log-concave measures. More recently, Cordero-Erausquin and Eskenazis [135] developed a general local framework for the concavity properties of weighted marginals, yielding functional forms of the inequality in rotationally invariant settings.

There are also several more general conjectured inequalities for the class of log-concave measures, involving different types of convex combinations of origin-symmetric convex sets in $\mathbb{R}^n$. The strongest among them is the log-Brunn–Minkowski conjecture of Böröczky, Lutwak, Yang, and Zhang [96], which asserts that the Euclidean volume of the geometric mean of two origin-symmetric convex sets can be bounded below by the geometric mean of their volumes. If true, this inequality would imply the dimensional Brunn–Minkowski inequality (10.3) and the B-inequality with γ_n replaced by any symmetric log-concave measure μ. In particular, the planar case of (10.40) can be derived for any symmetric log-concave measure μ on $\mathbb{R}^2$ and all symmetric convex sets K, L in $\mathbb{R}^2$ by combining [384, Proposition 1], [96, Theorem 1.7], and a result of Saroglou [499, Theorem 3.1]. Recent developments in this area are discussed in [340, 544, 130, 471, 95, 263], and [337].

An analytic approach of Kolesnikov and Livshyts [336], building on earlier ideas of Kolesnikov and Milman (see [339] and [340]), reduces (10.40) to coercivity estimates for the diffusion operator

$$Lu = \Delta u - \langle \nabla u, \nabla \psi \rangle$$

associated with μ. The key idea is that Brunn–Minkowski-type inequalities can be derived from infinitesimal perturbations of convex bodies, expressed via support functions and linearization along Minkowski combinations. This leads to second-order inequalities involving curvature and weighted boundary integrals, which are ultimately controlled by spectral-type quantities of the form

$$\int_K \left(\|\nabla^2 u\|^2 + \langle \nabla^2 \psi \, \nabla u, \nabla u \rangle\right) d\mu.$$

Using this framework, Livshyts [382] proved a version of (10.40) for all even log-concave measures, with exponent $c_n = n^{-4-o_n(1)}$. This universal estimate has recently been improved by Eskenazis, Giannopoulos, and Tziotziou [163]: for every even log-concave probability measure on $\mathbb{R}^n$, every pair of symmetric convex bodies $K, L \subset \mathbb{R}^n$, and $\lambda \in [0, 1]$

$$\mu(\lambda K + (1-\lambda)L)^{c_n} \geqslant \lambda \mu(K)^{c_n} + (1-\lambda)\mu(L)^{c_n},$$

where $c_n \geqslant \frac{c}{n^3 \log n}$.

10.5.2 Functional *M* position and reverse Brunn–Minkowski

The fact that functions in an appropriate position satisfy a form of the reverse Brunn–Minkowski inequality was proved by Klartag and Milman [298]. Many further results, including connections to mixed volumes, appear in Klartag's work [285]. The framework of covering numbers was extended to functions by Artstein-Avidan and Slomka in [25] (see also the earlier work [24] of Artstein-Avidan and Raz). Given measurable $f, g : \mathbb{R}^n \to [0, \infty)$, the functional covering number of f by g is defined by

$$N(f,g) = \inf\{\mu(\mathbb{R}^n) : \mu * g \geqslant f\},$$

where the infimum runs over non-negative Borel measures μ satisfying

$$(\mu * g)(x) = \int_{\mathbb{R}^n} g(x-t)\, d\mu(t) \geqslant f(x) \quad \text{for all } x \in \mathbb{R}^n.$$

Let $LC_g(\mathbb{R}^n)$ denote the class of geometric log-concave functions; these are the upper semi-continuous log-concave functions $f : \mathbb{R}^n \to [0,\infty)$ with $f(0) = \|f\|_\infty = 1$. If $f = e^{-\varphi} \in LC_g(\mathbb{R}^n)$, its Legendre dual is $f^* = e^{-\mathcal{L}\varphi}$, where

$$\mathcal{L}\varphi(x) = \sup_{y\in\mathbb{R}^n}\{\langle x,y\rangle - \varphi(y)\}$$

is the Legendre transform of φ. In the functional setting, the dual function f^* plays a role analogous to that of the polar body in classical convex geometry. The natural analogue of the Euclidean ball is the gaussian

$$g(x) = \exp\left(-\frac{1}{2}|x|^2\right) \quad \text{for which } g^* = g.$$

Artstein-Avidan and Slomka [26] established the existence of a functional version of Milman's M-position for geometric log-concave functions: If $f : \mathbb{R}^n \to [0,\infty)$ is a geometric log-concave function then there exists $T \in GL_n$ such that $\widetilde{f} = f \circ T$ satisfies $\int \widetilde{f} = (2\pi)^{n/2}$ and

$$\max\{N(\widetilde{f},g),\, N(\widetilde{f}^*,g),\, N(g,\widetilde{f}),\, N(g,\widetilde{f}^*)\} \leqslant C^n$$

for an absolute constant $C > 0$.

Giannopoulos and Tziotziou [211] established the existence of a regular functional M-position, in the sense of Pisier, for geometric log-concave functions. Recall that Pisier [463] proved that for any $0 < \alpha < 2$ and any symmetric convex body $K \subset \mathbb{R}^n$ there exists a linear image $\widetilde{K}$ such that

$$\max\{N(\widetilde{K}, tB_2^n),\, N(B_2^n, t\widetilde{K}),\, N(\widetilde{K}^\circ, tB_2^n),\, N(B_2^n, t\widetilde{K}^\circ)\} \leqslant \exp(c(\alpha)\, n/t^\alpha)$$

for every $t \geqslant c(\alpha)^{1/\alpha}$, where $c(\alpha) = O((2-\alpha)^{-\alpha/2})$ as $\alpha \to 2^-$. A convex body satisfying the above is said to be in an α-regular M-position.

Define the homothetic dilation of a log-concave function f by

$$(t \odot f)(x) = f(x/t), \quad t > 0.$$

It is shown in [211] that if $f : \mathbb{R}^n \to [0,\infty)$ is an isotropic geometric log-concave function, then for every $t \geqslant 1$

$$\max\{N(f, t\odot g), N(g, t\odot f^*)\} \leqslant \exp\left(\frac{\gamma_n n}{t}\right)$$

and

$$\max\{N(f^*, t\odot g), N(g, t\odot f)\} \leqslant \exp\left(\frac{\delta_n n}{t}\right),$$

where $\gamma_n \leqslant c(\log n)^2$ and $\delta_n \leqslant c\log n$.

Thus, the isotropic position yields a universal functional M-position whose regularity exponent is arbitrarily close to 1.

10.5.3 Functional John position

The functional analogue of the John and Löwner ellipsoids has attracted considerable attention, and several results in this direction are known. These include works by Alonso-Gutiérrez, González-Merino, Jiménez, and Villa [16], Li, Schütt, and Werner [373], Ivanov and Tsiutsiurupa [268], and Ivanov and Naszódi [267]. The latter use this position to obtain a new Helly-type theorem for log-concave functions: given a (possibly very large) family of log-concave functions $f_i, i \in I$ on $\mathbb{R}^n$, and defining for every subset $\sigma \subset I$ the pointwise minimum $f_\sigma = \min\{f_i : i \in \sigma\}$, there exists a subset of I of cardinality $3n + 2$ such that the integral of f_σ is bounded above by the integral of f_I times a constant $c(n)$, where $c(n)$ is of order $\exp(cn\log n)$.

10.5.4 Moment measure of a log-concave function

Let $f = e^{-\psi} : \mathbb{R}^n \to [0,\infty)$ be a log-concave function with finite positive integral. The moment measure μ_f of f is the Borel measure on $\mathbb{R}^n$ defined as the push-forward of fdx under $\nabla\psi$. Equivalently,

$$\int_{\mathbb{R}^n} g(y)\,d\mu_f(y) = \int_{\mathbb{R}^n} g(\nabla\psi(x))f(x)\,dx \tag{10.41}$$

for every Borel measurable function g such that $g \in L^1(\mu_f)$ or g is non-negative.

Cordero-Erausquin and Klartag [136] characterized those measures ν that arise as moment measures of log-concave functions f on $\mathbb{R}^n$ with finite positive integrals. Under the assumption that f is essentially continuous, meaning that f is upper semi-continuous and the set of its discontinuity points has zero measure for the $(n-1)$-dimensional Hausdorff measure, they showed that the moment measure μ_f has the following two properties: it is centered and not supported on a hyperplane. In particular, its first moment is finite

$$\int_{\mathbb{R}^n} |y|\, d\mu_f(y) = \int_{\mathbb{R}^n} |\nabla\psi(x)| f(x)\, dx < +\infty. \tag{10.42}$$

These necessary conditions are also sufficient. The main result of [136] establishes a bijection between essentially continuous functions $f = e^{-\psi}$ (modulo translations of ψ) and finite Borel measures on $\mathbb{R}^n$ that are centered and not supported on a hyperplane: Suppose that ν is a Borel measure on $\mathbb{R}^n$ such that

(i) $0 < \nu(\mathbb{R}^n) < +\infty$.
(ii) The measure ν is not supported by a lower-dimensional subspace.
(iii) The barycenter of ν lies at the origin; in particular, ν has a finite first moment.

Then, there exists an essentially continuous log-concave function $f = e^{-\psi}$ such that ν is the moment measure of the log-concave function f. Moreover, the function ψ is uniquely determined up to translation.

10.5.5 Surface area measures of a log-concave function

Recall that for every convex body K in $\mathbb{R}^n$ there exists a Borel measure $S(K,\cdot)$ on S^{n-1}, the surface area measure of K, such that

$$\lim_{t\to 0^+} \frac{|K+tL|-|K|}{t} = \int_{S^{n-1}} h_L(\xi)\, dS(K,\xi) \tag{10.43}$$

for every convex body L in $\mathbb{R}^n$. Choosing $L = B_2^n$ we see that $S(K, S^{n-1})$ equals the surface area of K.

A natural question in “functional convexity” is to extend the notion of the surface area measure of a convex body to the setting of functions in $LC(\mathbb{R}^n)$. For any log-concave function $f : \mathbb{R}^n \to \mathbb{R}$ with $0 < \int f < \infty$, Rotem considered in [484] and [485] two surface area measures of f. Write $f = e^{-\psi}$, where $\psi : \mathbb{R}^n \to (-\infty, \infty]$ is a convex function. Then, the first surface area measure μ_f of f is a measure on $\mathbb{R}^n$, defined as the push-forward

$$\mu_f = (\nabla\psi)_*(f dx).$$

Note that, in the terminology of Section 10.5.4, μ_f is precisely the moment measure of f. The second surface area measure of f is a measure ν_f on the sphere S^{n-1}, defined as the push-forward

$$\nu_f = (n_{K_f})_*(f d\mathcal{H}^{n-1}|_{\partial K_f}),$$

where $K_f = \overline{\{x : f(x) > 0\}}$ is the support of f, and n_{K_f} denotes the Gauss map $n_{K_f} : \partial K_f \to S^{n-1}$.

For example, if $f(x) = e^{-|x|^2/2}$, then $\mu_f = e^{-|x|^2/2}\,dx$ and $\nu_f \equiv 0$, because $\partial K_f = \partial\mathbb{R}^n = \varnothing$. On the other hand, if $f = \mathbb{1}_K$ for some convex body K in $\mathbb{R}^n$, then $\mu_{\mathbb{1}_K} = |K|\,\delta_0$ and $\nu_{\mathbb{1}_K} = S_K$, the usual surface area measure. The pair (μ_f, ν_f) is the pair of surface area measures of f.

It is useful to observe that we do not assume any regularity for the definitions of μ_f and ν_f. Since $\psi = -\log f$ is a convex function, it is differentiable almost everywhere with respect to the Lebesgue measure on the set $K_f = \overline{\{x : \psi(x) < \infty\}}$. Therefore, the push-forward $(\nabla\psi)_*(f dx)$ is well defined. Similarly, since K_f is a closed convex set, its boundary ∂K_f is a Lipschitz manifold, and hence the Gauss map n_{K_f} is defined $\mathcal{H}^{n-1}$ almost everywhere and the push-forward is again well defined.

Motivated by (10.43), for any $f, g \in \mathrm{LC}_n$ one may consider the first variation of the integral of f in the direction of g, defined by

$$\delta(f,g) = \lim_{t\to 0^+} \frac{\int f \star (t\cdot g) - \int f}{t}. \tag{10.44}$$

Under some additional regularity assumptions, $\delta(f,g)$ was studied by Colesanti and Fragalà in [133]. The same first variation appears in the work of Cordero-Erausquin and Klartag [136] as a step in their proof of their Minkowski-type existence theorem. Rotem dropped the additional hypotheses in [484] and proved that if $f = e^{-\psi}, g = e^{-\varphi} \in LC(\mathbb{R}^n)$, $0 < \int f < \infty$ and $\nu_f = 0$, then

$$\delta(f,g) = \int_{\mathbb{R}^n} h_g \, d\mu_f \tag{10.45}$$

with no regularity assumptions, where

$$h_g = \mathcal{L}\varphi$$

is the Legendre transform of φ, which is the functional analogue of the support function of g. Note that since f is log-concave and upper semi-continuous, it is only discontinuous at points $x \in \partial K_f$ such that $f(x) \neq 0$. Therefore, the condition $\nu_f = 0$ is equivalent to the statement that f is essentially continuous. In his subsequent work [485], Rotem obtained a very general version of (10.45): If $f, g \in LC(\mathbb{R}^n)$ such that $0 < \int f < \infty$, then

$$\delta(f,g) = \int_{\mathbb{R}^n} h_g \, d\mu_f + \int_{S^{n-1}} h_{K_g} \, d\nu_f.$$

Bibliography

[1] M. Agranovsky, *On polynomially integrable domains in Euclidean spaces*, in: Complex analysis and dynamical systems, 1–21, Trends Math., Birkhäuser/Springer, Cham, 2018.

[2] M. Agranovsky, *On algebraically integrable domains*, in: Functional analysis and geometry: Selim Grigorievich Krein centennial, 33–44, Contemp. Math. 733, Amer. Math. Soc., Providence, RI, 2019.

[3] M. Agranovsky, *Locally polynomially integrable surfaces and finite stationary phase expansions*, J. Anal. Math. 141 (2020), 23–47.

[4] M. Agranovsky, *Domains with algebraic X-ray traansform*, Anal. Math. Phys. 12 (2022), 60, 17 pp.

[5] M. Agranovsky, J. Boman, A. Koldobsky, V. Vassiliev and V. Yaskin, *On algebraically integrable bodies*, in Harmonic analysis and convexity, A. Koldobsky and A. Volberg, eds, De Gruyter, Berlin/Boston, 2023, 1–35.

[6] M. Agranovsky, A. Koldobsky, D. Ryabogin and V. Yaskin, *An analogue of polynomially integrable bodies in even dimensions*, J. Math. Anal. Appl. 529 (2024), paper 127071, 12 pp.

[7] R. Ahlswede and D. E. Daykin, *An inequality for the weights of two families of sets, their unions and intersections*, Z. Wahrscheinlichkeitstheor. Verw. Geb. 43 (1978), 183–185.

[8] A. Akopyan, A. Hubard and R. Karasev, *Lower and upper bounds for the waists of different spaces*, Topol. Methods Nonlinear Anal. 53 (2019), no. 2, 457–490.

[9] M. Alexander, M. Henk and A. Zvavitch, *A discrete version of Koldobsky's slicing inequality*, Israel J. Math. 222 (2017), 261–278.

[10] I. Aliev, *On the volume of hyperplane sections of a d-cube*, Acta Math. Hungar. 163 (2021), 547–551.

[11] M. Angeles Alfonseca, F. Nazarov, D. Ryabogin and V. Yaskin, Analysis and geometry near the unit ball: proofs, counterexamples and open questions, in Harmonic analysis and convexity, A. Koldobsky and A. Volberg, eds, De Gruyter, Berlin/Boston, 2023, 445–468.

[12] M. Angeles Alfonseca, F. Nazarov, D. Ryabogin and V. Yaskin, *A solution to the fifth and eighth Busemann-Petty problems in a small neighborhood of the Eucllidean ball*, Adv. Math. 390 (2021), paper 107920.

[13] M. Angeles Alfonseca, F. Nazarov, D. Ryabogin, A. Stancu and V. Yaskin, *Two results on the homothety conjecture for convex bodies of flotation on the plane*, Rev. Un. Mat. Argentina 69 (2026), no. 1, 93–108.

[14] M. Angeles Alfonseca, D. Ryabogin, A. Stancu and V. Yaskin, *On flotation, stability and related questions: a survey*, in New Probes into Discrete and Convex Geometry, J. Pach and G. Toth, eds, Bolyai Society Mathematical Studies, Vol. 32, Springer-Verlag, Berlin, in press.

[15] D. Alonso-Gutiérrez, E. Lucas and J. Yepes Nicolás, *On Rogers-Shephard-type inequalities for the lattice point enumerator*, Commun. Contemp. Math. 25 (2023), no. 8, Paper no. 2250022.

[16] D. Alonso-Gutiérrez, B. González Merino, C. H. Jiménez and R. Villa, *John's ellipsoid and the integral ratio of a log-concave function*, J. Geom. Anal., 28 1182–1201, no. 2, 2018.

[17] D. Alonso-Gutiérrez, B. González Merino, C. H. Jiménez and R. Villa, *Rogers-Shephard inequality for log-concave functions*, J. Funct. Anal. 271 (2016), no. 11, 3269–3299.

[18] D. Alonso-Gutiérrez, S. Artstein-Avidan, B. González Merino, C. H. Jiménez and R. Villa, *Rogers-Shephard and local Loomis-Whitney type inequalities*, Math. Ann. 374 (2019), 1719–1771.

[19] G. Ambrus, *Critical central sections of the cube*, Proc. Amer. Math. Soc. 150 (2022), no. 10, 4463–4474.

[20] M. Anttila, K. M. Ball and I. Perissinaki, *The central limit problem for convex bodies*, Trans. Amer. Math. Soc. 355 (2003), 4723–4735.

[21] V. I. Arnold, *Arnold's problems*, 2nd edition, Springer-Verlag, Berlin, 2004.

[22] V. I. Arnold and V. A. Vassiliev, *Newton's Principia read 300 years later*, Notices Amer. Math. Soc. 36 (1989), 1148–1154.

[23] S. Artstein-Avidan and V. D. Milman, *Hidden structures in the class of convex functions and a new duality transform*, J. Eur. Math. Soc. 13 (2011), no. 4, 975–1004.

https://doi.org/10.1515/9783111360508-011

[24] S. Artstein-Avidan and O. Raz, *Weighted covering numbers of convex sets*, Adv. Math. 227 (2011), 730–744.

[25] S. Artstein-Avidan and B. A. Slomka, *A note on Santaló inequality for the polarity transform and its reverse*, Proc. Amer. Math. Soc. 143 (2015), 1693–1704.

[26] S. Artstein-Avidan and B. A. Slomka, *Functional Covering Numbers*, J. Geom. Anal. 31 (2021), 1039–1072.

[27] S. Artstein-Avidan, A. Giannopoulos and V. D. Milman, *Asymptotic Geometric Analysis, Vol. I*, Mathematical Surveys and Monographs, 202, American Mathematical Society, Providence, RI, 2015.

[28] S. Artstein-Avidan, A. Giannopoulos and V. D. Milman, *Asymptotic Geometric Analysis, Vol. II*, Mathematical Surveys and Monographs., 262, American Mathematical Society, Providence, RI, 2021. xxxvii+645 pp.

[29] S. Artstein-Avidan, B. Klartag and V. D. Milman, *On the Santaló point of a function and a functional Santaló inequality*, Mathematika 54 (2004), 33–48.

[30] S. Artstein, V. D. Milman and S. J. Szarek, *Duality of metric entropy*, Annals of Math. 159 (2004), 1313–1328.

[31] S. Artstein-Avidan, K. Einhorn, D. Florentin and Y. Ostrover, *On Godbersen's conjecture*, Geom. Dedicata 178 (2015), 337–350.

[32] S. Artstein-Avidan, B. Klartag, C. Schütt and E. Werner, *Functional affine-isoperimetry and an inverse logarithmic Sobolev inequality*, J. Funct. Anal. 262 (2012), no. 9, 4181–4204.

[33] S. V. Astashkin, K. V. Lykov and M. Milman, *Majorization revisited: Comparison of norms in interpolation scales*, Math. Ann. 388 (2024), no. 4, 3373–3426.

[34] H. Auerbach, *Sur un probléme de M. Ulam concernant léquilibre des corps flottants*, Studia Math. 7 (1938), 121–142.

[35] D. Bakry and M. Ledoux, *Lévy-Gromov's isoperimetric inequality for an infinite dimensional diffusion generator*, Invent. Math. 123 (1996), 259–281.

[36] D. Bakry, I. Gentil and M. Ledoux, *Analysis and geometry of Markov diffusion operators*, Volume 348 of Grundlehren Math. Wiss., Springer, Cham 2014.

[37] K. M. Ball, *Isometric problems in ℓ_p and sections of convex sets*, Ph.D. dissertation, Trinity College, Cambridge, 1986.

[38] K. M. Ball, *Cube slicing in $\mathbb{R}^n$*, Proc. Amer. Math. Soc. 97 (1986), 465–473.

[39] K. M. Ball, *Some remarks on the geometry of convex sets*, Geometric aspects of functional analysis (1986/87), 224–231 Lecture Notes in Math. 1317, Springer-Verlag, Berlin-Heidelberg-New York, 1988.

[40] K. M. Ball, *Logarithmically concave functions and sections of convex sets in $\mathbb{R}^n$*, Studia Math. 88 (1988), 69–84.

[41] K. M. Ball, *Volumes of sections of cubes and related problems*, Geometric aspects of functional analysis (1987–88), 251–260, Lecture Notes in Math., 1376, Springer, Berlin-New York, 1989.

[42] K. M. Ball, *Shadows of convex bodies*, Trans. Amer. Math. Soc. 327 (1991), 891–901.

[43] K. M. Ball, *Normed spaces with a weak-Gordon-Lewis property*, Functional analysis (Austin, TX, 1987/1989), 36–47, Lecture Notes in Math. 1470, Springer, Berlin, 1991.

[44] K. M. Ball, *Volume ratios and a reverse isoperimetric inequality*, J. London Math. Soc. (2) 44 (1991), no. 2, 351–359.

[45] K. M. Ball, *Ellipsoids of maximal volume in convex bodies*, Geom. Dedicata 41 (1992), no. 2, 241–250.

[46] K. M. Ball, *Convex geometry and functional analysis*, Handbook of the geometry of Banach spaces, Vol. I, 161–194, North-Holland, Amsterdam, 2001.

[47] K. M. Ball and V. H. Nguyen, *Entropy jumps for isotropic log-concave random vectors and spectral gap*, Studia Math. 213 (2012), 81–96.

[48] W. Banaszczyk, *Inequalities for convex bodies and polar reciprocal lattices in $\mathbb{R}^n$*, Discrete Comput. Geom. 13 (1995), no. 2, 217–231.

[49] W. Banaszczyk, *Inequalities for convex bodies and polar reciprocal lattices in $\mathbb{R}^n$. II. Application of K-convexity*, Discrete Comput. Geom. 16 (1996), no. 3, 305–311.

[50] W. Banaszczyk, A. Litvak, A. Pajor and S. Szarek, *The flatness theorem for nonsymmetric convex bodies via the local theory of Banach spaces*, Math. Oper. Res. 24 (1999), no. 3, 728–750.

[51] I. Bárány and P. Frankl, *How (not) to cut your cheese*, Amer. Math. Monthly 128 (2021), no. 6, 543–552.

[52] I. Bárány and P. Frankl, *Cells in the box and a hyperplane*, J. Eur. Math. Soc. (JEMS) 25 (2023), no. 7, 2863–2877.

[53] I. Bárány and Z. Füredi, *Approximation of the sphere by polytopes having few vertices*, Proc. Amer. Math. Soc. 102 (1988), 651–659.

[54] D. L. Barrow and P. W. Smith, *Classroom Notes: Spline Notation Applied to a Volume Problem*, Amer. Math. Monthly 86 (1979), no. 1, 50–51.

[55] F. Á. Bartha, F. Fodor and B. González Merino, *Central diagonal sections of the n-cube*, Int. Math. Res. Not. IMRN 2021, no. 4, 2861–2881.

[56] F. Barthe, *Measures unimodales et sections des boules* B_p^n, C.R. Acad. Sci. Paris 321 (1995), 865–868.

[57] F. Barthe, *On a reverse form of the Brascamp-Lieb inequality*, Invent. Math. 134 (1998), 335–361.

[58] F. Barthe, *Extremal properties of central half-spaces for product measures*, J. Funct. Anal. 182 (2001), no. 1, 81–107.

[59] F. Barthe, K.-J. Böröczky and M. Fradelizi, *Stability of the functional forms of the Blaschke-Santaló inequality*. Monatsh. Math. 173 (2014), no. 2, 135–159.

[60] F. Barthe, M. Fradelizi and B. Maurey, *A short solution to the Busemann-Petty problem*, Positivity 3 (1999), 95–100.

[61] F. Barthe and D. Cordero-Erausquin, *Invariances in variance estimates*, Proc. London Math. Soc. 106 (2013), 33–64.

[62] F. Barthe and B. Klartag, *Spectral gaps, symmetries and log-concave perturbations*, Bull. Hellenic Math. Soc. 64 (2020), 1–31.

[63] F. Barthe and A. Koldobsky, *Extremal slabs in the cube and the Laplace transform*, Adv. Math. 174 (2003), 89–114.

[64] F. Barthe and A. Naor, *Hyperplane projections of the unit ball of* ℓ_p^n, Discrete Comput. Geom. 27 (2002), 215–226.

[65] F. Barthe, O. Guédon, S. Mendelson and A. Naor, *A probabilistic approach to the geometry of the* ℓ_p^n*-ball*, Ann. Probab. 33 2005, no. 2, 480–513.

[66] A. Barvinok, *Combinatorics, Geometry and Complexity of Integer Points*, Lecture Notes available from http://www.math.lsa.umich.edu/~barvinok/latticenotes669.pdf.

[67] V. Bayle and C. Rosales, *Some isoperimetric comparison theorems for convex bodies in Riemannian manifolds*, Indiana Univ. Math. J. 54 (2005), 1371–1394.

[68] M. Beck and S. Robins, *Computing the Continuous Discretely. Integer-point enumeration in polyhedra*, Undergraduate Texts in Mathematics. Springer, New York, 2007.

[69] J. Bennett and T. Tao, *Adjoint Brascamp-Lieb inequalities*, Proc. Lond. Math. Soc. (3) 129 (2024), no. 4, Paper no. e12633.

[70] Y. Benyamini and Y. Gordon, *Random factorization of operators between Banach spaces*, J. d'Analyse Math. 39 (1981), 45–74.

[71] M. Berger, *Geometry I*, Translated from the French by M. Cole and S. Levy. Universitext. Springer-Verlag, Berlin, 1987, xiv+428 pp.

[72] L. Berwald, *Verallgemeinerung eines Mittelswertsatzes von J. Favard, für positive konkave Functionen*, Acta Math. 79 (1947), 17–37.

[73] U. Betke and W. Weil, *Isoperimetric inequalities for the mixed area of plane convex sets*, Arch. Math. (Basel) 57 (1991), no. 5, 501–507.

[74] U. Betke, M. Henk and J. M. Wills, *Successive-minima-type inequalities*, Discrete Comput. Geom. 9 (1993), no. 2, 165–175.

[75] R. Bhatia, *Matrix analysis*, Graduate Texts in Mathematics, 169. Springer-Verlag, New York, 1997.

[76] P. Bizeul, *The slicing conjecture via small ball estimates*, Preprint (https://arxiv.org/abs/2501.06854).

[77] P. Bizeul and B. Klartag, *Distances between non-symmetric convex bodies: optimal bounds up to polylog*, Preprint (https://arxiv.org/abs/2510.20511).
[78] W. Blaschke, *Über affine Geometrie VII: Neue Extremaigenschaften von Ellipse und Ellipsoid*, Ber. Vergh. Sächs Akad. Wiss. Leipzig, Math.-Phys. Kl. 69 (1917), 306–318, Ges. Werke 3, 246–258.
[79] S. G. Bobkov, *Isoperimetric and analytic inequalities for log-concave probability measures*, Ann. Prob. 27 (1999), 1903–1921.
[80] S. G. Bobkov, *On concentration of distributions of random weighted sums*, Ann. Probab. 31 (2003), no. 1, 195–215.
[81] S. G. Bobkov, *Convex bodies and norms associated to convex measures*, Probab. Theory Related Fields 147 (2010), no. 1-2, 303–332.
[82] S. G. Bobkov and C. Houdré, *Isoperimetric constants for product probability measures*, Ann. Probab. 25 (1997), no. 1, 184–205.
[83] S. G. Bobkov and A. Koldobsky, *On the central limit property of convex bodies*, Geometric aspects of functional analysis, 44–52, Lecture Notes in Math., 1807, Springer, Berlin, 2003.
[84] S. G. Bobkov, A. Colesanti and I. Fragalá, *Quermassintegrals of quasi-concave functions and generalized Prékopa-Leindler inequalities*, Manuscripta Mathematica, 143, (2014), no. 1-2, 131–169.
[85] S. G. Bobkov, B. Klartag and A. Koldobsky, *Estimates for moments of general measures on convex bodies*, Proc. Amer. Math. Soc. 146 (2018), no. 11, 4879–4888.
[86] S. Bochner, *Lectures on Fourier integrals*, with an author's supplement on monotonic functions, Stieltjes integrals, and harmonic analysis, Annals of Mathematics Studies, vol. 42. Princeton University Press, Princeton, NJ, 1959, 333 pp.
[87] E. D. Bolker, *A class of convex bodies*, Trans. Amer. Math. Soc. 145 (1969), 323–345.
[88] B. Bollobás and A. Thomason, *Projections of bodies and hereditary properties of hypergraphs*, Bull. London Math. Soc. 27 (1995), 417–424.
[89] J. Boman, *A hypersurface containing the support of a Radon transform must be an ellipsoid I*, J. Geom. Anal. 31 (2021), 2726–2741.
[90] J. Boman, *A hypersurface containing the support of a Radon transform must be an ellipsoid II*, J. Inverse Ill-Posed Probl. 29 (2021), 351–367.
[91] C. Borell, *Complements of Lyapunov's inequality*, Math. Ann. 205 (1973), 323–331.
[92] C. Borell, *Convex measures on locally convex spaces*, Ark. Mat. 12 (1974), 239–252.
[93] C. Borell, *Convex set functions in d-space*, Period. Math. Hungar. 6 (1975), 111–136.
[94] C. Borell, *The Ehrhard inequality*, C. R. Math. Acad. Sci. Paris 337 (2003), no. 10, 663–666.
[95] K. J. Böröczky and P. Kalantzopoulos, *Log-Brunn-Minkowski inequality under symmetry*, Trans. Amer. Math. Soc. 375 (2022), no. 8, 5987–6013.
[96] K. J. Böröczky, E. Lutwak, D. Yang and G. Zhang, *The log-Brunn-Minkowski inequality*, Adv. Math. 231 (2012), no. 3-4, 1974–1997.
[97] A. A. Borovkov and S. A. Utev, *On an inequality and a related characterization of the normal distribution*, Teor. Veroyatn. Primen. 28 (1983) 209–218.
[98] S. Boucheron, G. Lugosi and P. Massart, *Concentration inequalities. A nonasymptotic theory of independence*. With a foreword by Michel Ledoux. Oxford University Press, Oxford, 2013.
[99] J. Bourgain, *On high-dimensional maximal functions associated to convex bodies*, Amer. J. Math. 108 (1986), 1467–1476.
[100] J. Bourgain, *On the L^p-bounds for maximal functions associated to convex bodies*, Israel J. Math. 54 (1986), 257–265.
[101] J. Bourgain, *Geometry of Banach spaces and harmonic analysis*, Proceedings of the International Congress of Mathematicians (Berkeley, CA, 1986), Amer. Math. Soc., Providence, RI, 1987, 871–878.
[102] J. Bourgain, *On the distribution of polynomials on high-dimensional convex sets*, Lecture Notes in Math. 1469 (1991), 127–137.
[103] J. Bourgain, *On the Busemann-Petty problem for perturbations of the ball*, Geom. Funct. Anal. 1 (1991), 1–13.

[104] J. Bourgain and V. D. Milman, *New volume ratio properties for convex symmetric bodies in* $\mathbb{R}^n$, Invent. Math. 88 (1987), 319–340.

[105] J. Bourgain and G. Zhang, *On a generalization of the Busemann-Petty problem*, Convex geometric analysis (Berkeley, CA, 1996), 65–76, Math. Sci. Res. Inst. Publ., 34, Cambridge Univ. Press, Cambridge, 1999.

[106] J. Bourgain, B. Klartag and V. D. Milman, *Symmetrization and isotropic constants of convex bodies*, Geometric aspects of functional analysis, 101–115, Lecture Notes in Math., 1850, Springer, Berlin, 2004.

[107] J. Bracho, L. Montejano and D. Oliveros, *Carousels, Zindler curves and the floating body problem*, Period. Math. Hungar. 49 (2004), 9–23.

[108] M-C. Brandenburg, J. de Loera, Yu Luo and C. Meroni, *Critical moments of slices and slabs of the cube (and other polyhedral norms)*, Preprint (https://arxiv.org/abs/2603.25643).

[109] H. J. Brascamp and E. H. Lieb, *On extensions of the Brunn-Minkowski and Prékopa-Leindler theorems, including inequalities for log-concave functions, and with an application to the diffusion equation*, J. Funct. Anal. 22 (1976), no. 4, 366–389.

[110] S. Brazitikos and D.-M. Liakopoulos, *On a version of the slicing problem for the surface area of convex bodies*, Trans. Amer. Math. Soc. 375 (2022), 5561–5586.

[111] S. Brazitikos, A. Giannopoulos and D.-M. Liakopoulos, *Uniform cover inequalities for the volume of coordinate sections and projections of convex bodies*, Adv. Geom. 18 (2018), no. 3, 345–354.

[112] S. Brazitikos, S. Dann, A. Giannopoulos and A. Koldobsky, *On the average volume of sections of convex bodies*, Israel J. Math. 222 (2017), 921–947.

[113] S. Brazitikos, A. Giannopoulos, P. Valettas and B. Vritsiou, *Geometry of isotropic convex bodies*, Mathematical Surveys and Monographs, 196. American Mathematical Society, Providence, RI, 2014. xx+594 pp.

[114] J. Bretagnolle, D. Dacunha-Castelle and J. L. Krivine, *Lois stables et espaces* L_p, Ann. Inst. H. Poincaré Probab. Statist. 2 (1966), 231–259.

[115] P. Brzezinski, *Volume estimates for sections of certain convex bodies*, Math. Nachr. 286 (2013), no. 17-18, 1726–1743.

[116] Y. D. Burago and V. A. Zalgaller, *Geometric Inequalities*, Springer Series in Soviet Mathematics, Springer-Verlag, Berlin-New York (1988).

[117] H. Busemann, *A theorem on convex bodies of the Brunn-Minkowski type*, Proc. Nat. Acad. Sci. U.S.A. 35 (1949), 27–31.

[118] H. Busemann, *Volumes and areas of cross-sections*, Amer. Math. Monthly 67 (1960), 248–250.

[119] H. Busemann and C. M. Petty, *Problems on convex bodies*, Math. Scand. 4 (1956), 88–94.

[120] H. Busemann and E. G. Straus, *Area and Normality*, Pacific J. Math. 10 (1960), 35–72.

[121] P. Buser, *A note on the isoperimetric constant*, Ann. Sci. École Norm. Sup. 15 (1982), 213–230.

[122] A. M. Caetano, *Weyl numbers in sequence spaces and sections of unit balls*, J. Funct. Anal. 106 (1992), no. 1, 1–17.

[123] U. Caglar, M. Fradelizi, O. Guédon, J. Lehec, C. Schütt and E. Werner, *Functional versions of* L_p*-affine surface area and entropy inequalities*, Int. Math. Res. Not. IMRN 2016, no. 4, 1223–1250.

[124] A. Carbery, *Radial Fourier multipliers and associated maximal functions*, Recent Progress in Fourier Analysis, North-Holland Math. Studies 111 (1985), 49–56.

[125] G. D. Chakerian and P. Filliman, *The measures of the projections of a cube*, Studia Sci. Math. Hungar. 21 (1986), no. 1-2, 103–110.

[126] G. Chasapis, A. Giannopoulos and D. Liakopoulos, *Estimates for measures of lower dimensional sections of convex bodies*, Adv. Math. 306 (2017), 880–904.

[127] G. Chasapis, H. König and T. Tkocz, *From Ball's cube slicing inequality to Khinchin-type inequalities for negative moments*, J. Funct. Anal. 281 (2021), no. 9, Paper no. 109185, 23 pp.

[128] G. Chasapis, P. Nayar and T. Tkocz, *Slicing* ℓ_p*-balls reloaded: stability, planar sections in* ℓ_1, Ann. Probab. 50 (2022), no. 6, 2344–2372.

[129] J. Cheeger, *A lower bound for the smallest eigenvalue of the Laplacian*, Problems in Analysis (Papers dedicated to Salomon Bochner, 1969), Princeton Univ. Press, Princeton (1970), 195–199.

[130] S. Chen, Y. Huang, Q. Li and J. Liu, *The L_p-Brunn-Minkowski inequality for $p < 1$*, Adv. Math. 368 (2020), 107166, 21 pp.

[131] Y. Chen, *An almost constant lower bound of the isoperimetric coefficient in the KLS conjecture*, Geom. Funct. Anal. 31 (2021), 34–61.

[132] A. Colesanti, *From the Brunn-Minkowski inequality to a class of Poincaré-type inequalities*, Commun. Contemp. Math. 10 (2008), no. 5, 765–772.

[133] A. Colesanti and I. Fragalà, *The first variation of the total mass of log-concave functions and related inequalities*, Adv. Math. 244 (2013), 708–749.

[134] A. Colesanti, G. V. Livshyts and A. Marsiglietti, *On the stability of Brunn-Minkowski type inequalities*, J. Funct. Anal. 273 (2017), no. 3, 1120–1139.

[135] D. Cordero-Erausquin and A. Eskenazis, *Concavity principles for weighted marginals*, Preprint (https://arxiv.org/abs/2506.16941).

[136] D. Cordero-Erausquin and B. Klartag, *Moment measures*, J. Funct. Anal. 268 (2015), no. 12, 3834–3866.

[137] D. Cordero-Erausquin and B. Maurey, *Some extensions of the Prékopa–Leindler inequality using Borell's stochastic approach*, Stud. Math. 238 (2017), no. 3, p. 201–233.

[138] D. Cordero-Erausquin and L. Rotem, *Improved log-concavity for rotationally invariant measures of symmetric convex sets*, Ann. Probab. 51 (2023), no. 3, 987–1003.

[139] D. Cordero-Erausquin, M. Fradelizi, G. Paouris and P. Pivovarov, *Volume of the polar of random sets and shadow systems*, Math. Ann. 362 (2015), no. 3-4, 1305–1325.

[140] D. Cordero-Erausquin, M. Fradelizi and B. Maurey, *The (B) conjecture for the Gaussian measure of dilates of symmetric convex sets and related problems*, J. Funct. Anal. 214 (2004), no. 2, 410–427.

[141] D. Dadush, *Integer programming, lattice algorithms, and deterministic volume estimation*, Georgia Institute of Technology, USA, 2012.

[142] D. Dadush, *On approximating the covering radius and finding dense lattice subspaces*, STOC'19-Proceedings of the 51st Annual ACM SIGACT Symposium on Theory of Computing, 1021–1026, ACM, New York, 2019.

[143] N. Dafnis and G. Paouris, *Small ball probability estimates, ψ_2-behavior and the hyperplane conjecture*, J. Funct. Anal. 258 (2010), 1933–1964.

[144] N. Dafnis and G. Paouris, *Estimates for the affine and dual affine quermassintegrals of convex bodies*, Illinois J. of Math. 56 (2012), 1005–1021.

[145] S. Dann and M. Zymonopoulou, *Sections of convex bodies with symmetries*, Adv. Math. 271 (2015), 112–152.

[146] S. Dann, G. Paouris and P. Pivovarov, *Bounding marginal densities via affine isoperimetry*, Proc. London Math. Soc. 113 (2016), 140–162.

[147] W. J. Davis, V. D. Milman, and N. Tomczak-Jaegermann, *The distance between certain n-dimensional Banach spaces*, Israel J. Math. 39 (1981), 1–15.

[148] A. De, I. Diakonikolas and R. A. Servedio, *A robust Khintchine inequality, and algorithms for computing optimal constants in Fourier analysis and high-dimensional geometry*, SIAM J. Discrete Math. 30 (2016), no. 2, 1058–1094.

[149] P. Diaconis and D. Freedman, *Asymptotics of graphical projection pursuit*, Ann. of Stat. 12 (1984), 793–815.

[150] H. Dirksen, *Sections of the regular simplex–volume formulas and estimates*, Math. Nachr. 290 (2017), no. 16, 2567–2584.

[151] H. Dirksen, *Hyperplane sections of cylinders*, Colloq. Math. 147 (2017), no. 1, 145–164.

[152] F. Dorrek and F. Schuster, *Projection functions, area measures and the Alesker-Fourier transfor*, J. Funct. Anal. 273 (2017), 2026–2069.

[153] N. Dunford and J. T. Schwartz, *Linear operators, Part I: General theory*, Interscience, New York, 1958.

[154] R. Durrett, *Stochastic Calculus: A Practical Introduction*, Probability and Stochastics Series. CRC Press, Boca Raton, FL, 1996. x+341 pp.
[155] D. Edmunds and H. Melkonian, *Behaviour of L_q norms of the sinc_p function*, Proc. Amer. Math. Soc. 147 (2019), no. 1, 229–238.
[156] A. Ehrhard, *Symétrisation dans l'espace de Gauss*, Math. Scand. 53 (1983), no. 2, 281–301.
[157] R. Eldan, *Thin shell implies spectral gap up to polylog via a stochastic localization scheme*, Geom. Funct. Anal. 23 (2013), 532–569.
[158] R. Eldan and B. Klartag, *Approximately gaussian marginals and the hyperplane conjecture*, Proceedings of the Workshop on "Concentration, Functional Inequalities and Isoperimetry", Contemp. Math. 545 (2011), 55–68.
[159] R. Eldan and J. Lehec, *Bounding the norm of a log-concave vector via thin-shell estimates*, Geometric aspects of functional analysis, 107–122, Lecture Notes in Math., 2116, Springer, Cham, 2014.
[160] S. Ellmeyer and G. Hofstätter, *Busemann-Petty type problems in complex vector spaces*. Indiana Univ. Math. J. 75 (2026), no. 2, 525–551.
[161] A. Eskenazis, *On extremal sections of subspaces of L_p*, Discrete Comput. Geom. 65 (2021), no. 2, 489–509.
[162] A. Eskenazis and G. Moschidis, *The dimensional Brunn-Minkowski inequality in Gauss space*, J. Funct. Anal. 280, (2021), no. 6, 108914.
[163] A. Eskenazis, A. Giannopoulos and N. Tziotziou, *Functional perimeter and the dimensional Brunn-Minkowski inequality for log-concave measures*, Preprint (arXiv:2605.02747).
[164] A. Eskenazis, P. Nayar and T. Tkocz, *Gaussian mixtures: entropy and geometric inequalities*, Ann. of Prob. 46 (2018), no. 5, 2908–2945.
[165] A. Eskenazis, P. Nayar and T. Tkocz, *Resilience of cube slicing in ℓ_p*, Duke Math. J. 173 (2024), no. 17, 3377–3412.
[166] M. Fathi, *A sharp symmetrized form of Talagrand's transport-entropy inequality for the Gaussian measure*, Electron. Commun. Probab. 23 (2018), Paper no. 81.
[167] K. J. Falconer, *Applications of a result on spherical integration to the theory of convex sets*, Amer. Math. Monthly 90 (1983), 690–693.
[168] W. Fenchel (ed.), *Proceedings of the Colloquium on Convexity, 1965*, Kobenhavns Univ. Mat. Inst., 1967.
[169] T. Figiel and N. Tomczak-Jaegermann, *Projections onto Hilbertian subspaces of Banach spaces*, Israel J. Math. 33 (1979), 155–171.
[170] P. Filliman, *The largest projections of regular polytopes*, Israel J. Math. 64 (1988), no. 2, 207–228.
[171] P. Filliman, *Exterior algebra and projections of polytopes*, Discrete Comput. Geom. 5 (1990), no. 3, 305–322.
[172] P. Filliman, *The extreme projections of the regular simplex*, Trans. Amer. Math. Soc. 317 (1990), no. 2, 611–629.
[173] S. R. Finch, *Mathematical constants*, Encyclopedia of Mathematics and its Applications 94, Cambridge University Press, Cambridge, 2003.
[174] A. Fish, F. Nazarov, D. Ryabogin and A. Zvavitch, *The unit ball is an attractor of the intersection body operator*, Adv. Math. 226 (2011), 2629–2642.
[175] B. Fleury, *Concentration in a thin Euclidean shell for log-concave measures*, J. Funct. Anal. 259 (2010), 832–841.
[176] D. Florentin and A. Segal, *A Santaló-type inequality for $\mathcal{J}$ transform*, Communications in Contemporary Mathematics 23 (2021), 1 1950090.
[177] M. Fradelizi, *Sections of convex bodies through their centroid*, Arch. Math. (Basel) 69 (1997), no. 6, 515–522.
[178] M. Fradelizi, *Hyperplane sections of convex bodies in isotropic position*, Beiträge Algebra Geom. 40 (1999), no. 1, 163–183.
[179] M. Fradelizi and M. Meyer, *Some functional forms of Blaschke-Santaló inequality*. Math. Z. 256 (2007), no. 2, 379–395.

[180] M. Fradelizi and M. Meyer, *Increasing functions and inverse Santaló inequality for unconditional functions*, Positivity 12 (2008), no. 3, 407–420.

[181] M. Fradelizi and M. Meyer, *Some functional inverse Santaló inequalities*, Adv. Math. 218 (2008), no. 5, 1430–1452.

[182] M. Fradelizi and M. Meyer, *Functional inequalities related to Mahler's conjecture*. Monatsh. Math. 159 (2010), no. 1-2, 13–25.

[183] M. Fradelizi and E. Nakhle, *The Functional Form of Mahler's Conjecture for Even Log-Concave Functions in Dimension* 2, Int. Math. Res. Not. IMRN (2022), rnac120.

[184] M. Fradelizi, N. Gozlan and S. Zugmeyer, *Transport proofs of some functional inverse Santaló inequalities*, High dimensional probability IX—the ethereal volume, 123–142, Progr. Probab., 80, Birkhäuser/Springer, Cham, 2023.

[185] M. Fradelizi, M. Meyer and A. Zvavitch, *Volume product*, Harmonic analysis and convexity, 163–222, Adv. Anal. Geom., 9, De Gruyter, Berlin, 2023.

[186] M. Fradelizi, Y. Gordon, M. Meyer and S. Reisner, *The case of equality for an inverse Santaló functional inequality*, Adv. Geom. 10 (2010), no. 4, 621–630.

[187] A. Freyer and M. Henk, *Bounds on the lattice point enumerator via slices and projections*, Discrete Comput. Geom. 67 (2022), no. 3, 895–918.

[188] A. Freyer and M. Henk, *Polynomial bounds in Koldobsky's discrete slicing problem*, Proc. Amer. Math. Soc. 152 (2024), no. 7, 3063–3074.

[189] A. Freyer and E. Lucas, *Interpolating between volume and lattice point enumerator with successive minima*, Monatsh. Math. 198 (2022), no. 4, 717–740.

[190] D. Galicer, J. Haddad and J. Singer, *Extensions of the Busemann-Petty problem for arbitrary measures*, Preprint (https://arxiv.org/abs/2512.13555).

[191] R. J. Gardner, *Intersection bodies and the Busemann-Petty problem*, Trans. Amer. Math. Soc. 342 (1994), 435–445.

[192] R. J. Gardner, *On the Busemann-Petty problem concerning central sections of origin-symmetric convex bodies*, Bull. Amer. Math. Soc. 30 (1994), 222–226.

[193] R. J. Gardner, *A positive answer to the Busemann-Petty problem in three dimensions*, Annals of Math. 140 (1994), 435–447.

[194] R. J. Gardner, *Geometric Tomography*, Second edition. Encyclopedia of Mathematics and its Applications 58, Cambridge University Press, Cambridge, 2006.

[195] R. J. Gardner, *Geometric tomography: Update to the second edition*, https://www.wwu.edu/faculty/gardner/research.html.

[196] R. J. Gardner, *The Brunn-Minkowski inequality*, Bull. Amer. Math. Soc. (N.S.) 39 (2002), 355–405.

[197] R. J. Gardner, *The dual Brunn-Minkowski theory for bounded Borel sets: dual affine quermassintegrals and inequalities*, Adv. Math. 216 (2007), 358–386.

[198] R. J. Gardner and P. Gronchi, *A Brunn-Minkowski inequality for the integer lattice*, Trans. Amer. Math. Soc. 353 (2001), no. 10, 3995–4024.

[199] R. J. Gardner, P. Gronchi and C. Zong, *Sums, projections, and sections of lattice sets, and the discrete covariogram*, Discrete Comput. Geom. 34 (2005), no. 3, 391–409.

[200] R. J. Gardner and P. Gritzmann, *Discrete tomography: determination of finite sets by X-rays*, Trans. Amer. Math. Soc. 349 (1997), 2271–2295.

[201] R. J. Gardner and P. Gritzmann, *Uniqueness and complexity in discrete tomography*, in: Discrete Tomography: Foundations, Algorithms and Application, G. T. Herman and A. Kuba, eds, Birkhäuser, Boston, MA, 1999, pp. 85–113.

[202] R. J. Gardner, A. Koldobsky and Th. Schlumprecht, *An analytic solution to the Busemann-Petty problem on sections of convex bodies*, Annals of Math. 149 (1999), 691–703.

[203] R. J. Gardner, D. Ryabogin, V. Yaskin and A. Zvavitch, *A problem of Klee on inner section functions of convex bodies*, J. Differential Geom. 91 (2012), no. 2, 261–279.

[204] R. J. Gardner and A. Zvavitch, *Gaussian Brunn-Minkowski inequalities*, Trans. Amer. Math. Soc. 362 (2010), no. 10, 5333–5353.

[205] I. M. Gelfand and G. E. Shilov, *Generalized functions, vol.1: Properties and operations*, Academic Press, New York, 1964.

[206] I. M. Gelfand and N. Ya. Vilenkin, *Generalized functions, vol.4: Applications of harmonic analysis*, Academic Press, New York, 1964.

[207] A. Giannopoulos, *A note on a problem of H. Busemann and C. M. Petty concerning sections of symmetric convex bodies*, Mathematika 37 (1990), 239–244.

[208] A. Giannopoulos and A. Koldobsky, *Variants of the Busemann-Petty problem and of the Shephard problem*, Int. Math. Res. Not. (IMRN), (2017), no. 3, 921–943.

[209] A. Giannopoulos and A. Koldobsky, *Volume difference inequalities*, Trans. Amer. Math. Soc. 370 (2018), 4351–4372.

[210] A. Giannopoulos and E. Milman, *M-estimates for isotropic convex bodies and their L_q-centroid bodies*, Geometric aspects of functional analysis, 159–182, Lecture Notes in Math., 2116, Springer, Cham, 2014.

[211] A. Giannopoulos and N. Tziotziou, *Regular functional covering numbers*, International Mathematics Research Notices 2026 (2026), no. 5, Paper: rnag035.

[212] A. Giannopoulos and M. Papadimitrakis, *Isotropic surface area measures*, Mathematika 46 (1999), 1–13.

[213] A. Giannopoulos, M. Hartzoulaki and G. Paouris, *On a local version of the Aleksandrov-Fenchel inequality for the quermassintegrals of a convex body*, Proc. Amer. Math. Soc. 130 (2002), 2403–2412.

[214] A. Giannopoulos, A. Koldobsky and P. Valettas, *Inequalities for the surface area of projections of convex bodies*, Canad. J. Math. 70 (2018), no. 4, 804–823.

[215] A. Giannopoulos, A. Koldobsky and A. Zvavitch, *Inequalities for the Radon transform on convex sets*, Int. Math. Res. Not. (IMRN) (2022), no. 18, 13984–14007.

[216] A. Giannopoulos, A. Koldobsky and A. Zvavitch, *Inequalities for sections and projections of convex bodies*, in "Harmonic analysis and convexity", 223–256, Adv. Anal. Geom., 9, De Gruyter, Berlin, 2023.

[217] A. Giannopoulos, P. Stavrakakis, A. Tsolomitis and B-H. Vritsiou, *Geometry of the L_q-centroid bodies of an isotropic log-concave measure*, Trans. Amer. Math. Soc. 367 (2015), no. 7, 4569–4593.

[218] M. Giertz, *A note on a problem of Busemann*, Math. Scand. 25 (1969), 145–148.

[219] O. Giladi, H. Huang, C. Schütt and E. M. Werner, *Constrained convex bodies with extremal affine surface areas*, J. Funct. Anal. 279 (2020), no. 3, 108531.

[220] D. Gilbarg and N. Trudinger, *Elliptic Partial Differential Equations of Second Order*, 2nd edition Classics in Mathematics, Springer-Verlag, Berlin Heidelberg, 2001.

[221] N. Glover, T. Tkocz and K. Wyczesany, *Stability of polydisc slicing*, Mathematika 69 (2023), no. 4, 1165–1182.

[222] P. R. Goodey, E. Lutwak and W. Weil, *Functional analytic characterization of classes of convex bodies*, Math. Z. 222 (1996), 363–381.

[223] P. R. Goodey, V. Yaskin and M. Yaskina, *Fourier transforms and the Funck-Hecke theorem in convex geometry*, J. London Math. Soc. (2) 80 (2009), 388–404.

[224] P. R. Goodey, V. Yaskin and M. Yaskina, *A Fourier transform approach to Christoffel's problem*, Trans. Amer. Math. Soc. 363 (2011), 6351–6384.

[225] P. Goodey and W. Weil, *Intersection bodies and ellipsoids*, Mathematika 42 (1995), 295–304.

[226] P. Goodey and G. Zhang, *Inequalities between projection functions of convex bodies*, Amer. J. Math. 120 (1998) 345–367.

[227] Y. Gordon and M. Meyer, *On the Minima of the Functional Mahler Product*, Houston J. Math. 40 (2014), no. 2, 385–393.

[228] V. Gorev and E. Kosov, *Functional analogs of the Shephard, Busemann-Petty and Milman problems*, J. Anal. Math. 157 (2025), no. 2, 511–536.

[229] E. A. Gorin and Yu. Favorov, *Generalizations of the Khinchin inequality, (Russian)* Teor. Veroyatnost. i Primenen. 35 (1990), no. 4, 762–767; translation in Theory Probab. Appl. 35 (1990), no. 4, 766–771 (1991).

[230] N. Gozlan, *The deficit is the Gaussian log-Sobolev inequlaity and inverse Santaló inequalities*, Int. Math. Res. Not. IMRN 17 (2022), 13396–13446.

[231] N. Gozlan and C. Léonard, *Transport inequalities. A survey*, Markov Process. Related Fields 16 (2010), no. 4, 635–736.

[232] N. Gozlan, C. Roberto, P.-M. Samson and P. Tetali, *Transport proofs of some discrete variants of the Prékopa-Leindler inequality*, Annali della Scuola Normale Superiore di Pisa. Classe di Scienze. Serie V 22 (2021), 1207–1232.

[233] W. Gregory and A. Koldobsky, *Inequalities for the derivatives of the Radon transform on convex bodies*, Israel J. Math. 246 (2021), 261–280.

[234] E. L. Grinberg, *Isoperimetric inequalities and identities for k-dimensional cross-sections of convex bodies*, Math. Ann. 291 (1991), 75–86.

[235] E. Grinberg, *Comparing volumes by concurrent cross-sections of complex lines: a Busemann Petty type problem*, Positivity 22 (2018), no. 5, 1297–1301.

[236] E. Grinberg and G. Zhang, *Convolutions, transforms, and convex bodies*, Proc. London Math. Soc. (3) 78 (1999), no. 1, 77–115.

[237] H. Groemer, *Geometric applications of Fourier series and spherical harmonics*, Cambridge University Press, New York, 1996.

[238] G. Grimmett, *Percolation*, 2nd edition, Grundlehren der Mathematischen Wissenschaften, vol. 321, Springer, 1999, xiv+444 pages.

[239] G. Grimmett, *Probability on graphs*, Institute of Mathematical Statistics Textbooks, vol. 8, Cambridge University Press, 2018, xi+265 pages.

[240] P. M. Gruber, *Convex and discrete geometry*, vol. 336 of Grundlehren der Mathematischen Wissenschaften. Springer, Berlin, 2007.

[241] B. Grünbaum, *Partitions of mass-distribution and of convex bodies by hyperplanes*, Pacific Journal of Mathematics 10 (1960), 1257–1261.

[242] B. Grünbaum, *On some properties of convex sets*, Colloq. Math. 8 (1961), 39–42.

[243] Q. Y. Guan, *A note on Bourgain's slicing problem*, Preprint (https://arxiv.org/abs/2412.09075).

[244] O. Guédon and E. Milman, *Interpolating thin-shell and sharp large-deviation estimates for isotropic log-concave measures*, Geom. Funct. Anal. 21 (2011), 1043–1068.

[245] U. Haagerup, *The best constants in the Khintchine inequality*, Studia Math. 70 (1981), no. 3, 231–283.

[246] J. Haddad, *Distance inequalities for generalized k-intersection bodies*, Preprint. (https://arxiv.org/abs/2405.10223).

[247] J. Haddad and A. Koldobsky, *Radon transforms with small derivatives and distance inequalities for convex bodies, Pure Appl.* Funct. Anal. 10 (2025), no. 5, 1245–1253.

[248] J. Haddad, C. H. Jiménez and M. A. Montenegro, *Asymmetric Blaschke-Santaló functional inequalities*, J. Funct. Anal. 278 (2020), no. 2, 108319.

[249] H. Hadwiger, *Radialpotenzintegrale zentralsymmetrischer rotationkörper und ungleichheitsaussagen Busemannscher art*, Math. Scand. 23 (1968), 193–200.

[250] H. Hadwiger, *Gitterperiodische punktmengen und isoperimetrie*, Monatsh. Math. 76 (1972), 410–418.

[251] D. Halikias, B. Klartag and B. Slomka, *Discrete variants of Brunn-Minkowski type inequalities*, Ann. Fac. Sci. Toulouse Math. (6) 30 (2021), no. 2, 267–279.

[252] G. Hargé, *A convex/log-concave correlation inequality for Gaussian measure and an application to abstract Wiener spaces*, Probability theory and related fields, 130 (2004), 415–440.

[253] X. He, T. Tkocz and K. Wyczesany, *A Sharp Gaussian Tail Bound for Sums of Uniforms*, Israel J. Math. 269 (2025), no. 2, 985–1000.

[254] S. Helgason, *The Radon transform*, Second edition, Progress in Mathematics, 5, Birkhäuser Boston, Inc., Boston, MA, 1999.

[255] S. Helgason, *Groups and geometric analysis*, Academic Press, New York, 1984.
[256] M. Henk, *Successive minima and lattice points*, Rendi. Circ. Matematico Palermo, Serie II, Suppl. 70 (2002), 377–384.
[257] D. Hensley, *Slicing the cube in* $\mathbb{R}^n$ *and probability (bounds for the measure of a central cube slice in* $\mathbb{R}^n$ *by probability methods)*, Proc. Amer. Math. Soc. 73 (1979), 95–100.
[258] M. A. Hernández Cifre, D. Iglesias and J. Yepes Nicolás, *On a discrete Brunn-Minkowski type inequality*, SIAM J. Discrete Math. 32 (2018), 1840–1856.
[259] S. Hoehner, *Extremal general affine surface areas*, J. Math. Anal. Appl. 505 (2022), no. 2, Paper no. 125506.
[260] G. C. Hofstatter and F. E. Schuster, *Blaschke Santaló Inequalities for Minkowski and Asplund Endomorphisms*, Int. Math. Res. Not. IMRN (2023), no. 2, 1378–1419.
[261] J. Hosle, *On the comparison of measures of convex bodies via projections and sections*, Int. Math. Res. Not. IMRN (2021), no. 17, 13046–13074.
[262] J. Hosle, *The isomorphic section-projection problem for convex bodies*, Preprint, https://arxiv.org/abs/2607.10881.
[263] J. Hosle, A. Kolesnikov and G. Livshyts, *On the* L_p*-Brunn-Minkowski and dimensional Brunn-Minkowski conjectures for log-concave measures*, J. Geom. Anal. 31 (2021), no. 6, 5799–5836.
[264] D. Iglesias, J. Yepes Nicolás and A. Zvavitch, *Brunn-Minkowski inequalities for the lattice point enumerator*, Adv. Math. 370 (2020), 107193.
[265] G. Ivanov, *Tight frames and related geometric problems*, Canad. Math. Bull. 64 (2021), no. 4, 942–963.
[266] G. Ivanov, *On the volume of projections of the cross-polytope*, Discrete Math. 344 (2021), no. 5, Paper no. 112312.
[267] G. Ivanov and M. Naszódi, *Functional John Ellipsoids*, J. Funct. Anal. 282 (2022), no. 11, Paper no. 109441.
[268] G. Ivanov and I. Tsiutsiurupa, *Functional Löwner Ellipsoids*, J. Geom. Anal. 31 (2021), no. 11, 11493–11528.
[269] G. Ivanov and I. Tsiutsiurupa, *On the volume of sections of the cube*. Anal. Geom. Metr. Spaces 9 (2021), no. 1, 1–18.
[270] J. Jakimiuk and H. König, *Maximal sections of the unit ball of* $\ell_p^n(\mathbb{C})$ *for* $p > 2$. Studia Math. 283 (2025), 177–201.
[271] F. John, *Extremum problems with inequalities as subsidiary conditions*, Studies and Essays Presented to R. Courant on his 60th Birthday, January 8, 1948, 187–204, Interscience Publishers, New York, 1948.
[272] Z. Kabluchko and D. Zaporozhets, *Intrinsic volumes of Sobolev balls with applications to Brownian convex hulls*, Trans. Amer. Math. Soc. 368 (2016), no. 12, 8873–8899.
[273] N. Kalton and A. Koldobsky, *Intersection bodies and* L_p*-spaces*, Adv. Math. 196 (2005), 257–275.
[274] R. Kannan and L. Lovász, *Covering minima and lattice-point-free convex bodies*, Ann. of Math. (2) 128 (1988), no. 3, 577–602.
[275] R. Kannan, L. Lovász and M. Simonovits, *Isoperimetric problems for convex bodies and a localization lemma*, Discrete Comput. Geom. 13 (1995), 541–559.
[276] M. Kanter, *Unimodality and dominance for symmetric random vectors*, Trans. Amer. Math. Soc. 229 (1977), 65–85.
[277] B. S. Kashin, *The widths of certain finite-dimensional sets and classes of smooth functions*, Izv. Akad. Nauk SSSR Ser. Mat. 41 (1977), 334–351.
[278] R. Kerman, R. Olhava and S. Spektor, *An asymptotically sharp form of Ball's integral inequality*, Proc. Amer. Math. Soc. 143 (2015), no. 9, 3839–3846.
[279] C. G. Khatri, *On certain inequalities for normal distributions and their applications to simultaneous confidence bounds*, Ann. Math. Statist. 38 (1967), 1853–1867.
[280] A. Khintchine, *Uber dyadische Brüche*, Math. Z. 18 (1923), no. 1, 109–116.
[281] A. Khintchine, *A quantitative formulation of the approximation theory of Kronecker, Izvestiya Akad.* Nauk SSSR. Ser. Mat., 12 (1948), 113–122.

[282] J. Kim, V. Yaskin and A. Zvavitch, *The geometry of p-convex intersection bodies*, Adv. Math. 226 (2011), no. 6, 5320–5337.

[283] B. Klartag, *An isomorphic version of the slicing problem*, J. Funct. Anal. 218 (2005), 372–394.

[284] B. Klartag, *On convex perturbations with a bounded isotropic constant*, Geom. Funct. Anal. 16 (2006), 1274–1290.

[285] B. Klartag, *Marginals of geometric inequalities*, Geometric aspects of functional analysis, 133–166, Lecture Notes in Math., 1910, Springer, Berlin, 2007.

[286] B. Klartag, *A central limit theorem for convex sets*, Invent. Math. 168 (2007), 91–131.

[287] B. Klartag, *Power-law estimates for the central limit theorem for convex sets*, J. Funct. Anal. 245 (2007), 284–310.

[288] B. Klartag, *A Berry-Esseen type inequality for convex bodies with an unconditional basis*, Probab. Theory Related Fields 145 (2009), 1–33.

[289] B. Klartag, *High-dimensional distributions with convexity properties, European Congress of Mathematics*, Eur. Math. Soc., Zurich (2010), 401–417.

[290] B. Klartag, *Logarithmic bounds for isoperimetry and slices of convex sets*, Ars Inven. Anal. (2023), Paper no. 4.

[291] B. Klartag and A. Koldobsky, *An example related to the slicing inequality for general measures*, J. Funct. Anal. 274 (2018), 2089–2112.

[292] B. Klartag and J. Lehec, *Poisson processes and a log-concave Bernstein theorem*, Stud. Math. 247 (2019), no. 1, 85–107.

[293] B. Klartag and J. Lehec, *Bourgain's slicing problem and KLS isoperimetry up to polylog*, Geom. Funct. Anal. 32 (2022), no. 5, 1134–1159.

[294] B. Klartag and J. Lehec, *Isoperimetric inequalities in high-dimensional convex sets*, Preprint (https://arxiv.org/abs/2406.01324).

[295] B. Klartag and J. Lehec, *Affirmative resolution of Bourgain's slicing problem using Guan's bound*, Geom. Funct. Anal. 35 (2025), no. 4, 1147–1168.

[296] B. Klartag and J. Lehec, *Thin-shell bounds via parallel coupling*, Preprint (https://arxiv.org/abs/2507.15495).

[297] B. Klartag and G. V. Livshyts, *The lower bound for Koldobsky's slicing inequality via random rounding*, Geometric aspects of functional analysis. Vol. II, 43–63, Lecture Notes in Math. 2266, Springer, Cham, 2020.

[298] B. Klartag and V. D. Milman, *Geometry of log-concave functions and measures*, Geom. Dedicata 112 (2005), 169–182.

[299] B. Klartag and V. D. Milman, *Rapid Steiner symmetrization of most of a convex body and the slicing problem*, Combin. Probab. Comput. 14 (2005), 829–843.

[300] B. Klartag and R. Vershynin, *Small ball probability and Dvoretzky theorem*, Israel J. Math. 157 (2007), 193–207.

[301] V. L. Klee, *Research Problems: Is a body spherical if its HA-measurements are constant?*, Amer. Math. Monthly 76 (1969), 539–542.

[302] H. Knothe, *Contributions to the theory of convex bodies*, Mich. Math. J. 4 (1957), 39–52.

[303] A. Koldobsky, *Generalized Lévy representation of norms and isometric embeddings into L_p-spaces*, Ann. Inst. H. Poincare, Prob. et Stat., ser. B 28 (1992), 335–353.

[304] A. Koldobsky, *An application of the Fourier transform to sections of star bodies*, Israel J. Math. 106 (1998), 157–164.

[305] A. Koldobsky, *Intersection bodies, positive definite distributions and the Busemann-Petty problem*, Amer. J. Math. 120 (1998), 827–840.

[306] A. Koldobsky, *Intersection bodies in $\mathbb{R}^4$*, Adv. Math. 136 (1998), 1–14.

[307] A. Koldobsky, *A generalization of the Busemann-Petty problem on sections of convex bodies*, Israel J. Math. 110 (1999), 75–91.

[308] A. Koldobsky, *Positive definite distributions and subspaces of L_{-p} with applications to stable processes*, Canad. Math. Bull. 42 (1999), 344–353.

[309] A. Koldobsky, *A functional analytic approach to intersection bodies*, Geom. Funct. Anal. 10 (2000), 1507–1526.

[310] A. Koldobsky, *Fourier analysis in convex geometry*, Mathematical Surveys and Monographs, 116. American Mathematical Society, Providence, RI, 2005. vi+170 pp.

[311] A. Koldobsky, *Stability in the Busemann-Petty and Shephard problems*, Adv. Math. 228 (2011), 2145–2161.

[312] A. Koldobsky, *Stability of volume comparison for complex convex bodies*, Arch. Math. (Basel) 97 (2011), 91–98.

[313] A. Koldobsky, *A hyperplane inequality for measures of convex bodies in $\mathbb{R}^n$, $n \leqslant 4$*, Discrete Comput. Geom. 47 (2012), 538–547.

[314] A. Koldobsky, *Stability and separation in volume comparison problems*, Math. Model. Nat. Phenom. 8 (2013), 156–169.

[315] A. Koldobsky, *A $\sqrt{n}$ estimate for measures of hyperplane sections of convex bodies*, Adv. Math. 254 (2014), 33–40.

[316] A. Koldobsky, *Estimates for measures of sections of convex bodies, GAFA Seminar Volume, B. Klartag and E. Milman, editors*, Lect. Notes in Math. 2116 (2014), 261–271.

[317] A. Koldobsky, *Slicing inequalities for measures of convex bodies*, Adv. Math. 283 (2015), 473–488.

[318] A. Koldobsky, *Stability inequalities for projections of convex bodies*, Discrete Comput. Geom. 57 (2017), 152–163.

[319] A. Koldobsky, *Functions positively associated with integral transforms*, Preprint. (https://arxiv.org/abs/2503.23310).

[320] A. Koldobsky and M. Lifshits, *Average volume of sections of star bodies*, Geom. Aspects of Funct. Analysis, Lecture Notes in Math. 1745 (2000), 119–146.

[321] A. Koldobsky and D. Ma, *Stability and slicing inequalities for intersection bodies*, Geom. Dedicata 162 (2013), 325–335.

[322] A. Koldobsky and A. Pajor, *A remark on measures of sections of L_p-balls*, Geometric aspects of functional analysis, 213–220, Lecture Notes in Math., 2169, Springer, Cham, 2017.

[323] A. Koldobsky and V. Yaskin, *The interface between convex geometry and harmonic analysis*, CBMS Regional Conference Series in Mathematics, 108. Published for the Conference Board of the Mathematical Sciences, Washington, DC; by the American Mathematical Society, Providence, RI, 2008.

[324] A. Koldobsky and A. Zvavitch, *An isomorphic version of the Busemann-Petty problem for arbitrary measures*, Geom. Dedicata 174 (2015), 261–277.

[325] A. Koldobsky and A. Zvavitch, *Isomorphic Busemann–Petty for arbitrary measures: the sharp order*, Preprint, arXiv:2608.02098, 2026.

[326] A. Koldobsky and M. Zymonopoulou, *Extremal sections of complex ℓ_p-balls, $0 < p \leqslant 2$*, Studia Math. 159 (2003), no. 2, 185–194.

[327] A. Koldobsky, H. König and M. Zymonopoulou, *The complex Busemann-Petty problem on sections of convex bodies*, Adv. Math. 218 (2008), no. 2, 352–367.

[328] A. Koldobsky, A. Merkurjev and V. Yaskin, *On polynomially integrable convex bodies*, Adv. Math. 320 (2017), 876–886.

[329] A. Koldobsky, G. Paouris and M. Zymonopoulou, *Isomorphic properties of intersection bodies*, J. Funct. Anal. 261 (2011), 2697–2716.

[330] A. Koldobsky, G. Paouris and M. Zymonopoulou, *Complex intersection bodies*, J. London Math. Soc. 88 (2013), 538–562.

[331] A. Koldobsky, G. Paouris and A. Zvavitch, *Measure comparison and distance inequalities for convex bodies*, Indiana Univ. Math. J. 71 (2022), no. 1, 391–407.

[332] A. Koldobsky, D. Ryabogin and A. Zvavitch, *Fourier analytic methods in the study of projections and sections of convex bodies*, Fourier analysis and convexity, 119–130, Appl. Numer. Harmon. Anal., Birkhäuser Boston, Boston, MA, 2004.
[333] A. Koldobsky, D. Ryabogin and A. Zvavitch, *Projections of convex bodies and the Fourier transform*, Israel J. Math. 139 (2004), 361–380.
[334] A. Koldobsky, M. Roysdon and A. Zvavitch, *Comparison problems for Radon transforms*, Adv. Math. 434 (2023), Paper no. 109336.
[335] A. Koldobsky, V. Yaskin and M. Yaskina, *Modified Busemann-Petty problem on sections of convex bodies*, Israel J. Math. 154 (2006), 191–207.
[336] A. Kolesnikov and G. Livshyts, *On the Gardner-Zvavitch conjecture: symmetry in inequalities of Brunn-Minkowski type*, Adv. Math. 384 (2021), Paper no. 107689.
[337] A. Kolesnikov and G. Livshyts, *On the local version of the Log-Brunn–Minkowski conjecture and some new related geometric inequalities*, Int. Math. Res. Not. IMRN 2022, no. 18, 14427–14453.
[338] A. V. Kolesnikov and E. Milman, *Brascamp-Lieb-type inequalities on weighted Riemannian manifolds with boundary*, J. Geom. Anal. 27 (2017), no. 2, 1680–1702.
[339] A. V. Kolesnikov and E. Milman, *Poincaré and Brunn-Minkowski inequalities on the boundary of weighted Riemannian manifolds*, Amer. J. Math. 140 (2018), no. 5, 1147–1185.
[340] A. V. Kolesnikov and E. Milman, *Local L^p-Brunn–Minkowski inequalities for $p < 1$*, Mem. Amer. Math. Soc. 277 (2022), no. 1360, v+78 pp.
[341] H. König, *Non-central sections of the simplex, the cross-polytope and the cube*, Adv. Math. 376 (2021), Paper no. 107458.
[342] H. König, *On hyperplane sections and projections in ℓ_p^n*, Banach J. Math. Anal. 18 (2024), Paper no. 64.
[343] H. König, *On maximal hyperplane sections of the unit ball of ℓ_p^n for $p > 2$*, Adv. Oper. Theory 10 (2025), Paper no. 15.
[344] H. König, *Non-central sections of the regular n-simplex*, Pure Appl. Funct. Anal. 10 (2025), 1287–1321.
[345] H. König and A. Koldobsky, *Volumes of low-dimensional slabs and sections in the cube*, Adv. in Appl. Math. 47 (2011), no. 4, 894–907.
[346] H. König and A. Koldobsky, *Minimal volume of slabs in the complex cube*, Proc. Amer. Math. Soc. 140 (2012), no. 5, 1709–1717.
[347] H. König and A. Koldobsky, *On the maximal measure of sections of the n-cube*, Geometric analysis, mathematical relativity, and nonlinear partial differential equations, 123–155, Contemp. Math., 599, Amer. Math. Soc., Providence, RI, 2013.
[348] H. König and A. Koldobsky, *On the maximal perimeter of sections of the cube*, Adv. Math. 346 (2019), 773–804.
[349] H. König and M. Rudelson, *On the volume of non-central sections of a cube*, Adv. Math. 360 (2020), Paper no. 106929.
[350] L. Kryvonos and D. Langharst, *Weighted Minkowski's existence theorem and projection bodies*, Trans. Amer. Math. Soc. 376 (2023), no. 12, 8447–8493.
[351] E. Kuwert, *Note on the isoperimetric profile of a convex body*, Geometric Analysis and Nonlinear Partial Differential Equations, Springer, Berlin (2003), 195–200.
[352] S. Kwapień, R. Latala and K. Oleszkiewicz, *Comparison of moments of sums of independent random variables and differential inequalities*, J. Funct. Anal. 136 (1996), no. 1, 258–268.
[353] M. Lafi and A. Zvavitch, *Measure comparison problems for dilations of convex bodies*, Canad. Math. Bull. 68 (2025), 270–285.
[354] D. Langharst, M. Roysdon and A. Zvavitch, *General measure extensions of projection bodies*, Proc. Lond. Math. Soc. (3) 125 (2022), 1083–1129.
[355] D. G. Larman and C. A. Rogers, *The existence of an origin-symmetric convex body with central sections that are unexpectedly small*, Mathematika 22 (1975), 164–175.
[356] R. Latala, *A note on the Ehrhard inequality*, Stud. Math. 118 (1996), no. 2, 169–174.

[357] R. Latala, *On some inequalities for Gaussian measures*, in: Proceedings of the International Congress of Mathematicians, vol. II, Beijing, 2002, Higher Ed. Press, Beijing, 2002, pp. 813–822.

[358] R. Latala and K. Oleszkiewicz, *On the best constant in the Khinchin-Kahane inequality*, Studia Math. 109 (1994), no. 1, 101–104.

[359] R. Latala and K. Oleszkiewicz, *Gaussian measures of dilatations of convex symmetric sets*, Ann. Probab. 27 (1999), no. 4, 1922–1938.

[360] M. Ledoux, *A simple analytic proof of an inequality by P. Buser*, Proc. Am. Math. Soc. 121 (1994), 951–959.

[361] M. Ledoux, *The concentration of measure phenomenon*, Mathematical Surveys and Monographs, 89. American Mathematical Society, Providence, RI, 2001.

[362] M. Ledoux and M. Talagrand, *Probability in Banach spaces. Isoperimetry and processes*, Springer-Verlag, Berlin, 1991.

[363] Y. T. Lee and S. S. Vempala, *Eldan's stochastic localization and the KLS hyperplane conjecture: an improved lower bound for expansion*, 58th Annual IEEE Symposium on Foundations of Computer Science—FOCS 2017, 998–1007, IEEE Computer Soc., Los Alamitos, CA, 2017.

[364] Y. T. Lee and S. S. Vempala, *Eldan's stochastic localization and the KLS conjecture: Isoperimetry, concentration and mixing*, Ann. of Math. (2) 199 (2024), no. 3, 1043–1092.

[365] J. Lehec, *The symmetric property (τ) for the Gaussian measure*, Ann. Fac. Sci. Toulouse Math. (6) 17 (2008), no. 2, 357–370.

[366] J. Lehec, *Partitions and functional Santaló inequalities.*, Archiv der Mathematik 92 (2009), no. 1, 89–94.

[367] J. Lehec, *A direct proof of the functional Santaló inequality*, C. R. Acad. Sci. Paris, Ser. I 347 (2009), 55–58.

[368] L. Leindler, *On a certain converse of Hölder's inequality II*, Acta Sci. Math. Szeged 33 (1972), 217–223.

[369] P. Lévy, *Théorie de l'addition de variable aléatoires*, Gauthier-Villars, Paris, 1937.

[370] D. R. Lewis, *Finite dimensional subspaces of L_p*, Studia Math. 63 (1978), 207–212.

[371] D. R. Lewis, *Ellipsoids defined by Banach ideal norms*, Mathematika 26 (1979), 18–29.

[372] J. Li, *Rényi entropy power inequality and a reverse*, Studia Math. 242 (2018), no. 3, 303–319.

[373] B. Li, C. Schütt and E. Werner, *The Löwner Function of a Log-Concave Function*, J. Geom. Analysis 31 (2021), 423–456.

[374] A. Lichnerowicz, *Géométrie des groupes de transformations*, Volume III of Travaux et Recherches Mathématiques. Dunod, Paris, 1958.

[375] C. Lin, YD. Liu and G. Xiong, *A positive solution to the generalized Busemann–Petty problem with subspace dimensions 2 and 3*, Preprint, arXiv:2608.05621, 2026.

[376] J. E. Littlewood, *On bounded bilinear forms in an infinite number of variables*, Quart. J. Math., Oxford Ser. 1 (1930), 164–174.

[377] D.-M. Liakopoulos, *Reverse Brascamp-Lieb inequality and the dual Bollobás-Thomason inequality*, Arch. Math. (Basel) 112 (2019), no. 3, 293–304.

[378] D.-M. Liakopoulos, *Inequalities for the quermassintegrals of sections of convex bodies*, Indiana Univ. Math. J. 74 (2025), no. 2, 385–406.

[379] A. E. Litvak, V. D. Milman and G. Schechtman, *Averages of norms and quasi-norms*, Math. Ann. 312 (1998), 95–124.

[380] R. Liu and T. Tkocz, *A note on the extremal non-central sections of the cross-polytope*, Adv. in Appl. Math. 118 (2020), Paper no. 102031.

[381] G. V. Livshyts, *An extension of Minkowski's theorem and its applications to questions about projections for measures*, Adv. Math. 356 (2019), Paper no. 106803.

[382] G. V. Livshyts, *A universal bound in the dimensional Brunn-Minkowski inequality for log-concave measures*, Trans. Amer. Math. Soc. 376 (2023), no. 9, 6663–6680.

[383] G. Livshyts, G. Paouris and P. Pivovarov, *On sharp bounds for marginal densities of product measures*, Israel J. Math. 216 (2016), no. 2, 877–889.

[384] G. Livshyts, A. Marsiglietti, P. Nayar and A. Zvavitch, *On the Brunn–Minkowski inequality for general measures with applications to new isoperimetric-type inequalities*, Trans. Amer. Math. Soc. 369 (2017), no. 12, 8725–8742.
[385] L. H. Loomis and H. Whitney, *An inequality related to the isoperimetric inequality*, Bull. Amer. Math. Soc. 55 (1949), 961–962.
[386] L. Lovász, *How to compute the volume?*, Jber. d. Dt. Math.-Verein, Jubiläumstagung 1990, (1990), 138–151.
[387] L. Lovász and M. Simonovits, *Random walks in a convex body and an improved volume algorithm*, Random Structures and Algorithms 4 (1993), 359–412.
[388] G. Lozanovskii, On some Banach lattices, Siberian Math. J. 10 (1969), 419–431.
[389] E. Lutwak, *Dual mixed volumes*, Pacific J. Math. 58 (1975), 531–538.
[390] E. Lutwak, *A general isepiphanic inequality*, Proc. Amer. Math. Soc. 90 (1984), 415–421.
[391] E. Lutwak, *On some affine isoperimetric inequalities*, J. Differential Geom. 23 (1986), 1–13.
[392] E. Lutwak, *Intersection bodies and dual mixed volumes*, Advances in Math. 71 (1988), 232–261.
[393] E. Lutwak, *Extended affine surface area*, Adv. Math. 85 (1991), no. 1, 39–68.
[394] E. Lutwak, *Selected affine isoperimetric inequalities*, in Handbook of convex geometry, Vol. A, B, 151–176, North-Holland, Amsterdam, 1993.
[395] E. Lutwak and G. Zhang, *Blaschke-Santaló inequalities*, J. Differential Geom. 47 (1997), 1–16.
[396] E. Lutwak, D. Yang, and G. Zhang, *L_p-John ellipsoids*, Proc. London Math. Soc. 90 (2005), 497–520.
[397] K. Mahler, *Ein Minimalproblem für konvexe Polygone*, Mathematica B (Zutphen) B7 (1938), 118–127.
[398] R-D. Malikiosis, *A discrete analogue for Minkowski's second theorem on successive minima*, Adv. Geom. 12 (2012), no. 2, 365–380.
[399] A. Manui, C. S. Ndiaye and A. Zvavitch, *On the volume of sums of anti-blocking bodies*, Preprint (https://arxiv.org/abs/2409.14214).
[400] E. Markessinis, G. Paouris and Ch. Saroglou, *Comparing the M-position with some classical positions of convex bodies*, Math. Proc. Cambridge Philos. Soc. 152 (2012), 131–152.
[401] A. Marsiglietti, *On the improvement of concavity of convex measures*, Proc. Amer. Math. Soc. 144 (2016), no. 2, 775–786.
[402] B. Maurey, *Theoremes de factorization pour les operateurs lineaires a valeurs dansles espaces L_p*, Asterisque 11 (1974), Societe Math. de France.
[403] B. Maurey, *Some deviation inequalities*, Geom. Funct. Anal. 1 (1991), no. 2, 188–197.
[404] L. Mayrhofer, J. Schade and S. Weltge, *Lattice-free simplices with lattice width $2d - o(d)$*, Integer Programming and Combinatorial Optimization: 23rd International Conference, IPCO 2022, Eindhoven, The Netherlands, June 27–29, 2022, Proceedings, 375–386, Springer-Verlag, Berlin, Heidelberg, 2022.
[405] V. G. Maz'ya, *Classes of domains and imbedding theorems for function spaces*, Dokl. Acad. Nauk SSSR (Engl. transl. Soviet Math. Dokl., 1 (1961) 882–885) 3 (1960), 527–530.
[406] V. G. Maz'ya, *The negative spectrum of the higher-dimensional Schrödinger operator*, Dokl. Akad. Nauk SSSR 144 (1962), 721–722.
[407] P. McMullen, *Volumes of projections of unit cubes*, Bull. London Math. Soc. 16 (1984), no. 3, 278–280.
[408] J. Melbourne and C. Roberto, *Quantitative form of Ball's Cube slicing in R^n and equality cases in the min-entropy power inequality*, Proc. Amer. Math. Soc. 150 (2022), no. 8, 3595–3611.
[409] J. Melbourne and C. Roberto, *Transport-majorization to analytic and geometric inequalities*, J. Funct. Anal. 284 (2023), no. 1, Paper no. 109717.
[410] M. Meyer, *A volume inequality concerning sections of convex sets*, Bull. London Math. Soc. 20 (1988), no. 2, 151–155.
[411] M. Meyer and A. Pajor, *Sections of the unit ball of ℓ_q^n*, J. Funct. Anal. 80 (1988), 109–123.
[412] M. Meyer and A. Pajor, *On the Blaschke-Santaló inequality*, Arch. Math. 55 (1990) 82–93.
[413] E. Milman, *Dual mixed volumes and the slicing problem*, Adv. Math. 207 (2006), 566–598.
[414] E. Milman, *Generalized intersection bodies*, J. Funct. Anal. 240 (2006), no. 2, 530–567.

[415] E. Milman, *Generalized intersection bodies are not equivalent*, Adv. Math. 217 (2008), 2822–2840.
[416] E. Milman, *On the role of convexity in isoperimetry, spectral gap and concentration*, Invent. Math. 177 (2009), no. 1, 1–43.
[417] E. Milman, *On the mean width of isotropic convex bodies and their associated L_p-centroid bodies*, Int. Math. Res. Not. IMRN 2015, no. 527, 3408–3423.
[418] E. Milman and A. Yehudayoff, *Sharp Isoperimetric Inequalities for Affine Quermassintegrals*, J. Amer. Math. Soc. 36 (2023), no. 4, 1061–1101.
[419] E. Milman, S. Shabelman and A. Yehudayoff, *Fixed and periodic points of the intersection body operator*, Invent. Math. 241 (2025), 509–558.
[420] V. D. Milman, *Geometrical inequalities and mixed volumes in the local theory of Banach spaces*, Astérisque 131 (1985), 373–400.
[421] V. D. Milman, *Random subspaces of proportional dimension of finite dimensional normed spaces: approach through the isoperimetric inequality*, Lect. Notes in Math. 1166 (1985), 106–115.
[422] V. D. Milman, *Inegalité de Brunn-Minkowski inverse et applications á la théorie locales des espaces normés*, C.R. Acad. Sci. Paris 302 (1986), 25–28.
[423] V. D. Milman and A. Pajor, *Isotropic position and inertia ellipsoids and zonoids of the unit ball of a normed n-dimensional space*, in: Geometric Aspects of Functional Analysis, J. Lindenstrauss and V. D. Milman, Lecture Notes in Mathematics 1376, Springer, Heidelberg, 1989, pp. 64–104.
[424] V. D. Milman and L. Rotem, *Mixed integrals and related inequalities*, J. Funct. Anal. 264 (2013), no. 2, 570–604.
[425] V. D. Milman and G. Schechtman, *Asymptotic Theory of Finite Dimensional Normed Spaces*, Lecture Notes in Math. 1200 Springer, Berlin, 1986.
[426] V. D. Milman and G. Schechtman, *Global versus Local asymptotic theories of finite-dimensional normed spaces*, Duke Math. J. 90 (1997), 73–93.
[427] H. Minkowski, *Volumen und Oberfläche*, Math. Ann. 57 (1903), 447–495.
[428] H. Minkowski, *Gesammelte Abhandlungen*, Vol. 2, 131–229, Teubner, Leipziig, 1911.
[429] J. Moody, C. Stone, D. Zach and A. Zvavitch, *A remark on the extremal non-central sections of the unit cube*, Asymptotic geometric analysis, 211–228, Fields Inst. Commun., 68, Springer, New York, 2013.
[430] S. Moriguti, *A lower bound for a probability moment of any absolutely continuous distribution with finite variance*, Ann. Math. Statistics 23 (1952), 286–289.
[431] S. Myroshnychenko, K. Tatarko and V. Yaskin, *Unique determination of ellipsoids by their dual volumes*, Int. Math. Res. Not. IMRN 2022, no. 527, 13569–13589.
[432] S. Myroshnychenko, K. Tatarko and V. Yaskin, *Answers to questions of Grünbaum and Loewner*, Adv. Math. 461 (2025), paper 110081.
[433] S. Myroshnychenko, C. Tang K. Tatarko and T. Tkocz, *Stability of simplex slicing*, Discrete Comput. Geom. 2026, in press, arXiv:2403.11994.
[434] A. Naor and D. Romik, *Projecting the surface measure of the sphere of l_p^n*, Ann. Inst. H. Poincaré Probab. Statist. 39 (2003), no. 2, 241–261.
[435] P. Nayar and T. Tkocz, *A note on a Brunn-Minkowski inequality for the Gaussian measure*, Proc. Amer. Math. Soc. 141 (2013), no. 11, 4027–4030.
[436] P. Nayar and T. Tkocz, *Extremal sections and projections of certain convex bodies: a survey*, Harmonic analysis and convexity, 343–390, Adv. Anal. Geom. 9, De Gruyter, Berlin, 2023.
[437] F. L. Nazarov and A. N. Podkorytov, *Ball, Haagerup, and distribution functions*, Complex analysis, operators, and related topics, 247–267, Oper. Theory Adv. Appl., 113, Birkhäuser, Basel, 2000.
[438] F. Nazarov, D. Ryabogin and A. Zvavitch, *Non-uniqueness of convex bodies with prescribed volumes of sections and projections*, Mathematika 59 (2013), 213–221.
[439] F. Nazarov, D. Ryabogin and A. Zvavitch, *An asymmetric convex body with maximal sections of constant volume*, J. Amer. Math. Soc. 27 (2014), 43–68.
[440] F. Nazarov, M. Sodin and A. Volberg, *The geometric Kannan-Lovász-Simonovits lemma, dimension-free estimates for the distribution of the values of polynomials, and the distribution of the zeros of random analytic functions*, Algebra i Analiz, 14 (2002), no. 2, 214–234.

[441] I. Newton, *Philosophiae naturalis principia mathematica*, London, 1687.
[442] E. M. Nikishin, *Resonance theorems and superlinear operators*, Uspeki Mat. Nauk. 25 (1970), 129–191.
[443] D. M. Oberlin and E. M. Stein, *Mapping properties of the Radon transform*, Indiana Univ. Math. J. 31 (1982), no. 527, 641–650.
[444] R. O' Donnell, *Analysis of Boolean functions*, Cambridge University Press, New York, 2014.
[445] B. Oksendal, *Stochastic Differential Equations: An Introduction with Applications*, Sixth edition. Universitext. Springer-Verlag, Berlin, 2003. xxiv+360 pp.
[446] K. Oleszkiewicz, *Comparison of moments via Poincaré-type inequality*, Advances in stochastic inequalities (Atlanta, GA, 1997), 135–148, Contemp. Math., 234, Amer. Math. Soc., Providence, RI, 1999.
[447] K. Oleszkiewicz, *On p-pseudostable random variables, Rosenthal spaces and ℓ_p^n ball slicing*, Geometric Aspects of Functional Analysis, 2001-2002, 188–210, Lecture Notes in Math. 1807, Springer-Verlag, Berlin-Heidelberg, 2003.
[448] K. Oleszkiewicz and A. Pelczynski, *Polydisc slicing in $\mathbb{C}^n$*, Studia Math. 142 (2000), 281–294.
[449] Y. Ollivier and C. Villani, *A curved Brunn–Minkowski inequality on the discrete hypercube, or: what is the Ricci curvature of the discrete hypercube?*, SIAM J. Discrete Math. 26 (2012), no. 3, 983–996.
[450] A. Pajor and N. Tomczak-Jaegermann, *Remarques sur les nombres d'entropie d'un opérateur et de son transposé*, C.R. Acad. Sci. Paris 301 (1985), 743–746.
[451] G. Paouris, *On the isotropic constant of non-symmetric convex bodies, Geom. Aspects of Funct. Analysis. Israel Seminar 1996-2000*, Lect. Notes in Math., 1745 (2000), 239–244.
[452] G. Paouris, *Concentration of mass in convex bodies*, Geom. Funct. Analysis 16 (2006), 1021–1049.
[453] G. Paouris, *Small ball probability estimates for log-concave measures*, Trans. Amer. Math. Soc. 364 (2012), 287–308.
[454] G. Paouris and P. Pivovarov, *Small-ball probabilities for the volume of random convex sets*, Discrete Comput. Geom. 49 (2013), 601–646.
[455] M. Papadimitrakis, *On the Busemann-Petty problem about convex, origin-symmetric bodies in $\mathbb{R}^n$*, Mathematika 39 (1992), 258–266.
[456] L. E. Payne and H. F. Weinberger, *An optimal Poincaré inequality for convex domains*, Arch. Rat. Mechanics and Analysis 5 (1960), 286–292.
[457] C. M. Petty, *Surface area of a convex body under affine transformations*, Proc. Amer. Math. Soc. 12 (1961), 824–828.
[458] C. M. Petty, *Projection bodies*, Proc. Coll. Convexity (Copenhagen 1965), 234–241, Kobenhavns Univ. Mat. Inst., 1965.
[459] R. E. Pfiefer, *The Extrema of Geometric Mean Values*, PhD dissertation, Department of Mathematics, University of California, Davis, CA, 1982.
[460] R. E. Pfiefer, *Maximum and minimum sets for some geometric mean values*, J. Theoret. Probab. 3 (1990), no. 2, 169–179.
[461] G. Pisier, *Sur les espaces de Banach K-convexes*, Séminaire d'Analyse Fonctionnelle 79/80, Ecole Polytechniques, Palaiseau, Exp. 11 1980.
[462] G. Pisier, *Holomorphic semi-groups and the geometry of Banach spaces*, Ann. of Math. 115 (1982), 375–392.
[463] G. Pisier, *A new approach to several results of V. Milman*, J. Reine Angew. Math. 393 (1989), 115–131.
[464] G. Pisier, *The Volume of Convex Bodies and Banach Space Geometry*, Cambridge Tracts in Mathematics 94 (1989).
[465] G. Pólya, *Berechnung eines bestimmten integrals*, Math. Ann. 74 (1913), 204–212.
[466] B. Pooley, *Upper bounds on the cross-sectional volumes of cubes and other problems*, MSc dissertation, University of Warwick, 2012.
[467] L. Pournin, *Shallow sections of the hypercube*, Preprint (2021), Israel J. Math. 255 (2023), no. 2, 685–704.
[468] L. Pournin, *Local extrema for hypercube sections*, J. Anal. Math. 152 (2024), no. 2, 557–594.

[469] A. Prékopa, *On logarithmic concave measures and functions*, Acta Sci. Math. Szeged 34 (1973), 335–343.
[470] J. Prochno, C. Thäle and N. Turchi, *Geometry of ℓ_p^n-balls: classical results and recent developments*, High dimensional probability VIII – the Oaxaca volume, 121–150, Progr. Probab., 74, Birkhäuser/Springer, Cham, 2019.
[471] E. Putterman, *Equivalence of the local and global versions of the L^p-Brunn-Minkowski inequality*, J. Funct. Anal. 280 (2021), no. 527, Paper no. 108956.
[472] S. Rabinowitz, *A theorem about collinear lattice points*, Utilitas Math. 36 (1989), 93–95.
[473] S. T. Rachev and L. Rüschendorf, *Approximate independence of distributions on spheres and their stability properties*, Ann. Probab. 19 (1991), no. 3, 1311–1337.
[474] C. R. Rao, *Linear statistical inference and its applications*. Second edition, Wiley Series in Probability and Mathematical Statistics. John Wiley & Sons, New York-London-Sydney, 1973.
[475] O. Regev, *A note on Koldobsky's lattice slicing inequality*, arXiv e-prints, arXiv:1608.04945.
[476] O. Regev and N. Stephens-Davidowitz, *A reverse Minkowski theorem*, STOC'17-Proceedings of the 49th Annual ACM SIGACT Symposium on Theory of Computing, 941–953, ACM, New York, 2017.
[477] V. Reis and T. Rothvoss, *The subspace flatness conjecture and faster integer programming*, 2023 IEEE 64th Annual Symposium on Foundations of Computer Science—FOCS 2023, 974–988, IEEE Computer Soc., Los Alamitos, CA, 2023.
[478] M. Ritoré and J. Yepes Nicolás, *Brunn-Minkowski inequalities in product metric measure spaces*, Adv. Math. 325 (2018) 824–863.
[479] I. Rivin, *Surface area and other measures of ellipsoids*, Adv. in Appl. Math. 39 (2007), no. 4, 409–427.
[480] C. A. Rogers and G. C. Shephard, *The difference body of a convex body*, Arch. Math. 8 (1957), 220–233.
[481] C. A. Rogers and G. C. Shephard, *Convex bodies associated with a given convex body*, J. London Math. Soc. 33 (1958), 270–281.
[482] L. C. G. Rogers and D. Williams, *Diffusions, Markov processes, and martingales. Vol. 2: Itô calculus*, Reprint of the second (1994) edition. Cambridge Mathematical Library. Cambridge University Press, Cambridge, 2000. xiv+480 pp.
[483] B. A. Rogozin, *An estimate for the maximum of the convolution of bounded densities*, Teor. Veroyatnost. i Primenen. 32 (1987), no. 1, 53–61.
[484] L. Rotem, *Surface area measures of log-concave functions*, J. Anal. Math. 147 (2022), no. 1, 373–400.
[485] L. Rotem, *The anisotropic total variation and surface area measures*, Geometric aspects of functional analysis, 297–312, Lecture Notes in Math., 2327, Springer, Cham, 2023.
[486] T. Royen, *A simple proof of the Gaussian correlation conjecture extended to some multivariate gamma distributions*, Far East J. Theoret. Stat. 48 (2014), no. 2, 139–145.
[487] B. Rubin, *The lower dimensional Busemann-Petty problem for bodies with the generalized axial symmetry*, Israel J. Math. 173 (2009), 213–233.
[488] B. Rubin, *Comparison of volumes of convex bodies in real, complex, and quaternionic spaces*, Adv. Math. 225 (2010), 1461–1498.
[489] B. Rubin and G. Zhang, *Generalizations of the Busemann-Petty problem for sections of convex bodies*, J. Funct. Anal. 213 (2004), 473–501.
[490] M. Rudelson, *Distances between non-symmetric convex bodies and the MM^*-estimate*, Positivity 4 (2000), 161–178.
[491] W. Rudin, *Functional analysis*, Mc Graw Hill, 1973.
[492] I. Z. Ruzsa, *Sum of sets in several dimensions*, Combinatorica 14 (1994), 485–490.
[493] I. Z. Ruzsa, *Sets of sums and commutative graphs*, Studia Sci. Math. Hungar. 30 (1995), 127–148.
[494] D. Ryabogin, *A negative answer to Ulam's Problem 19 from the Scottish Book*, Ann. of Math. 195 (2022), 1111–1150.
[495] D. Ryabogin and A. Zvavitch, *Analytic methods in convex geometry*, IM PAN Lecture Notes, Vol. 2, Warsaw 2014.

[496] S. G. Samko, A. A. Kilbas, and O. I. Marichev, *Fractional integrals and derivatives*, Gordon and Breach Sci. Publ., Amsterdam, 1993.

[497] L. A. Santaló, *Un invariante afin para los cuerpos convexos del espacio de n dimensiones*, Portugaliae Math. 8 (1949), 155–161.

[498] C. Saroglou, *Remarks on the conjectured log-Brunn-Minkowski inequality*, Geom. Dedicata 177 (2015), 353–365.

[499] C. Saroglou, *More on logarithmic sums of convex bodies*, Mathematika 62 (2016), no. 3, 818–841.

[500] G. Schechtman and J. Zinn, *On the volume of the intersection of two L_p^n balls*, Proc. Amer. Math. Soc. 110 (1990), no. 1, 217–224.

[501] G. Schechtman and A. Zvavitch, *Embedding subspaces of L_p into ℓ_N^p*, $0 < p < 1$, Math. Nachr. 227 (2001), 133–142.

[502] M. Schmuckenschläger, *Volume of intersections and sections of the unit ball of l_p^n*, Proc. Amer. Math. Soc. 126 (1998), no. 5, 1527–1530.

[503] R. Schneider, *Zu einem problem von Shephard über die projektionen konvexer Körper*, Math. Z. 101 (1967), 71–82.

[504] R. Schneider, *Functional equations connected with rotations and their geometric applications*, Enseign. Math. (2) 16 (1970), 297–305.

[505] R. Schneider, *Convex Bodies: The Brunn-Minkowski Theory*, Second expanded edition. Encyclopedia of Mathematics and Its Applications 151, Cambridge University Press, Cambridge, 2014.

[506] R. Schneider and W. Weil, *Zonoids and selected topics*, Convexity and its Applications, P. M. Gruber and J. M. Wills, eds, Birkhauser, Boston, Mass., 1983.

[507] R. Schneider and W. Weil, *Stochastic and integral geometry*, Probability and its Applications, Springer-Verlag, Berlin, 2008.

[508] G. C. Shephard, *Shadow systems of convex bodies*, Israel J. Math. 2 (1964), 229–306.

[509] G. C. Shephard, *Combinatorial properties of associated zonotopes*, Canadian J. Math. 26 (1974), 302–321.

[510] Z. Sidák, *On multivariate normal probabilities of rectangles: their dependence on correlation*, Ann. Math. Statist. 39 (1968), 1425–1434.

[511] N. Skarmogiannis, *On a multi-integral norm defined by weighted sums of log-concave random vectors*, Proc. Amer. Math. Soc. 151 (2023), no. 10, 4355–4370.

[512] B. A. Slomka, *A remark on discrete Brunn-Minkowski type inequalities via transportation of measure*, Israel J. Math. 261 (2024), no. 2, 791–807.

[513] I. Soprunov and A. Zvavitch, *Bezout inequality for mixed volumes*, Int. Math. Res. Notices (2016), no. 527, 7230–7252.

[514] J. Spingarn, *An inequality for sections and projections of a convex set*, Proc. Amer. Math.Soc. 118 (1993), 1219–1224.

[515] E. Stein, *The development of square functions in the work of A. Zygmund*, Bull. Amer. Math. Soc. 7 (1982), 359–376.

[516] P. Sternberg and K. Zumbrun, *On the connectivity of boundaries of sets minimizing perimeter subject to a volume constraint*, Commun. Anal. Geom. 7 (1999), 199–220.

[517] V. N. Sudakov, *Gaussian random processes and measures of solid angles in Hilbert spaces*, Soviet Math. Dokl. 12 (1971), 412–415.

[518] V. N. Sudakov, *Typical distributions of linear functionals in finite-dimensional spaces of high dimension*, Soviet Math. Dokl. 19 (1978), 1578–1582.

[519] S. J. Szarek, *On the best constants in the Khinchin inequality*, Studia Math. 58 (1976), no. 2, 197–208.

[520] M. Talagrand, *Transportation cost for Gaussian and other product measures*, Geom. Funct. Anal. 6 (1996), no. 3, 587–600.

[521] C. Tang, *Simplex slicing: an asymptotically sharp lower bound*, Adv. Math. 451 (2024), Paper no. 109784.

[522] T. Tao and V. Vu, *John-type theorems for generalized arithmetic progressions and iterated sumsets*, Adv. Math. 219 (2008), no. 2, 428–449.

[523] T. Tao and V. Vu, *Additive Combinatorics*, Cambridge Studies in Advanced Mathematics, 105. Cambridge University Press, Cambridge, 2010.

[524] M. Tointon, *New bounds in the discrete analogue of Minkowski's second theorem*, Discrete Anal. (2024), Paper no. 7.

[525] B. Tomaszewski, *A simple and elementary proof of the Kchintchine inequality with the best constant*, Bull. Sci. Math. (2) 111 (1987), no. 1, 103–109.

[526] N. Tziotziou, *Inequalities for sections and projections of log-concave functions*, J. Geom. Analysis, 36 (2026), Paper no. 145.

[527] N. Tziotziou, *Sections and projections of the outer and inner regularizations of a convex body*, Mathematika, in press (https://arxiv.org/abs/2511.02974).

[528] S. Ulam, *A collection of mathematical problems*, Interscience Tracts in Pure and Applied Mathematics, New York-London, 1960, Problem 19.

[529] P. S. Urysohn, *Mittlere Breite und Volumen der konvexen Körper im n-dimensionalen Raume*, Mat. Sb. USSR 31 (1924), 313–319.

[530] J. D. Vaaler, *A geometric inequality with applications to linear forms*, Pacific J. Math. 83 (1979), 543–553.

[531] P. L. Várkonyi, *Neutrally floating objects of density 1/2 in three dimensions*, Stud. Appl. Math. 130 (2013), 295–315.

[532] V. Vassiliev, *Newton's lemma XXVIII on integrable ovals in higher dimensions and reflection groups*, Bull. London Math. Soc. 47 (2015), 290–300.

[533] V. Vassiliev, *Lacunas and local algebraicity of volume functions*, J. Singul. 18 (2018), 350–357.

[534] V. Vassiliev, *Algebroidally integrable bodies*, Arnold Math. J. 6 (2020), 291–309.

[535] C. Vernicos and D. Yang, *A centro-projective inequality*, C. R. Math. Acad. Sci. Paris 357 (2019), no. 8, 681–685.

[536] H. von Weizsäcker, *Sudakov's typical marginals, random linear functionals and a conditional central limit theorem*, Probab. Theory Related Fields 107 (1997), 313–324.

[537] S. Wang, G. Xiong and K. Yang, *A complete solution to the Grünbaum–Loewner centroid problems*, Preprint, arXiv:2606.19865, 2026.

[538] S. Webb, *Central slices of the regular simplex*, Thesis (Ph.D.)–University of London, University College London, United Kingdom, 1996.

[539] S. Webb, *Central slices of the regular simplex*, Geom. Dedicata 61 (1996), no. 1, 19–28.

[540] F. Wegner, *Floating bodies of equilibrium*, Stud. Appl. Math. 111 (2003), 167–183.

[541] F. Wegner, *Floating bodies in equilibrium in 2D, the tire track problem and electrons in a parabolic magnetic field*, (https://arxiv.org/abs/physics/0701241).

[542] W. Weil, *Centrally symmetric convex bodies and distributions*, Israel J. Math. 24 (1976), 352–367.

[543] W. Weil, *Centrally symmetric convex bodies and distributions II*, Israel J. Math. 32 (1979), 173–182.

[544] Y. Yang and D. Zhang, *The log-Brunn-Minkowski inequality in* $\mathbb{R}^3$, Proc. Amer. Math. Soc. 147 (2019), no. 10, 4465–4475.

[545] V. Yaskin, *The Busemann-Petty problem in hyperbolic and spherical spaces*, Adv. Math. 203 (2006), no. 2, 537–553.

[546] V. Yaskin, *An extension of polynomial integrability to dual quermassintegrals*, Int. Math. Res. Not. IMRN 2020, no. 527, 3978–3990.

[547] V. Yaskin, *On a generalization of Busemann's intersection inequality*, J. Funct. Anal. 287 (2024), Paper no. 110561.

[548] V. Yaskin, Private communication.

[549] V. Yaskin and B. Zawalski, *On separably integrable symmetric convex bodies*, Adv. Math. 441 (2024), Paper no. 109527.

[550] D. Ye, L_p *Geominimal surface areas and their inequalities*, Int. Math. Res. Not. IMRN 9 (2015), 2465–2498.

[551] G. Zhang, *Centered bodies and dual mixed volumes*, Trans. Amer. Math. Soc. 345 (1994), 777–801.

[552] G. Zhang, *Intersection bodies and Busemann-Petty inequalities in* $\mathbb{R}^4$, Annals of Math. 140 (1994), 331–346.

[553] G. Zhang, *Sections of convex bodies*, Amer. J. Math. 118 (1996), 319–340.

[554] G. Zhang, *A positive answer to the Busemann-Petty problem in four dimensions*, Annals of Math. 149 (1999), 535–543.

[555] A. Zvavitch, *The Busemann-Petty problem for arbitrary measures*, Math. Ann. 331 (2005), 867–887.

[556] A. Zvavitch, *Gaussian measure of sections of dilates and translations of convex bodies*, Adv. in Appl. Math. 41 (2008), no. 2, 247–254.

[557] M. Zymonopoulou, *The complex Busemann-Petty problem for arbitrary measures*, Archiv der Math. 91 (2008), 436–449.

Index

https://doi.org/10.1515/9783111360508-012

Author Index

https://doi.org/10.1515/9783111360508-014

Advances in Analysis and Geometry

Volume 13
Keng Hao Ooi
Capacitary Calculus.
With Special Attention to Sobolev Multiplier Spaces and Their Preduals, 2026
ISBN 978-3-11-224232-2, e-ISBN 978-3-11-224233-9,
e-ISBN (ePUB) 978-3-11-224234-6

Volume 12
Hitoshi Tanaka, Hiroki Saito
Introduction to Potential Theory. Maximal Operators and Weights, 2025
ISBN 978-3-11-072596-4, e-ISBN 978-3-11-072603-9,
e-ISBN (ePUB) 978-3-11-072610-7

Volume 11
W. A. Zúñiga-Galindo
p-Adic Analysis. Stochastic Processes and Pseudo-Differential Equations, 2025
ISBN 978-3-11-157833-0, e-ISBN 978-3-11-157868-2,
e-ISBN (ePUB) 978-3-11-157904-7

Volume 10
Boris N. Apanasov
Dynamics of Discrete Group Action, 2024
ISBN 978-3-11-078403-9, e-ISBN 978-3-11-078410-7,
e-ISBN (ePUB) 978-3-11-078413-8

Volume 9
Alexander Koldobsky, Alexander Volberg (Eds.)
Harmonic Analysis and Convexity, 2023
ISBN 978-3-11-077537-2, e-ISBN 978-3-11-077538-9,
e-ISBN (ePUB) 978-3-11-077543-3

Volume 8
Suzanne Lenhart, Jie Xiao (Eds.)
Potentials and Partial Differential Equations. The Legacy of David R. Adams, 2023
ISBN 978-3-11-079265-2, e-ISBN 978-3-11-079272-0,
e-ISBN (ePUB) 978-3-11-079278-2

www.degruyterbrill.com